GEOTECHNICAL AND ENVIRONMENTAL APPLICATIONS
OF KARST GEOLOGY AND HYDROLOGY

PROCEEDINGS OF THE EIGHTH MULTIDISCIPLINARY CONFERENCE ON SINKHOLES
AND THE ENGINEERING AND ENVIRONMENTAL IMPACTS OF KARSTS
LOUISVILLE / KENTUCKY / 1-4 APRIL 2001

GEOTECHNICAL AND ENVIRONMENTAL APPLICATIONS OF KARST GEOLOGY AND HYDROLOGY

Edited by

BARRY F.BECK & J.GAYLE HERRING

P.E.LaMoreaux & Associates, Inc., Oak Ridge, Tennessee

Taylor & Francis
Taylor & Francis Group
LONDON AND NEW YORK

Sponsored jointly by

P.E. LaMoreaux & Associates, Inc.,
The GeoInstitute of the American Society of Civil Engineers, and
The Association of Ground Water Scientists and Engineers of the National Ground Water Association

Cover photo

Cover photo by Antonio Barsotti, Office of Civil Protection, Community of Camaiore, Italy: A sinkhole under repair in Camaiore, October 1995. The sinkhole was approximately 30 m in diameter and 13 m deep. It totally engulfed two houses and damaged five others surrounding the collapse. A 300-m circle of cracks outlined the area of subsidence. A local street was also destroyed. The limestone in this area is more than 100 m below ground surface. Camaiore is located on a graben indented into the mountains to the north. This structural low is infilled with Pliocene-Quaternary sediments. This is a tectonically active area near the Appenine Fault. Seismic activity was noted in the adjacent mountains approximately a week before the collapse. This, and another similar recent sinkhole near Grosseto in southern Tuscany, may be related to tectonism and deep-seated water circulation rather than the more classic, near-surface karst. These two sinkholes—in Camaiore and Grosseto—were the impetus for an international meeting on the phenomenon: *La Voragini Catastrofiche (Sinkholes)—Un Nuovo Problema Per La Toscana*. The meeting was convened in Grosseto, Italy, in March 2000; it was sponsored by the City and Province of Grosseto, the Region of Tuscany, and *Banca di Credito Cooperativo Della Maremma Grossetana*. I was privileged to see these areas and participate in this conference as a guest of Maria Sargentini, Luigi Micheli and the Region of Tuscany to whom I extend my thanks for their fascinating geology, good wine, good food, and warm hospitality.

A Note of Editorial Clarification

The purpose of these multidisciplinary conferences is to share professional information on the application of karst science to practical problems. As such, we feel that it is our job as editors to insure that the information presented herein is both professional and practical.

As we all realize in today's global environment, high-quality professional work is not confined to those who speak, or write, English fluently. Therefore, this is not one of the criteria for papers that you will read in this volume. Gayle and I hope that we have insured that the English language in these papers communicates the high-quality professional information on karst that we all wish to share, and from which we want to learn. However, because both the editors and the authors have limited time to invest in this endeavor, we have not made an exhaustive effort to render a text in fluent, grammatical English. We trust our readers will understand and will see the good, useful information in these papers.

Finally, I feel it necessary to add a note regarding the graphic quality of the illustrations in this current volume. As editors, we have asked that authors submit original figures or photographic reproduction thereof. However, in the interest of saving time it has become necessary to accept computer transmitted images. Unfortunately, the quality of the latter is not the same as the former. This is not in any way the fault of the printing process. The publisher does the best job possible with the "camera ready" copy he is provided by the author. In almost all cases, the final print quality faithfully reproduces the quality of the illustrations which the author(s) provided.

Barry F. Beck, Editor-in-Chief

Published by Taylor & Francis
2 Park Square, Milton Park, Abingdon, Oxon, OX14 4RN
270 Madison Ave, New York NY 10016

Transferred to Digital Printing 2007

ISBN 90 5809 190 2

Publisher's Note
The publisher has gone to great lengths to ensure the quality of this
reprint but points out that some imperfections in the original may be apparent

Table of contents

IV Evaluating the Risk of and Remediating the Damage from Sinkhole Collapse

V Groundwater Flow and Contaminant Movement in Karst

VI Treatment of Environmental Problems in Karst

VII Using Geophysics in Karst Investigations

VIII Field Trip Guide

I Keynote Address

Geotechnical and Environmental Applications of Karst Geology and Hydrology, Beck & Herring (eds)
© *2001 Taylor & Francis, ISBN 90 5809 190 2*

Tunnelling and mining in karstic terrane: an engineering challenge

PAUL G.MARINOS   National Technical University of Athens, Faculty of Civil Engineering, Athens, 10682, Greece
marinos@central.ntua.gr

ABSTRACT

Although limestone and most carbonate rocks exhibit good geotechnical behavior, when karstic, they may induce hazards during tunnelling operations, which may evolve into huge problems. Groundwater is the main source of problems and so is the crossing of voids and caverns, either empty, aquiferous or filled. In order to estimate the probability of encountering such conditions and be prepared to face them, a thorough hydrogeological study should complement the traditional site investigation program. This study has to consider a broader area embracing the whole hydrogeological basin of the karstic aquifer with background knowledge of the tectonic and paleogeographic evolution. In this paper a series of hydrogeological models are discussed depending on the internal karstic geometry of the aquifer and the position of the tunnel, either in the transfer or the inundation zone. Each model is associated with its own tunnelling particularities in terms of hazards and countermeasures. The crossing of big limestone mountains is discussed through a case history since it is very likely that the interior of similar mountains is not affected by karstification and in order to present a number of deviations from the persisting hydrogeologic regime. A discussion on the solutions to be engineered in order to cross big karstic cavities is also presented. For mining in karst, two case histories on the efforts to achieve effective dewatering illustrate the scale and the size of such difficult operations, along with the associated environmental implications.

INTRODUCTION
Tunnelling can be a high risk business

Groundwater is often the main source of problems in tunnel construction associated with stability and safety issues. Groundwater control during both construction and operation of the tunnel, is one of the most challenging problems faced by tunnel designers and contractors. Drainage facilities from the headings may be required and when the necessary invert grades are not available, the additional trouble and expense of pumping are unavoidable. Water can affect roof and face stability and in appreciable quantity will impede construction. If the host ground is soft and prone to erosion the risk is further increased.

Seepages, or leakages, into underground works from the surrounding aquifer can also affect the surrounding ground and adjacent facilities. Depending on local geology, hydrogeology and geotechnical parameters of the material, a severe environmental impact may be expected. In the opposite case, i.e. when leakages from underground works to the aquifer are possible, the hazard of groundwater contamination has to be considered.

Mining works due to the extension of the underground void space involved and to the absence of lining, can affect more seriously the surrounding hydrogeological conditions and increase the risk either for the development of instability or for the depletion of the groundwater resources.

The crossing of voids and caverns, either empty, aquiferous or filled with erodable material causes difficulties and the solutions that should be engineered, are often site specific.

Hence, although limestone and carbonate rocks in general exhibit a good geotechnical behavior, when karstic, they may induce all the aforementioned problems in tunnelling operations[*]. Many large engineering projects involving tunnels are currently under construction in countries where limestones are a very common geological formation. The design of underground excavations in these materials requires knowledge of the geological and hydrogeologic model in which these excavations are carried out.

INTERACTION WITH GROUNDWATER; GENERAL CONSIDERATIONS

The interaction of tunnelling and mining works with groundwater can be summarized as follows:

During construction
-   Inflows of water in the underground space, affecting normal construction procedures and possibly induce face and roof stability.
-   Sudden inflows associated with specific and localized geological features, e.g. faults, crushed zones, big karstic conduits etc.

---

[*] In the following text the term limestone refers also to all carbonate rocks that undergo karstification.

- Decline in yields of springs, decrease of groundwater discharge to wells.
- Development of sinkholes in susceptible areas due to piping or internal erosion.
- Acceleration of dissolution of soluble sediments (e.g. gypsum).
- Unacceptable settlements, where compressible fine-grained soils or heavily fractured rock masses are present, due to the increase of effective stresses by lowering of the groundwater table.
- Temporary contamination of groundwater occurring at lower elevations, by infiltration of polluting substances used for the construction.

During operation

- Infiltration of used chemically and organically contaminated waters from road or rail tunnels can affect the quality of the groundwater if the tunnels are crossing the non saturated zone.
- Rise of piezometric levels by the obstruction of groundwater flow by lined tunnels; the rise is effective when the tunnel is located at a shallow depth under a shallow water table and can affect the built environment (foundation, basements) and/or mobilize contaminants in case of saturation.
- Influence of the hydrostatic head on the lining of the tunnel.
- Tunnel collapse by wide fluctuation in hydrostatic pressure associated with normal operation of hydraulic unlined tunnels.
- In the case of water conveyance tunnels with lining deficiencies, the relation between the head of the waters flowing in the tunnel and the head of the surrounding aquifer can cause:
  - Inflow of eventually polluted waters in the tunnel and/or development of all the related and abovementioned risks (internal head lower than the head of the aquifer). Underground excavations containing fluids such as petroleum products at near-atmospheric pressures can be left unlined if the rock quality is high and if the excavation is below the water table since the fluids are contained by inward seepage of groundwater
  - leakages from sewer tunnels can contaminate the surrounding aquifer (interior head higher than head of the aquifer); leakage is a major concern when tunnels carry high-pressure water with toxic ingredients. Such fluids must be contained by an impervious liner.

INVESTIGATION

General considerations

It is essential to have accurate preconstruction assessment of groundwater conditions. No major underground engineering operation should be initiated before a comprehensive knowledge about the loads and flow regime of groundwater is established.

In the case of a tunnelling project close to the surface or in urban areas a good number of investigation techniques suitable to provide direct information and measurements are available (on piezometric heads, permeabilities, discharges). Geophysical investigation is often of great assistance. Unforeseeable conditions are thus very constrained.

In case of long tunnels at greater depths (in mountainous areas) the investigation possibilities are rather limited due to high cost. In such areas the investigation is mainly based on classical hydrogeological studies, procedures and techniques and covers a broader area for getting all necessary data and all geological boundary conditions. A study of this caliber must be based also on some kind of geological judgment.

This procedure must include:
- identification and classification of aquifer media (lithological and structural mapping)
- distinction of hydrogeologic units and water tables
- definition of hydrogeologic basins (underground catchment areas) and of the discharge areas
- delineation of water budgets
- study of springs: location, elevation, flow dynamics and discharge rates
- compilation of piezometric maps
- evaluation of hydraulic parameters both locally around the tunnel and in the broader area and basin (permeability, transmissivity, storativity)
- conclusions in the form of a report on the hydrogeological and geometrical boundary conditions for each aquifer and evaluation of heads and inflows relative to the underground construction. The report must provide also approaches for likely zones of sudden inrush hazard, such as fault zones.

Particularities in karstic rock masses

The particular or even unique hydrogeological features in a karstic environment demand special attention as there is an increased risk for water inflows and for environmental problems. Tunnelling in limestone terrane may thus be a challenge for both geologists and engineers owing to:
- high coefficient of infiltration from meteoric water.
- very high permeability; often non linear underground flow.
- preservation of high values of permeability at greater depths.
- potential of development of large hydrogeological basins, which may extend far beyond the boundaries of the corresponding geographic - hydrological basins of the considered area, involving, thus, greater quantities of groundwater.
- development of a non uniform, heterogeneous pattern of flow paths; depending on the post-tectonic and paleogeographic evolution of the area, preferential flow conduits and karstic tubes could be developed with a capacity to transmit water at large discharge rates; these conduits drain the surrounding jointed or finely fractured rock mass of low or medium permeability.
- groundwater flow in a flooding manner throughout the transfer ("unsaturated") zone.

- potential crossing of large underground cavities filled eventually with earth materials, with the possibility also to carry a column of perched ground water.

POTENTIAL HYDROGEOLOGICAL MODELS TO BE ENCOUNTERED

During the first stage of investigation in a limestone terrane it is crucial to understand the karstic pattern around the tunnel by means of a detailed hydrogeological study[*]. Such hydrogeological study should include a paleogeographic evaluation of vertical movements and changes of the geographic base level related to past locations of springs, in order to assess the depth of karstification inside the limestone mountain and the geometry of the karstic base level. This level is not necessarily restricted at the present elevation of the surficial springs. Thus, the geological reconnaissance in a broader area is a prerequisite for the investigation regarding tunnelling in karstic terrane.

Dye tracing testing and follow up of the route of major underground flow axes, i.e. between sinkholes (ponors) and springs, greatly assists the understanding of the delay of underground flow and is thus elucidating as to the presence of potential branching of the large karstic conduits or a general dispersion of flow to several directions. In this same rationale, the study of the distribution and the hydrographs of springs is always the most reliable tool for understanding the internal structure and geometry of a karstic aquifer, since it reflects the hydrodynamics of the interior of the karstic mass.

The question of whether concentrated or dispersed inflows are to be expected is of great concern since the former may threaten tunnelling operations. A detailed structural analysis of the hydrogeologic basin will define zones of possibly very high permeability (i.e. faults, or systematic bending zones).

Finally, the position of groundwater levels and fluctuations in the investigative boreholes, must be recorded at all times since they reflect the thansmissivity of the whole karstic mass. In the case of tunnelling in mountainous areas, pumping tests from wells, even if feasible, are not as helpful as for tunnels in low relief terrain. In those cases, packer tests restricted in the zone around the tunnel controlling the inflows, is a common practice.

Table 1 intends to provide the main hydrogeological models in a limestone environment. The answer on the most probable model to be crossed will facilitate the appropriate design of the tunnel and the provision of the methods and equipment necessary to face the hazards associated with the karstic conditions to be encountered.

Table 1: Potential hydrogeological models in limestone environment. Note that in some cases (e.g. platform karst) the inundation zone may be insignificant or transient. Carbonate rocks with substantial primary porosity can be considered of the finely-jointed type presented in this table. Few climatic type of karstification may produce patterns different from those above.

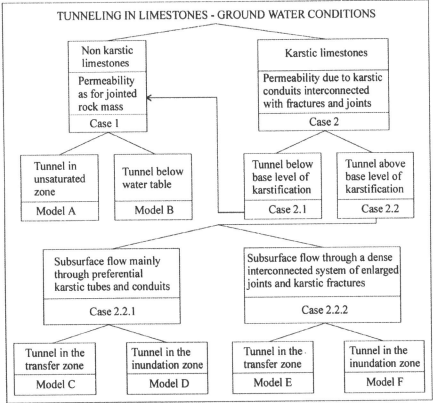

Case1: Groundwater issues are considered as for a jointed or fractured rock mass. Permeability is generally low and decreases dramatically with depth. Exceptions may occur in fault zones.
- Model A: Tunnel will cross a completely dry limestone mass; no risk for floods
- Model B: Tunnel will encounter medium to insignificant flow, depending on the frequency and aperture of joints or fractures.

Case 2: Dramatic difference in behavior compared with other aquiferous media; presence of high permeabilities, large discharges.
- *Case 2.1:* The rock mass surrounding the tunnel has never been exposed to underground erosion due to the paleogeographic evolution of the area or its isolation from infiltration and flow to outlets. In low relief morphology, the past geographic base level of the area to be crossed has never been lower than that of the tunnel. However in large mountainous masses the interior of the

---

[*] The reader can get insight on karstic processes in some excellent recent publications (Breznik, 1998, Milanović, 2000 and White, 1999).

mountain could have escaped karstification and the base level of karst lies at much higher elevations than the present level of the springs. Tunnels with such conditions will comply with either model A or B.
- *Case 2.2:* The size of the problems and risks depend on the internal geometry of the karstic system. Two options are possible:
  - *Case 2.2.1* when the underground flow is mainly concentrated and governed by distinct preferential large karstic tubes and conduits or,
  - *Case 2.2.2:* when flow is guided by a more homogeneous interconnected system of karstic fractures and enlarged joints. The latter is usually the case of well-bedded limestone in areas characterized by a long lasting persistence of an extended flat geographic base level. The former is often the case where a continuous downward underground erosion persists as the geographic base level was progressing towards lower elevations or where the lowest geographic level was restricted to a confined zone.

Model C: The tunnel is in the transfer zone of a selectively highly karstified mass. It will cross dry limestones but if located at depth the hazard for personnel and equipment from sudden inrushes and flooding will be high when storms occur in the catchment area. The stability of the tunnel might also be endangered. Erosion of loose filling material may result to a mud flow into the tunnel. Probing ahead should be a common practice. Contamination of the underlying "water table" is a real risk.

Model D: The tunnel is in the inundation zone and will drain moderate quantities of ground water between karstic conduits. These quantities are fed by water stored in fractures between these conduits. Upon encounter of the conduits, considerable increase of inflow will be experienced and violent inrush or flooding of the tunnel cannot be excluded. Probing ahead during construction is an absolute need. Predrainage techniques with site specific character should be applied in order to assist the crossing of the conduit. A quasi-permanent drainage of the karstic aquifer will last almost all of the construction period. The water resources of the area will be affected. Ground water discharges from the limestone mass between karstic conduits can be approached by the graph of Fig. 1. This estimation does not apply for the discharges of the conduits themselves.

Model E: The tunnel is in the transfer zone of a dense interconnected system of slightly karstified joints and fractures of moderate aperture. It will cross a mass with dripping waters or small amounts of transient water during wet periods. There is no risk for floods as the infiltration is widely dispersed inside the karstic mass.

Model F: The tunnel, being in the inundation zone, will drain, almost permanently significant or very significant quantities of ground waters during the construction, imposing the need for appropriate draining equipment. Violent inrushes should be restricted. Special design arrangements are to be implemented (i.e. diversion of waters to the sides of the tunnel). A drainage umbrella in front of the face should reduce the head and control inflows during the excavation (Fig. 6). Stability problems may occur only if the limestone is brecciated. Groundwater resources can be seriously affected.

Figure 1: Estimation of water inflow in a 10 m diameter tunnel for steady flow condition. This graph can be applied in the inundation zone of a limestone aquifer for estimating maximum values before transient flow is established and in sections between two main karstic conduits. It does not apply to discharges through the conduits themselves. These conduits may recharge their fractured-jointed limestone environment simulating steady flow conditions.

Some Case Histories

Any tunnel in a karstic environment offers experiences or incidents regarding facing of groundwater problems. However little is published in scientific or technical literature. The karst commission of the International Association of Engineering Geology in a report published by L. Calambert, in 1975, succeeded at that time to collect a number of cases, some of which are presented followingly.

Several cases are reported in Spain, where large quantities of water flooded a number of tunnels (Yagüe, A. 1975, *in* Calembert, 1975). In the Talave tunnel, quantities in the order of 1000 l/sec owing to structural features, faults or synclinal zones, were diminished only after several months. Large quantities of water were also encountered in a tunnel in Asturia, where all works were ceased until a drawdown of the water table was achieved. An important amount of fill material was also eroded and filled partially both tunnels.

An interesting case is reported during the construction of the highway tunnels of Gran Sasso (Calembert, 1975). One of the tunnels came upon a thrust fault with a heavily sheared zone 25m of thickness. The roof of the fault was Cretaceous limestone with karstic conduits communicating with the surface where a high-yield aquifer was present. An inrush of 900 l/sec along with eroded material lasted for 5 days until the real discharge of the faulted zone occurred with 4-6000 l/sec and a peak of 20000 l/sec (!) filling, additionally, the tunnel with more than 30000 m³ of debris such as sands and limestone blocks. The works were called off for many months.

In Turkey (Erguvanli, K., 1974, *in* Calembert, 1979) during the construction of a 7 km tunnel north of Tarsus, a localized discharge of karstic waters in the order of 250 l/sec caused a considerable delay of the works. A number of cases in Switzerland are briefly described in the same report with the countermeasures being mainly drainage; freezing techniques or cases of isolation are also reported (tunnels of Mont Dore and of Simplon).

TUNNELLING IN LARGE LIMESTONE MASSIF

The potential of coming across large cavities at great depths (more than several hundreds of meters) seems to be limited. However, small active conduits have been reported as was the case in France described by Petiteville, P. and Toulemont, M., (1974 *in* Calembert, 1975) where such conduits were found at a depth of about 1000 m.

The encounter of such karstic features partially filled or empty, under a thick cover, has a strongly accidental character that no method of investigation from the surface can trace. Recently, the application of geophysical methods from the face has been developed but with little success due to inherent limitations and because results are influenced by the tunnelling equipment and the steel support. Anyway in all cases where geophysics can be applied, corroboration must follow through exploratory boring from the face. Thus, often the best solution is to simply probe ahead immediately if there is suspicion for the presence of such karstic features.

An insight to the karstic conditions inside big mountains was gained while crossing a large karstic mountain at great depth in central Greece (Marinos, 1992). This experience can be easily utilized for other big massifs, as for instance, those occurring around the Mediterranean or other undergone the same geological evolution.

The question regarding the prevailing state in the interior of the karstified mountains at great depths and far beyond the areas where springs appear is often open. Such an issue arose during the construction of the Giona tunnel for the Mornos-Athens water-supplying aqueduct concerning both construction and operation. The tunnel now traverses this mountain for a length of 14.6 km, parallel to the coast, at an altitude of 377 m, beneath a cover of 1700 m, and with the central section 14-20 km from the coast of the Corinthian Gulf, where the groundwaters of the mountain are discharged through big coastal springs (Fig. 2).

At the beginning of the project, the intense karstification of the surface of the mountain and the drainage towards the low points of the coastline led the designers to a first hasty hypothesis that the tunnel would pass through karstic limestone, but more or less above the karstic water table due to the gentle hydraulic gradient expected for such a karstic environment (Fig. 3A). This is the case described by model C in Table 1. The tunnel was thus expected to be within the transfer (or conveyance) zone of groundwaters with high risk for sudden inflows during floods but with no permanent underground water. This water table would be considerably lower in areas of high permeability and of unobstructed discharge to the coast.

A few investigative drillings, although not deep enough, provided however some indications that the limestone could not be karstified at depth, but, merely, finely fissured. In such a case (model B in Table 1), the water table could lie considerably above the level of the tunnel and obviously with low-yield inflows in the tunnel (Fig. 3B).

Finally the karstic and hydrogeological conditions of the interior of the mountain appear more composite and to a certain extent, are a combination of models B and C (Fig. 4 and 5). The interior of the carbonate mountain does not appear karstified; karstification seems to stop at a depth of a few hundred meters and creates a karstic zone, which proceeds in stages parallel to the surface of the mountain. The paleogeographic development of the area, with gradual surface erosion and leveling due to successive faulting and changes of sea level, contributes to the formation of such an underground karstic geometry. Beneath the karstic zone, the limestones are not karstified but appear finely and tightly jointed, hence leading to low permeability.

The water table exhibits low gradients only at the karstic zone (in the lateral envelop of the mountain and behind the springs) but the gradients become steeper towards the low permeability interior of the mountain. The classical concept of the karst base level does not apply except for the areas below these peripheral parts of the mountain and evidently the mountain remains non karstified in its central part.

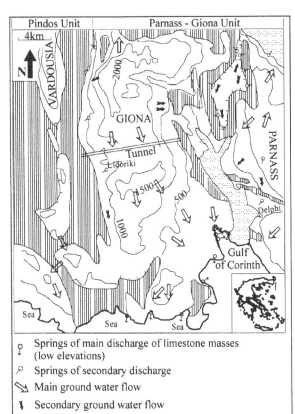

Springs of main discharge of limestone masses (low elevations)

Springs of secondary discharge

Main ground water flow

Secondary ground water flow

Limestone   Flysch   Alluvium   Thrust

Figure 2: Hydrogeological map of Giona karstic mountain and the crossing of the tunnel of the Mornos-Athens aqueduct (Marinos, 1992)

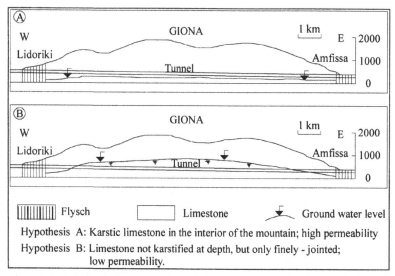

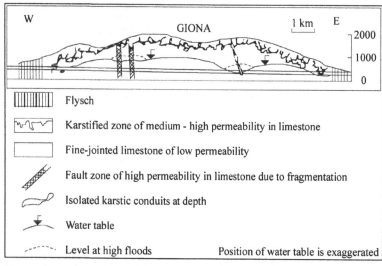

The water table in the outer karstified parts of the mountain is below the position of the tunnel and rises above it in the non karstified central areas. In those areas, the limestone is of low to very low permeability and the flow can be thought of as that in a poor porous medium. Drainage in the tunnel is barely perceptible mainly in the form of "transpiration", wet sidewalls or drip flows.

However, a few deviations from the general regime of the mountain's interior were found in the form of very limited zones of high permeability. Throughout its length, for more than 11 km in the interior of the mountain, the tunnel crossed just two karstic conduits, which were developed most probably in fault zones (Fig. 4). These conduits constitute no more than an exception and do not change the general non karstic characteristics of the interior of the mountain. These barely wide conduits were crossed at 9.8 and 6.5 km from the western entrance of the tunnel. The voids

Figure 3: Schematic sections of two hypothesis of groundwater development in Giona Mountain, Central Greece.

were bridged by fill and concrete slabs to allow boring by the Tunnel Boring Machine in use. When the first conduit was crossed, water was released under pressure but then the discharge quickly declined to small amounts. The second conduit was partially filled with clay, sand and gravel without water, but with clear indications of underground flow. Following a heavy storm, a flood reached the tunnel with a delay of only 8 hours. The water drained away from the tunnel within about a week. The active hydrogeological role of these conduits as a zone of transfer of infiltrated waters to an underlying inundated section was verified when they could not drain the water discharged into them by the tunnel; given that it was the season of high rainfall, the water table was elevated close to the level of the tunnel in that area.

These karstic tubes comprise axes of preferential isolated drainage according to a model similar of C or D (Table 1) but with a restricted extension inside the mountain. An additional result of their presence is a local significant lowering of the high, but low-yield water table prevailing inside the massif.

Given the information gathered during the tunnel construction, the hydrogeological description of the interior of the mountain is only complete if one takes also into account the presence and role of faults with or without a mylonitized zone. These fault zones do not bear karstic features or voids along their discontinuities, but few of them induced water problems, especially in the sections between 9.3 and 10 km from the East side (Fig. 4). In total, more than 400 l/sec of water entered the tunnel, 150 l/sec of which were contributed by a single fault through its fairly narrow mylonitic zone.

Figure 4: Underground hydraulic regime of Giona Mountain, Central Greece (Marinos, 1992).

The water-bearing faults increase the underground hydraulic heterogeneity of the interior of the mountain. These faults are fed by the karstic and highly permeable portions of the surface of the mountain. Hence, according to the geometry of the faults and their discharge capacity, the resulting water column can maintain a significantly raised water table despite the high permeability of these zones. This column recharges the surrounding finely-jointed limestone during wet periods and drains it during the dry periods.

Leakage from this hydraulic tunnel (piezometric head of 80 m) was impossible for most part of the mountain and the water table applies a high hydrostatic load that the lining was designed to withstand. On the contrary, leakage from the tunnel was possible at the endmost parts the tunnel, where the karstic zone was crossed. In these parts, a tighter grouting program had to be applied, as the karstic water level lied lower than the tunnel.

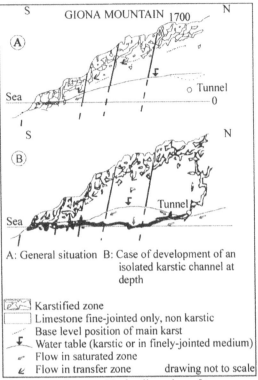

A: General situation   B: Case of development of an
                           isolated karstic channel at
                           depth

Karstified zone	
	Limestone fine-jointed only, non karstic
	Base level position of main karst
	Water table (karstic or in finely-jointed medium)
	Flow in saturated zone
	Flow in transfer zone drawing not to scale

Figure 5: Underground hydraulic regime of two cross sections of Giona Mountain (Marinos, 1992). (A) Model B according to Table 1, (B) Model C according to Table 1.

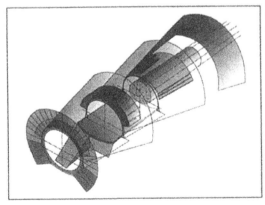

Figure 6: Driving a tunnel through an important water bearing zone with predrainage through embracing drainage umbrellas (sketch from "Geodata", Torino, personal communication).

CONFRONTING THE PROBLEM OF WATERS

Groundwater in tunnelling can be faced mainly with the following generally regarded operations (general information can be obtained from Anonymous, 1992 and Bauer, 1994):
-   lowering groundwater level by controlled drainage or dewatering via pumping, thus reducing both head of water pressure and discharge into the tunnel
-   grouting

Methods such as freezing, ground control by slurry, compressed air or earth pressure balance boring machines cannot be applied in highly permeable karstic limestone. Usually, drainage is more effective and often cheaper than any other operation. Predrainage prior to tunnel construction is probably the most commonly used water control method. The technique basically involves the lowering of the water table by drilling a series of wells or boreholes at either side of the projected tunnel. Drainage can be achieved from within the tunnel itself when dewatering from the surface is impossible. This can be done through drain holes from the face or from a long systematic drainage umbrella embracing the tunnel (Fig. 6), or even though the construction of small side pilot drainage galleries.

In the case of grouting in limestone, the primary goal is to reduce permeability. Modern practice is to drill a 360° array of grouting holes forwarded subhorizontally, then blast out and seal a section of tunnel inside this completed grout curtain. This also largely deals with the hazard of catastrophic inrush, i.e. a flooded cavity should be first encountered by a narrow bore drill hole that can be sealed off quickly. Grouting anyhow is difficult in large openings or under high pressure of water.

Such an example is described by Ford and William (1992), "the cooling water intake tunnel for an atomic power station, in Ontario, was an 8m diameter tunnel extending 600 m from shore beneath Lake Huron. It followed a corallian limestone formation just below the lake bed. Grouting forward proceeded in 20 m sections and the tunnel was cut in 8 m sections i.e. there was 60% overlap of successive grout curtains. However a cavity was encountered that could not be grouted because it was too large. It was sealed off and the tunnel was then deflected around it without serious difficulty, but at substantial extra cost". It is obvious that this deflection is not always possible (e.g. for traffic tunnels). Thus most extended dewatering methods have to be applied.

Dewatering can have undesired side effects on adjacent properties, the tunnel itself and the environment, such as (see also Powers, 1985):
-   ground settlement due to consolidation of compressible soils filling big karstic cavities up to the surface as an effect of increased effective stresses from water table lowering. Fortunately, such ground settlement cannot take place when limestones cover the ground as the rock is not compressible
-   development of sinkholes
-   depletion of adjacent groundwater and/or surface water supplies
-   salt water intrusion
-   expansion of contamination plumes
-   release of contaminated water into the environment

THE CASE OF CONFRONTING KARSTIC GROUND WATER IN MINES

Grouting is not feasible in the extracting galleries of a mine or in an open cast exploitation. Here, a more elaborate strategy is to dewater the mine zone entirely, i.e. maintain a huge cone of depression around it for as long as the mine is worked.

Case from a mine in Poland

A good example of the method is provided by the development of lead/zinc mines at Olkusz, Poland (Wilk, *in* Ford and William, 1992). "The ores are contained in filled dolines and cavities in a dolomite paleokarst at a depth of 200 to 300 m below a plain of

Quaternary sediments that is in hydrologic contact with the bedrocks. Potentially, this was a very hazardous situation. An area of 500 km² was surveyed about the potential mine. It contained 70 natural springs and 600 wells. A further 1700 exploration boreholes were drilled. Piezometers were installed in 300 wells and boreholes for carrying out pumping tests. From the latter it was estimated that 300 x 10⁶ m³ of groundwater would have to be pumped to establish the cone of depression for the mine. The cone was pumped via vertical wells plus drainage audits with high capacity pumps that were cut beneath each extraction level before ore extraction began. By these means, maximum local inrushes of water were held to 1.5 m³/sec, i.e. within the capacity of the pumps".

A case in Greece: prediction of groundwater table lowering for lignite mining and environmental implications

Extensive lignite deposits in the Ptolemais basin, in the Macedonia region of North Greece, are being exploited for electric power generation in thermal plants. The lignite-bearing horizons in the area are of about 50-70 meters thick and are covered by 100-200 meters of sterile overburden. Large-scale open-cast mining techniques are employed for the removal and disposal of the overburden and the excavation of the coal. Earth-removing operations are coupled with extensive groundwater table lowering in the alluvia of the basin using deep wells along the periphery of the mines.

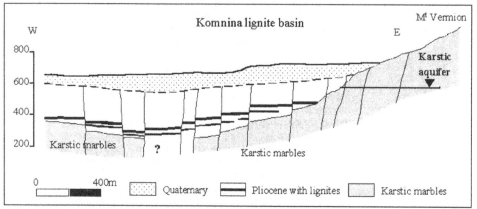

Figure 7: Typical East-West section of the eastern boundary of the Komnina basins, Ptolemais, Northern Greece (Kavvadas and Marinos, 1994).

The coal deposits are almost horizontally bedded, but a series of normal faults results in a progressive deepening of the beds and a corresponding increase of the thickness of the overburden towards the east-southeast rim of the basin (Fig. 7). Because of the continuously increasing demand for electric power in Greece, the Public Power Corporation, which owns and operates the lignite mines, plans to develop new mines in the Komnina Field, adjacent to the eastern boundary of the basin. Open cast coal exploitation in this field will require excavations down to 300 meters below the ground surface. The significant increase in the depth of the excavation is not the only challenge in planning the development of this new field. Mining operations will require lowering of the groundwater table by about 260 meters for the entire life of the field (estimated to be 10-15 years). In the planned Komnina Field, the increased depth of the lignite deposits and mainly the proximity of the field to the eastern boundary of the Ptolemais basin, which consists of highly karstified water-bearing marbles, make groundwater table lowering very costly. Furthermore, the effects of such an extensive operation will extend to a significant distance from the mines, altering the hydrological budget of a region, which is already stressed by over-exploitation for intensive use in irrigated agriculture and industry. More specifically, the long-term groundwater table lowering might also effect the hydrological budget of lake Vegoritis, some nine kilometers to the north of the Komnina Field, with possible consequences to its ecosystem (Fig. 8). It was thus necessary to study the nature of the karst aquifer along the eastern boundary of the Ptolemais basin, and assess the feasibility of dewatering and its effects on the hydrological budget of the region.

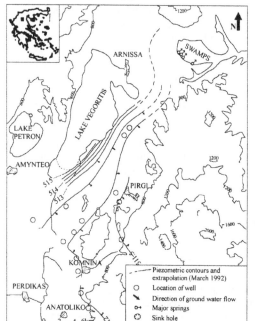

Figure 8: Iso-piezometric contours in the karstic aquifer to the east of the Komnina sub-basin in March, 1992 (Kavvadas and Marinos, 1994).

Hydrological and hydrogeological investigations in the well developed karstic marbles of western Mount Vermion revealed the existence of a single quite homogeneous unconfined aquifer having a very low hydraulic gradient (0.02-0.05%) towards the southeast and a high transimissivity (0.01-0.1 m²/s) (Fig. 8). The high transmissivity of the aquifer poses severe problems to the feasibility of the required extensive and sustained groundwater table lowering.

A two-dimensional regional flow model based on the finite element method was developed and used to study the behavior of the aquifer (Kavvadas and Marinos, 1994). Model parameters were estimated from

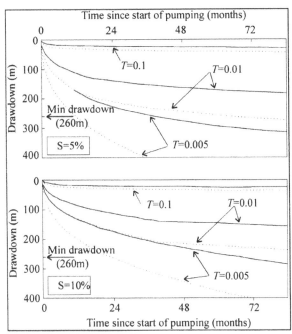

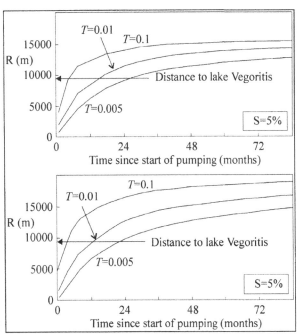

Figure 9: Predicted evolution with time of the average drawdown at the periphery of the mines, for various values of the transmissivity (T) of the aquifer. Pumping intensities: 10.000 m³/h (continuous line), and 15.000 m³/h (dotted line). Storativity: 5% (upper diagram), 10% (lower diagram).

Figure 10: Predicted evolution with time of the average radius of of dewatering (defined as the distance at which drawdown is 0.5 m), for various values of the transmissivity (T). Storativity: 5%. Pumping intensities: 10.000 m³/h (continuous line: upper diagram) and 15.000 m³/h (dotted line: lower diagram).

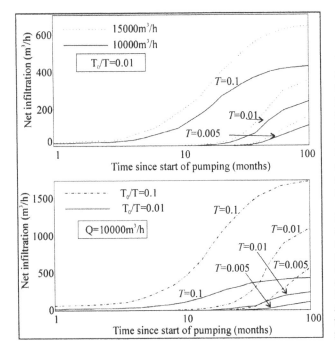

Figure 11: Predicted infiltration rates from lake Vegoritis to the karstic aquifer as a function of the time since the beginning of pumping. Upper diagram: a zone of alluvial deposits having transmissivity T_o = 0.01T, exists at the interface between the lake and the aquifer. Pumping intensities at the Komnina field: 10.000 m³/h (continuous line), and 15.000 m³/h (dotted line). Lower diagram: a zone of alluvial deposits having transmissivity T_o = 0.01T (continuous line) and T_o = 0.1T (dashed – dotted line) exists at the interface between the lake and the aquifer. Pumping intensity at the Komnina field: 10.000 m³/h.

field measurements in the aquifer, experience with similar karstic aquifers elsewhere in Greece, as well as knowledge of the steady-state response of the aquifer. Darcy limitations were marginally accepted due to the homogeneity of the karstic net, the uniformity of the piezometric surface and the low hydraulic gradients. The predictions of the regional flow model show that for reasonable values of the aquifer transmissivity (0.01 to 0.1 m²/s), the required groundwater table lowering cannot be achieved and maintained without extremely high cost (Fig. 9). It was also found that within a period of one to two years, the hydrological budget of lake Vegoritis will be significantly affected by the dewatering, with induced losses from the lake being large, compared to their present estimated values (Fig. 10, 11). Thus large-scale dewatering schemes around the proposed Komnina field should be combined with artificial recharge of the lake with part of the water pumped at Komnina. It is thus concluded that the deep lignite-bearing horizons along the eastern boundary of the Ptolemais basin could be only partially exploited at present, and a zone of sufficient thickness has to be left intact along the margin of the basin to act as a seal between the karstic aquifer and the mine. However, the stability of the seal as well as the possibility of piping through sandy seams interbedded with lignite horizons would still require partial groundwater table lowering in the karstic aquifer to reduce the hydrostatic pressures acting on the seal.

GEOTECHNICAL ISSUES WHEN TUNNELLING IN LIMESTONE ROCK MASS

The rock mass itself

With the exception of the problem associated with the karstic characteristics, limestone and all other carbonate rocks in general, exhibit a good geotechnical behavior and a friendly tunnelling response. They exhibit reasonably good resistance to drilling or boring with reduced wear of excavation tools. The strength of a limestone rockmass can never reach low levels such as those of a squeezing ground, even when brecciated. Limestone breccias always exhibit good frictional values; however, support is sometime necessary with light steel sets or lattice girders, beyond rock bolts and shotcrete.

From another point of view, when the rock is at great depths or under high horizontal stresses, it cannot generate typical bursting instability, as is the case of hard rocks, since it is not a brittle material[*]. Any mild spalling problems in tunnels can be satisfactorily coped with rock bolting and reinforced shotcrete.

The case of voids and karstic caverns

The meeting of caverns and big karstic conduits may be associated with the following problems, often very difficult to overcome:
- bridging the void, if empty
- tunnelling through a geotechnically weak fill material
- confronting water inrush associated with mud flow if the void is water bearing and filled partially or totally with earth materials (as discussed earlier).

In the case of urban tunnels with a thin cover, the occurrence of these voids can effectively be investigated with a drilling program assisted by geophysical testing. In most shallow depths the georadar can give reliable information. In deeper levels cross-hole tomography could be the best choice. In tunnels close to the surface an associated risk is the collapse of an adjacent cavern after an earthquake; filling these voids prior to the completion of the tunnel is an additional task to be undertaken. In the case of deep tunnel through a mountain and given there are clear indications that such cavities are present, the only reliable method is probing ahead, as was previously mentioned.

Figure 12: Typical appearance of a small karstic void partially filled with clay and silt; Dodoni tunnel, northwestern Greece, 2000.

The Dodoni tunnel in northern Greece, with a length of 3.3 km and 12 m in diameter, is currently (fall – winter 2000) being driven in a limestone sequence with well developed bedding and possible local intercalations of siltstones or cherts a few cm or dm of thickness. The limestone encountered so far has behaved well and this behavior is expected to continue. However, significant overbreaks have occurred at some locations and these overbreaks were due to instability of the fill in karstic cavities (Fig. 12). Karstic solution features may indeed be observed in outcrops on the surface of the mountain ridge crossed by the tunnel under a cover of at least 100 m. These features indicate that karstic processes were active inside the limestone ridge.

Two major collapses occurred related to the presence of sinkholes at the surface with outcropping chimneys almost 100 m of height. The voids were filled with clayey material and pieces of broken rock and were prominently wet. The main collapse had a diameter of approximately 1.5 m in the tunnel and 3 m on the surface (Fig. 13), leading to 1200 m^3 of material falling into the tunnel.

In order to detect karstic cavities, pockets filled with soft and broken material, shear zones and gouge-filled faults, it was recommended that routine probe drilling ahead of the tunnel face should be carried out (Hoek and Marinos, 2000, experts' unpublished report to "Egnatia Highway S.A."). Typically, such probe holes are percussion drilled using the normal jumbo. Ideally, the probe hole should always be kept one tunnel diameter ahead of the advancing face and the most convenient way to achieve this is by drilling long holes (30 to 50 m) during maintainance shifts or at weekends. As in all karstic voids, because of the irregular and unpredictable shape and location of weak zones, it is recommended that at least three probe holes should be drilled from the face at 10, 12 and 2 o'clock positions. These holes are believed to have the highest probability of detecting the most dangerous zones. During

---

[*] In terms of mechanical properties typical bursting situation can usually be met in hard, strong and brittle rock, e.g. having an unconfined compressive strength higher than 100 MPa and a modulus of deformation greater than 4 GPa.

Figure 13: Collapse of the filling of a karstic chimney crossed by Dodoni Tunnel. The collapse outcropped on the surface about 100 m. over the tunnel.

drilling, a supervising geologist or engineer should be present and, together with the driller, should watch for rapid changes in drill penetration rates, nature of the chippings and color of the return drilling water. An experienced driller will usually be able to detect changes in the drill performance and to give a reliable prediction of the nature of the ground ahead of the face.

When a significant weak zone is detected additional probe holes should be drilled to define the extent and shape of the zone as accurately as possible. In exceptional cases, one or two cored holes may be required to determine the nature of the filling material.

As a general rule grouting of the filling material within the cavity is a primry consideration in order to improve its cohesive strength. However, it has to be realized that the effects of such grouting are highly unpredictable, depending on the nature of the filling materials.

The support measures to be used depend upon the nature and the extent of the weak zone. When a weak zone (e.g. a karstic cavity or a filled pocket of limited extent) is to be dealt with, the use of forepoles to bridge the cavity should be considered. These forepoles play an entirely different role from those used to pre-support the face in squeezing ground. Their function is to form a roof over the tunnel through which the weak filling material of the cavity cannot pass. Hence, depending on the volume of material to be supported, the forepoles should be reasonably light (say 75 mm diameter tubes) and they should be as closely spaced as possible. They should also be long enough to ensure that they are securely socketted in good limestone on either side of the cavity. The number of forepoles to be installed should be limited to the number required to form an effective barrier under the cavity. It is not necessary to implement a complete support system, with an extensive forepole umbrella and additional support measures, such as that used in squeezing ground.

When the probe drilling detects a continuous feature of significant size, the approach has to be quite different from that described above. In such a case, the rock mass on either side of the cavity will be most probably weaker than the surrounding limestone and the zone may be 10 m thick or more, depending on the orientation of the void. In such a case, it is prudent to implement the full forepoling solution, similar to that used in squeezing ground (Fig. 14).

One further possibility needs to be considered and that is the case of a large empty karstic void. Such a void will generally require bridging and backfilling. The nature of the backfill will depend on the location of the void relative to perimeter of the tunnel. If water is associated with the void, drainage holes have to be foreseen as described earlier (see also Fig. 6)

Encountering a vertical karst channel, which is the most common case in the transfer zone of the aquifer (model C in Table 1), unpredictable concentrated water pressure may load the tunnel lining. In order to prevent possible damage, forced drainage of the channel towards lower elevations has to be secured (Fig. 15).

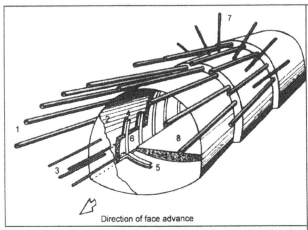

Direction of face advance

Figure 14: Full face excavation through weak ground under the protection of a forepole umbrella. The final concrete lining is not included in this figure (Hoek, 2000). The method can be applied in cases of large karstic caverns or large chimneys filled with cohesive soil under substantial load. Note that it is not always necessary to implement all the components shown in this figure.

1. Forepoles – typically 75 or 114 mm diameter pipes, 12 m long installed every 8 m to create a 4 m overlap between successive forepole umbrellas.
2. Shotcrete – applied immediately behind the face and to the face, in cases where face stability is a problem. Typically, this initial coat is 25 to 50 mm thick.
3. Grouted fiberglass dowels – Installed to reinforce the rock immediately ahead of the face. These dowels are usually 6 to 12 m long and are spaced on a 1 m x 1 m grid.
4. Steel sets – Installed as close to the face as possible and designed to support the forepole umbrella and the stresses acting on the tunnel.
5. Invert struts – Installed to control floor heave and to provide a footing for the steel sets.
6. Shotcrete – Typically steel fiber reinforced shotcrete applied as soon as possible to embed the steel sets to improve their lateral stability and also to create a structural lining.
7. Rockbolts as required. In very poor quality ground it may be necessary to use self-drilling rockbolts in which a disposable bit is used and is grouted into place with the bolt.
8. Invert lining – Either shotcrete or concrete can be used, depending upon the end use of the tunnel.

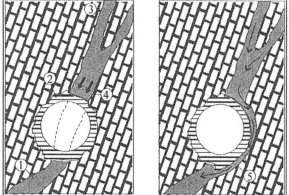

Figure. 15: Crossing a big karst conduit by tunnelling (Milanović, 2000). 1. Karst, 2. Tunnel, 3. Water level in the channel, 4. Lining under concentrated pressure, 5. Drainage around the tunnel pipe.

P. Milanović (2000) describes the way the tunnel Učka in Croatia faced the case of the approach of the tunnel to a large cavernous zone (Fig. 16).

"The total length of investigated karst channels is more than 1,300 m. The largest cave hole is 175 m long and 70 m wide while the cave hall close to the tunnel is 60 m long, 40 m wide and 55 m high. In the lower part of the cave system permanent groundwater flow is present. Summer flow rates vary from 10 to 30 l/s. After rainy storms the flow abruptly increases and exceed 1000 l/s (Hudec et al, 1980). Since the stability of the tunnel is endangered, artificial support is needed. Total plugging of the cave could not be carried out because of the existing permanent water flow through the cave. The use of reinforce concrete arch structures were rejected as complicated and very expensive. Since heavy mechanical equipment could not be used in the cave space, a simplified solution was applied to support the potentially unstable rock mass between the tunnel and the roof of the cave. The roof was supported by well compacted and stabilized fill (using water jet) which composed of fine-grained limestone aggregate, partially strengthened by addition of cement (150 kg of cement per 1 m³ of fill). The strengthened zone forms the concrete "skin" around filled aggregate, and over the natural cone of limestone blocks. The average thickness of this concrete zone was 1 m. The space between strengthened aggregate and the cavern roof was filled by concrete. Along the entire contact zone between concrete and cave roof, 24 mm anchors have been installed. The entire supported structure was constructed without grouting treatment."

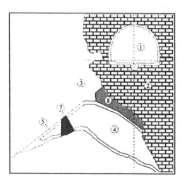 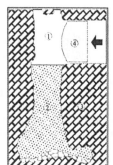 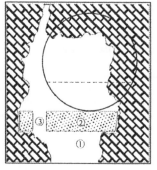 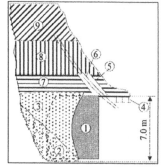

Figure 16: Driving the tunnel over a big karstic cavern. Longitudinal (a) and perpendicular (b) cross-sections (from Hudec et al, 1980). 1. Tunnel, 2. Limestone, 3. Cavern, 4. Fill – aggregate, 5. Fill – reinforced with cement, 6. Stone wall, 7. Retaining wall (reinforced fill), 8. Concrete, 9. Tunnell floor.

Figure 17: Cavern treatment by concrete filling for a TBM drive (Milanović, 2000). 1. Cavern; 2. Part of cavern filled by concrete; 3. Limestone; 4.TBM.

Figure 18: Cavern rehabilitation by bridging for a TBM drive (Milanović, 2000). 1. Cavern; 2. Concrete slab; 3. Aeration – drainage opening.

Figure 19: Filling of a cavern from inside the tunnel (Milanović, 2000). 1. TBM; 2. Cave clayey deposits; 3. Inert material – sand; 4. Tunnel support; 5. Shaft; 6. Pipe connected with concrete pump; 7, 8, 9. Stages of concreting.

When tunnel boring machines (TBM) are to be used, local realignment of the tunnel axis in order to avoid voids is not an option and usually a stoppage is imposed in order to backfill or bridge the void (Fig. 17, 18). If backfilling of the karstic cavern should be carried out from within the tunnel (Fig. 19) care should be given not to obstruct the cutter head with the concrete operations. When naturally filled, the voids have to be crossed by conventional tunnelling since the TBM, being usually of an open type, cannot bore the fill which could ravel through the cutter head of the machine.

SOME CONTRACTUAL CONSIDERATIONS

Successful groundwater control during underground construction is probably as dependent on the form of the contractual documents as it is on technical details of any particular method (Cuertin, 1989).

Indeed, groundwater control is always a high risk activity and the way that the risk is to be shared is regulated in the contractual documents. This sharing is obviously dependent on the results of a quality site investigation program and on the sound understanding of the most probable predominant karstic model.

Sharing risks associated with unpredictable events can substantially improve the success of a contract both in terms of cost as well as of schedule control. Where the overall financial and contractual arrangements permit, it may be possible for all parties to agree on some form of "Risk Sharing Package".

Table 2: Example of a "Risk Sharing Package" for a tunnel in karstic limestone (Hoek and Palmieri, 1998, slightly simplified)

Tunnel length: 5020 m; finished diameter: 3.5 m; concrete lining to be provided. Geology: Miocene Limestone and Jurassic Dolomitic Limestone, cover 80 to 200 m.	
Description	*Extent (m)*
Massive to slightly jointed	2600
Closely jointed	1650
Weakly cemented	670
Fault zones, karstic cavities	100
Risk description	*Risk sharing*
Rock mass quality along the route	Since the main risk is associated with large karstic cavities and the average rock quality is fair to good, deviations from the assumed distribution can be included in the Contractor's risk.
Presence of groundwater	Inflows into the tunnel are within the Contractor's risk up to the following limits: a) 20 l/s at the tunnel face; b) 50 l/s at the tunnel portal; c) head of water not to exceed 50 m.
Karstic cavities	Limits to the Contractor's risk: a) cavity zone not exceeding the tunnel span, say 4 m; b) water inflows not exceeding 20 l/s and decreasing to less than 20 l/s at the unconfined state; c) delays caused by the occurrence of cavities do not exceed 30 days

An example of a Risk Sharing Package for a tunnel in karstic limestone is given by Hoek and Palmieri (1989) and is presented in Table 2. In this particular project, the general geological conditions indicated a potential for sizeable karstic cavities. Site investigations had revealed some information, however the incremental investigation cost, which aimed at fully identifying them was not considered financially justifiable.

Hence, a set of limits was derived and agreed upon by both parties to the contract, based upon experience in the construction of similar tunnels. All of the indicators shown in Table 2 can be measured by simple quantitative site observations.

The provisions listed in Table 2 represent a reasonable risk package and it is probable that any international arbitrator would classify anything in excess of the limits defined in this package as Force Majeur conditions.

CONCLUSIONS

Tunnelling and mining in karst terrane require a thorough hydrogeological knowledge over a broader area. Lack of this knowledge may result to a design which will not be able to face problems or hazards that may occur during construction with probably dramatic consequences on the completion of the operation. Judgment and engineered solutions should always assist any decision at all stages during design and construction.

ACKNOWLEDGMENTS

Particular acknowledgments are due to Nikos Sirtariotis as well as Achilleas Papadimitriou, Maria Benissi and Jasmine Athanasiou for their assistance in the editing of this paper.

REFERENCES

Anonymous, 1992, Solving water problems: World Tunnelling, April issue, p. 154-160

Bauer, G., 1994, How to control groundwater in tunnelling projects: Tunnels and Tunnelling, June issue, p. 55-57

Breznik, M., 1998, Storage Reservoirs and Deep Wells in Karst Regions: Balkema publ., 251 p.

Calembert, L., 1975, Engineering Geological problems in karstic regions. Bulletin of IAEG, No 12, p. 39-82

Ford, D.C., and William, P.W., 1992, Karst Geomorphology and Hydrology: Chapman & Hall, p. 534-536

Guertin, J.D.Jr., 1989, Water control: in «Underground Structures. Design and Instrumentation». R.S. Sinha editor, Elsevier, p. 321-371

Hoek, E., 2000, Big Tunnels in Bad Rock: 2000 Terzaghi Lecture, Seattle, October 2000. To be published in the ASCE journal of Geotechnical and Geoenvironmental Engineering

Hoek, E., and Palmieri, A., 1999, Geotechnical risks on large civil engineering projects: 8th International IAEG Congress, Vancouver, Balkema Publ., p. 79-88

Hudec, M., Bozicević, S., Bleiwess, R., 1980, Support of cavern roof near tunnel Učka: 5th Yugoslav Symposium for Rock Mechanics and Underground Works, Split

Kavvadas, M.J., and Marinos, P.G., 1994, Prediction of groundwater table lowering for lignite open cast mining in a karstic terrain, in Western Macedonia, Greece: Quarterly Journal of Engineering Geology, No 27, p. 41-55

Marinos, P.G, 1992, Karstification and groundwater hydraulics of the interior of large calcareous massifs: The case of Giona mountain in Central Greece: In "Hydrogeology of selected Karst Regions", H. Pale and K. Bac editors, IAH International Contributions to Hydrogeology, Verlag Heinz Heise or Balkema Publ., Vol. 13, p. 241-247

Marinos, P.G, 1996, Hydrogeological problems related to tunnelling and mining works: Ingegneria e Geologia degli Arquiferi, No 6, p. 45-52.

Milanović, P., 2000, Geological Engineering in karst: Zebra Publ. Belgrade (zebra@EUnet.yu), 347p.

Powers, J.P., 1985, Dewatering; avoiding its unwanted side effects: Technical Committee on groundwater control of the underground technology research council of the ASCE Technical Council on Research, American Society of Civil Engineers, New York, NY

White, W.B., 1999, Karst Hydrology: Recent developments and open questions: Proceedings 7th Conference, "Hydrogeology and Engineering Geology of Sinkholes and Karst, B. Beck et al edit., Balkema publ. p. 3-20

II Sinkhole Formation

Geotechnical and Environmental Applications of Karst Geology and Hydrology, Beck & Herring (eds)
© 2001 Taylor & Francis, ISBN 90 5809 190 2

Geological and geotechnical context of cover collapse and subsidence in mid-continent US clay-mantled karst

TONY COOLEY   Consulting Hydrogeologist/Geotechnical Engineer, Lexington, KY 40517, USA, tlcool0@pop.uky.edu

ABSTRACT

This paper presents a synthesis of geologic and geotechnical concepts to present a unified model of conditions controlling the development of cover-collapse sinkholes and associated ground subsidence. The geotechnical characteristics of the overlying clay mantle and occurrence of the associated cover-collapse features are not random, but rather are directly tied to the underlying water flow routes and their development through time. The clay mantle and underlying epikarst are two components of a single system, each of the components influencing the other. This paper brings together these two aspects in terms of the author's personal experience and observations as a geologist, geotechnical engineer, hydrogeologist, and caver. A summary of the basic model follows.

Much of the clay mantle and pinnacled upper surface of the epikarst forms while surface drainage still prevails. At this stage, the karst underdrains are insufficiently developed to transport soils, though some subsidence into cutters occurs due to dissolutional rock removal. Soil arches and macropore flow routes associated with cutters have developed by this stage. As competent deep conduits extend into the area by headward linking, the cutters with the most favorable drains are linked to the conduits first and act as attractors for the development of a tributary, laterally integrated drainage system in the epikarst. Once the most efficient cutter drains become competent to transport soils, the depressed top-of-rock and ground surfaces characteristic of dolines develop. A given doline underdrain is likely to have multiple tributary drains from adjacent cutters which vary in soil transport competence.

Soil stiffness in the clay mantle over the limestone varies as a result of the pattern of stresses imposed as the underlying rock surface is lowered by dissolution and later as soil piping locally removes soils. In the absence of karst, these soils would have developed a laterally uniform, stiff to very stiff consistency. Where soil near the soil/bedrock interface is locally removed, however, the weight of the materials overlying this void is transferred to abutment zones on the pinnacles by soil arches. Local soil loading in the abutment areas of these arches would increase at least on the order of 50% in the case of an isolated cavity. In some cases, multiple closely spaced cutters whose soil arches have narrow, laterally constrained abutment zones bearing on the intervening pinnacles may produce substantially higher soil abutment stresses. If the clays in the abutment zones don't fail, they would respond to this increase in stress by consolidating: stiffening and decreasing in volume. The cutters spanned by the soil arches accumulate raveled soils that are "underconsolidated", the soft zones noted between pinnacles by Sowers. A simple integral of stresses analysis makes it obvious, however that no continuous soft zone exists. It is the transfer of load to the pinnacles through the stiffened abutment soils that allows these locally soft areas to exist. Soil stiffness profiles from borings substantiate this pattern.

Cover-collapse features develop where soil transport through cutter drains is sufficient to remove the soils from beneath these arched areas. Two types of collapse have been observed: Type 1 collapses have an upward-stoping open void whose rubble pile is removed by transport as fast as it is generated; producing a deep, steep-sided final collapse. In some cases, multiple voids in clusters can form with narrow abutments separating them. Large collapses may involve a progressive failure of several members of a cluster, including intervening pillars. Type 2 features are soil-filled voids limited in their rate of upward growth by the rate of soil removal, have little open void space, and migrate to the ground surface as a column of soft soils, finally producing a shallow depression. The Type 2 features have geotechnical significance due to their effect on settlement under imposed loads. A single underdrain system may service both types of features, the behavior of particular voids being dependent on the relative efficiencies of their drains. This behavior can also change with time, as backfilling of the underdrains with soil or flushing out of the soil filling can occur with changes in hydrologic or erosional regimes.

INTRODUCTION

This paper presents a synthesis of karst and soil mechanics theory that addresses the factors controlling cover collapse and subsidence in clay-mantled U.S. mid-continent fluvio-karst developed on Paleozoic carbonates. It is based on both theory and the author's consulting and research experience in Alabama and Kentucky, with limited observations from other areas. The model is presented only in brief outline without much of my supporting documentation because of paper length limitations for this conference. An expanded version of this model will be available in a future publication.

The physical conditions of the clay mantle and development of cover collapse features are a direct result of ground water flow and soil transport into and through the underlying epikarst. While difficult or impractical to predict with precision, the locations where cover-collapse features occur are not random. There is a geologic context to collapse and subsidence processes. To be meaningful, finite-element or other engineering modeling must incorporate geologic input on the epikarst and soil mantle. Any mathematical model constrains the possible failure modes by its selection of algorithms; unless all realistic failure modes and material behaviors have been recognized for incorporation into a model, the model output is unreliable at best. In this paper, I will use the term "soil" in a soil-mechanics manner, rather than as used by agronomists, to refer to what is alternately called the regolith.

THE LINKING MECHANISM FOR FORMATION OF KARST UNDERDRAINAGE

In Paleozoic mid-continent carbonates, flow occurs through a triple porosity system. The primary component of the porosity is the intergranular flow through the rock matrix itself and is negligible for Paleozoic carbonates. The secondary component occurs along naturally occurring discrete discontinuities such as bedding planes and joints that predate weathering of the rock, though the lower stress field near the ground surface may slightly dilate them. Flow in this system follows rules for fracture flow, being dominated by the degree of opening of the planar features, called aperture, and the interconnectivity of these features. The secondary component controls flows prior to development of karst. The tertiary component of porosity, that due to enlargement of selected flow routes by dissolution of rock bounding these routes, is the essence of karst. While involving a small portion of the rock mass, the tremendous increase in transmissibility of dissolutionally enhanced flow routes can dominate flow through a rock, once it develops, and in some cases results in water velocities sufficient to transport soils.

Mid-continent carbonates were originally covered with clastic rocks and were uncovered by fluvial erosion. A soil cover developed under a surface drainage regime that incorporated clastic sediments as well as insoluble residue of the carbonates. Until an epikarst developed, most ground water flow occurred through these soils, with only a minor portion of flow moving through the secondary porosity system of the bedrock. The upper portion of the bedrock was attacked by dissolutionally aggressive groundwater, but the rate of reaction drops extremely rapidly after a threshold saturation of the solute is reached, so until karst underdrainage arrived at a location, little dissolution occurred deeper within the rock.

The arrival of karst underdrainage and its elaboration was successfully modeled by the Ewers network linking model (Ewers, 1982). Figure 1 shows the basic concept as applied to elaboration of a karst drain system on a vertical plane through a system with secondary porosity flow routes along joints and bedding planes. A similar model was applied for the creation and extension of the original conduit system. The transmissivity of the individual fractures is very sensitive to the aperture of the individual features, as discussed in White (1988). Dissolutional enlargement of the upper portion of a flow route increases its transmissivity, even though soil infilling may limit this increase. The low transmissivity of the remainder of the flow route largely controls the flow through the feature so nearly all of the available head loss occurs along the unenhanced portion of the route. In Figure 1a, the available head is that between water in the soil mantle and the head of an underlying conduit drain. For illustration purposes, it is assumed that the resistance to flow is uniform along the length of any given fracture, but differs between fractures. This is not an essential assumption and would not apply to the real world, it just simplifies illustration. Along the portion of the fracture that has been dissolutionally enlarged, the transmissivity has been greatly increased. Thus, in Figure 1a, nearly all of the head is expended in flowing through the unenhanced portion of the flow route.

Flow routes differ in their original transmissivity. Provided water is commonly available in the soils, the flow routes with higher original transmissivities will transmit more solute and it will move faster. This extends the dissolutionally enhanced portion of the route farther along such routes than in the less transmissive routes. By reducing the length of the remaining unenhanced portion of the route, this increases the gradient along this portion of the route and further increases its advantage over other routes. When the competing routes are interconnected, as by the bedding planes of Figure 1, the high head of the enhanced portion actually retards further development of adjacent competing flow routes. As seen in Figure 1b, the original head of the soil water is carried down with little change to the start of the unenhanced portion of the flow route. This is greater than the head in the adjacent flow routes, so some flow actually diverges from the main flow and reduces the gradient in the adjacent portion of these routes.

In Figure 1c, the favored flow route has connected to the low head of the conduit. At this point, flow will greatly increase through the connected feature and a cone of depression in the overlying soils will decrease the head available to continue propagating the competing flow routes. In addition, the head distribution in the connected flow route is similar to that of the conduit and now acts as an attractor for flow through the bedding planes from adjacent flow routes. Lateral linking results in the capture of the flow of these adjacent flow routes, ultimately producing a laterally integrated system as shown in Figure 1d. Because this is on a vertical plane, the lower, beheaded portion of the captured flow routes still have "sump" areas that will retain water and provide for some continued flow and downward extension of these adjacent flow routes, though at a much reduced pace.

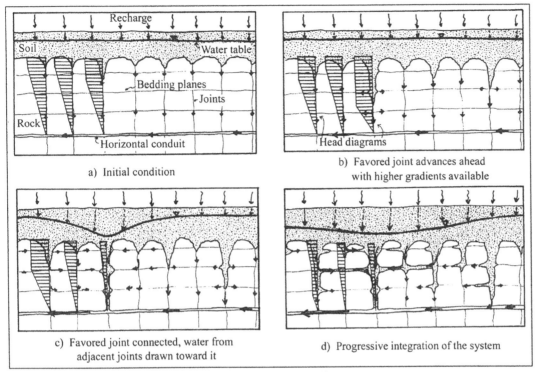

a) Initial condition

b) Favored joint advances ahead with higher gradients available

c) Favored joint connected, water from adjacent joints drawn toward it

d) Progressive integration of the system

Figure 1. Ewers network linking model applied to development of epikarst

SOILS AND MACROPORES

The soil mantle consists of soils transported in while surface drainage persisted and residual soils produced by the removal of the soluble fraction of the carbonate rock. Sowers (1996) notes that lowering of carbonate rock surfaces due to dissolutional weathering occurs at a rate of a few tens of millimeters per thousand years. The residual soils formed as the insoluble residue of this process would accrete to the bottom of the soil mantle at a rate on the order of a few millimeters per thousand years. This residue consists of a mixture of clay, silt, sand, and chert fragments. It would form under the loading of the overlying soils and would be in equilibrium with it; there is an abundance of time for consolidation (in the soil mechanics sense). Local zones of soft "underconsolidated" clays exist whose Overconsolidation Ratios (OCR's) are less than 1 if the overburden loading is assumed to equal the full weight of overlying materials. The OCR is defined as the ratio of the "preconsolidation pressure" as estimated from the shape of the one-dimensional consolidation test results on undisturbed soil samples to the estimated effective stress loading at the sample location. A truly underconsolidated soil would have excess pore pressures due to carrying some of the load in the pore fluids. These soft zones are not really underconsolidated, they never have experienced significantly greater loads than their present loads and are normally consolidated at those loads. The soft soils form in low spots in the bedrock surface sheltered by soil arches bearing on abutments on the adjacent higher portions of the rock. Soils in these abutment areas are "overconsolidated" as a result of the higher effective stresses in these locations.

A simple freebody equilibrium analysis shows that the integral of the vertical stresses on any horizontal surface must equal the weight of overlying materials provided the freebody extends beyond the limits of any lateral transfer of loading by vertical shears. The presence of soft "underconsolidated" zones requires the presence of "overconsolidated" soil arch abutment areas; no continuous soft layer could exist at the soil/bedrock interface. Further, "pinnacle punching" as described by Sowers (1996) is a realistic failure mode for surficial loading by fill or rapid natural deposition of sediment, but not as an explanation for the natural pattern of soft soils in the low portions of the soil/bedrock interface with firm to stiff soils on the pinnacle tops. The soil cover and bedrock surface did not form separately and then come together with the pinnacles impaling through a continuous soft layer of soil; the two components formed concurrently maintaining equilibrium throughout their history except for brief moments of cover collapse of a soil arch.

The stiffness of the clay mantle does often show a very stiff crust in the upper 3 to 6 meters (10 to 20 feet). OCR values of 2 or 3 are not uncommon in this upper soil. Desiccation loading is a major factor in this stiffness, though Sowers (1996) does mention age-related bonding of the clays with cements. Clay consolidation occurs in response to effective stress loading. When soil pore water pressures become negative, this increases the effective stresses beyond the simple weight of the overlying sediment. Roots, particularly tree roots, are a major factor in this because they can extend of depths of at least 6 to 9 meters (20 to 30 feet) and exert suctions equivalent to a negative head of more than 6 meters (20 feet) of water. I have observed tree roots in split spoon samples of fissured clays from the depths mentioned and have measured soil suctions with tensiometers near a root that went off-scale at the suction noted.

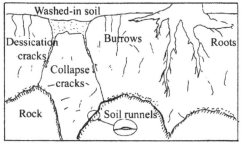

Figure 2. Macropore types

Figure 2 shows the type of macropores and other features that affect water flow through the soils. Water flow in the clayey mantle is strongly influenced by macropores during periods of prolonged wet weather or very intense storms. During drier weather when upper soils have negative soil porewater pressures, macropores are less important as water is drawn out of the walls of the macropore to satisfy matrix water demand. An intense storm with ponding or perching of water can locally overwhelm the matrix soil suction demand by channeling water down a macropore faster than it can be drawn laterally into the soils bounding the macropore even during a dry period. It is the wet weather and periods of high flow that are of most significance to cover collapse development so the above qualification will generally not be important to this discussion.

At depths to about 1 to 2 meters (3 to 6 feet) macropores are commonly abundant. These are typically animal burrows (worms and insects through mammals), root holes, desiccation cracks, and human artifacts, though they may include collapse cracks and swallets related to the underlying karst. A former collapse filled by a landowner with brush, trash, and old barbed wire is a major macropore that will be present at some sites. This network of shallow macropores is commonly laterally interconnected and lateral flow through the shallow macropore system should be considered in addition to overland flow in conveying water toward depressions such as dolines. Washed-in soils filling low areas such as former collapses and possibly partially extending down former collapse cracks are also features in this upper zone. Water perching at the base of such washed-in soils and above the residual soils can initiate flows down macropores connected to that interface.

Deeper macropores would mostly consist of tree roots, tree-root induced desiccation cracking, and collapse and soil deformation cracking. Such collapse cracking would include the network of openings between raveled soils within the collapse limits. These deeper macropores would mostly be vertically oriented. The enhancement of soil bulk permeabilities attributable to macropores would primarily be in the vertical direction at depths below the upper 2 meters. Were an excavation made into such a clay, the head available to produce flow into such an excavation may well be less than the bottom of the excavation. Even though the head in adjacent soils is well above the excavation floor, the lateral permeability is too low to promote inflow from that direction. The water flow route would have to be downward to the soil/bedrock interface, laterally along the runnel features there, and then back up lateral macropores into the excavation floor or walls. The head losses of such a route may be too great for this flow to occur.

At the top of rock itself, lateral flows would be conveyed by soil runnels. These are soil roofed and rock floored features that commonly are slightly sculpted into the rock by dissolution while the soil roof may or may not have been arched by erosion. Where I have observed these directly at my Frankfort, KY., research site, the clay soil is moist but firm or stiff essentially to the top of rock. Then immediately at the top of rock, water inflows occur as the power or hand auger actually contacts the rock. Often it is only as the last few centimeters (inch or so) of soil is removed that inflows occur. Once water begins to come in, water levels in the borings have risen as much as 30 or 40 centimeters (1 to 1.3 feet) in 10 minutes to an hour, indicating sumps occur along the flow route. It would be expected that such macropores would not be constrained to continuously downstream sloping channels like a surface channel. As a roofed feature, pressure flow like a pipe would not be surprising. Those I have examined using a garden trowel blade and 12 volt auto taillight bulb on the end of a hand auger rod have not had an observable height, being more like cracks in the vertical dimension. I could not observe the width of such a feature, but as the runnel could not bear load, its width would be limited by a necessity for soil arching to span it. Where a runnel crosses through a closely spaced boring array at the Frankfort research site, it also follows a slightly depressed area of the soil/bedrock interface. It may be difficult for runnels to go through soil abutment areas with their higher stresses. Dr. Ralph Ewers (personal communication) has seen runnels in the walls of washed out collapses that had a height of about 5 centimeters (2 inches), most of it in the soil with a slight channel into the rock, and a width of less than 30 centimeters (1 foot).

In addition to runnels, lag accumulations of the coarser fraction of the clay mantle are sometimes encountered at the soil/bedrock interface, especially in the throats of active drains capable of transporting soils. This is commonly a chert and sand concentration as the finer-grained soils are depleted by piping erosion from the flowing water.

At present, I must speculate that the vertical macropore system and runnels are directly connected as I have not yet done the soil staining and excavation work necessary to demonstrate this. Considering the low matrix permeability of the clay, it is judged likely that the macropores would be a significant factor in delivering water to the soil/bedrock interface and that runnels would develop in response to the availability of this water. A headward linking mechanism as discussed previously for karst development would also work for connecting the high head macropores to a discharge area along the bedrock surface.

RELATIONSHIP OF COVER COLLAPSE AND SUBSIDENCE TO DEVELOPMENT OF THE MACROPORE SYSTEM

Figure 3 illustrates a simple model for the development of an epikarst and associated macropore system and its influence on cover collapse and subsidence. The initial case, Figure 3a, would occur after the clastic rocks have been eroded off of the underlying carbonates and surface drainage prevails. The water table would be shallow because of the absence of significant underdrainage and the low bulk permeability of the clay. The soil mantle would be a combination of transported and locally derived materials with some accretion of insoluble residue to the bottom of the mantle. Ground water would aggressively attack the bedrock interface, lowering this. Some cutter development would occur due to the limited deeper circulation of water that would eventually produce a connection to the conduit. As most flow of ground water must be lateral in the soils and soil/bedrock interface at this point, it is also likely that a runnel system would develop and probably follow and enlarge the cutters. A laterally integrated drainage system at the soil/bedrock interface with possibly some small conduits slightly below would be expected with discharge to nearby low points. The more distant

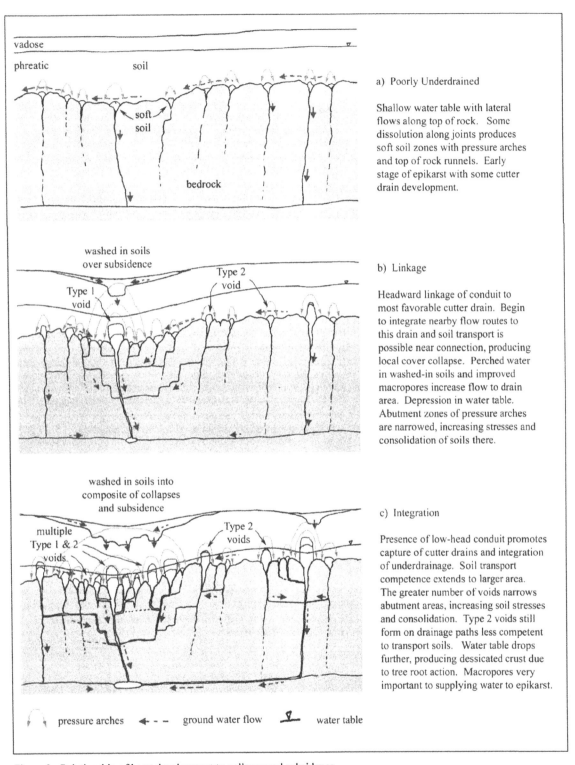

vadose

phreatic   soil

soft
soil

bedrock

a) Poorly Underdrained

Shallow water table with lateral
flows along top of rock. Some
dissolution along joints produces
soft soil zones with pressure arches
and top of rock runnels. Early
stage of epikarst with some cutter
drain development.

washed in soils
over subsidence

Type 1
void

Type 2
void

b) Linkage

Headward linkage of conduit to
most favorable cutter drain. Begin
to integrate nearby flow routes to
this drain and soil transport is
possible near connection, producing
local cover collapse. Perched water
in washed-in soils and improved
macropores increase flow to drain
area. Depression in water table.
Abutment zones of pressure arches
are narrowed, increasing stresses and
consolidation of soils there.

washed in soils into
composite of collapses
and subsidence

multiple
Type 1 & 2
voids

Type 2
voids

c) Integration

Presence of low-head conduit promotes
capture of cutter drains and integration
of underdrainage. Soil transport
competence extends to larger area.
The greater number of voids narrows
abutment areas, increasing soil stresses
and consolidation. Type 2 voids still
form on drainage paths less competent
to transport soils. Water table drops
further, producing dessicated crust due
to tree root action. Macropores very
important to supplying water to epikarst.

pressure arches   ◄ - -   ground water flow   ▽   water table

Figure 3. Relationship of karst development to collapse and subsidence

base level discharges would have little influence at this stage. Removal of the rock differentially would produce areas of non-support
that would accumulate raveled soft clays with transfer of loads to abutment areas by soil arches. Macropores would form to connect to
the soil/bedrock interface due to broad irregularities in the soil/bedrock interface that are too large-scale to be bridged by soil arches.

If a significantly lower head discharge area exists at a distance, this may extend into the area by headward linking according to the Ewers network linking model. This will occur along the most favorable local flow route and once this connection is made, the low head of the discharge area would be transferred to this area. This head differential occurs because the natural hydraulic gradients in soils are typically on the order of a few percent while gradients in karst conduit flow systems are typically a few tenths of a percent. Thus a conduit working headward from a discharge area by a series of sort headward links to potential sources of water would commonly gain in head advantage relative to a water table in the soil and non-underdrained rock.

Once a connection occurs, as in Figure 3b, the availability of the low head allows the connected drain to begin to capture the flows of nearby cutter drains. The system becomes more integrated and soil transport becomes possible through the main drain. Removal of the soft soils at some locations in the cutters provides space for further upward raveling of the pockets of soft soils formed under the soil arches. If the soil removal is very efficient, the pile of soil under the upward raveling roof of the new void leaves an empty or water-filled cavity above itself. The upward propagation of this cavity, and possibly its widening, continues until a cover collapse of the type commonly described occurs when the cavity roof thinned by raveling finally fails in shear. This has a final collapse form with vertical or overhanging side walls and is commonly deep. I refer here to this as a Type 1 collapse.

In adjacent areas, the size of the connections and the hydraulic gradients are less favorable such that these routes have lesser soil transport competence. Soil removal allows the upward raveling of the cavity roof, but is not fast enough to produce a cavity. The pile of raveled soils stays close to the roof, though a boring into such a feature may experience a water loss or gain as it crosses the roof into the rubble pile. I call these Type 2 features. I became aware of the existence of such features when drilling geophysical anomalies based on seismic reflection diffraction patterns. These borings repeatedly revealed columns of soft soils without a significant open void, while offset borings outside the anomalies did not encounter them. The diffraction pattern was a response of the seismic waves to the roof of a cavity. A soil arch forms in response to the width of the area of non-support. The height of the cavity doesn't matter. A cavity roof would form even if the raveled soils were in contact with the roof as this material would not be able to provide support to the roof once the soil strains required for formation of the soil arch occurred, though the seismic waves may not identify such a feature.

I then recognized the likely existence of such features at other sites. Propagation of such features to the ground surface also explained several surface depressions about 3 to 5 meters (10 to 15 feet) in diameter and 0.3 to 0.6 meter (one or two feet) deep which had well defined edges and significant edge curvatures. My first interpretation of washed-in-sediment-filled conventional collapses was unsatisfactory because the microrelief of the area consisted of very gently sloping flat ground, abrupt but curving edge of the feature, and then uniform bottom. Not even minor notching of the edge or rills in the surrounding soils were present. My alternate interpretation is that the raveling feature came to the surface with only about half a meter (foot or two) of void between roof and rubble pile with final failure as possibly a sag mediated by a root mat or a shallow collapse with minor erosional modification of the lip but no significant wash-in of surrounding soils. The depression might also reflect further downward movement of the soft soil plug after it reached the ground surface due to continued erosional removal of its base with continued soil transport through the drain.

Another feature of Figure 3b is the general sag of the ground in the vicinity of the drain produced by consolidation of the soils in the abutments. This results as the local stresses increase because of enlargement and proliferation of locations where soil transport removes soils at the base of the mantle. Washed-in soils collect in this depression as well as filling any Type 1 collapse features. These soils provide a perched aquifer above the residual soils for maintaining concentrated flow to the drain area. An accelerated dissolution of the upper bedrock surface also occurs because of the greater water conveyed to the depression by overland and shallow macropore flows. This further contributes to the ground surface depression and deformation cracking of the clay mantle.

Figure 3c simply shows a continuation of the process. As integration increases, the number of entryways into the rock competent to transport soils increases, narrowing the abutment areas between adjacent voids. If clay mantle thickness is great enough, a pressure arch spanning several voids may form with the intervening soil pillars on the narrow pinnacles acting like yield pillars in mining engineering pressure arch theory (Lucas and Adler, 1973). Progressive failures of multiple cavities due to failure of one pillar then are possible, creating collapses larger than individual voids.

A modification of the above scenarios is that formerly active flow routes can be plugged with deposited soils and require the development of new connections. Cavers have noted backfilling and re-excavating of caves on both a pre-historic and historic time scale. I have documented a situation where a plugged former bedrock depression supported about half a meter of head differentiation upward from the rock without having an effect on the water table in the overlying mantle.

REFERENCES

Ewers, R.O. (1982) An analysis of solution cavern development in the dimensions of length and breadth. PhD thesis, Geography, McMaster university, Hamilton, Ontario

Lucas, J.R. and Adler, L. (editors) (1973) Section 13, Roof and Ground Control, in SME Mining Engineering Handbook, Volume 1, A. Cummins and I.A. Given editors, Society of Mining Engineers of AIMMPE, New York

Sowers, G. F. (1996) Building on Sinkholes, ASCE Press, American Society of Civil Engineers, New York 202 p.

White, W.B. (1988) Geomorphology and Hydrology of Karst Terrains, Oxford University Press, New York, 464 p.

Geotechnical and Environmental Applications of Karst Geology and Hydrology, Beck & Herring (eds)
© 2001 Taylor & Francis, ISBN 90 5809 190 2

A comparison of human-induced sinkholes between China and the United States

YONGLI GAO & E.CALVIN ALEXANDER, JR.   Dept. of Geology & Geophysics, University of Minn.,
Minneapolis, MN 55455, USA, gaox0011@tc.umn.edu, alexa001@tc.umn.edu

MINGTANG LEI   The Institute of Karst Geology, Guilin, Guangxi 541004, China,
mingtang@mailbox.gxnu.edu.cn

ABSTRACT

Human-induced sinkholes have become a substantial problem in many countries due to economic development. This paper compares the human impact on sinkhole development between the world's largest developing country, China, and the world's largest developed country, the United States.

The number of human-induced sinkholes in China has dramatically increased since the 1970s. Water-pumping sinkholes, mine-drainage sinkholes, and reservoir-induced sinkholes are the three major types of human-induced sinkholes in China. The most disastrous sinkhole cases are related to the mining industry. Human-induced sinkholes in the United States are divided into two types: sinkholes induced by lowering water level due to pumpage and sinkholes resulting from construction (Newton, 1984). Sinkholes resulting from groundwater withdrawal are the dominant type of human-induced sinkholes in most southeastern karst regions of the United States such as Florida, Alabama, and South Carolina. Although there is a significant number of sinkholes caused by groundwater withdrawal, the number of sinkholes caused by construction has increased in other parts of the United States such as Tennessee, Kentucky, Pennsylvania, Georgia, Missouri, North Carolina, Minnesota, and Virginia.

Due to limited water resources, higher population density, an abundance of karst lands, and poor construction management, human-induced sinkholes are more disastrous in China. Some management methods used to investigate and treat sinkholes in the United States are applicable in China and may help prevent sinkhole development and reduce the damages caused by human-induced sinkholes. Discoveries and approaches made in China will be mutually beneficial to the management and evaluation of human-induced sinkholes in both China and the United States.

INTRODUCTION

During the past three decades, problems caused by human-induced sinkholes in karst regions have increasingly attracted public and regulatory attention as well as scientific interest. The world's karst areas have long been recognized as fragile environments that are vulnerable to groundwater contamination and degradation by man's activities. Sinkholes are common features in a karst terrain and serve as direct links between the surface and the underlying aquifers. Surface contamination typically enters the aquifer systems on time-scales that range from days down to seconds. Contamination can then quickly spread through the groundwater system along solution channels and pollute the aquifer and surface waters via springs. An additional hazard associated with sinkholes is the danger of the catastrophic surface collapse. Within seconds or over a period of a few hours to days, large sinkholes can develop and destroy surface structures.

Induced sinkholes are those caused or accelerated by human activities. The number of human-induced sinkholes in China has dramatically increased since the 1970s due to economic development (Liu, 1997). According to the most recent statistics, sinkholes have appeared in 26 provinces in China and more than 70% of those sinkholes are related to human activities (Xing and Que, 1997). The majority of the sinkholes in China are in southern China including the provinces of Guangxi, Guizhou, Hunan, Jiangxi, Sichuan, Yunnan, Hubei, and Guangdong. Water-pumping sinkholes, mine-drainage sinkholes, and reservoir-induced sinkholes are the three major types of human-induced sinkholes in China.

In the U.S., most investigations and studies dealing with human-induced sinkholes started since the 1950s (Newton, 1984). Human-induced sinkholes were categorized as collapses induced by lowering groundwater levels and collapses induced by raising groundwater levels (Aley and others, 1972). Newton (1984) divided human-induced sinkholes as two types: those induced by lowering water level due to pumpage and those caused by construction. Diversion of drainage and impoundment of water account for most sinkholes resulting from construction (Newton, 1986). Subsequent workers have documented several other types of human activities that induce sinkholes but water pumping, water impoundments, and construction/drainage changes are prominent in most lists.

This paper compares distributions, causes, and assessments of human-induced sinkholes between China and the United States. The purpose of this study is to compare and contrast the influence of human activities on sinkhole formation in China and the United States in the hope that this comparison may encourage the exchange of ideas about how effectively to prevent and reduce damages caused by human-induced sinkholes.

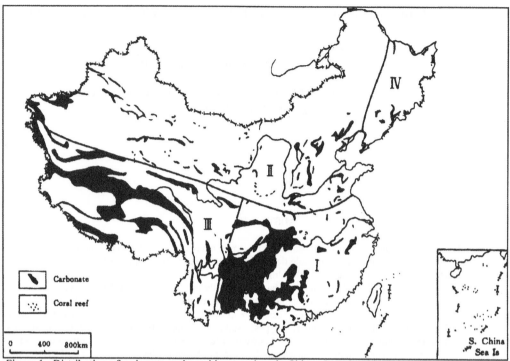

Figure 1. Distribution of carbonate rocks and karst regions in China. I – subtropical karst; II – arid, semi-arid karst; III – high mountain karst; IV – temperate semi-humid karst (Yuan, 1998)

METHODOLOGY

Since water pumping, drainage diversion and water impoundment are the three major human activities causing sinkholes in both China and the United States, sinkholes induced by these three activities were compared in this paper. Mingtang Lei provided statistics of human-induced sinkholes in China based on data gathered in the most recent national sinkhole database developed by the Institute of Karst Geology. To our knowledge no national tabulation of human-induced sinkholes has been made in the United States. Detailed investigations on human-induced sinkholes have been published in Alabama (Newton, 1986) and Missouri (Aley and others, 1972; Williams and Vineyard, 1976). Therefore, statistics of human-induced sinkholes in the United States were based on the Alabama (Newton, 1986) and Missouri (Aley and others, 1972; Williams and Vineyard, 1976) examples. Although other karst regions of the U.S. have distributions that differ in detail, we believe the major trends are similar.

RESULTS AND DISCUSSION
Karst regions in China and the United States

About 20% of the earth's land surface is karst (White and others, 1995). Karst lands in China are mainly developed in areas where carbonate rocks at or near surface, more than 30% of Mainland China is karst. The total area of karst lands in China is about 3,443,000 km^2 (Yuan and others, 1998). Figure 1 shows the distribution of karst and carbonate areas in Mainland China. Most of the carbonate rocks are pre-Triassic except some Jurassic and Cretaceous carbonate rocks in Tibet. Subtropical karst in southern China is the most unique and important karst region in China (region I in Figure 1). The famous tower karst is developed in this region and most of the sinkholes are distributed in southern China.

Davies and Legrand (1972) reported that nearly 15% of the continental United States contains soluble rocks at or near surface. That corresponds to about 1,183,000 km^2 of karst in the U.S., i.e. China has about three times as much karst as the U.S. Carbonate karst lands cover more than 40% of the contiguous United States east of Tulsa, Oklahoma (White and others, 1995). Figure 2 displays the major karst regions in the United States (Davies and Legrand, 1972). Limited water resources, higher population density, an abundance of karst lands, and poor construction management caused human-induced sinkholes to be more disastrous in China.

China's population in 2000 was about 1.26×10^9 people while the comparable U.S. figure was about 2.76×10^8 people (U.S. Census Bureau, 2000). China has about 4.6 times as many people as the U.S. In both countries the density of people living in most karst regions is higher than their average population densities. In both countries the karst areas are desirable and habitable regions. Human-induced sinkhole development has tracked modern energy-intensive and material-intensive human activities. Induced sinkholes in U.S. karst regions began to require study and management as early as the 1950s. In China, despite its larger karst areas and higher population, it was not until industrial activities intensified in the 1970s that induced sinkholes were recognized as a significant problem.

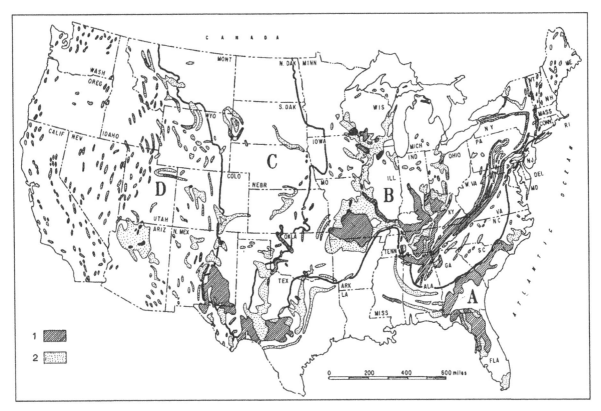

Figure 2. Distribution of soluble rocks and karst regions in the United States. A – Atlantic and Gulf Coastal Plain region; B – east-central region of Paleozoic and other old rocks; C – Great Plains region; D – western mountain region; 1 – karst areas; 2 – carbonate and sulfate rocks at or near surface (Davies and Legrand, 1972)

Human-induced sinkholes in China and the United States

Table 1 lists a summary of all different types of sinkholes in China. Water-pumping sinkholes, mine-drainage sinkholes, and reservoir-induced sinkholes are the three major types of human-induced sinkholes recorded in the national sinkhole database developed by the Institute of Karst Geology in China (Table 1). Mine-drainage sinkholes are the most disastrous sinkholes since many mineral deposits are located in or close to karst aquifers (Zhou, 1997). In July 1970, in the Zhangcun mine, Heshan Coal Basin, Guangxi, a flood of karst water into the mine dropped the overlying groundwater level 24m. An area of about 2000 m^2 subsided on the surface, and a number of other collapses occurred afterwards (Yuan, 1987). Water pumping is the second most important activity causing catastrophic sinkholes in China. Karst

Table 1. Types of sinkhole sites and sinkholes in China, based on sinkholes database from the Institute of Karst Geology in China

Types of Sinkhole	# of sites	% sites	# of sinkholes	% sinkholes
Unknown	151	10.5%	3047	6.8%
Natural	300	20.9%	2798	6.2%
Other	203	14.1%	1634	3.6%
Mine-drainage	203	14.1%	30974	69.0%
Water-Pumping	444	30.9%	4922	11.0%
Water-Impoundment	137	9.5%	1529	3.4%

aquifers are the primary fresh water resources in many parts of China and are increasingly over pumped. For example, in the past 20 years, large-scale pumping has lowered water levels 10-20m in many karst areas of northern China (Lin and Liao, 1999). Nearly 70% of the sinkholes in China were induced by mine drainage and these sinkholes clustered in 14% of all the sinkhole sites. On the other hand, water-pumping induced sinkholes were widely spread over 39% of all the sinkhole sites in China but account for only 11% of the total number of sinkholes.

Table 2 and Table 3 list the three major types of human-induced sinkholes in Alabama and Missouri. In Alabama, the majority (more than 80%) of human-induced sinkholes were caused by water pumping, only three drainage-diversion induced sinkholes were recorded before 1981 (Table 2). In Missouri, the majority (more than 70%) of human-induced sinkholes were caused by drainage diversion, water-pumping induced sinkholes (less than 10%) are significantly less than those in Alabama (Table 3). Troester and Moore (1989) stated that about 9% of groundwater withdrawals (25 million cubic meters per day) were from carbonate aquifers in the U.S. in 1985. In Florida, heavy pumping of water from wells at an unusually rapid pace is a major triggering mechanisms associated

Table 2. Distribution of human-induced sinkholes in Alabama before 1981

Types of Sinkhole	# of sinkholes	age
Water Pumping	1157	1950s-1970s
Drainage Diversion	3	1976
Water Impoundment	164	1950s-1981

Source: Newton , 1986

Table 3. Distribution of human-induced sinkholes in Missouri before 1976

Types of Sinkhole	# of sinkholes	age
Water Pumping	7	1930s-1970s
Drainage Diversion	62	1870s-1972
Water Impoundment	18	1870s-1969

Sources: Aley and others, 1972; Williams and Vineyard, 1976

with sinkhole development (Currin and Barfus, 1989). Hundreds of sinkholes formed within six hours during the development of a newly drilled irrigation well on February 25, 1998, in west central Florida (Tihansky, 1999). Some human-induced sinkhole examples in other parts of the U.S. are as follows: 1) about 70% of the sinkholes in the Warren County, Kentucky were induced primarily by drainage diversion associated with farming and urban development (Crawford and others, 1999); 2) quarry dewatering was believed to be associated with sinkhole formation in the North Carolina coastal plain since 1994 (Strum, 1999); 3) seepage from the pipes for storm water storage and conveyance was a likely cause of subsequent sinkhole formation at three commercial sites in eastern Tennessee (Scarborough, 1995); 85% of the 72 sinkholes along eastern Tennessee highways were mainly induced by drainage diversion along road ditch lines (Moore, 1987); and 4) failures of three waste water treatment (WWTF) lagoons induced 18 sinkholes in Minnesota from 1974 to 1992 (Alexander and Book, 1984; Jannik and others, 1991; Alexander and others, 1993).

Cooperation of study on karst and sinkholes between China and the United States

Scientists and researchers from both China and the U.S. have started to work together to solve karst and sinkhole problems since the 1980s. In May 1983, scientists from both countries organized a symposium on karst environment for the 149th session of AAAS in Detroit, Michigan, with sinkhole as one of its main points of discussion. From 1983 to 1999, seven international conferences on sinkholes were held in the U.S., Chinese scientists took active part in these meetings. Starting from 1990, the National Natural Science Foundation of China and the Ministry of Geology and Mineral Resources of China have sponsored three International Geological Correlation Programs (IGCP 299, IGCP 379, and IGCP 448) about global karst correlation. Scientists from the U.S. have made significant contributions to these programs. Some management methods used to investigate and treat sinkholes in the United States are applicable in China and may help prevent sinkhole development and reduce the damages caused by human-induced sinkholes. In the 1990s, Chinese institutions began to import Ground Penetrating Radar (GPR) techniques from the U.S. to investigate sinkholes. Remote Sensing (RS), Geographical Information System (GIS), and Global Positioning System (GPS), have been widely used in both countries to process and manage the information related to sinkholes. Discoveries and approaches made in China will be mutually beneficial to the management and evaluation of human-induced sinkholes in both China and the United States. Chines new nation-wide sinkhole database has proved to be an exceptionally effective and comprehensive approach to evaluate, predict and manage sinkholes in different regional scales (Lei and others, 1997; Ru and others, 1997).

CONCLUSION

China has about three times as much karst terrain as does the contiguous 48 United States. In both countries the karst areas are desirable and habitable regions. Induced sinkholes in the U.S. karst regions began to require study and management as early as the 1950s. In China, despite its larger karst areas and higher population, it was not until industrial activities intensified in the 1970s that induced sinkholes were recognized as a significant problem. Water-pumping sinkholes, mine-drainage sinkholes, and reservoir-induced sinkholes are the three major types of human-induced sinkholes in China. The most disastrous sinkhole cases are related to the mining industry. In China, mine-drainage induced sinkholes clustered in 14% of all the sinkhole sites; meanwhile, water-pumping induced sinkholes were widely spread over 39% of all the sinkhole sites in China but account for only 11% of the total number of sinkholes. Sinkholes resulting from groundwater withdrawal are the dominant type of human-induced sinkholes in most southeastern karst regions of the United States such as Florida, Alabama, and South Carolina. Although there is a significant number of sinkholes caused by groundwater withdrawal, the number of sinkholes caused by construction has increased in other parts of the United States such as Tennessee, Kentucky, Pennsylvania, Georgia, Missouri, North Carolina, Minnesota, and Virginia. Water impoundment and diversion of drainage account for most of the construction-induced sinkholes. Interdisciplinary collaborations between scientists from both countries and all over the world would significantly improve the study of reducing and preventing damages caused by human-induced sinkholes.

REFERENCES

Alexander, E.C., Jr., and Book, P.R., 1984, Altura Minnesota lagoon collapses: *in* Beck, B. F., ed., Sinkholes: their geology, engineering and environmental impact, Proceedings of the First Multidisciplinary Conference on Sinkholes: Orlando, Florida., 15-17 October, A.A. Balkema, Rotterdam, p. 311-318.

Alexander, E.C., Jr., Broberg, J.S., Kehren, A.R., Graziani, M.M., and Turri, W.L., 1993, Bellechester Minnesota lagoon collapses: *in* Beck, B. F., ed., Applied karst geology, Proceedings of the Fourth Multidisciplinary Conference on Sinkholes and the Engineering and Environmental Impacts of Karst: Panama City, Florida., 25-27 January, A.A. Balkema, Rotterdam, p. 63-72.

Aley, T.J., Williams, J.H., and Massello, J.W., 1972, Groundwater contamination and sinkhole collapse induced by leaky impoundments in soluble rock terrain: Missouri Geological Survey and Water Resources, Engineering Geology Series, No. 5, 32p

Crawford, N.C., Lewis, M.A., Winter, S.A., and Webster, J.A., 1999, Microgravity technique for subsurface in investigations of sinkhole collapses for detection of groundwater flow paths through karst aquifers: *in* Beck, B.F., Pettit A.J., and Herring J.G., eds., Hydrogeology and Engineering Geology of Sinkholes and Karst, Proceedings of the Seventh Multidisciplinary Conference on Sinkholes and the Engineering and Environmental Impacts of Karst: Harrisburg-Hershey, Penn., 10-14 April, A.A. Balkema, Rotterdam, p. 203-218.

Currin, J.L., and Barfus, B.L., 1989, Sinkhole distribution and characteristics in Pasco County, Florida: *in* Beck, B. F., ed., Engineering and Environmental Impacts of Sinkholes and Karst, Proceedings of the Third Multidisciplinary Conference on Sinkholes and the Engineering and Environmental Impacts of Karst: St.Petersburg Beach, Florida., 2-4 October, A.A. Balkema, Rotterdam, p. 97-106.

Davies, W.E., and Legrand, H.E., 1972, Karst of the United States: *in* Herak, M., and Stringfield, V.T., eds., Karst, Important Karst Regions of the Northern Hemisphere: Elsevier Pub Co., Amsterdam, p. 466-505.

Jannik, N.O., Alexander, E.C., Jr., and Landherr, L.J., 1991, The sinkhole collapse of the Lewiston, Minnesota waste water treatment facility lagoon: Proceedings of the Third Conference on Hydrogeology, Ecology, Monitoring, and Management of Groundwater in Karst Terranes, Nashville, TN, 4-6, December, USEPA and NGWA, p. 715-724.

Lei, M., Jiang, X., and Li, Y., 1997, Karst collapse information system of Guilin: The Chinese Journal of Geological Hazard and Control, Vol.8, Supplement, p. 78-82.

Lin, X., and Liao, Z., 1999, Agricultural impact on karst water resources in China: *in* Drew, D., and Hötzl, H., eds., Karst Hydrogeology and Human Activities: A.A. Balkema, Rotterdam, Netherlands, p. 63-66.

Liu, C., 1997, Research on distributive regularity of karst collapses area in China: The Chinese Journal of Geological Hazard and Control, Vol.8, Supplement, p. 11-17.

Moore, H.L., 1987, Sinkhole development along 'untreated' highway ditchlines in East Tennessee: *in* Beck, B. F., and Wilson, W.L., eds., Karst Hydrogeology: Engineering and Environmental Applications, Proceedings of the Second Multidisciplinary Conference on Sinkholes and the Environmental Impacts of Karst: Orlando, Florida., 9-11 February, A.A. Balkema, Rotterdam, p. 115-119.

Newton, J.G., 1984, Review of induced sinkhole development: *in* Beck, B. F., ed., Sinkholes: Their Geology, Engineering and Environmental Impact, Proceedings of the First Multidisciplinary Conference on Sinkholes: Orlando, Florida., 15-17 October, A.A. Balkema, Rotterdam, p. 3-9.

Newton, J.G., 1986, Development of Sinkholes Resulting from Man's Activities in the Eastern United States: U.S.G.S. Circular 968, 54 p.

Ru, J., Zhou, L., Ma, Z., and Mo., Y., 1997, GIS for evaluation & prediction of karst collapse in China: The Chinese Journal of Geological Hazard and Control, Vol.8, Supplement, p. 78-82.

Scarborough, J.A., 1995, Risk and reward: Pipes and sinkhole in East Tennessee: *in* Beck, B.F., ed., Karst Geohazards, Proceedings of the Fifth Multidisciplinary Conference on Sinkholes and the Engineering and Environmental Impacts of Karst: Gatlinburg, Tenn., 2-5 April, A.A. Balkema, Rotterdam, p. 349-354.

Strum, S., 1999, Topographic and hydrogeologic controls on sinkhole formation associated with quarry dewatering: *in* Beck, B.F., Pettit A.J., and Herring J.G., eds., Hydrogeology and Engineering Geology of Sinkholes and Karst, Proceedings of the Seventh Multidisciplinary Conference on Sinkholes and the Engineering and Environmental Impacts of Karst: Harrisburg-Hershey, Penn., 10-14 April, A.A. Balkema, Rotterdam, p. 63-66.

Tihansky, A.B., 1999, Sinkholes, West-Central Florida: *in* Galloway, D., Jones, D.R., and Ingebritsen, S.E., eds., Land subsidence in the United States: U.S.G.S. Circular 1182.

Troester, J. W., and Moore, J.E., 1989, Karst hydrogeology in the United States of American: Episodes, Vol. 12, no. 3, P.172-178.

U.S. Census Bureau (2000) IDB -- Rank Countries by Population, <http://www.census.gov/cgi-bin/ipc/idbrank.pl>, downloaded Dec. 2000, updated 10 May 2000.

White, W.B., Culver, D.C., Herman, J.S., Kane, T.C., and Mylroie, J.E., 1995, Karst Lands: American Scientist, Vol. 83, P.450-459.

Williams, J.H., and Vineyard, J.D., Geological indicators of catastrophic collapse in karst terrain in Missouri: National Academy of Science, Transportation Research Record, 612, P.31-37.

Xing, L. and Que, L., 1997, The distribution and harm of the land collapse in China: The Chinese Journal of Geological Hazard and Control, Vol.8, Supplement, p. 23-28.

Yuan, D.,1987, Environmental and engineering problems of karst geology in China: *in* Beck, B. F., and Wilson, W.L., eds., Karst Hydrogeology: Engineering and Environmental Applications, Proceedings of the Second Multidisciplinary Conference on Sinkholes and the Environmental Impacts of Karst: Orlando, Florida., 9-11 February, A.A. Balkema, Rotterdam,, p. 1-11.

Yuan, D., Li, B., and Liu, Z., 1998, Karst of China: *in* Yuan, D., and Liu, Z., eds., Global karst correlation: Science Press, Beijing, p. 167-177.

Zhou, W., 1997, The formation of sinkholes in karst mining areas in China and some methods of prevention: Environmental Geology, 31 (1/2), May, P.50-58.

Geotechnical and Environmental Applications of Karst Geology and Hydrology, Beck & Herring (eds)
© 2001 Taylor & Francis, ISBN 90 5809 190 2

Karstic and pseudokarstic sinkholes in Hawaii

WILLIAM R.HALLIDAY   Hawaii Speleological Survey, Nashville, TN 37205, USA, bnawrh@webtv.net

ABSTRACT

Innumberable pseudokarstic sinkholes exist in Hawaii. Most are the result of subsidence or collapse into lava tubes of various dimensions. In comparison to those of classical karst terrains, they are dry, short-lived and tend to form collapse trenches to a much greater degree than their calcareous counterparts. Karstic sinkholes in Hawaii are few and are only minor features. Collapse causes intermittent nuisances in the Moiliili District of Honolulu, and other sections of the city may be at some risk. However, the only catastrophic dewatering on record was the result of excavational breach of a karstic conduit below sea level, not sinkhole collapse. Because of the long isolation of Hawaii geology and archaeology from the karstic mainstream, the archaeological literature is full of references to other supposed sinkholes in elevated reef limestone. According to Mylroie's classification, these actually consist of pit caves, rudimentary flank margin caves, and one "banana hole". The AGI definition of the term should be clarified to exclude such features.

INTRODUCTION

Sinkholes in Hawaii do not conform to the currently accepted definition of this term (Jackson, ed., 1997). This is because:

1. Most occur in pseudokarstic terrains, and
2. Many of them are not funnel-shaped.

Readers of the Hawaiian archaeological literature receive the impression that innumerable karstic sinkholes are present in elevated reef limestones. This is a fallacy. Apparently, it resulted from long isolation of Hawaiian geology and archaeology from the karstic mainstream.

KARSTIC SINKHOLES IN HAWAII

The largest karstic sinkhole in Hawaii is an isolated feature on the south shore of the island of Kauai. Located in a high calcarenite hill, it is a sheer-walled vertical shaft about 27x33 m in diameters. It is the largest orifice of the Grove Farm Sinkhole Cave system and is a slightly atypical cenote; only intermittently is a little water present in debris at its base (Halliday, 1997). Its only risk is to unwary hikers.

In the Moiliili District of Honolulu, catastrophic dewatering caused sudden development of numerous small collapse sinkholes in 1934. This resulted in extensive but localized damage. After recharge, gradual lowering of the water table caused gradual to sudden development of depressions and a few small sinkholes in the same area. A single small example has been recorded in nearby Waikiki. All except the latter appear to have been related to dewatering of a clearly-defined expanse of elevated reef limestone drained by a single dendritic conduit system. At most, these phenomena have impacted individual small buildings. More commonly, they disrupted King Street, Kueilei Lane and other streets in the area. Occasionally, a car has had to be hauled out of a brand-new sinkhole. The 1934 dewatering was the result of accidental rupture of the karstic conduit about 7 m below sea level by construction activities. Other parts of Honolulu downslope from swallets at the volcanic/limestone interface may be risk of similar catastrophic dewatering (Halliday, 1998a). To date, however, this is the only example of a karstic conduit system identified in Hawaii.

Although the archaeological literature contains references to a large number of sinkholes in reef limestone, most of these actually are small pit caves, as defined by Mylroie and Carew (1995), one to four meters deep. These are commonly impacted by man, rather than the reverse. One "banana hole" (as defined by Mylroie and Carew) has been found, and a few small flank margin caves have vertical entrances which also could be mistaken for sinkholes. A sizeable karstic cavity destroyed in 1978 was variously termed "an unmodified wet sink-cave", "a flooded sink", and "a wet sink (cave)". It appears to have been a flank margin cave at the water table, half-full of water (Halliday, 2000).

PSEUDOKARSTIC SINKHOLES

Characteristically, sinkholes in Hawaii are pseudokarstic collapse features. Because most are the result of roof collapse into sizeable lava tube caves, only the smallest examples are funnel-shaped; most lava tube caves are sinuous conduit features or braided or dendritic complexes thereof, and sinkholes resulting from roof collapse reflect these patterns. In well-developed volcanic pseudokarstic terrain like northern Puna district, Hawaii Island, bulldozer operators properly fear thin-roofed sections of lava tube caves. Less dramatic subsidences and collapses are well-known to road maintenance crews bur rarely cause major problems, and lava tubes much too small to qualify as caves can and do cause some of these. Collapses and subsidences into large and small lava tubes divert contaminated surface water underground for varying distances, flowing in a conduit which characteristically is only partly obstructed by stream fill. Because of their pseudokarstic nature, however, pseudokarstic sinkholes tend to be much drier than their karstic counterparts.

The ultimate stage of collapse of roofs of lava tube caves consists of secondary lava trenches. These are rarely considered to be sinkholes but a sequence obviously exists. Some are difficult to distinguish from primary lava trenches. When a secondary lava trench is heavily eroded (e.g., in comparatively steep rainforest terrain), it is difficult to distinguish them from ordinary gullies of intermittent streams.

SINKHOLES ASSOCIATED WITH TECTONIC FISSURES

In volcanic terrains in Hawaii, sinkholes also occur along the course of linear tectonic features such as The Great Crack of Kilauea Volcano. Some are much more than 100 m long and at least 25 m deep. Because of an echelon cracking of brittle lavas, some of these sinkholes have a more complex form than might be expected. Except during an eruption close to an inhabited place, these features are of interest primarily to volcanologists and road maintenance crews.

PIT CRATERS AS VOLCANIC SINKHOLES

Because of controversies surrounding the definition and origin of "pit craters" (Halliday, 1998b), some rimless vertical volcanic structures may ultimately qualify as volcanic collapse sinkholes. Unlike Graciosa Island (Azores, Portugal), no circular-roofed cavities large enough to form pit craters by roof collapse have been found in Hawaii, however. Wood Valley Pit Crater in Kau District, Hawaii Island, however, appears to have been formed by stoping upward from a rift tube. It is believed to be unique in Hawaii if not the world.

SUMMARY

Parts of Honolulu downslope from swallets at the volcanic/limestone interface may be at risk of catastrophic dewatering like that of 1934. Most of the so-called sinkholes in elevated reef limestones in Hawaii, however, actually are small pit caves which are more at risk from humans than the reverse. The largest karstic sinkhole in the state is a cenote-like structure in a calcarenite hill. Pseudokarstic sinkholes associated with lava tube caves are much commoner and of more concern. Other types of volcanic sinkholes are primarily of scientific interest.

REFERENCES AND BIBLIOGRAPHY

Favre, G. 1993. Some observations on Hawaiian pit craters and relations to lava tubes. Proceedings, 3rd International Symposium on Vulcanospeleology, Bend, Oregon, 1982. Seattle, International Speleological Foundation, p. 37.

Halliday, W.R. 1966. Terrestrial pseudokarst and the lunar topography. National Speleological society Bulletin, 28 (4): 169.

Halliday, W.R. 1997. Karsts of Oahu and other Hawaiian islands. Geo2; Newsletter of the Cave Geology and Geography Section of the National Speleological society, 25 (1): 96.

Halliday, W.R., 1998a. "Pit craters", lava tubes, and open vertical volcanic conduits in Hawaii: a problem in terminology. International Journal of Speleology, 27B(1/4):113.

Halliday, W.R., 1998b. History and status of the Moiliili Karst, hawaii. Journal of Cave and Karst Studies, 60(3):141.

Halliday, W.R., 2000. Initial observations of karstification of elevated carbonate reefs in semi-arid sections of Hawaii. Proceedings, Karst 2000 Symposium, Marmaris, Turkey (in press).

Jackson, Julia A. 1997. Glossary of Geology, 4th Edition. Alexandria, VA., American Geological Institute, p. 595.

Mylroie, J.E. and J.L. Carew. 1995. Geology and karst geomorphology of San Salvador Island, Bahamas. Carbonates and Evaporites, 19(3):193.

Geotechnical and Environmental Applications of Karst Geology and Hydrology, Beck & Herring (eds)
© *2001 Taylor & Francis, ISBN 90 5809 190 2*

Sinkhole distribution of the Valley and Ridge Province, Virginia

DAVID A.HUBBARD, JR.   Virginia Division of Mineral Resources, Charlottesville,
VA 22903, USA, dhubbard@geology.state.va.us

ABSTRACT

The major karst in Virginia is developed within the Valley and Ridge province west of the Blue Ridge Mountains. It is a covered karst formed by the dissolution of folded and faulted Paleozoic carbonate rocks ranging in age from Cambrian to Mississippian. A mapping project spanning much of the period 1980-2000 incorporated the use of stereoscopic viewing of winter scenes of low altitude (4,000 m), panchromatic, aerial photography for sinkhole location. Approximately 27,300 square kilometers were viewed and a total of 48,807 sinkholes were located and plotted. Subsidence features observed ranged in size from 10 m to the Dungannon polje, a large alluviated closed-depression 6.8-km in length. Relative patterns of sinkhole development show a range of relationships with folds and faults, but the most important aspect of the study is the relative degree of karst development as expressed by the distribution of sinkholes. The greater the degree of relative sinkhole development, the greater the potential for karst associated hazards including subsidence, sinkhole flooding, and groundwater contamination associated with land-use modifications. Three types of pseudokarst features may cause interpretation problems: constructed closed-contour basins such as ponds and road bounded drainages; sag ponds and landslide blocked drainages; mining features such as pits.

INTRODUCTION

Virginia's most significant karstlands are developed in Paleozoic limestones and dolostones ranging from Cambrian to Mississippian age in the Valley and Ridge physiographic province. These karstlands extend over portions of 26 western counties and are bounded by the clastic, volcanic, and plutonic rocks of the Blue Ridge province to the east and by the Appalachian Plateaus province to the west in West Virginia and in southwestern Virginia. Karst is a terrain that forms by the interaction of water and a soluble host rock. Characteristics include solutional patterns on rock referred to as *karren*, subsidence features called *sinkholes*, solutional bedrock conduits termed *caves*, and *subsurface drainage*. As is typical throughout most of the Appalachians, soils and sediments cover Virginia's karst and much of the host rock and its karren are not readily observable. The closed-contour subsidence features developed in these cover materials and collectively referred to as sinkholes, in the U.S., are recognizable through the lush vegetation, which easily obscures many karst features. Although much of the Valley and Ridge karst is easily inferred by the complex patterns of sinkholes developed over it's folded and faulted carbonate rocks, some of the karst is sparsely marked or unmarked by sinkholes.

The mapping of Virginia's karst began in 1980 in the northern third (Hubbard, 1983; see area N in Figure 1) of the Valley and Ridge province and continued southward through the central (Hubbard, 1988; see area C in Figure 1) and southern portions of the province project area. The project's field mapping is complete and the map of the final third of the Old Dominion's Valley and Ridge province karst currently is in preparation (see area S in Figure 1).

The remote sensing mapping method produced a total of 17,568 observed sinkholes over the first two thirds of the project area (Hubbard, 1984). But mapping in the final third of the project area proved far more complex a proposition than the two previous map areas prompting a status report at the Fourth Sinkhole Conference (Hubbard, 1993).

PROJECT METHODS

Sinkhole locations were determined by stereoscopic viewing of cloudless winter and early spring scenes of low altitude (4,000 m), panchromatic aerial photography and plotted on 1:24,000 scale topographic maps. Sinkholes mapped ranged in size from about 10 meters to the 6.8-kilometer long Dungannon polje. Questionable sinkholes were field checked. Features not recognized by remote sensing, but observed during field checks were not added to the database to minimize bias. The use of remote sensing methods with the same low altitude aerial photography used in the generation of 7.5-minute topographic maps results in significantly more sinkhole locations than are recorded by topographic mapping because of contour interval limitations. Shallow sinkholes at elevations in between the elevations contoured on individual maps are not shown on those topographic maps. This problem is compounded in mountainous areas where 12 m contour intervals are used instead of 6 m intervals.

Detailed geologic information from published and manuscript maps on file at the Virginia Division of Mineral Resources (formerly the Virginia Geological Survey) also was transferred to these topographic maps. Geological map units were combined into non-carbonate, limestone, or limestone and dolostone map units. Sinkhole and geologic data were subsequently photographically reduced for regional compilation at a 1:250,000 scale on three separate maps. The project area for all three maps totaled approximately 27,300 square kilometers and included all or portions of 254 7.5-minute topographic maps.

METHODOLOGY AND MAPPING DIFFICULTIES

Not all sinkholes are recognizable by remote sensing methods. Sinkholes smaller than about 10 m were not consistently differentiable using the equipment settings and photography employed in this project. Many larger sinkholes, especially low profile features undifferentiated by vegetative differences, were not seen on aerial photography, but found during field checks. This was especially the case in wooded areas. Problems of aircraft tilt created slope anomalies on some frames that made some shallow sinkholes unrecognizable and some low slope areas appear as sinkholes.

The most formidable problems occurred in the southern map area and related to availability and quality of aerial photography and the quality of topographic maps. The southern mapping was delayed until about 1991, when new photographic coverage was available for two counties. Smaller gaps in coverage were eventually filled using some late spring and fall scenes. The extreme variability in quality of 1:24,000-scale (7.5-minute) topographic map coverage was an unforeseen problem that was not discovered until the southern mapping had commenced. Most of Virginia is covered by U.S. Geological Survey (USGS) generated topographic maps. These maps are consistently of uniform high quality. In southwest Virginia, much of the coverage is of Tennessee Valley Authority (TVA) maps, which are highly variable in quality. Some of these TVA maps contain an unusual number of topographic errors involving the inconsistent use of hatchures. The worst-case example is the Damascus map on which about 10 percent of the karst features observed or depicted on the map and not observed are in error. Some sinkhole features depicted on the map are hills, some topographic highs depicted on the map are sinkholes, and some other sinkholes shown on the map are not real. This problem only occurs in the southern third of the karst mapping area.

One problem inherent in the regional maps concerns scale and resolution. On 1:24,000 scale maps, the standard line width represents about 12 m. The representation of the minimum sized sinkhole of 10 m is actually 12 m. This same 10 m feature will be depicted as a 125 m sinkhole on the 1:250,000 scale map. At this scale, major highways are shown as 244 m wide and major streams and rivers pose similar problems. Direct translations place some sinkhole features on the wrong side of a road or creek. Corrections were made for northern and central maps, but the density of sinkholes on the southern map may not allow sinkholes to be refitted.

Throughout the mapping program some non-karst closed-contour depressions have been problematical. An old pond may look like a sinkhole. A drainage truncated by a fill for an old abandoned rail line or a road may look like a small blind valley or sinking stream. Old sag ponds or landslide truncated drainages also look like subsidence features. One of the more significant type of features are old mine pits, especially those associated with secondary and residual lead-zinc ores mined in the late 1800s in Wythe and Pulaski Counties and iron and manganese mining up until the early 1900s. These are carbonate-associated ores and the prospect and mine pits reclaimed by vegetation are within karst and resemble sinkholes. In some of the lead-zinc mine pits, bedrock pinnacles were left exposed. The iron and manganese ores were deposited as acidic waters rich in dissolved iron and manganese drained onto carbonate rock, were buffered during the dissolution of the carbonate bedrock and deposited ores. Many of these ores may have occurred within karst depressions at insurgencies. At other sites the depressions appear to only represent mining activities. In some instances it appears as though mining created a depression which subsequently engaged local paleokarst and began to pirate surface drainage. Because the ores are associated with carbonate bedrock subject to karst processes, such a change in surface drainage pattern can result in karst subsidence and pirating of surface drainage.

SINKHOLE DISTRIBUTION

The distribution of sinkholes varies along the Valley and Ridge province in a remarkable fashion. A total of 48,807 sinkholes have been mapped over 254 standard (7.5-minute) topographic maps for an average of 192.1 sinkholes per map. In the northern map area, 6674 sinkholes were mapped on 88 topographic (7.5-minute) maps. A simple statistical average of 75.8 sinkholes per topographic map compares to a maximum of 501 sinkholes mapped on the Harrisonburg Quadrangle. In the central map area, 10,894 sinkholes were located on 81 topographic maps. The statistical average of 134.5 sinkholes per map compares to the high of 776 sinkholes recorded on the Radford North Quadrangle. In the southern map area, 31,239 sinkholes were marked on 85 maps. The statistical average of 367.5 sinkholes per map compares to a maximum of 1350 sinkholes mapped on the Crockett Quadrangle (Figure 1). Even compared to the combined total of 17,568 sinkholes mapped in the northern and central portions of the province, the 31,239 sinkholes mapped in the southern third of the project area reveals an exceptional enhancement in the relative degree of karstification from north to south across the state of Virginia. It is quite obvious from the simple statistics for each of the three map areas that the sinkhole distribution across Virginia's Valley and Ridge province is not random.

A viewing of any of the three regional sinkhole maps reveals there are three obvious factors in the distribution of karst as expressed by sinkholes: lithology, structure, and hydraulic gradients. On a regional scale, the lithological differentiation is rather limited and first and foremost defines sinkholes as occurring where carbonate rocks are present. A quadrangle map with 20 percent carbonate rock will generally host fewer sinkholes than a map containing 80 percent carbonate rock. On the draft of the southern map, linear zones of sinkholes demonstrate preferential development along specific lithologies within limestone or limestone and dolostone map units.

Preferential sinkhole development along fold and fault structures was discussed with respect to the northern and central regional karst maps (Hubbard, 1984). Sinkhole development at the nose of plunging anticline folds in carbonate rocks probably is related to tensile fracture patterns. The channeling effects of non-karst rocks bounding these carbonates are thought to intensify the solutioning

Concentrations of sinkholes adjacent to deeply incised segments of rivers and major tributaries appear to indicate the importance of hydraulic gradient. Enhanced sinkhole development occurs along buffs of the Shenandoah River on the northern map, the New River on the central map, and Copper Creek on the southern map.

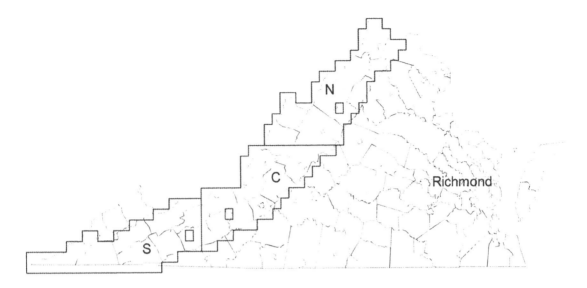

Figure 1. Layout of the Virginia karst regional maps. The northern regional map (published as Hubbard, 1983) is labeled N; the central regional map (published as Hubbard, 1988) is labeled C; the southern regional map area (drafting is underway) is labeled S. The individual 7.5-minute maps outlined in each regional map contained the highest number of sinkholes recorded for each region and are: the Harrisonburg Quadrangle with 501 sinkholes in N; the Radford North Quadrangle with 776 sinkholes in C; and the Crockett Quadrangle with 1350 sinkholes in S. Richmond is the State Capital.

KARST HAZARDS

There are three Virginia karst hazards which are related to sinkholes: subsidence, karst flooding, and groundwater contamination. Subsidence traditionally has been the hazard of greatest concern to karstland residents. The formation of sinkholes and other subsidence features may occur slowly over time or catastrophically. The location and occurrence of sinkholes is keyed to two major processes active in karst, to the movement of water into the subsurface aquifer and the dissolution of the host bedrock. Most sinkholes form at locations where water moves through epikarst drains or conduits to recharge the karst aquifer. Some sinkholes may mark the sink points of permanent surface streams. Other sinkholes are the locations were the surface runoff of precipitation sinks. Many sinkholes are not obviously linked to surficial waters, but mark the points were water moving along the soil/bedrock interface sinks into solution enlarged bedrock fractures or partings at epikarst drains. Unfortunately, not all epikarst drains are marked by subsidence features in covered karst. These unmarked drains are potential sites of subsidence or collapse. A small percentage of cover karst sinkholes represent the sites of cave collapses, where the bedrock ceilings of caves and their overlying soils and sediments have catastrophically collapsed. These bedrock collapses are the rarest of sinkholes (Sowers, 1996). The more typical catastrophic collapse type of sinkhole is related to changes in the local hydrology, whereby changes in the amount or pattern of surfical drainage or water table fluctuations are perpetuated by natural or humankind processes (Kemmerly, 1980). Drought followed by a storm event is one natural trigger of sinkhole formation and the natural analog of water-well development or excessive pumping induced sinkhole formation. The increased drainage resulting from an exceptional storm events may trigger sinkhole formation. Intense storm triggered sinkholes may be a natural analog to collapses triggered by increased runoff from developmental landscape modifications. However, the same magnitude of increase in runoff from pavements and other landscape alterations to surficial drainage may occur during much lower intensity storms than at unaltered sites. In addition to triggering subsidence, developmental landscape modifications may increase the incidence of karst or sinkhole flooding.

There are two types of karst or sinkhole flooding: ponding and epiphreatic flooding. Sinkhole ponding occurs when the amount of water draining to the sinkhole exceeds the capacity of the solution-enlarged fractures or conduit that underdrains the sinkhole. This type of flooding is typically aggravated during and following site development as additional drainage demand is placed on the sinkhole by increased drainage from paved surfaces, changes to the drainage pattern, and partial clogging of the drain by development associated siltation. Epiphreatic flooding results from increased hydraulic heads in the conduit system that underdrains a sinkhole. This problem is characteristically in response to increased drainage at other sinkholes draining into the same underdraining conduit system, but usually up-gradient from the affected sinkhole. The net result is that a high hydraulic head in the underdraining fracture or conduit system functionally converts the sinkhole to an estavelle. During exceptional storm events, the sinkhole shifts from an insurgence to a resurgence and temporarily disgorges groundwater.

associated siltation. Epiphreatic flooding results from increased hydraulic heads in the conduit system that underdrains a sinkhole. This problem is characteristically in response to increased drainage at other sinkholes draining into the same underdraining conduit system, but usually up-gradient from the affected sinkhole. The net result is that a high hydraulic head in the underdraining fracture or conduit system functionally converts the sinkhole to an estavelle. During exceptional storm events, the sinkhole shifts from an insurgence to a resurgence and temporarily disgorges groundwater.

Because sinkholes are underdrained by epikarst drains or conduits formed by solution enlargement of bedrock fractures or partings, little filtering occurs along the flowpath from the sinkhole to the groundwater aquifer. Both the epikarst drains and the larger conduits to which they drain are the natural plumbing directly linking surficial waters and their contaminants to karst groundwater. Sinkholes represent some of the recharge points for the local karst aquifer. They also represent classical dump sites which are linked by primary flow-paths to the aquifer. With the exception of base level surface drainage, most of the available water in mature karst is groundwater. The more mature the karst the more developed the natural subsurface plumbing and the more susceptible the groundwater system is to liquid or waterborne contaminants. A logical assumption is that the more developed the karst, the greater the density of sinkholes, and the greater the proportion of karst hazards. An important caution is that not all highly developed karst is densely covered with sinkholes. In many karst areas with well developed multiple mile long cave systems, the degree of cave development appears to vary inversely to the degree of sinkhole development.

DISCUSSION OF INTENDED USES VS ACTUAL USES OF THE VIRGINIA KARST MAPS

The three 1:250,000 scale karst maps of sinkhole locations and carbonate rock boundaries for the Valley and Ridge province of Virginia were planned as an accurate, but relative depiction of the major Virginia karstlands. Of the characteristic features found in covered karst, sinkholes are the easiest features to observe and map. Unfortunately, it would have taken approximately 250 man-years to accurately map all of the sinkholes in Virginia's Valley and Ridge province. The use of low altitude photography and remote sensing techniques has yielded a series of useful products that depict a relative degree of the karst development in the state's Valley and Ridge province despite the fact that only a fraction of the existing sinkholes are mapped. As with most geologic hazards and geotechnical evaluations, site specific studies are necessary to characterize karst sites for development. Karst is notorious for the "out of sight - out of mind" processes that have created and are actively modifying the terrain, despite inadequate understanding of these processes by many of the individuals developing this unstable terrain.

Some years ago I made a presentation about the nature of the two published Virginia karst maps and explained how they didn't record all sinkholes and at best should be used to relatively scale the degree of development of karst processes and hazards from one area to another. A subsequent speaker remarked in his presentation that he had consulted the appropriate karst map and found no sinkholes were recorded in the area, in which his project was sited, so he concluded there were no karst problems to be encountered at the site! This is not an isolated example of the misuse of these maps. I've been shown site evaluations citing these karst maps as characterizing the karst present at sites. The largest group of users of these karst maps wants the information in digital form so that the sinkhole locations can be scaled and plotted in site evaluation maps and other consulting report maps. An alarming trend is the lack of a single example of a site evaluation or environmental report that has cited one of the regional karst maps and noted any additional karst features found during an on-site evaluation.

REFERENCES

Hack, J.T., 1965, Geomorphology of the Shenandoah Valley, Virginia and West Virginia, and origin of the residual ore deposits: U.S. Geology Survey Professional Paper 400-B, p. 387-390.

Hubbard, D.A.., Jr., 1983, Selected karst features of the northern Valley and Ridge province, Virginia: Virginia Division of Mineral Resources Publication 44, single sheet.

Hubbard, D.A.., Jr., 1984, Sinkhole distribution in the central and northern Valley and Ridge province, Virginia: in Beck, B.F., ed., Sinkholes: Their Geology, Engineering & Environmental Impact, Proceedings of the First Multidisciplinary Conference on Sinkholes, Orlando, FL. A.A. Balkema, Rotterdam, p. 75-78.

Hubbard, D.A.., Jr., 1988, Selected karst features of the central Valley and Ridge province, Virginia: Virginia Division of Mineral Resources Publication 83, single sheet.

Hubbard, D.A.., Jr., 1993, Status report on the Virginia karst mapping program: in Beck, B.F., ed., Applied Karst Geology, Proceedings of the Fourth Multidisciplinary Conference on Sinkholes and the Engineering and Environmental Impacts of karst, Panama City, FL. A.A. Balkema, Rotterdam, p. 281-284.

Hubbard, D.A.., Jr. and Holsinger, J.R., 1981, Karst development in Rye Cove, Virginia: in Beck, B.F., ed., Proceedings of the Eighth International Congress of Speleology: National Speleological Society, Huntsville, Alabama, p. 515-517.

Kemmerly, P.R., 1980, Sinkhole collapse in Montgomery County, Tennessee: Tennessee Division of Geology, Environmental Geology Series No. 6, 42 p.

Sowers, G.F., 1996, Building on sinkholes: Design and construction of foundations in karst terrain: ASCE Press, New York, 202 p.

Geotechnical and Environmental Applications of Karst Geology and Hydrology, Beck & Herring (eds)
© 2001 Taylor & Francis, ISBN 90 5809 190 2

Topographic, geologic, and hydrogeologic controls on dimensions and locations of sinkholes in thick covered karst, Lowndes County, Georgia

JAMES A.HYATT & ROY WILSON   Dept. of Environmental Earth Science, Eastern Connecticut State University, Willimantic, CT 06226, USA, hyattj@easternct.edu, wilsonr@easternct.edu

JEFFREY S.GIVENS   Dept. of Geography, University of Tennessee, Knoxville, TN 1379961, USA, jsgivens@usit.net

PETER M.JACOBS   Dept. of Geography and Geology, University of Wisconsin Whitewater, Whitewater, WI53190, USA, jacobsp@mail.uww.edu

ABSTRACT

The dimensions and spatial locations of 649 sinkholes occurring in Lowndes County, Georgia, are analyzed in relation to possible geologic and hydrogeologic controls. The southern Georgia study area consists of 1320 km^2 of covered karst that formed within 0-70 m of Miocene through modern sediment cover overlying Oligocene Suwannee Limestone. Sinkhole locations and size characteristics were digitized from topographic maps and geologic data consists of digital files of elevation, soils, geology, and hydrogeology. Multiple one-way analysis of variance and cross-area tabulation statistical techniques identify trends between sinkhole sizes, locations, and a variety of topographic (elevation, slope), geologic (soil and overburden type, overburden thickness), and hydrogeologic (modeled potentiometric head) variables. These analyses indicate that the type of overburden is more important to conditioning sinkhole size than is soil infiltration capacity. Cross-area tabulation identifies several spatial trends that indicate two fundamentally different sets of sinkholes, one occurring predominantly in low elevation fluvial settings and the other in upland Miocene-Pliocene marine sediment. Low elevation sinkholes are smaller, and are more tightly clustered than upland sinkholes. Low elevation sinkholes occur in thinner and sandier sediments within steeper terrain than is the case for upland sinkholes. Elevation of the potentiometric surface in relation to the top of covered limestone bedrock is more important in explaining the locations of tightly clustered sinkholes than is the difference in head between unconfined surface aquifers and the Floridian aquifer.

INTRODUCTION

Sinkholes are common throughout covered karst regions in the Gulf Coastal plain of the southeastern United States. In many locations they form by cover collapse and cover subsidence mechanisms whereby overburden is piped into dissolutional openings in underlying carbonate bedrock (Beck and Sinclair, 1986). However, in many covered karst locations sinkholes may not directly overly subsurface openings in bedrock (e.g. Varrington and Lindquist, 1987). This is because piping and the migration of void space through insoluble overburden is influenced by topography, overburden properties, and by the prevailing hydrogeology (Palmquist, 1979; Wilson and Beck, 1992). For example, recent work in south Georgia suggests that locally enhanced hydraulic gradients associated with fluvial landforms have influenced the locations and forms of sinkholes (Hyatt et al. 1999). In central Florida, Whitman and Gubbels (1999) report that variations in head within surface and deep aquifers may influence the locations of sinkholes.

Although sinkholes are widespread in Georgia, there have been fewer studies examining their spatial distribution and hydrogeologic controls than in other southeastern states. Previous studies of sinkholes in southern Georgia have focused largely on thinly covered karst of the Dougherty Plain near Albany Georgia (Herrick and LeGrande, 1964; Beck and Arden, 1984; Brook and Alison, 1983; Hyatt and Jacobs, 1996). In this paper we present the first study of sinkholes in south Georgia from a region of thick covered karst outside the Dougherty Plain. We examine the locations, dimensions (area and perimeter), elevations, overburden thicknesses, soil conditions, and hydrogeologic settings for 649 sinkholes in Lowndes County, Georgia (Figure 1). The purpose of this work is to identify relationships between sinkhole dimensions, locations, and a variety of topographic, geologic, and hydrogeologic controls in order to better understand the spatial variability of sinkholes in south Georgia.

PHYSIOGRAPHY, SURFICIAL GEOLOGY, AND HYDROGEOLOGY

Lowndes County Georgia extends northward from the Georgia-Florida State line encompassing 1,320 km^2 of low relief terrain bounded to the west by the Withlacoochee and Little Rivers, and to the East by the Alapaha River (Figure 1). Sinkholes are most concentrated in the southern third of the county within the Lake Park Karst District physiographic region (Huddlestun, 1997), particularly at sites near the Withlacoochee River. The central and northern thirds of the county consist of a series of minimally incised, flat upland Pleistocene marine terraces of the Bacon Terrace District (Clark and Zisa, 1976). Sinkholes within this region are more widely spaced than in the Lake Park Karst District, although sinkholes are locally abundant adjacent to high order streams.

Lowndes County is underlain by >200 m of Cenozoic carbonate and clastic sedimentary rocks and sediments that host several aquifers important to the development of sinkholes (Table 1). Oligocene carbonates of the Suwannee Limestone are poorly exposed,

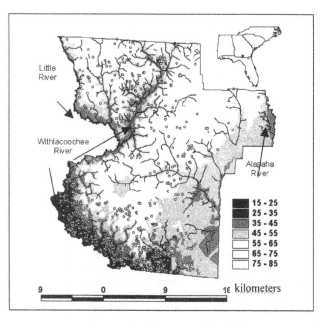

Figure 1. Inset showing the location of Lowndes County Georgia with associated digital topography (shading), sinkhole centroid locations (dots), and major streams (lines).

cropping out only within incised sections of the Withlacoochee River where surface water directly recharges the underlying upper Floridan aquifer system. Recharge through sinkholes in the bed of the Withlacoochee River varies between 1.4 and 8.5 m^3s^{-1} in response to changing flow conditions in the river (Krause, 1979; McConnell and Hacke, 1993). Elsewhere the Suwannee Limestone is unconformably overlain by up to 70 m of sandy to argillaceous marine sediments that include the Miocene Jumping Gully, Statenville, and Cooswatchie Formations, and the Pliocene Miccosukee Formation (Figure 2a). These formations enclose two small and unnamed confined aquifers that are primarily used for agricultural water supply (Table 1). Miocene exposures are largely restricted to the southwest portion of the county, whereas the Pliocene Miccosukee Formation, which contains thick beds of clay, is widely exposed in upland surfaces (Figure 2a) (Huddlestun, 1997). Sandy Quaternary alluvium occurs as extensive flood plain and river terrace deposits along major stream valleys. In addition several organic-rich Quaternary Carolina Bay swamp deposits discontinuously overly upland deposits of the Miccosukee Formation.

Analysis of water well logs by Krause (1979) reveals that post-Oligocene sediments that cover limestone bedrock are thickest along the NW-SE trending drainage divide that separates tributaries of the Withlacoochee and Alapaha Rivers (Figure 2b). Furthermore, despite abundant Quaternary flood plain and river terrace deposits along the Withlacoochee, Little and Alapaha Rivers (Figure 2a), the thickness of sediment cover above limestone bedrock decreases towards major stream valleys. This implies efficient evacuation of sediment likely both by fluvial erosion, and by subsurface flushing into the underlying Suwannee Limestone.

Table 1. Hydrogeologic Setting in Lowndes County with emphasis on units with surficial exposure within the study site. Adapted from McConnell and Hacke (1993) and Huddlestun (1997).

Age	Unit / Formation (map symbol, Fig. 2)	Lithology	Hydrogeologic Unit
Quaternary.	Alluvim Flood Plain (QALfp)	Fine, well-sorted, massive sand in larger streams. Sand, mud & organics in smaller streams.	Surficial Aquifer Restricted to stream valleys. Domestic water in some areas.
	Alluvium River Terrace (QALrt)	Fine to medium, variably sorted, structureless to crudely stratified sand.	
	Alluvium Carolina Bay (Qcb)	Organic-rich muds to muddy peat, swamp deposits.	Confining Unit Removed in some areas by stream erosion
Pliocene	Miccosukee Fm. (Pm)	Sand, variably argillaceous, some scattered thick clay beds; sand fine and well-sorted; prominently bedded with thin to medium white clay laminae.	Unnamed Aquifer Domestic and irrigation water source at some sites.
Miocene (Hawthorn Group)	Coosawatchie Fm. (Mhcm)	Clay, finely sandy, siliceous, with some clay intraclasts. No carbonates.	Confining Unit
	Statenville Fm. (Mhs)	Sand, clay, variably dolomitic. Lower part prominently bedded with thin to thick layers of sand, clay & dolostone. Upper part massive argillaceous sand to sandy clay.	Unnamed Aquifer Exposed in river valleys.
			Confining Layer
	Jumping Gully Fm. (Mhjg)	Sand to clay, siliceous with chert nodules. Variably dolomitic and phosphatic. Lower part mainly thinly bedded clay; upper part argillaceous fine sand with chert.	
Oligocene	Suwannee Limestone	Granular textured, calcarenitic limestone with scattered fossiliferous beds. Crudely stratified in outcrop, but massive where unweathered.	Upper Floridan Aquifer Major source of groundwater. Recharged directly through sinkholes in Withlacoochee River

Potentiometric head values for the upper Floridan aquifer in Lowndes County vary both spatially and temporally. Figure 2c depicts the distribution of average annual potentiometric head derived from May and November 1975 head data (Krause, 1979). Note that head values decrease toward the south and west, and that a potentiometric mound occurs in the southern part of the county where

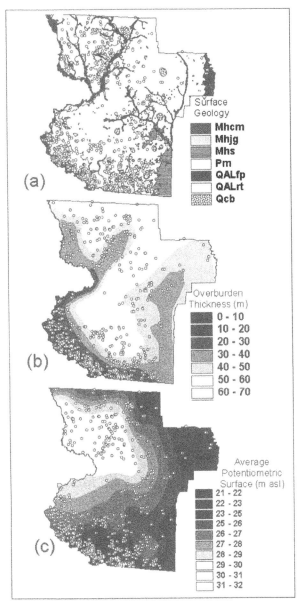

Figure 2. Overlays showing sinkhole centroid locations and (a) the surficial geology of Lowndes County Georgia adapted from Huddlestun (1997) and identified in Table 1; (b) overburden thickness derived from analysis by Krause (1979) of water well records; (c) average annual potentiometric surface for upper Floridan aquifer calculated by averaging May and November 1975 contoured potentiometric surfaces from Krause (1979).

surface water is known to recharge the Floridan aquifer through sinkhole lakes. Although the general configuration of the average potentiometric surface in Figure 2c is consistent with more recent data (e.g. McConnell and Hacke, 1993), head values have undoubtedly varied through time. As can be seen from water well records for a well in Valdosta near the center of Lowndes County (Figure 3) potentiometric head values have indeed varied by about 5 m annually and by as much as 11 m since 1957. We wish to emphasize, however, that the average head values for 1975 in Figure 3, which correspond to the surface in Figure 2c, fall approximately in the middle of the long-term range of head elevations. Thus, we believe this surface to be a good estimate of long-term average potentiometric elevations about which annual variations have occurred. We return to this point later in the paper.

DATA SETS

Our analysis examines the dimensions (area and perimeter) and the spatial locations (centroids) of 649 closed depressions identified from contour lines on portions of 15 USGS 1:24,000 map sheets. We exclude depressions that are clearly related to human activities, and assume that all 649 depressions are indicative of locations where

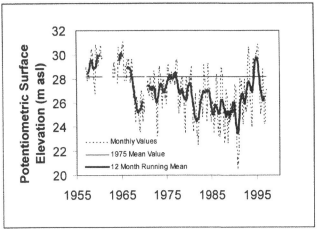

Figure 3. 40-year record for potentiometric head from observation well 19E009 in Valdosta Georgia. Horizontal line denotes average head value in 1975, the year that corresponds to potentiometric data in Figure 2c.

overburden has been transported into deeper openings in bedrock. We do recognize that such a definition of a sinkhole is non-genetic and may in fact include depressions with polygenetic origins. The dimensions and centroid locations of all sinkholes, so defined, were calculated using ArcView geographic information system software.

Sinkhole locations are compared with digital topography, surficial geology, soil type, and potentiometric head (hydrogeology) maps. Much of this original data is presented as contour maps that we digitized and converted to triangulated integrated network (TIN) files. TIN files were converted to raster grids with 30x30 m grid resolution and clipped to conform to the limits of the study site.

Topography data files were derived from USGS digital elevation model data downloaded from the Georgia GIS clearinghouse (http://gis.state.ga.us/). Surficial geology boundaries were digitized from a map by Huddlestun (1997) while overburden thicknesses were digitized from a map by Krause (1979). Digital county soil polygon files were obtained from the South Georgia Regional Development Center in Valdosta Georgia. Spatial hydrologic data, discussed at length in a subsequent section, are based on county-wide potentiometric head maps for the Floridan aquifer (Krause, 1979).

APPROACH TO DATA ANALYSIS

We analyze the sizes of sinkholes following techniques described by Hyatt et al. (1999), whereas our analysis of the locations of sinkholes follows that of Whitman et al. (1999). Important differences between the underlying assumptions for these analyses follow.

Our analysis of sinkhole dimensions assumes that the size of individual sinkholes is spatially independent of other sinkholes, and that error terms for log-transformed sinkhole dimensions are normally distributed. Normality is confirmed by a one sample Kolmogorov-Smirnov (KS) test between log transformed sinkhole dimensions and the standard normal curve. Establishing spatial independence of sinkhole sizes is more difficult, and we offer several qualitative arguments that support this assumption. Most importantly, our analysis of sinkhole dimensions does not rely on specific proximity criteria for grouping sinkholes (particularly in relation to overburden and soil groupings discussed subsequently). We recognize that some authors have identified sites where large parent sinkholes alter surface hydrology and promotes the development of smaller daughter sinkholes nearby (e.g. Drake and Ford 1972, Kemmerly 1982). This multigenerational mechanism for sinkhole formation clearly violates assumptions of spatial independence and, if operative in Lowndes County, would invalidate our use of ANOVA techniques in subsequent sections of this paper. However, analysis of sinkhole distributions at sites similar to Lowndes County, albeit in thinner overburden, have not found evidence of multiple generations of spatially dependent sinkholes (Hyatt et al., 1999). Thus, although we do not specifically test for spatial association in this paper, for reasons outlined above, we feel justified in assuming that the sizes of sinkholes in Lowndes County are spatially independent of one another.

In contrast, our analysis of the spatial distribution of sinkholes does NOT assume spatial independence. In keeping with Whitman et al. (1999), we use cross-area tabulation techniques to identify interesting spatial relationships between sinkhole proximity and a variety of topographic, geologic, and hydrogeologic variables. However, because we cannot assume spatial independence we refrain from assigning levels of significance to our spatial findings.

RESULTS AND ANALYSIS

Sinkhole dimensions in relation to overburden and soil types

Several studies have reported relationships between the type and thickness of overburden and the locations and sizes of sinkholes within covered karst (e.g. Palmquist, 1979). For example, Hyatt et al. (1999) report that sinkholes on the Dougherty Plain near Albany Georgia are smaller at low elevations in sandy Quaternary alluvium than are sinkholes at higher elevation where clay and sandy clay sediments dominate. Analysis of the dimensions of sinkholes in Lowndes County reveal similar trends (Table 2).

Mean area and mean perimeter values for sinkholes in Lowndes County are smallest for sinkholes with centroids located in sandy flood plain and river terrace deposits. Mean dimensions are 1.3 to 2.4 times larger for sinkholes within finer-grained upland Miocene and Pliocene deposits, and 2.8 to 10.3 times larger for sinkholes within Quaternary Carolina Bay swamp deposits (Table 2). Furthermore, multiple one-way analysis of variance (MANOVA) tests performed on log-transformed and normally distributed sinkhole dimensions (Table 3) indicate that differences between overburden groups are highly significant ($p<0.001$). In contrast, clear and consistent differences between mean dimensions for sinkholes grouped by hydrologic soil group are not evident. This suggests that, despite substantial differences in soil group infiltration capacity, a factor that conceivably could influence piping of sediment into subsurface voids, soil group is less important in conditioning the size of sinkholes than is the type of overburden.

Finally, it has widely been reported that sinkholes are more frequent in areas where cover is thin. While subsequent spatial analyses will confirm this trend for sinkholes in Lowndes County, Figure 4 indicates that thickness is not a good predictor of sinkhole area or perimeter.

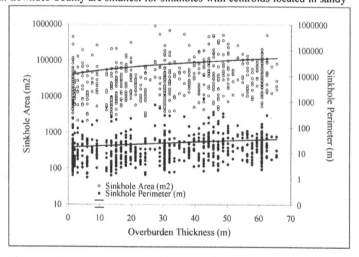

Figure 4. Comparison of sinkhole areas (open) and sinkhole perimeters (closed) with thickness of overburden. Note that both y-axes are logarithmic. Trend lines show a slight, but statistically insignificant, increase in size with increasing overburden thickness.

Spatial analysis of sinkholes locations using cross-area tabulation

We quantify spatial relationships between sinkhole locations and other variables using a cross-area tabulation technique recently applied to the study of sinkholes in central Florida (Whitman et al., 1999). Cross-area tabulation determines the area of overlap between maps of sinkhole location and maps of explanatory variables (e.g. topography, geology, hydrogeology). The following provides a brief explanation of this technique as applied to relationships between sinkhole locations and digital elevation classes. Interested readers are referred to excellent discussion by Whitman et al. (1999) and Bonham-Carter (1997) for a more detailed explanation.

The following cross-area tabulation calculates the area of overlap between a map of sinkhole proximity (Figure 5) and a map of classified digital elevation (Figure 1). Sinkhole proximity is a measure of the distances from each 30 m raster cell in the study site to intervals and converted to a series of polygons of equal proximity (Figure 5). The resulting proximity map identifies regions where sinkholes are closely spaced (low values on map) as well as regions where sinkholes are more widely spaced (high values on map).

Table 2. Summary statistics for sinkholes grouped by overburden type.

Class[a]	Age & Description[b]	Median	Mean	Std	Range	n
		Sinkhole Perimeters (m)				
Group 1	Quaternary flood plain & river terrace	238	395	633	65 to 7,252	317
Group 2	Miocene-Pliocene Sandy-Clay	385	529	559	56 to 4,729	308
Group 3	Quaternary Carolina Bay	940	1,103	598	651 to 2,913	13
		Sinkhole Areas (m^2)				
Group 1	Quaternary flood plain & river terrace	3,237	10,582	26,830	181 to 325,118	317
Group 2	Miocene-Pliocene Sandy-Clay	9,664	25,501	56,421	248 to 421,152	308
Group 3	Quaternary Carolina Bay	64,274	108,515	147,043	31,200 to 582,154	13

a – Overburden classes exclude 11 sinkholes that were classified as water filled. Groupings of the remaining overburden classes were determined as follows: First log-transforming perimeter and area values to achieve a normal distribution (significance confirmed at $p < 0.01$ by a KS nonparametric comparison with the standard normal curve). Sinkholes were grouped by overburden class (Figure 2), with the single sinkhole within Mhcm deposits being combined with the spatially adjacent Pm overburden class. MANOVA analyses were performed using log-transformed dimensions and groups that did not differ significantly from one another were combined. This resulted in groups 1 through 3 presented in this table.
b - Group description reflect combined overburden types in Figure 2a: 1 (QALfp & QALrt), 2 (Mhcm, Mhjg, Mhs, Pm), 3 (Qcb).

Table 3. Comparison of means matrix for sinkhole perimeters and areas grouped by overburden type and hydrologic soil group. Test results are from multiple comparison of mean t-tests conducted by multiple one-way ANOVA. Results are reported as indicating a significant difference (*SD*) or no significant difference (ND) for classes compared in the matrix. Probability values for each result are provided parenthetically.

MANOVA Comparisons For Sinkhole Dimensions Grouped by Overburden and Soil Type				
(a) MANOVA Results For Overburden Grouping[a]				
Overburden Class	Overburden Group 1 (Quaternary flood plain & river terrace)		Overburden Group 2 (Miocene & Pliocene Sandy-Clay)	
	Log Perimeter	Log Area	Log Perimeter	Log Area
2	*SD (<0.001)*	*SD (<0.001)*	---	---
3	*SD (<0.001)*	*SD (<0.001)*	*SD (<0.001)*	*SD (<0.001)*
(b) Soil Hydrologic Grouping[b]				
Group	Hydrologic Soil Group A		Hydrologic Soil Group B	Hydrologic Soil Group C
B	*SD (0.018)*	*SD (0.006)*	--- ---	--- ---
C	ND (0.212)	ND (0.244)	ND (0.537) ND (0.314)	--- ---
D	ND (0.910)	*SD (0.039)*	ND (0.095) ND (0.236)	ND (0.351) ND (0.170)

a - Overburden groupings are as follows: Group 1 combines sinkholes within Pm and Mhjg surficial geologic units, Group 2 combines sinkholes within QALfp and QALrt units, and Group 3 consists of sinkholes within Qcb. These groupings are based on combining sinkhole classes for individual units that did not differ significantly from one another.
b - Soil Classes correspond to regrouping 24 soil types in the Lowndes County Soil Survey into their soil hydrologic groups.

Next the reclassified sinkhole proximity map is overlaid on top of a classified map of digital elevation (derived from Figure 1). The intersections between these maps define a number of new polygons with various combinations of sinkhole proximity values and digital elevation values. The area contained within these polygons is totaled for each combination of proximity and digital elevation and is recorded as a cross-area tabulation table. Columns in the cross-area tabulation table correspond to sinkhole proximity classes, whereas rows refer to digital elevation classes. The contents of each cell in the table contains the total area in the study site with proximity and elevation values equal to the corresponding column and row classes.

Cross-area tabulations are more easily evaluated graphically. Figure 6 depicts the results of a cross-area tabulation between sinkhole proximity (Figure 5) and digital elevation (Figure 1) with contours identifying areas of similar cross-tabulated area. Large values for the observed cross-area tabulation (Figure 6a) indicate pairings of sinkhole proximity and digital elevation as classes that are areally extensive within the study site. In contrast, small values in Figure 6a indicate combinations of sinkhole proximity and digital elevation that do not account for much area in the study site. For example, Figure 6a indicates that the bulk of the study site (shaded region) is located within 4 km of a sinkhole, and that the elevations for these regions vary substantially (25 to 75 m asl). Upon closer examination, however, it is evident that sites where sinkholes are close together or clustered (i.e. <1.5 km proximity) are most extensive for two elevation ranges centered on 47 and 60 m asl.

Although observed cross-area tabulation results are informative, they can be difficult to interpret directly. A better approach is to compare observed cross-area tabulations (Figure 6a) with cross-area tabulations that would be expected if the variables under consideration were spatially independent of one another (Figure 6b). Techniques for calculating expected cross-area tabulations are summarized in the figure caption for Figure 6. The difference between observed and expected cross-area tabulations are hereafter referred to as residual plots (Figure 6c).

41

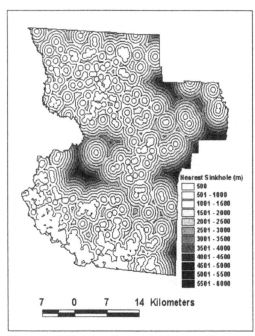

Figure 5. Sinkhole proximity map showing distance to nearest sinkhole classified in 500 m intervals.

Figure 6c reveals two regions where sinkhole proximity and digital elevation are more strongly related than would be expected if these variables were independent of one another. Strongly positive residual values occur at low elevations (27 to 40 m asl) for closely spaced sinkholes (proximity <1 km), and to a lesser extent at elevations around 60 m for more widely separated sinkholes (proximity 1 to 3.5 km). These positive regions indicate greater than expected clustering of sinkholes (0.5 km proximity) at the elevation of most flood plains and river terraces, and greater than expected clustering of sinkholes with 2.5 km proximity, at the elevation of many of the upland Pliocene-Miocene surfaces.

Spatial trends with slope, distance to major rivers, and overburden thickness

As explained above, cross-area tabulation provides a means for identifying greater than expected overlap between sinkhole proximity and a variety of geologic variables. We continue our exploratory analysis by performing cross-area tabulations between sinkhole proximity and a variety of topographic and geologic factors that may contribute to sinkhole development. The results of these analyses are presented as residual plots in Figure 7.

Topographic slope may be important to the development of sinkholes because it can enhance local hydraulic gradients thereby increasing the likelihood of piping. Figure 7a shows residual cross-area tabulations for proximity and average topographic slope (where slope is calculated as the average pixel-to-pixel slope within a 300 m x 300 m rectangular search window stepped across the study site). Most sites in Lowndes County have gentle slopes as is evident from the distribution of positive residual values at the bottom of Figure 7a. However, it is interesting to note that the most

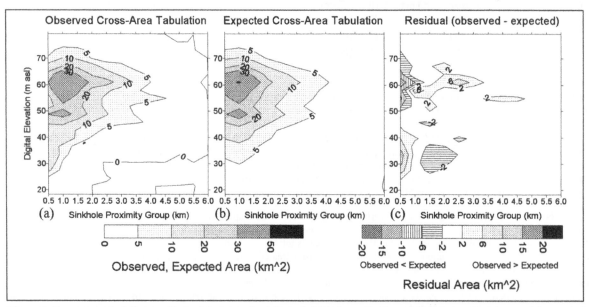

Figure 6. Cross-area tabulation results for (a) observed pairings of sinkhole proximity and digital elevation classes, (b) expected cross-area tabulations if sinkhole proximity and digital elevation classes are spatially independent, and (c) residual values calculated as the difference between observed and expected cross-area tabulations. Expected values are calculated as follows. The cross-area tabulation matrix T between two maps (A and B) yields matrix area elements T_{ij}, where i refers to class 1, 2, ... n for map A and j refers to class 1, 2, ... m for map B. The sum of each i-th row, and j-th column in the matrix is referred to as a margin total (T_i and T_j). If maps A and B are independent, then expected values T^*_{ij}, are calculated as the product of respective margin totals divided by the grand total area of the matrix (Bonham-Carter, 1997).

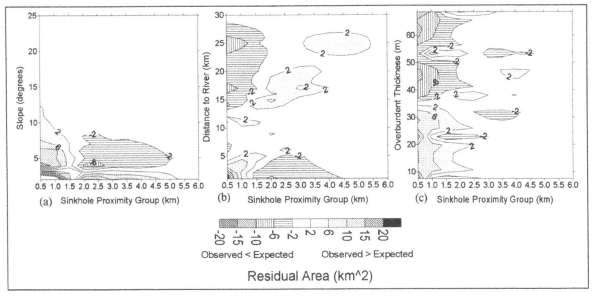

Figure 7. Residual cross-area tabulation results comparing sinkhole proximity with (a) slope class, (b) distance to major rivers, and (c) overburden thickness. See text for discussion.

positive residual values occur in areas that are near sinkholes (<1.5 km proximity) and have slope values between 3 and 7, distinctly steeper than is the case for more widely spaced sinkholes.

Previous discussion has suggested that sinkholes are more numerous near major river valleys. This qualitative observation is confirmed from cross-area tabulations between sinkhole proximity and the distance to major rivers (Withlacoochee, Little, and Alapaha Rivers) (Figure 7b).

Lastly, unlike its relationship with sinkhole size (Figure 3), overburden thickness does appear to be important in determining the spatial location of closely spaced sinkholes (Figure 7c). Large positive residual regions only occur for sites where sinkholes are tightly clustered (proximity < 1.5 km) in thin overburden (< 30 m). Residual values for sites where sinkholes are more widely separated (e.g. upland Miocene-Pliocene surfaces) are near zero indicating that overburden thickness is unrelated to the location of sinkholes.

Spatial Relationships between sinkholes and hydrogeology.

Several authors have recently argued that the location of sinkholes (Whitman et al., 1999), as well as the development of sinkhole lakes (Motz, 1998; Kindinger et al., 1999), are influenced by differences in hydrostatic head, associated with surface aquifers, and potentiometric head within the deeper Floridan Aquifer System. They argue

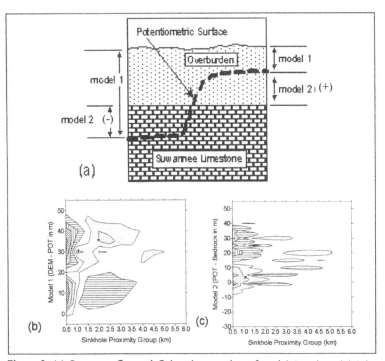

Figure 8. (a) Summary figure defining the meaning of model 1 and model 2 in relation to the elevation of the ground surface (which we use as a first-order proxy for the position), of the water table), the elevation of the average potentiometric surface, and the elevation of the top of the Suwannee Limestone. Also included are cross-area tabulation results for comparisons of sinkhole proximity and (b) the results of model 1, and (c) the results of model 2. See text for discussion.

that downward seepage and the transport of overburden into the subsurface is enhanced where head differences are large (model 1 in Figure 8a). Further, where head differences are small there is less leakage from surface aquifers downward into deeper bedrock aquifers and therefore the likelihood of sinkhole collapse is diminished. Cross-tabulation comparisons of sinkhole proximity and head difference by Whitman generally confirm this relationship for thick covered karst sites in central Florida. However, as is evident from

Figure 8b, similar analysis of sinkhole distributions in Lowndes County Georgia do not display the same trend. In fact, we find the opposite, with the most positive residual regions in cross-area tabulation occurring for closely spaced sinkholes at low head difference values (5 to 15 m).

Figure 8 also presents residual cross-area tabulation results for a second model (Figure 8a) that examines elevation differences between of the average county-wide potentiometric surface (Figure 2c) and the top of the Suwannee Limestone. The top of bedrock corresponds to the top of the Floridan aquifer and represents the beginning elevation of karstic voids, into which overburden is piped. This elevation is calculated as digital elevation (Figure 1) minus the thickness of overburden (Figure 2b).

We believe that model 2 is important to sinkhole development in Lowndes County for several reasons. First, piping of overburden into subsurface voids in the Suwannee Limestone is expected to be most effective where the potentiometric surface is below the top of the limestone bedrock (i.e. negative values for model 2). Secondly, cross-area tabulation results (Figure 8c) indicate reasonably strong positive residuals for closely spaced sinkholes at sites where the potentiometric surface is below to within 10 m above bedrock. This indicates that sinkholes cluster more strongly than would be expected if model 2 were unimportant, and this clustering preferentially occurs where low model 2 values dominate. Furthermore, it is important to recall that although we use an average potentiometric surface, water well data indicate variation about this surface by approximately 5 m seasonally and as much as 11 m since 1957 (Figure 3). Finally, as can be seen from Figure 8c sinkholes visibly cluster within regions of low to moderate model 2 values.

SUMMARY

The forgoing analysis, while identifying several individual trends, repeatedly points to two fundamentally different groups of sinkholes in Lowndes County (Figure 9). Sinkholes in the first group are smaller on average (Tables 1, 2), and are more closely spaced (Figure 5) than are sinkholes in group 2. In addition, sinkholes in group 1 occur predominantly at lower elevations (Figure 6), in sandy Quaternary alluvium (Figure 2), along major river valleys where topographic slope is steeper and overburden is thinner (Figure 7) than for sites where sinkholes in group 2 are found. The second group of sinkholes is more prevalent in upland terrain where more clay-rich Miocene-Pliocene sediments are exposed. Although our present analysis does not determine the relative importance of independent variables, we firmly believe, as has been pointed out elsewhere (Whitman et al., 1999), that the elevation of potentiometric head for the Floridan

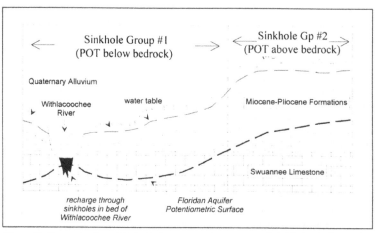

Figure 9. Conceptual model depicting two major groups of sinkholes in Lowndes County and associated topographic, geologic, and hydrologic controlling variables.

aquifer plays a critical role in controlling the location and size of sinkholes. As is implied in Figure 9, we believe that the proximity of the time-averaged potentiometric surface is more frequently at or even below the top of limestone bedrock in low-lying river valleys than it is below upland terrain. This in combination with the prevalence of sandy alluvium in river valleys, a sediment that is easily piped, contributes to the observed differences in sinkhole characteristics and spatial locations in Lowndes County, Georgia.

ACKNOWLEDGEMENTS

We thank the following individuals and institutions for logistical and financial support relating to this study: Eastern Connecticut State University and the American Association of University Professors for grants supporting conference travel. Phaedra Durost and Tracey Gaylord (ECSU) for assistance in digitizing maps. Valdosta State University (Department of Physics Astronomy, and Geosciences) for support while the first author was in Georgia. Chris Strom of the South Georgia Regional Development Center for access to soil data files. We also appreciate public access to data supplied on the Georgia GIS clearing house web site.

REFERENCES

Beck, B. F. and Arden, D. D., 1984. Karst hydrogeology and geomorphology of the Dougherty Plain, Southwest Georgia. Southeastern Geological Society Guidebook No. 26, Southeastern Geologic Society, Tallahassee, 59 pp.

Beck, B. F., and Sinclair, W. C., 1986. Sinkholes in Florida, an introduction. Florida Sinkhole Research Institute, University of Central Florida. Report 85-86-4, 16 pp.

Bonham-Carter, G. F. 1997. Geographic information systems for geoscientists, modelling with GIS. Peragamon. 398 pp.

Brook, G. A., and Allison, T. L., 1983. Fracture mapping and ground subsidence susceptibility modelling in covered karst terrain: Dougherty County, Georgia. In: P. H. Dougherty (editor), Environmental Karst, GeoSpeleo Publications, Cincinnati, pp. 91-108.

Clark, W. Z., and Zisa, A. C., 1976. Physiographic map of Georgia. Georgia Geologic Survey, Statewide Map 4.

Drake, J. J., and Ford, D. C., 1972. The analysis of growth patterns of two-generation populations: the example of karst sinkholes. Canadian Geographer XVI:381-384.

Herrick, S. M., and LeGrande, H. E., 1964. Solution subsidence of a limestone terrane in southwest Georgia. International Association of Scientific Hydrology, Bulletin 9: 25-36.

Huddlestun, P. F. 1997. Geologic atlas of the Valdosta area. Georgia Department of Natural Resources & U.S. Environmental Protection Agency, Geologic Atlas 10. 25 pp.

Hyatt, J. A., and Jacobs, P. M., 1996. Distribution and morphology of sinkholes triggered by flooding following Tropical Storm Alberto at Albany, Georgia, USA. Geomorphology 17:305-316.

Hyatt, J. A., Wilkes, H. P., and Jacobs, P. M. 1999. Spatial relationships between new and old sinkholes in covered karst, Albany, Georgia, USA. In Proceedings of the seventh multidisciplinary conference on sinkholes and the engineering and environmental impacts of karst; hydrogeology and engineering geology of sinkholes and karst. Beck, F. B., Pettit, A. J., and Herring, J. G. (eds). A. A. Balkema, Rotterdam. pp. 37-44.

Kemmerly, P. R., 1982. Spatial analysis of a karst depression population: clues to genesis. Geological Society of America Bulletin 93:1078-1086.

Kindinger, J. L., Davis, J. B., and Flocks, J. G. 1999. Geology and evolution of lakes in north-central Florida. Environmental Geology, 38: 301-321.

Krause, R. E., 1979. Geohydrology of Brooks, Lowndes, and western Echols counties, Georgia. U.S. Geological Survey Water-Resources Investigations Report 78-117, 49 pp.

McConnell, J. B., and Hacke, C. M. 1993. Hydrogeology, water quality, and water-resources development potential of the upper Floridan Aquifer in the Valdosta area, south-central Georgia. U.S. Geological Survey, Water-Resources Investigations Report 93-4044, 44 pp.

Motz, L. H. 1998. Vertical leakage and vertically averaged vertical conductance for karst lakes in Florida. Water Resources Research, 34:159-167.

Palmquist, R., 1979. Geologic controls on doline characteristics in mantled karst. Z. Geomorph. N.F. Suppl. Bd 32, pp. 90-106.

Varrington, D. V., and Lindquist, R. C. 1987. Thickly mantled karst of the Interlachen, Florida aeaa. Proceedings of the 2nd multidisciplinary conference on sinkholes and the environmental impacts of Karst.

Whitman, D., and Gubbels, T. 1999. Applications of GIS technology to the triggering phenomena of sinkholes in Central Florida. In Proceedings of the seventh multidisciplinary conference on sinkholes and the engineering and environmental impacts of karst; hydrogeology and engineering geology of sinkholes and karst. Beck, F. B., Pettit, A. J., and Herring, J. G. (eds). A. A. Balkema, Rotterdam. pp. 67-73.

Whitman, D., Gubbels, T., and Powell, L. 1999. Spatial interrelationships between lake elevations, water tables, and sinkhole occurrence in central Florida; a GIS approach. Photogrammetric Engineering and Remote Sensing, 65: 1169-1178.

Wilson, W., and Beck, B. 1992. Hydrogeologic factors affecting new sinkhole development in the Orlando area, Florida. Ground Water, 30: 918-930.

Wilson, W. L., McDonald, K. M., Barfus, B. L., and Beck, B. F., 1987, Hydrogeologic factors associated with recent sinkhole development in the Orlando area, Florida, Florida Sinkhole Research Institute, University of Central Florida, Report No. 87-88-4, 109 pp.

Geotechnical and Environmental Applications of Karst Geology and Hydrology, Beck & Herring (eds)
© 2001 Taylor & Francis, ISBN 90 5809 190 2

Subsidence rates of alluvial dolines in the central Ebro Basin, Northeastern Spain

M.A.SORIANO & J.L.SIMÓN   Dept. de Geología, Universidad de Zaragoza, 50009 Zaragoza, Spain,
asuncion@posta.unizar.es

ABSTRACT

The central sector of the Ebro basin is filled with Paleogene and Neogene materials (clastic, evaporite and carbonate facies). Alluvial dolines develop on Quaternary terraces and pediments overlying Neogene evaporites. These kinds of landforms are very active at present. Along with dissolution, the mechanisms that cause doline development are dragging and collapse of the materials.

Maps of dolines prepared by analyzing aerial photographs of different years (from 1946 till 1993) clearly show important differences through this period of time, caused by human influence and natural processes. The main urbanization of the area took place in the seventies, and since then a great number of existing buildings and infrastructures have been seriously damaged as a consequence of karst processes. This damage was at great economic loss. Damage inventory includes: (1) *fractures* in walls, floors, roads and water pipes, (2) *tilting* in buildings and (3) *soil subsidence* in floors and roads. From the dates of construction and repair of some of the buildings and pavements of that area, we establish approximate subsidence rates that oscillate between 12 mm/year to 110 mm/year. We decided to monitor several of these dolines developed in urban areas to determine in a more detailed way the subsidence behavior.

Monitoring of three dolines located on different streets of this area took place from September 1993 to November 1997. A water level system (with an error of 2-3 mm) was utilized for measuring the evolution of these depressions. During these four years, the mean subsidence rates obtained were 64.5, 39 and 21 mm/y, respectively. Assuming the initial level of roads being almost horizontal and taking into account the whole time since these areas were urbanized, the subsidence rate is 92, 32.5-40 and 24 mm/y, which fits with the previous data from repaired zones.

INTRODUCTION

The central part of the Ebro Basin, northeast Spain, shows many examples of active alluvial dolines generated by dissolution of underlying gypsum. The presence of dolines causes many problems both in farming activities and civil engineering. Farmers lose arable lands, and they attempt to replace it either by filling the depressions each year or by building drainage channels to eliminate the excess of moisture in the soil. However, the main problems appear in civil engineering, especially since the 1970s, when a large number of factories were built in this zone on terrain where karstic depressions were previously filled. No special measures for construction were taken at that time. The generation of dolines is still an active process, and both subsidence and sudden collapse (as a completely isolated event or as recurrent events affecting the same or neighboring sites) frequently occur all around this zone. As a consequence, important damage appears in many civil engineering constructions, i.e., fractures in walls, floors and ceilings of buildings (in some cases causing almost total destruction), subsidence or collapse in roads, and breaking of water supply systems, etc.

As it is well known, the solubility of gypsum is higher than the solubility of calcium carbonate (in pure water at 20°C they are 2.531 gr/l and 1.5 mg/l, respectively, as shown by Klimchouk, 1996a). This fact causes variation in the size of landforms developed on gypsum that can be appreciated on a human scale. There are different karst landforms on gypsum with a great variability in size and shape appearing under different climatic conditions. (Calaforra, 1996; Forti, 1996; Klimchouk, 1996b and Sauro, 1996). Alluvial dolines are generated when soils or other superficial deposits overlying karstic rocks fall into conduits enlarged by dissolution, thus causing the sinking of the surface. Their development is conditioned by: (1) mobilization of material by groundwater (often including the creation of subsoil cavities), which may take place by either dissolution of a soluble substratum or piping of the cover; and (2) cave-in of the alluvial cover, developed by either sudden collapse or slow subsidence (Beck, 1986; White et al., 1986; Yuan, 1988; Benito et al., 1995; Buttrick and van Schalkwyk, 1998).

The Ebro basin (Figure 1a) was filled with Paleogene and Neogene sediments including clastic, evaporite (mainly gypsum and halite) and carbonate facies (Quirantes, 1978). During the Quaternary, several nested levels of alluvial terraces and pediments were formed. Dolines often develop where the alluvial deposits overlie Neogene evaporites that were deposited around the depocentre of the Neogene basin. The climate in this area is semiarid with an average annual rainfall in Zaragoza around 300 mm, most of it by precipitation during spring and autumn; although summer storms are also frequent. The average monthly maximum temperature is 24-25°C in July, and the minimum is 5-6.5°C in January.

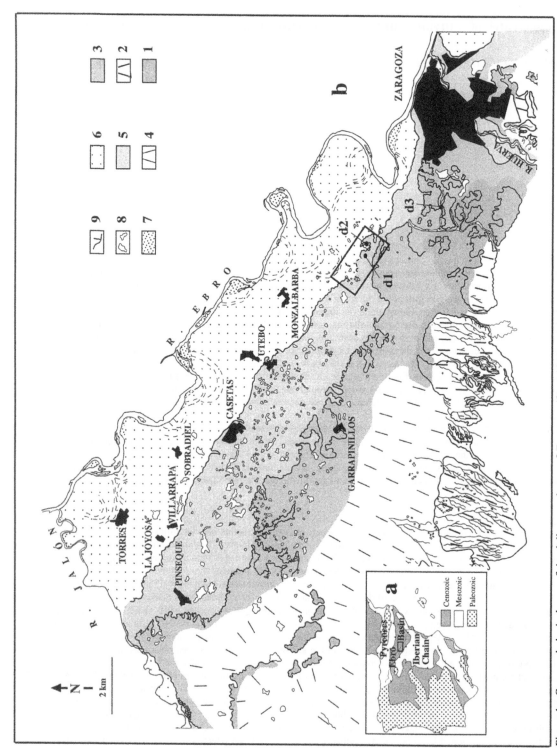

Figure 1.- Geomorphological map of the dolines northwest of Zaragoza (from 1957 aerial photographs). 1: Terrace T$_4$. 2: Pediment related to T$_4$. 3: Terrace T$_3$. 4: Pediment related to T$_3$. 5: Terrace T$_2$. 6: Terrace T$_1$. 7: Present-day river bars. 8: Dolines. 9: Gullies. The location of the three monitoring dolines is pointed with d1, d2 and d3. The square shows the location of the industrial areas "El Portazgo" and "Europa".

The main objective of this work is to determine the rate of subsidence of dolines in the central sector of the Ebro basin. For this purpose the methodology followed here is varied because different types of data are required. (1) A compilation of previous technical reports (data from mechanical drilling, geoelectrical profiles, groundwater wells and geological maps was obtained from 580 points or mall sections distributed over an area of 124 km^2), both from private and public enterprises, was made to learn in detail the geological characteristics of the zone. (2) Mapping of dolines from aerial photographs of several years (from 1947 to 1993) was attained, which permits changes in the distribution and evolution to be analyzed. (3) Interviews with people that live or work in the area were made with the purpose of checking the presence of dolines and learning when and where other events not identified in aerial photographs occurred. (4) Three active dolines located in urban areas were monitored for a little more than four years.

ALLUVIAL DOLINES: CHARACTERISTICS AND DISTRIBUTION.
The geomorphological context. A brief description.

In the central Ebro basin, close to the city of Zaragoza, eight terrace levels are recognized. The best developed are the last four (Figure 1b). All of them appear on the right bank of the Ebro River, and they cover the Tertiary materials. The terraces are mainly constituted of layers of gravels, sands and silts. The size of gravels is between 5 and 20 cm. Sedimentary structures such as planar and trough cross bedding and imbrication of gravels are very common. In most levels (except in the most recent, T$_1$) carbonate crusts are frequently developed at top. There are also six pediment levels with layers of gravels and silts. Alluvial dolines are developed preferentially on the three youngest terrace levels, which have a relative altitude to the Ebro river of 3-6 m (T$_1$), 10-14 m (T$_2$) and 29-34 m (T$_3$) (Soriano, 1990a). A small number of dolines are also detected in T4 and pediment levels 3 and 4.

Taking into account the data that we have collected at detailed scales (1:5.000-1:10.000), the most important factors that influence the development of dolines are (1) the irregular topography existing between the Miocene and the Quaternary, where *paleovalleys* favor the development of dolines; (2) the low thickness of the Quaternary cover; and (3) the low percentage of lutites in the Quaternary deposits (Simón et al, 1998). At less detailed scales (1:50.000) the maximum annual variation of the water level and the sulphate content in groundwater seem to be important factors in doline genesis (Soriano and Simón, 1995).

This information justifies the more detailed characteristics of the three main factors just mentioned. The limit between Quaternary terraces and the underlying Neogene shows a clear decrease in elevation towards the Ebro River. In detail, an irregular relief is shown with high and low areas that in many cases appear in a perpendicular direction to the present scarps of terraces. The thickness of T$_4$ varies between 30-50 m, T$_3$ between 25-40 m, T$_2$ between 8-20 m, and T$_1$ between 4-15 m. Consequently, the lowest thickness is registered in the two more recent levels (1 and 2), T$_2$ being in fact the level with the highest number of dolines on it. In relation to lutite percentage, the most relevant data is its high content in the lowest level of terrace, where the maximum is around 60% of the total amount of the deposit. In the other terraces the percentage is lower than 20%, except some parts in T$_2$ where the values can reach 40%. This fact can explain the low number of dolines in T$_1$.

Karst landforms and time variation.

The origin of these alluvial dolines is conditioned by dissolution of the underlying gypsum and dragging or collapse of the Quaternary layers present at top. From a morphological point of view (Cvijic, 1893 and Palmquist, 1979), basin, pan, funnel and well shaped types have been identified. Their diameters range from several meters to 100 m, and the depth from 1 to 20 m. There are also bigger depressions of around 1 km length that correspond to uvalas (Figure 1).

In this area there is a paleokarst, inactive at present, that had an important development at least during the sedimentation of the fourth level of terrace and pediment (from middle Pleistocene). Most of them show funnel and well shapes, and they were developed at the same time, as Quaternary materials were deposited (Soriano, 1990b; 1992).

Table 1: 1946 - Total number of dolines identified from aerial photographs of 1946. A - Number of dolines that remain. B - Number of dolines that vanish. C - New dolines. 1986 - Total number of dolines identified from aerial photographs of 1986 (modified from Soriano and Simón, 1995).

	1946	A	B	C	1986
T1	34	16	18	15	31
T2	147	46	101	99	145
T3	96	24	72	85	109
TOTAL	277	86	191	199	285

A high number of dolines have been recognized through time. The use of aerial photographs favors (if both the phenomenon and photograph scales permits) the identification of dolines in an area. By analyzing photographs of different years, it is possible to determine the evolution of dolines. The studied aerial photographs correspond to different years: 1946, 1957, 1970, 1982, 1986, 1987, 1988 and 1993. Not every one covers the same area, and there are differences in their quality and scale (1:43000, 1:32000, 1:20000, 1:18000, 1:18000, 1:3000, 1:20000 and 1:3500, respectively). Even in these conditions, it is possible to see that the number, shape and dimensions of dolines have changed throughout time (Soriano, 1992; Soriano and Simón, 1995). The two oldest series of photographs have the disadvantage of a small scale, with the consequent loss of detail, but have the advantage that in those years most of the land was dedicated to farming activities. Consequently, very few modifications in the original landform were made. The urbanization of a great part of this zone started at the beginning of the seventies, which is well reflected in the other series of photographs. The aerial photographs from years 1946, 1957, 1970 and 1986 were chosen to determine the variations in dolines (Soriano, 1992) because they cover the same area, and in the rest of the years there is only a partial overlap. From this analysis, changes are observed in each period. Some of the depressions disappeared by human filling, while a high number of new dolines were generated indicating the active character of the process. Nevertheless, the final balance is almost constant in the three terrace levels during these forty years, which could signify that the rate of doline generation and reactivation is similar to the rate of human filling (Soriano and Simón, 1995).

In urban areas built on old alluvial dolines, filled around 1970, it is more difficult to identify them, because buildings and pavement mask the original topography, and depressions can be directly observed only in a few cases. In such areas it is even more necessary that additional field surveys be completed to recognize doline activity and damage to building structures. The types of damage identified are the following:

(a) Fractures in walls: A high number of buildings of the area show this kind of damage. By studying their geometry, it is possible to identify the direction in which the subsidence center is located.

(b) Fractures and steps in floors and roads: It is usually easier to follow the fracture than in the previous case. Frequently they show curved patterns, whose concave sides point to the center of the sinking area. In some occasions, continuity between fractures in floors and walls have been observed.

(c) Soil subsidence: Sometimes sinking zones appear as deformation in the pavement without fractures at all. They are easily recognized by observing differences of several centimetres or decimetres in the floor or road surface, below its normal position.

(d) Tilting of buildings: This happens when the subsidence area affects directly one or more master pillars of the building, thus causing the tilt of all or only a part of the edifice (up to 8-10°). In most cases the structure must be destroyed to avoid danger to humans.

In some cases it is possible to determine the precise position of the subsidence or sinking center from the previous data. In the industrial area located close to Zaragoza a lot of damage is recognized, and when the sinking centers are well determined, most of them (around 90%) are located in relation with old dolines previously identified by aerial photographs. The sinking centers can be inside of the contour or just in the borders of these depressions. All of these facts show a clear relation between dolines and the response observed in built-up areas.

RATES OF SUBSIDENCE IN THE AREA.
Rates from collected information.
Direct observations in the area and interviews with people living or working there permits: (1) confirmation of research data, (2) updates of recent cases of subsidence which are revealed by damage that appears in farming areas and buildings, and (3) attainment of exact values of sinking in the zone through time. To achieve these objectives, after the study of aerial photographs, we visited all the places where dolines were mapped (Simón et al., 1998).

Some of the factory and office buildings affected by subsidence needed to be repaired several times since their construction (Polígonos Industriales "El Portazgo" and "Europa"). In these cases, interviews with people have been basic in determining the kind of damage. The type of reparations is varied, but an especially interesting one is the use of movable pillars that level the structure of the building and permit exact control of the difference in height at each moment. From data on three buildings the estimated subsidence rates are between 1.2 and 4 cm/year.

A lot of filled dolines show a similar activity after the urbanization of the zone. In streets and roads sinking damage was solved by occasional refilling and leveling with asphalt. In one of these streets, the thickness of filling has been estimated between 2.5 and 3 m. It was urbanized in 1972; consequently, when the interview was held in 1999, the subsidence rate was between 9 and 11 cm/year (considering the lowest and the highest value).

In another case, there is a sinking center related to the border of a bigger doline. From our experience and the oral information, it may be concluded that the sinking rate is increasing in the last years. This zone was paved in 1973, but in a field study made in 1991, no deformation of the pavement was observed. Nevertheless, fractures in a wall were detected from 1989. In addition, the difference in height measured in 1999 was 80 cm. This means that in these 10 years the rate is 8 cm/y; if we consider the data from 1973, the rate is 3.1 cm/y. The fractures appearing in the pavement were repaired in 1998 and some of them have a new opening of 6 cm in just one year (from 1998 to 1999).

The situation is different in cases of sudden collapse. This may appear as a completely isolated event, giving rise to an instantaneous sinking of several metres, or as recurrent moderate events affecting the same or neighboring sites. Several examples of the first case took place in several buildings of the Polígonos "El Portazgo" and "Europa". In one industrial building, a sudden collapse happened in the spring of 1999. Ten years ago several fractures in walls and floor were observed, but that year part of the floor of the building collapsed, which affected even the roof and walls. One week after the collapse the size of the doline was 10x10 m in surface and 6 m in depth. In the same zone, another doline affected a building in 1994. In this case the diameter was around 15 m, and the depth was at least 3 m (the presence of water in the doline makes its observation difficult). Several points in the motorway to the northwest of Zaragoza are good example of the second situation; these points have suffered repeated collapses since construction in 1976 in spite of rapid refilling and covering of the depressions. At one point (Autopista Vasco-Aragonesa, km 292.500) five events had been registered from 1976 to 1991: 1981, depth = 320 cm; 1983, unknown depth; 1984, 25 cm; 1984, 70 cm; 1991, 200 cm. The two main events took place in 1981 and 1991 after the two longest periods of inactivity.

Rates from monitoring dolines.
Due to the serious problems caused by the presence of dolines, in 1993 we started monitoring some of the most active ones to determine the speed of the process in each. Several topographic profiles were carried out across three dolines developed in urban areas (Figure 1). Two are located in the industrial areas just mentioned (points d1 and d2 of Figure 1b). The other is in the urban area of the city of Zaragoza (d3 in Figure 1b). For measuring the evolution of these dolines, we utilized a water-level device with a precision of 2-3 mm (it was tested at the beginning of the topographic survey). In each profile we measured the differences in height every 2 m laterally, and the measure was repeated twice during each survey in order to ensure that no significant error was introduced during the

Table 2. Data from doline monitoring (1 and 2 are mapped in figure 1 and 3 is in the city of Zaragoza). M.S. = Maximum sinking from the first till the last measure (September 93-November 97). Rate: the total period of measure is 4.2 years. T.A.S. = Total apparent sinking (it is assumed that the streets were more or less horizontal). T.U. = Time from the urbanization of the streets till 1997. A. Rate = Apparent rate (total apparent sinking through the period that the streets have been urbanized). M. Rate M.T. and m. Rate M.T. = Maximum and minimum rate between two periods of monitoring.

	1	2	3
M. S. (mm)	271	164	88
Rate (mm/y)	64.5	39	21
T. A. S. (mm)	1200	650-800	600
T. U. (y)	13	20	25
A. Rate (mm/y)	92	32.5-40	24
M. Rate M. T. (mm/y)	120	62	40
	(March-Dec.1994)	(April 96-Feb.97)	(Feb-Nov.1997)
m. Rate M. T. (mm/y)	40	21	4
	(Feb.-Nov.1997)	(Feb.-Nov.1997)	(March-Dec.1994)

process. The surveys started in September of 1993 and finished in November 1997. A partial summary of the results obtained during this period of time is represented in Table 2.

From these results, the continuous increase in depth in all dolines is obvious. Nevertheless, this value, and consequently the rate, is bigger for doline 1 than for the others. Supposing that the original surface of the streets was more or less horizontal in the three cases, the total apparent sinking was determined. This value is very similar for dolines 2 and 3 over 4.2 years, but it is again higher for doline 1. The same trend is observed between the values registered in successive measures; indeed, the minimum value found for doline 1 coincides with the highest of doline 3.

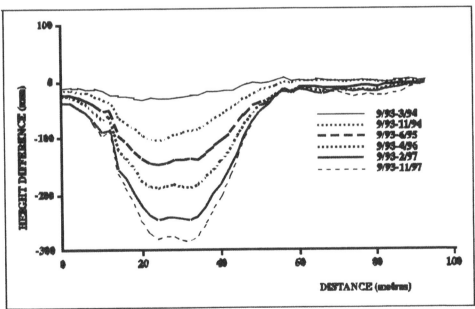

Figure 2. Profiles of the differences in height of the measures taken in doline 1 from September 1993 to November 1997.

In Figure 2 the differences between the data of the first measure (September 1993) and the successive measures are shown for doline 1. From these profiles a clear sinking area with minor irregularities is defined. The rest of the profile shows differences always less than 5 cm during this time with logical larger differences close to the borders of the main sinking area.

If we compare monitoring data with those obtained in previous observations and reports, it seems clear that there is a good correlation between them. The subsidence rates range from 12 to 110 mm/y, and the results of our monitoring are comprised between 24 and 92 mm/y. Even the highest rate of doline 1 does not surpass the maximum calculated rate. These two maximum values (90-110 and 92 mm/y) are measured in two dolines located very close to each other, probably on a more active dissolution zone at present. The rest of the data are, in general, less than 50 mm/y.

CONCLUSIONS

In the central part of the Ebro basin dissolution processes of Neogene gypsum and development of alluvial dolines on Quaternary deposits are frequent phenomena, developed at least from the middle Pleistocene. Variations of size and the number of dolines through time can be established with the aid of several series of aerial photographs.

Subsidence rates for some areas (located in industrial zones close to the city of Zaragoza) were established from data obtained from field work, the monitoring of three dolines from September 1993 to November 1997, and from information about repairs made to damage caused by the doline evolution. There is a good agreement between the sources of data. The minimum rate is 12 mm/y (data of special pillars in industrial buildings) and the maximum is 90-110 mm/y (refilling of a street in an industrial zone). The results of our monitoring (24-92mm/y) are clearly included in the former interval.

REFERENCES

Beck, B., 1986, A generalized genetic framework for the development of sinkholes and karst in Florida, U.S.A.: Environ. Geol. Water Sci., 8, pp. 5-18.

Benito, G, Pérez, P., Gutiérrez, M. and Sancho, C., 1995, Natural and human-induced sinkholes in gypsum terrain and associated environmental problems in NE Spain: Environmental Geology 25, pp.156-164.

Buttrick, D. and Schalkwyk van, A., 1998, Hazard and risk assessment for sinkhole formation on dolomite land in South Africa: Environmental Geology 36 pp. 170-178

Calaforra, J.M., 1996, Some examples of gypsum karren: In (eds) J. Fornós & A. Ginés: Karren landforms pp. 253-260. Palma de Mallorca.

Cvijic, J., 1893, The dolines: Translation of Geog. Abhandlungen 5, pp. 225-276. In Sweeting, M.M. (ed.) 1981 Karst Geomorphology. Hutchinson.

Forti, P., 1996, Erosion rate, crystal size and exokarst microforms: In (eds) J. Fornós & A. Ginés Karren landforms: pp. 261-276. Palma de Mallorca.

Klimchouk, A., 1996a, The dissolution and conversion of gypsum and anhydrite: Int. J. Speleol. 25, pp. 21-36.

Klimchouk, A., 1996b, The typology of gypsum karst according to its geological and geomorphological evolution: Int. J. Speleol. 25, pp. 49-60.

Palmquist, R., 1979, Geologic controls on doline characteristics in mantled karst: Z. Geomorph. Suppl. Bd., 32, pp. 90-106.

Quirantes, J., 1978, Estudio sedimentológico y estratigráfico del terciario continental de los Monegros: Zaragoza. Diputación Provincial de Zaragoza. 200 pp.

Sauro, U., 1996, Geomorphological aspects of gypsum karst areas with special emphasis on exposed karst: Int. J. Speleol. 25, pp. 105-114.

Simón, J.L., Soriano, M.A., Arlegui, L. and Caballero, J., 1998, estudio de riesgos de hundimientos kársticos en el corredor de la carretera de Logroño: (Unpublished report).Universidad de Zaragoza, 94 p.

Soriano, M.A., 1990a, Geomorfología del sector centro-meridional de la Depresión del Ebro: Zaragoza. Diputación Provincial de Zaragoza. 269 p.

Soriano, M.A., 1990b, Le karst du gypse du centre de la Dépression de l'Ebre (Espagne): Karstologia, 16, pp.39-45.

Soriano, M.A., 1992, Characteristics of the alluvial dolines developed because of gypsum dissolution in the central Ebro Basin: Z. fur Geomorphology Suppl. Bd , 85, pp. 59-72

Soriano, M.A. & Simón, J.L., 1995, Alluvial dolines in the central Ebro basin, Spain: a spatial and developmental hazard analysis:Geomorphology 11, pp. 295-309.

White, E.L.; Gert, A. & White, W.B., 1986, The influence of urbanization in sinkhole development in central Pennsylvania: Environ. Geol. Water Sci., 8, pp. 91-97.

Yuan, D., 1988, Environmental and engineering problems of karst geology in China: Environ. Geol. Water Sci., 12, pp. 79-87.

Geotechnical and Environmental Applications of Karst Geology and Hydrology, Beck & Herring (eds)
© *2001 Taylor & Francis, ISBN 90 5809 190 2*

Cover-collapse sinkhole formation and piezometric surface drawdown

THOMAS M.THARP   Dept. of Earth & Atmospheric Sciences, Purdue University, West Lafayette, IN 47907, USA,
ttharp1@purdue.edu

ABSTRACT

Piezometric surface drawdown is a common cause for formation of cover-collapse sinkholes in soil. When water pressure drops in a soil void at the rock contact, the pore pressure drop in the surrounding soil lags that in the opening. This may result in effective tensile stresses that produce hydraulic fracturing in the soil, leading to progressive fracturing and sloughing that propagates to the land surface. To model this phenomenon poroelastic stress and pore pressure equations are derived for a spherical opening under spherically symmetric loading. If the soil void is small relative to depth below surface the governing pore pressure and displacement relationships are nearly uncoupled. The two equations are solved by finite difference for each time step to find pore pressures, stresses and effective stresses in the soil around the void. In soils the poroelastic analysis yields higher tensile effective stresses than are found in a non-poroelastic analysis. Local susceptibility to formation of cover-collapse sinkholes is found to be strongly a function of soil compressibility, permeability and tensile strength. If these and other soil parameters are known, the piezometric surface drawdown conditions necessary to trigger sinkhole formation can be found for a particular locality.

INTRODUCTION

In formation of a cover-collapse sinkhole, a subsurface void forms in soil above an opening in the underlying rock. The void ultimately propagates to the surface, resulting in subsidence and, generally, formation of a surface pit. In terms of mechanics and causation, cover-collapse sinkholes probably fall into two classes dependent on depth to the water table. When the water table lies below the soil-rock interface, sinkholes form in response to a diverse set of triggering mechanisms and a number of different mechanical processes may be responsible for the failure (reviewed by Tharp, 1999). The more restricted class of failures that occur where the water table lies, at least initially, above the soil-rock interface are the subject of this paper. In areas with a high water table most induced sinkholes result from lowering of the water table (Newton, 1984; Brink, 1984; Metcalfe and Hall, 1984; Kiernan, 1989; Sowers, 1996). Reasons suggested for this association include loss of buoyant support, increased pore pressure gradients (and thus water velocities), and increased amplitude of water level variations (Newton, 1984).

It was noted by Tharp (1997, 1999) and Anikeev (1999) that high pore pressure gradients at the perimeter of the soil void could lead to failure by hydraulic fracturing. Tharp noted that once initiated this process will typically propagate rapidly upward. Recent appearances of cover-collapse sinkholes in regions with high water tables are common. In Florida, the increased frequency of sinkhole formation associated with the spring dry season (Benson and LaFountain, 1984; Currin and Barfus, 1989) may be associated with water table lowering. It is significant that sinkholes sometimes appear very quickly in response to groundwater lowering. Bengtsson (1987) observed that almost all sinkholes associated with a 3-day drawdown episode occurred within a day to a few days, Sowers (1975) describes two large sinkholes formed after 3 days of pumping, and Daoxian (1987) noted the occurrence of the first sinkholes within hours of the commencement of pumping.

INITIAL FORMATION OF SOIL CAVITY

Formation of a surface sinkhole is preceded by formation of a soil void that forms by movement of soil into an opening in the rock below. Openings in solutioned limestone, commonly 10-40 cm (4-16 in.) in diameter (White et al., 1984), are reported by White (1999) to be 5-20 cm (2-8 in.) across in central Pennsylvania. There is commonly a zone of very soft soil just above the rock (Wilson and Beck, 1988, Pazuniak, 1989; Siegel and Belgeri, 1995; Iqbal, 1995; Sowers, 1996). This soil just above the rock surface may be protected from compaction by rock pinnacles or overhangs and, being farthest from the surface, is least subject to consolidation by desiccation (Siegel and Belgeri, 1995). This soft zone has also been interpreted as soil that has sloughed (ravelled) into and filled previous soil voids (Sowers, 1975; Wilson and Beck, 1988). I assume that this soft soil will move easily into the opening in the rock to form the soil void. Tensile failure in the soft soil and perhaps overlying stronger soil will allow the roof of the soil void to assume a no-tension domal shape (Hodek et al., 1984) with all load carried in compression.

HYDRAULIC FRACTURING AROUND THE SOIL VOID

It is assumed for this analysis that the resulting soil void is submerged, with the water in the void in communication with the karst aquifer below. When the piezometric surface in the karst aquifer is drawn down the water pressure on the wall of the soil void is immediately reduced, reducing the normal radial compressive stress in the soil. The pore pressure in the soil around the opening exhibits a more gradual drop, with the result that high pore pressure gradients may occur. If pore pressure exceeds the sum of the small radial compressive stress near the soil void, plus tensile strength of the soil, hydraulic fracturing will occur. This will cause the innermost layer of soil around the soil void to slough off, exposing soil with even higher pore pressure to the same boundary conditions. Hydraulic fracturing and sloughing at this new interface will occur immediately. Once initiated this process of successive sloughing may rapidly propagate the soil void to the ground surface.

POROELASTIC ANALYSIS

Because the soil matrix deforms easily, a change in pore pressure causes deformation and a change in stresses in the soil. Deformation of the soil matrix also generally affects pore pressure. Because these effects can be significant in a soil, the effect of the drawdown of the piezometric surface is appropriately analyzed as a poroelastic problem. In order to simplify the analysis and to allow isolation of the effect of the critical parameters, a simplified geometry is assumed for the analysis. Rather than considering a three-dimensional domal soil void above a rigid rock interface, the soil void is modeled as spherical, with spherical symmetry in all stress and pore pressure boundary conditions to allow a one-dimensional representation of the problem. Because both stresses and pore pressures drop to near background levels only a few radii away from an opening, this simplified geometry can, with appropriate boundary conditions, reasonably approximate a soil void that is small relative to its depth below the surface. To accommodate spherical symmetry it is also assumed that horizontal and vertical stresses in the soil are equal. The spherical approximation of a hemispherical cavity may be relatively accurate because the soft layer at the soil-rock interface may act as a plane of nearly zero shear stress, which approximates the zero shear stress of the horizontal symmetry plane of the spherical problem.

A poroelastic solution must satisfy constitutive equations relating stresses and pore pressure to strains, force equilibrium equations, fluid continuity equations, and Darcy's Law governing flow of water through the porous medium. It is convenient to consider first the force equilibrium equation for spherically symmetric stresses:

$$\frac{d\sigma_r}{dr} - \frac{2(\sigma_\theta - \sigma_r)}{r} = 0 \tag{1}$$

where r is radial distance from the center of the sphere and σ_r and σ_θ are radial and tangential normal stresses. Stresses are stated in terms of strains as:

$$\sigma_r = 2G\varepsilon_r + 2G(v/(1-2v))\varepsilon - \alpha P \tag{2a}$$

$$\sigma_\theta = 2G\varepsilon_\theta + 2G(v/(1-2v))\varepsilon - \alpha P \tag{2b}$$

where ε_r and ε_θ and are strains in the radial and tangential directions, $\varepsilon = \varepsilon_r + 2\varepsilon_\theta$ is volumetric strain, P is pore pressure, G is shear modulus, v is Poisson's ratio and α is the Biot-Willis parameter (Wang, 2000). Substitution of (2) into (1) gives:

$$2G\frac{\partial\varepsilon_r}{\partial r} + 2G\left(\frac{v}{1-2v}\right)\frac{\partial\varepsilon}{\partial r} - \alpha\frac{\partial P}{\partial r} - 4G[(\varepsilon_\theta - \varepsilon_r)/r] = 0 \tag{3}$$

Strains for the spherically symmetric problem are:

$$\varepsilon_r = du/dr \tag{4a}$$

$$\varepsilon_\theta = u/r \tag{4b}$$

where u is radial displacement. Recognition that $(\varepsilon_\theta - \varepsilon_r)/r = (u/r - du/dr)/r = -d\varepsilon_\theta / dr$ allows simplification of (3) to:

$$\frac{\partial\varepsilon}{\partial r} = c_m\frac{\partial P}{\partial r} \tag{5}$$

Integrating (5) gives:

$$\varepsilon = c_m P + g(t) \tag{6}$$

where $g(t)$ is a function only of time t and $c_m = \alpha(1-2v)/G(2-2v)$. Wang (2000) gives the remaining steps in the derivation of the governing equation for P This is stated briefly to illustrate the principles of the derivation. The identity for ε (Wang, 2000):

$$\varepsilon = \sigma/K + \alpha P/K \tag{7}$$

is substituted into equation (6) to give:

$$\sigma + P(\alpha - Kc_m) = Kg(t) \tag{8}$$

where $\sigma = (\sigma_r + 2\sigma_\theta)/3$ is the mean normal stress, and $K=2G(1+v)/3(1-2v)$ is the bulk modulus. Equation (8) is solved for σ and substituted into equation (9) for transient porous medium flow (Wang, 2000):

$$(\alpha/KB)\left[B\frac{\partial\sigma}{\partial t} + \frac{\partial P}{\partial t}\right] = \frac{k}{\mu}\nabla^2 P \tag{9}$$

to yield:

$$S\frac{\partial P}{\partial t} - \frac{k}{\mu}\nabla^2 P = -\alpha\frac{\partial g}{\partial t} \tag{10}$$

where B is Skempton's coefficient (Wang, 2000), k is permeability , μ is viscosity of water, and $S = (\alpha/K)[1/B - \alpha(\alpha - Kc_m)/3K]$. The significance of equation (10) is that stress has been eliminated, except perhaps from the time dependent term g. If the spherical cavity is an infinite domain with stress and pore pressure perturbation only at the spherical cavity, $\partial g / \partial t$ must be zero (Wang, 2000), and the equation reduces to:

$$\frac{\partial P}{\partial t} - c \, \nabla^2 P = 0 \tag{11}$$

where $c = k/\eta S$. For the spherically symmetric case equation (11) expands as:

$$\frac{\partial P}{\partial t} - c \frac{1}{r^2}\left(2r \frac{\partial P}{\partial r} + r^2 \frac{\partial^2 P}{\partial r^2}\right) = 0 \tag{12}$$

Equation (12) is solved by implicit finite difference. With the pore pressure field known, the force equilibrium equation (3) may also be solved. Equations (4) are substituted into equation (3) to yield:

$$\frac{d^2 u}{dr^2} + \frac{2}{r}\frac{du}{dr} - 2\frac{u}{r^2} = \frac{\alpha}{2G\left[(1-\nu)/(1-2\nu)\right]}\frac{dP}{dr} \tag{13}$$

With appropriate boundary conditions equation (13) is solved for radial displacement u. Stresses σ_r and σ_θ are then found from equations (14) which result from substituting equations (4) into equations (2).

$$\sigma_r = 2G\frac{du}{dr} + 2G\frac{\nu}{1-2\nu}\left(\frac{du}{dr} + \frac{2u}{r}\right) - \alpha P \tag{14a}$$

$$\sigma_\theta = 2G\frac{u}{r} + 2G\frac{\nu}{1-2\nu}\left(\frac{du}{dr} + \frac{2u}{r}\right) - \alpha P \tag{14b}$$

ESTIMATION OF PARAMETERS

The parameters needed for a poroelastic analysis are not measured directly in a typical geotechnical investigation, but they can commonly be calculated with sufficient accuracy from the results of standard tests when the poroelastic medium is a soil. This is possible because in soils the poroelastic parameters are strongly a function of drained compressibility, which can be estimated with reasonable accuracy from the results of one-dimensional consolidation tests. The soil directly overlying the solutioned carbonate rock will commonly be a residual soil. To evaluate conditions for sinkhole formation in such a soil I will evaluate the poroelastic parameters for a typical silty clay residual soil, classified as CH in the Unified Soil Classification System. Geotechnical parameters of this soil at Oak Ridge, Tennessee were measured by Woodward–Clyde Consultants (1984) as part of the evaluation of a possible waste disposal site.

The elastic parameters G, K and ν appear directly in the poroelastic relationships and Young's modulus E is also useful. Only two of these four isotropic elastic parameters are independent, and once two are known the others can be calculated. Because Poisson's ratio ν varies over a relatively narrow range in soils, little accuracy is lost in the final parameter values by estimating ν rather than measuring it directly. The coefficient of earth pressure at rest K_o is related to ν by $K_o = \nu/(1-\nu)$. Because both approximate theories and empirical data suggest that $K_o = 1 - \sin \psi$ (Kezdi, 1974), where ψ is the angle of internal friction, ν may be estimated by:

$$\nu = (1 - \sin \psi)/(2 - \sin \psi) \tag{15}$$

This yields ν equal to 0.38 for the measured ψ value of 23°. With this value of ν, Young's modulus E can be calculated from results of a one-dimensional consolidation test by:

$$E = \left(\Delta \sigma_z / \Delta \varepsilon_z\right)\left(1 - 2\nu^2 /(1-\nu)\right) \tag{16}$$

where $\Delta \sigma_z$ and $\Delta \varepsilon_z$ are small increments of vertical stress and strain. The mean vertical stress for the increment was chosen to be appropriate for a depth of burial of 15m.

Another required parameter is K_v, the drained uniaxial incompressibility (Wang, 2000). It is the ratio of longitudinal stress to longitudinal strain with orthogonal strains zero. This is calculated directly from results of the one-dimensional consolidation test:

$$K_v = \Delta \sigma_z / \Delta \varepsilon_z \tag{17}$$

Bulk modulus K and shear modulus G are calculated from E and ν by the standard elastic relationships $K = E/3(1-2\nu)$ and $G = E/2(1+\nu)$. Other moduli are also needed to evaluate the parameters in the poroelastic equations. The bulk modulus of water K_f varies modestly with temperature, and handbook values for pure water are adequate for soil problems.

The unjacketed bulk compressibility K_s' is the compressibility measured when pore water pressure equals confining pressure. Because the drained bulk modulus K is small for soils, the exact value of K_s' need not be known. It is approximately equal to the bulk modulus of the mineral grains making up the soil, which may be estimated with acceptable accuracy from experimental single-crystal data tabulated by Simmons and Wang (1971). Single crystal data exist for quartz, but not for clay minerals. However, there are data for the micas muscovite, biotite and phlogopite. If bulk moduli of these phyllosilicates are assumed to be representative of those for clays, and if the soil is assumed to be about half quartz, a bulk modulus of 4.5×10^4 MPa (6.5×10^6 psi) is appropriate.

The last fundamental modulus required is K_ϕ, the unjacketed pore compressibility. It is a measure of the change in pore volume with change in confining pressure when pore pressure is equal to confining pressure (Wang, 2000). If the solid phase is composed of a single constituent, $K_\phi = K_s'$, but indirect experimental evidence (Wang 2000) suggests that K_ϕ is close to K_f for clayey sandstones. For a soil, the difference between the two is inconsequential and I assume that $K_\phi = K_s'$. The poroelastic parameters α, B and S are related to the fundamental parameters by (Wang, 2000):

$$\alpha = 1 - K/K_s' \tag{18}$$
$$B = (1/K - 1/K_s') / [1/K - 1/K_s' + \phi (1/K_f - 1/K_\phi)] \tag{19}$$
$$S = \alpha^2/K_v + \phi (1/K_f - 1/K_\phi) \tag{20}$$

The measured, estimated or computed values for all parameters for the residual soil analyzed are shown in Table 1. Because K and K_v are much smaller than K_s', K_ϕ and K_f, the parameters α and B are approximately 1.0, and $S \sim 1/K_v$. It is clear, as noted above, that the important terms are dominated by matrix compressibility and that rough estimates suffice for K_s', K_ϕ and K_f. Permeability for the soil is calculated from in situ hydraulic conductivity, which ranges from 1.3×10^{-9} m/sec (4.3×10^{-9} ft/sec) to 1.3×10^{-8} m/sec (4.3×10^{-8} ft/sec) (Woodward-Clyde Consultants, 1984). Assumed tensile strength of 2.36 Kpa (0.34 psi) is the mean for four clay soils measured by Nearing et al. (1991).

Table 1. Parameters estimated for a residual soil

Parameter	Value
ϕ	0.46
ν	0.38
E	1.25×10^1 MPa (1.81×10^3 psi)
G	4.52 MPa (656 psi)
K	1.73×10^1 MPa (2.51×10^3 psi)
K_v	2.34×10^1 MPa (3.39×10^3 psi)
K_s'	4.5×10^4 MPa (6.5×10^6 psi)
K_f	1.93×10^3 MPa (2.80×10^5 psi)
α	0.99961
B	0.99963
S	4.30×10^{-2} MPa^{-1} (2.96×10^{-4} psi^{-1})

RESULTS

Analyses were performed for a constant rate of head drop in the soil cavity. Radial stress drops as the water pressure on the spherical wall of the cavity drops, but pore pressure in the wall of the cavity decays more slowly as pore water flows toward the cavity. With continued drawdown radial stress plus tensile strength may eventually become less than pore pressure a short distance from the wall of the cavity. Effective tensile stress exceeds tensile strength at this point and hydraulic fracturing occurs. This removes a thin rind of soil from the wall of the soil cavity. This exposes a new soil surface with an even greater effective tensile stress, which also spalls. Once the first failure occurs, the failure will propagate rapidly upward until it reaches the surface or a shallow depth where pore pressure is little more than tensile strength.

It is convenient to portray the results of the analysis as total head drop to cause hydraulic fracturing (failure) versus rate of head drop. Figure 1A contrasts results for poroelastic analyses with non-poroelastic analyses. The non-poroelastic analyses are based on closed-form solutions for stress (Timoshenko and Goodier, 1970) and pore pressure (Carslaw and Jaeger, 1959), but without taking into account the effect of changing pore pressure on displacements and therefore stresses. The small difference between the poroelastic and non-poroelastic analyses at high rates of head drop results from the finite cell size of 1cm (0.4 in) in the finite difference poroelastic solution. At the most practically significant rates of head drop of 0.1 to 0.01 m/hr (0.3 – 0.03 ft/hr), the difference between the solutions is substantial. Figure 1A also shows the large effect of differences in hydraulic conductivity on failure conditions and Salvati et al. (2001) shows the effect of tensile strength. If a locality experiences cycles of high and low water table, a drawdown episode may occur when the soil is only partly saturated. This may be modeled by the simple expedient of changing the effective bulk

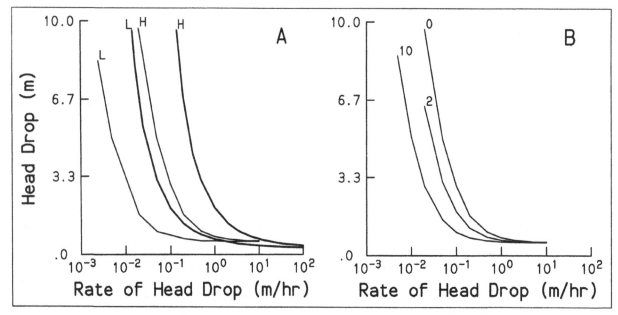

Fig. 1. Head drop at failure (hydraulic fracturing) versus rate of head drop within a soil void at a depth of 30 m (98 ft). A) Thin curves are for poroelastic analyses, thick curves for non-poroelastic analyses. L and H refer respectively to low hydraulic conductivity of 1.3×10^{-9} m/sec (4.3×10^{-9} ft/sec) and high hydraulic conductivity of 1.3×10^{-8} m/sec (4.3×10^{-8} ft/sec). B) Curves for air bubbles constituting 0, 2, and 10% of the fluid fraction in the soil, with hydraulic conductivity of 1.3×10^{-8} m/sec (4.3×10^{-8} ft/sec).

modulus of water. Fig. 1B shows the effect of pore volumes that are 0, 2 and 10% air filled. The effect of air solubility is ignored. Even without this effect, expansion of air bubbles slows the decay of pore pressure and allows failure for smaller drawdowns for a given rate of drawdown. The analyses shown assume a large soil void with a radius of 1m (3 ft). However, because radial stress gradient is strongly a function of void radius, whereas rate of pore pressure drop very near the opening is only weakly dependent, conditions for hydraulic fracturing are strongly size dependent. Smaller initial openings are less likely to fail.

CONCLUSIONS

Hydraulic fracturing is a plausible causative mechanism for the upward propagation of soil voids that results in cover-collapse sinkholes. Poroelastic analysis is necessary for an accurate assessment of the conditions for failure, and the parameters necessary for the analysis can be calculated from standard geotechnical test results, plus a measurement of tensile strength. The probability of failure is enhanced for low permeability or tensile strength and the presence of isolated air bubbles in the soil.

REFERENCES

Anikeev, A.V., 1999. Casual hydrofracturing theory and its application for sinkhole development prediction in the area of Novovoronezh Nuclear Power House-2 (NV NPH-2), Russia. Proc. 7th Multidisciplinary Conf. on Sinkholes and the Engineering and Environmental Impacts of Karst. Balkema, Rotterdam.

Bengtsson, T.O., 1987. The hydrologic effects from intense ground-water pumpage in East-Central Hillsborough County, Florida. In: Proc. 2nd Multidisciplinary Conf. on Sinkholes and the Environmental Impacts of Karst. Balkema. Rotterdam, pp. 109-114.

Benson, R.C., LaFountain, L.J., 1984. Evaluation of subsidence or collapse potential due to subsurface cavities. In: Proc. 1st Multidisciplinary Conf. on Sinkholes. Balkema. Rotterdam, pp. 201-215.

Brink, A.B.A., 1984. A brief review of the South African sinkhole problem. In: proc. 1st Multidisciplinary Conf. on Sinkholes. Balkema. Rotterdam, pp. 123-127.

Carslaw, H.S., Jaeger, J. C., 1959. Conduction of Heat in Solids. Clarendon Press, Oxford.

Currin, J.L., Barfus, B.L., 1989. Sinkhole distribution and characteristics in Pasco County, Florida. In: Proc. 3rd Multidisciplinary Conf. on Sinkholes and the Engineering and Environmental Impacts of Karst. Balkema, Rotterdam, pp. 97-106.

Daoxian, Y., 1987. Keynote address: Environmental and engineering problems of karst geology in China. In: Proc. 2nd Multi-disciplinary Conf. on Sinkholes and the Environmental Impacts of Karst. Balkema, Rotterdam, pp. 1-11.

Hodek, R.J., Johnson, A.M., Sandri, D.B., 1984. Soil cavities formed by piping. In: Proc. 1st Multidisciplinary Conf. on Sinkholes. Balkema, Rotterdam, pp. 249-254.

Iqbal, M.A., 1995. Engineering experience with limestone. In: Proc. 5th Multidisciplinary Conf. on Sinkholes and the Environmental Impacts of Karst. Balkema, Rotterdam, pp. 463-468.

Kezdi, A., 1974. Handbook of Soil Mechanics. Elsevier, Amsterdam, 249-252.

Kiernan, K., 1989. Sinkhole hazards in Tasmania. In: Proc. 3rd Multidisciplinary Conf. on Sinkholes and the Engineering and Environmental Impacts of Karst. Balkema, Rotterdam, pp. 123-128.

Metcalfe, S.J., Hall, L.E., 1984. Sinkhole collapse induced by groundwater pumpage for freeze protection irrigation near Dover, Florida, January 1977. In: proc. 1st Multidisciplinary Conf. on Sinkholes. Balkema, Rotterdam, pp. 29-33.

Nearing, M.A., Parker, S.C, Bradford, J.M. and Elliot, W.J., 1991. Tensile strength of thirty-three saturated repacked soils. Soil Sci. Soc. Am. V., 55: 1546-1551.

Pazuniak, B.L., 1989. Subsurface investigation response to sinkhole activity at an eastern Pennsylvania site. In: Proc. 3rd Multi-disciplinary Conf. on Sinkholes and the Engineering and Environmental Impacts of Karst. Balkema, Rotterdam, pp. 263-269.

Salvati, R., Tharp, T.M. Capelli, G., 2001. Conceptual model for geotechnical evaluation of sinkhole risk in the Latium region. In: Proc. 8th Multidisciplinary Conf. on Sinkholes and the Environmental Impacts of Karst (this volume).

Siegel, T.C., Belgeri, J.J., 1995. The importance of a model in foundation design over deeply-weathered, pinnacled, carbonate bedrock. In: Proc. 5th Multidisciplinary Conf. on Sinkholes and the Engineering and Environmental Impacts of Karst. Balkema, Rotterdam, pp. 375-382.

Simmons, G. and Wang, H., 1971. Single Crystal Elastic Constants and Calculated Aggregate Properties: A Handbook, MIT Press, Cambridge, MA.

Sowers, G.F., 1975. Failures in limestones in humid subtropics. J. Geotech. Engng. Div., ASCE 101, 771-787.

Sowers, G.F., 1996. Building on Sinkholes. ASCE Press, New York.

Tharp, T.M., 1997. Mechanics of formation of cover collapse sinkholes. Proc. 6th Multidisciplinary Conf. on Sinkholes and the Engineering and Environmental Impacts of Karst: 29-36, Balkema, Rotterdam.

Tharp, T.M., 1999. Mechanics of upward propagation of cover-collapse sinkholes. Engineering Geology, 52: 23-33.

Timoshenko, S.P.,Goodier, J.N., 1970. Theory of Elasticity, 3rd ed., McGraw-Hill, New York.

Wang, F., 2000. Theory of Linear Poroelasticity with Applications to Geomechanics and Hydrogeology. Princeton University Press, Princeton, NJ.

White, W.B., 1999. Karst hydrology: recent developments and open questions. Proc. 7th Multidisciplinary Conf. on Sinkholes and the Engineering and Environmental Impacts of Karst. Balkema, Rotterdam.

White, E.L., Aron, G., White, W.B., 1984. The influence of urbanization on sinkhole development in central Pennsylvania. In: proc. 1st multidisciplinary Conf. on Sinkholes. Balkema, Rotterdam, pp. 275-281.

Wilson, W.L., Beck, B.F., 1988. Evaluating sinkhole hazards in mantled karst terrane. Geotechnical Aspects of Karst Terrains. Geotech. Sp. Pub. No.14, ASCE, New York, pp. 1-24.

Woodward-Clyde Consultants, 1984. Subsurface characterization and Geohydrologic site evaluation West Chestnut Ridge Site. Woodward-Clyde Consultants, Wayne, New Jersey. NTIS ORNL/sub-83-64764.

III Regional Studies of Sinkholes and Karst

Geotechnical and Environmental Applications of Karst Geology and Hydrology, Beck & Herring (eds)
© *2001 Taylor & Francis, ISBN 90 5809 190 2*

Characterization and classification of Alfisols developed in the karstic land of northwestern Turkey

CUMHUR AYDINALP   Faculty of Agriculture, Dept. of Soil Science, Uludag University, 16384 Bursa, Turkey,
aydinalp@uludag.edu.tr

ABSTRACT

Evolution of Alfisols in northwestern Turkey in genetically isolated pockets on concave landforms is a unique phenomenon. The objective was to investigate the properties and genesis of these Alfisols. These soils are developed on limestone parent materials of Tertiary age and are found alongside Inceptisols in the karstic landforms. The soils are largely a product of microtopographic differences, which modify distribution of water across the landscape. These soils are deep, show A-Bt-C horizons which are chromic and have strong argillic B horizons. Textures of surface horizons are clay to clay loam, and subsurface textures are clay. The soils are slightly acidic to slightly alkaline. The minerals identified are chlorite and illite and are the major minerals followed by a lesser amount of chlorite-smectite. Quartz and plagioclase feldspar are present in trace amounts in all horizons. The soils are allocated to the class of Typic Rhodoxeralfs. The concavity of land surface and Mediterranean type of climate facilitate argillic horizon development. Keywords: Alfisols, clay illuviation, karstic topography, soil formation

INTRODUCTION

Alfisols are some of the most productive soils in the world. Soils with these properties are found on all continents and are estimated to cover about 13% of the world's land area (Rust, 1983). Alfisols cover 2.94% area of the Bursa province in northwestern Turkey. In this region these soils are found under conditions of Mediterranean climate. In the semi-arid Bursa region, these soils occur in isolated pockets under a xeric moisture regime where topographic position facilitates greater water percolation and negligible run-off and, therefore, greater effective precipitation. Alfisols in the Bursa province occur alongside weakly developed Inceptisols (Aydinalp, 1996) rather than in a chronosequence of Mollisols and Ultisols. This isolated development shows a unique property of soil formation induced by topography-climate interaction.

The objective of this research was to determine some important properties of these soils developed on karstic landforms and to classify these soils according to USDA Soil Taxonomy (1994). This would enhance our understanding of these soils' properties which will have an impact on their agricultural utilization.

MATERIALS AND METHODS
Description of study area

The study area was located on the west side of the Bursa province. The province, covering an area of 1.104.301 ha, is situated in northwestern Turkey between 28° 10′-30° 00′ N latitudes and 40° 40′-39° 35′ E longitudes (Fig. 1a and 1b). The soils were developed on limestone of karstic landforms under a Mediterranean climate, characterized by hot summers and mild winters. The mean annual precipitation and temperature are around 713.1 mm and 14.4 °C in the province. The soil temperature and moisture regimes are thermic and xeric respectively. Pedons representing five soils range in elevation from 250 to 400 m above mean sea level.

Methods

Five pedons of Alfisols were exposed. The morphology of the soils was described according to Soil Survey Manual (Soil Survey Division Staff, 1993). Soil samples collected from different horizons were air dried and passed through a 2 mm sieve. The soil samples were analyzed for

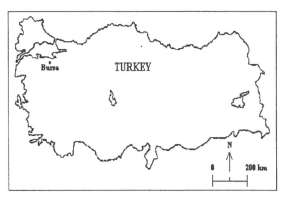

Fig. 1a. The location of the study area.

particle size distribution (Piper, 1950), pH (1:2.5 water, Jackson, 1958), organic carbon (Tinsley, 1950), total nitrogen (Bradstreet, 1965), calcium carbonate (Bascomb, 1961), EC (SCS, 1972), free iron oxide (SCS, 1972), CEC (American Society of Agronomy Method 57-3, 1965), and exchangeable cations (American Society of Agronomy Method 57-2, 1965). The clay minerals were identified by XRD in the <2μm size fraction using oriented samples, both untreated and treated for organic matter and iron removal. The latter were further subjected to conventional treatments: K saturation and heating to 100, 300 and 550 °C and Mg saturation followed by ethylene glycol solvation. The K saturated samples were also treated with CsCl-hydrazine-DMSO (Lim et al., 1981).

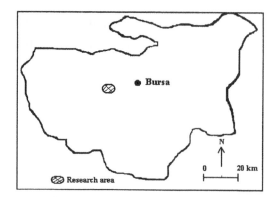

Fig. 1b. The location of the study area.

RESULTS AND DISCUSSION
Soil morphological properties
The morphological properties of the soils are given in Table 1.

Table 1. Morphology of the pedons

Horizon	Depth (cm)	Munsell colour, (moist)	Texture	Structure	Consistence (moist)	Roots	Boundary
Pedon 1							
Ah	0-25	7.5YR 5/4	cl	2f, abk	fi	c, m	c
Bt	25-50	2.5YR 4/6	c	3m, abk	fi	c, f	w
C	50-65	10Y 8/2	cl	1m, sbk	fi	-	g
Pedon 2							
Ah	0-30	7.5YR 5/3	cl	2f, sbk	fi	c, f	c
Bt	30-55	2.5YR 4/8	c	3m, abk	fi	c, f	w
C	55-70	10Y 7/2	cl	1m, sbk	fi	-	g
Pedon 3							
Ah	0-20	7.5YR 5/6	c	2m, sbk	fi	c, m	c
Bt	20-60	2.5YR 3/4	c	3m, abk	fi	f, f	w
C	60-90	10Y 7/1	cl	2m, abk	fi	f, f	w
Pedon 4							
Ah	0-20	7.5YR 5/3	c	2m, abk	fi	c, m	c
Bt	20-70	2.5YR 3/4	c	3m, abk	fi	c, f	w
C	70-90	10Y 6/1	cl	2m, sbk	fi	f, f	w
Pedon 5							
Ah	0-15	7.5YR 5/8	c	2f, sbk	fi	c, f	c
Bt	15-65	2.5YR 3/6	c	3m, abk	fi	f, f	w
C	65-95	10Y 7/2	cl	2m, abk	fi	f, f	g

Structure: 1 = weak, 2 = moderate, 3 = strong. Type: f = fine, m = medium. Class: abk = angular blocky, sbk = subangular blocky. Consistency: fi = firm. Roots: abundance: f = few, c = common; size: f = fine, m = medium. Boundary: distinctness: c = clear, g = gradual, w = wavy. Topography = concave

The upper horizon of pedons had 7.5YR hues, whereas the lower horizons of these pedons and all the B horizons had 2.5YR hues. The values ranged from 3 to 8 and chromas from 1 to 8. All the soils were well or moderately drained and as such devoid of any aquic features. The Bt horizon of all the pedons contained at least 3% ped and channel argillans. The orientation of argillans was moderate and continuous in all the pedons. All the soils were deep and had clay-enriched Bt horizons. The width of argillic horizon varied from 25 to 50 cm. In general the width of Bt horizons was lower in pedons 1 and 2 as compared with other pedons. The horizon boundaries were clear without any tonguing of A into B horizon or any form of pedoturbation, indicating that Bt horizons are essentially ordinary argillic horizons. Structurally, Bt horizons were stronger than A horizons, and the structure was either angular or subangular blocky in pedons. The stronger structure, due to increased abundance and thickness of clay films in argillic horizons, compared with surface horizons also reported from Ultisols of North Caroline (Southard and Buol, 1988). In comparison with other soils (Aydinalp, 1996) of the region, consistence is hard, requiring greater force tillage operations. Surface horizons were clay to clay loam, and, obviously, subsurfaces were harder. Soil texture did not show a relationship with rainfall or location of the pedons. Moderate to deep root systems were found in all pedons. This is in concurrence with annual rainfall—moderate rainfall being a stimulant for a deeper root system.

Table 2. The particle size distribution of Alfisols.

Horizon	Depth (cm)	Sand (%)	Silt (%)	Clay (%)	Texture
Pedon 1					
Ah	0-25	27.9	29.5	37.2	CL
Bt	25-50	21.1	25.0	47.7	C
C	50-65	29.8	31.2	32.4	CL
Pedon 2					
Ah	0-30	28.5	30.7	34.5	CL
Bt	30-55	23.9	26.3	43.0	C
C	55-70	29.3	37.1	28.7	CL
Pedon 3					
Ah	0-20	22.1	31.3	41.8	C
Bt	20-60	17.2	27.4	50.2	C
C	60-90	28.6	29.5	39.4	CL
Pedon 4					
Ah	0-20	24.0	31.8	42.4	C
Bt	20-70	13.9	28.5	55.1	C
C	70-90	29.0	29.7	38.7	CL
Pedon 5					
Ah	0-15	23.3	33.4	40.1	C
Bt	15-65	17.6	26.2	52.4	C
C	65-95	31.5	31.2	35.8	CL

Table 3. Some chemical properties of Alfisols.

Horizon	Depth (cm)	pH	EC (dS m^{-1})	Org. C (%)	Total N (%)	C/N	CaCO$_3$ (%)	CEC	Ca	Mg	K	Na	BS (%)	Free Fe$_2$O$_3$ (%)
									cmol (+) kg^{-1}					
Pedon 1														
Ah	0-25	6.4	0.34	1.12	0.10	11.2	—	35.4	29.2	1.7	1.4	2.3	100	2.15
Bt	25-50	7.0	0.28	0.84	0.08	10.5	—	39.1	30.1	3.9	2.3	2.0	100	2.72
C	50-65	7.4	0.25	0.55	0.06	9.2	1.3	30.0	23.2	2.7	1.9	1.4	100	1.83
Pedon 2														
Ah	0-30	6.3	0.30	1.03	0.08	12.9	—	31.7	25.2	1.9	1.6	2.1	100	2.24
Bt	30-55	7.2	0.23	0.67	0.07	9.6	—	36.3	27.7	3.6	2.0	1.8	100	2.50
C	55-70	7.3	0.21	0.49	0.06	8.2	1.8	27.5	21.5	2.4	1.3	1.1	100	1.65
Pedon 3														
Ah	0-20	6.5	0.24	1.25	0.11	11.4	—	38.8	32.1	2.1	1.3	2.4	100	3.40
Bt	20-60	7.3	0.20	0.97	0.09	10.8	—	41.5	33.0	3.5	2.4	1.5	100	3.73
C	60-90	7.6	0.18	0.71	0.08	8.9	2.1	32.3	25.4	3.1	1.8	1.3	100	2.12
Pedon 4														
Ah	0-20	6.2	0.22	1.08	0.09	12.0	—	37.5	31.3	2.0	1.5	1.9	100	3.23
Bt	20-70	7.5	0.19	0.76	0.07	10.8	—	42.8	35.3	3.3	1.9	1.5	100	3.97
C	70-90	7.6	0.17	0.52	0.06	8.7	1.1	33.1	26.7	2.7	1.6	1.2	100	2.01
Pedon 5														
Ah	0-15	6.9	0.27	1.38	0.13	10.6	—	36.9	30.4	1.8	1.6	2.0	100	2.85
Bt	15-65	7.0	0.25	0.85	0.09	9.4	—	40.7	32.8	3.2	2.1	1.7	100	3.18
C	65-95	7.7	0.23	0.40	0.05	8.0	1.9	31.5	24.6	2.8	1.9	1.5	100	1.95

Physical and chemical properties

The physical and chemical properties of the studied soils are presented in Tables 2 and 3. Texture classes ranged between clay and clay loam within all pedons. The pedons contain clay between 28.7% to 55.1% and significantly increase with depth.

The soil pH is slightly acidic to slightly alkaline, and values increased with depth. The EC values were low in all the pedons. The organic carbon content appeared to be influenced by the combined effects of rainfall and temperature—the high temperature and moderate rainfall resulting in low organic matter content. Within the profiles, organic carbon invariably decreased with depth,

Table 4. Semiquantitative abundance of the minerals in the clay fraction, (2 μm) estimated by the relative heights of the diffraction peaks.

Horizon	Depth (cm)	Chlorite	Illite	Chlorite-smectite	Quartz	Plag. Feldspar
Pedon 1						
Ah	0-25	***	***	**	*	*
Bt	25-50	***	***	**	*	*
C	50-65	**	*	**	*	*
Pedon 2						
Ah	0-30	***	***	**	*	*
Bt	30-55	***	***	**	*	*
C	55-70	*	*	**	*	*
Pedon 3						
Ah	0-20	***	***	**	*	*
Bt	20-60	***	***	**	*	*
C	60-90	*	**	*	*	*
Pedon 4						
Ah	0-20	***	***	**	*	*
Bt	20-70	***	***	**	*	*
C	70-90	**	*	*	*	*
Pedon 5						
Ah	0-15	***	***	**	*	*
Bt	15-65	***	***	**	*	*
C	65-95	*	**	*	*	*

*** major   ** minor   * trace

suggesting horizon development and stability of the land surfaces. Calcium carbonate is absent throughout the pedons. It suggests that $CaCO_3$ is completely leached down to C horizons. The presence or absence of $CaCO_3$ strongly influencing the grade of soil structures was found in all the pedons. Perhaps cohesive forces of plasma were augmented with the cementing properties of $CaCO_3$. The free iron oxide values range from 1.65% to 3.97% and increase with depth in all the pedons. The free iron oxide content also shows a parallel increase with the clay fraction of the soils. High cation exchange capacity of these soils is due to high clay content. The soils have 100% base saturation. As clay content increased from upper horizons to lower horizons, CEC increased the variation in the clay content being matched by variation in CEC. This implies that CEC is mainly governed by clay. Similar observations were made by Aydinalp (1998) in some Alfisols located on the east side of the region.

Clay Mineralogy

The clay mineralogy of all studied pedons is summarized in Table 4. Minerals in the clay fraction were estimated semiquantitatively by means of the relative intensities of the diffraction peak. In all horizons chlorite and illite are the major minerals followed by lesser amounts of chlorite-smectite. Quartz and plagioclase feldspar are present in trace amounts in all horizons. Apart from some variations in the intensities of different reflections, no differences between the five pedons developed in different locations of the karstic landscape.

Soil Classification

The soils are formed on karstic land in northwestern Turkey. These pedons were developed on limestone parent material under *Pinus brutia* and *Macchie*. The soils are classified according to their type and sequence of horizons. The ochric and argillic horizons were defined in all the pedons. The nomenclature and classification of pedons are given according to the USDA Soil Taxonomy (1994). The most advanced stage of soil development in these pedons is the genesis of an argillic horizon. The argillic horizon is indicated by: (i) clay enrichment (Table 2) in B horizons, and (ii) development of argillans by eluviation of silicate clays from A horizon and their illuviation in B horizons along root channels and soil profiles. It places these soils into the order of Alfisols (USDA Soil Taxonomy, 1994). Alfisols are morphologically characterized by the presence of an argillic horizon with high base saturation. These features illustrate the moderate maturity of the Alfisols in the studied area.

CONCLUSION

Alfisols form in the region with a xeric moisture regime. These soils invariably occur in concave surfaces within the landscape. The adjoining surfaces are characterized by the occurrence of Inceptisols (Aydinalp, 1996). Vegetation seems to have played a limited and different role in the formation of these Alfisols as these are developed under natural forest cover, and clay skins are common along the root channels. These Alfisols are slightly acidic to slightly alkaline in reaction. Heavy winter rains following hot and dry summers promote dispersion of clay particles upon wetting. Concavity of the landscape results in accumulation of runoff water from the adjoining areas. The rainfall distribution pattern and topography interact to facilitate lessivage leading to formation of Alfisols in association with Inceptisols.

REFERENCES

American Society of Agronomy., 1965, Methods of Soil Analysis Part I and II. Pub. Mad. USA ch.57-2&3.

Aydinalp, C., 1996, Characterization of the main soil types in the Bursa province, Turkey. Ph.D. thesis, University of Aberdeen, Aberdeen, U.K.

Aydinalp, C., 1998, Bursa ovasinda uc farkli teras duzeyinde olusmus kirmizi akdeniz topraklarinin (Terra-Rossa) genesisi ve siniflandirilması. J. of Faculty of Agriculture, Uludag University, 13: 31-41, Bursa.

Bascomb, C.L.A., 1961, A calcimeter for routine use on soil samples. Chem. & Ind, 45-1926.

Bradstreet, R.B., 1965, Kjeldahl Methods for Organic Nitrogen, Ac. Press.

Jackson, M.L, 1958, Soil Chemical Analysis, Prentice-Hall Inc., New Jersey.

Lim, C.H., Jackson, M.L., and Higashi, I., 1981, Intercalation of soil clays with dimethylsulfoxide. Soil Sci. Soc. Am. J., 45 (2): 433-436.

Piper, C.S., 1950, Soil and Plant Analysis, Adelaide.

Rust, R.C., 1983, Alfisols. In: Wilding, L.P., Smeck, N.E., Hall, G.F. (Eds.), Pedogenesis and Soil Taxonomy. II. The Soil Orders. Elsevier, Amsterdam, pp. 253-281.

Tinsley, J., 1950, The determination of organic carbon in soils by dichromate mixtures. Transactions of the Fourth International Meeting of the Society of Soil Science, 1: 161-164.

SCS, Soil Conservation Service, 1972, Soil Survey Laboratory Methods and Procedures for Collecting and Analysing Soil Samples. Soil Survey Invest. Report 1. USDA, Washington DC, USA.

Soil Survey Division Staff., 1993, Soil Survey Manual. USDA, Washington, D.C.

Southard, R.J., Boul, S.W., 1988, Subsoil blocky structure formation in some North Caroline Paleudults and Paleaquults. Soil Sci. Soc. Am. J. 52, 1069-1076.

USDA Soil Taxonomy, 1994, Keys to Soil Taxonomy, Sixth Edition.

Geotechnical and Environmental Applications of Karst Geology and Hydrology, Beck & Herring (eds)
© 2001 Taylor & Francis, ISBN 90 5809 190 2

Developing a GIS for the Jamaican Cockpit Country

M.SEAN CHENOWETH   Dept. of Geography, University of Wisconsin - Milwaukee, Milwaukee, WI 53211, USA
cheno@uwm.edu

MICHAEL J.DAY   Dept. of Geography, University of Wisconsin - Milwaukee, Milwaukee, WI 53211, USA
mickday@uwm.edu

ABSTRACT

The Jamaican Cockpit Country is the World's "type-example" of the cockpit style of polygonal karst. Although it has been the subject of extensive geomorphological, hydrological and biological studies, extensive interest in its future conservation status has only recently emerged with the suggestion that it be nominated as a U.N. Natural World Heritage Site and with the involvement of the World Bank in developing a management strategy. One critical element in the formulation of an appropriate management policy is the development of a reliable biogeomorphological database. In that context, this research aims to classify and quantify the cockpit karst landforms and identify associated vegetation patterns. A digital elevation model (DEM) and cartographic modeling will be used to delineate the complex karst landforms in a geographic information system (GIS). High resolution remote sensing imagery (IKONOS 4-meter multispectral) is used to identify vegetation patterns by landform, slope and aspect. Reference data on landforms and vegetation type were collected in the field and used for image classification and overall accuracy assessment. The results of this project will facilitate planning by the Jamaican Government and the World Bank in the creation of the Cockpit Country National Park.

INTRODUCTION

The Cockpit Country of west central Jamaica covers about 500 square kilometers (Figure 1). This carbonate karst landscape is composed of residual limestone hills, dry or underdrained valleys, enclosed depressions or cockpits, and caves. Soil cover is discontinuous with deep, fertile soil in the valleys and cockpits and little or no soil on hill slopes and hill summits. Forest cover tends to be denser & greener in the lowland soil covered areas and less dense, drier vegetation tends to occur on the hill slopes and summits.

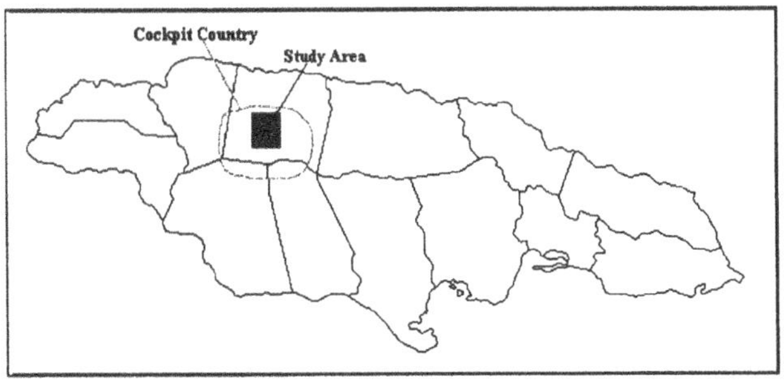

An intimate but spatially unquantified relationship exists between geomorphology and vegetation, with soil and climate as important mediating agents (Tinkler, 1985; Bauer, 1996). Vegetation, climate and landforms interact at a wide range of spatial and temporal scales, and there is scope for a greater understanding of the processes involved (Kirby et al., 1995). Digital remote sensing for geomorphological study and landscape interpretation has considerable potential and is currently underdeveloped (Pickup, 1990; Young & White, 1994; Giles, 1998).

Figure 1: Study area in the Jamaican Cockpit Country.

The Study Area

The study area consists of about 26 square kilometers and is located in Trelawny Parish adjacent to the north coast of Jamaica (Figure 1). Topographic map sheets 52c and 53a (1:12,500) published by the Survey Department of Jamaica (1975) essentially cover the study area, with sheet 53a adjacent and to the south of 52c. The northwest corner of map 52c is latitude N 18^0 23' 04", longitude W 77^0 41' 32" and the southeast corner of map 53a is latitude N 18^0 16' 28", longitude W 77^0 36' 19". The hamlet of Windsor is located on the north side of the study area and undeveloped cockpit landscape lies to the south.

Climate

The Cockpit Country is characterized by a tropical, seasonally dry climate (Koppen Aw). The average regional rainfall ranges from 1,900 to 3,800 mm annually (Proctor, 1983). The dry season is from December through March when precipitation falls below 100 mm per month.

Geology

The following geological summary is based upon the Geological Sheet 8 (Falmouth) published by the Geological Survey Department of Jamaica in 1974. The Swanswick and Troy/Claremont Formations, which dominate the study area, belong to the White Limestone Group, which is of Middle Eocene age. The Swanswick Formation is a foraminiferal limestone with a sparry calcite matrix with well-sorted comminuted debris of mollusks, corals and algae. Maximum thickness is about 100 meters. The Troy / Claremont Formation underlies the Swanswick Formation. The Troy Member is a well-bedded to massive recrystallized limestone or dolostone that is usually unfossiliferous. The Claremont Member is an evenly-bedded, bioclastic, mollusk-rich limestone with irregular alternating beds of compact, fine-grained foraminiferal limestones. Maximum thickness of the Troy/Claremont Formation is about 340 meters.

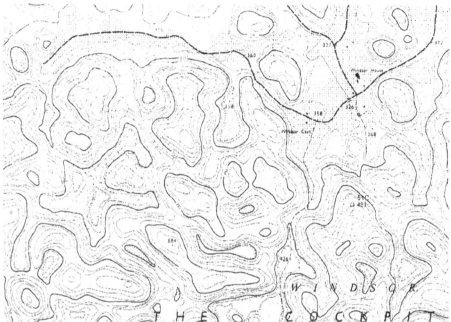

Figure 2: Portion of topographic map sheet 52c, near Windsor, Trelawny. Contour interval is 50 feet. Scale is 1:12,500.

Landforms and Land Cover

The Cockpit Country landforms are well-pronounced, and the overall landscape is a highly irregular combination of positive and negative landform elements (Figure 2). The cockpits, for which the terrain is named, are deep enclosed depressions surrounded by three or more residual limestone hills. In their bases is a thick cover of soil that supports large trees and dense vegetation. Many of the cockpits near populated places are used or have previously been used for subsistence agriculture, although this practice appears to be in decline. The term "old-ground" is used locally to describe the cockpits in which agriculture has been abandoned. Cockpit morphology ranges from complex star-shaped patterns to simple circular forms. Cockpits often are drained by one or more sinkholes or swallets, which are typically located at hillslope bases or within the center of the depression. In many cases broad, shallow channels convey ephemeral surface drainage to the swallets.

Corridors or passages are divides between hills that connect adjacent cockpits or valleys. Some corridors are flat-floored, narrow defiles, others rise steeply to tens of meters above the cockpit or valley floor. They often have a soil cover but can be quite rocky, and they represent important routes for travel through the Cockpit Country.

Saddles are less pronounced notches between adjacent limestone hills. They have either a soil cover or a thick root mat at the surface, typically underlain by limestone rubble. If soil is present, trees such as the Trumpet Tree and John Crow Bead are common. If soil is not present there is generally a less dense, scrubby vegetation cover.

Slopes are of two major types: talus and soil covered. Other authors have identified a wider variety of slope morphologies (Aub, 1969b; Day 1986, 1987) but our concern here is with ecological relationships rather than geomorphology *per se*. Most slopes are talus covered, and there is a suggestion that tree-tipping and bedrock disruption as a result of windthrow may play an important role in the development and maintenance of these slopes. Saddle slopes are generally extensions of soil-covered saddles, and they are usually narrowest at the top, widening slightly toward the base. At their edges there is often a sharp boundary between them and adjacent talus slopes. Colluvium is usually present at the base of saddle slopes.

There are two basic hill top morphologies, and these often relate to vegetation density. Some hilltops are dome shaped and very rocky and are covered by a flexible or spongy root mat typically about 10cm thick. Vegetation on these summits is notably more xeric than in other situations. Other hill summits are flatter in plan view. At the edges of these summits rocky, dry conditions are also characteristic but toward the center there is usually a soil-filled depression, which can support denser vegetation and larger trees.

Glades are broad areas between residual hills that can individually be categorized as either compound depressions (uvalas) or as dry or underdrained valleys. Their cross-profile may be flat, convex, concave, or undulating, and in plan view their shapes are either linear or sinuous. Many glades have thick soils and when left undisturbed they support larger trees and denser vegetation.

Vegetation

The Cockpit Country forest is remarkably pristine for the Caribbean and stands in sobering contrast to Jamaica's largely cultivated and eroded landscape (Pregill et al., 1991). Vegetation variation occurs as a reflection of landscape heterogeneity (Asprey,

1953). In the Cockpit Country, both soil and vegetation are unevenly distributed (Aub, 1969a,b). There are 101 vascular plant species endemic to the Cockpit Country and the majority have only been collected a few times or are known from a single location; each of these species should be considered endangered (Proctor, 1983).

Forest type and density is possibly related to landform and soil cover. Valley (glade, cockpit) vegetation is relatively consistent but the slopes may have more varied vegetation cover reflecting wetter or drier conditions (Asprey, 1953). Forest of the rocky or talus-covered hills represents the fundamental essence of the Cockpit Country vegetation (Proctor 1983). Slopes and ridges are mantled by talus interspersed with soil and vegetation litter (Harvey et al. 1988). Woodland on the rocky hills is more species-rich than that in the cockpit bases (Proctor, 1983). Vegetation on the hill summits is characteristically stunted, thinner and more xerophytic (Aub, 1969a,b).

Locally, commercial agriculture, particularly the cultivation of sugarcane and coffee, occurs within the valleys (glades) and subsistence farming, involving yams, bananas and other crops is practiced in cockpits, on saddle slopes and on soil-covered hill tops. Marijuana is grown illicitly in small patches.

PROBLEM STATEMENT
The essential problems at the core of this study are as follows:
1. There has been no previous integration of geomorphological and ecological information about the Cockpit Country.
2. Satellite imagery has not been utilized previously in studies of the area.
3. There currently exists no digital elevation model for the area.
4. There currently exists no Geographic Information System (GIS) for the area.
5. The above are necessary for the planning, creation and management of a proposed Cockpit Country National Park.

METHODS
Reference Data
An extensive ground survey was conducted in the study area over a period totaling two months in 1999 and 2000. Reference data or ground truth data was collected by visiting 140 preselected sites and recording attributes about the landforms, land cover, soil presence and latitude and longitude on a standard form. The precise location of each site was recorded with a Magellan 4000XL GPS receiver. This data will be used ultimately for supervised satellite image classification and accuracy assessment. Some of the data has high positional errors (30 to 80 meters) that are associated with the rugged terrain, the lack of an external antenna and the selective availability that was still in effect for most of the data collection period.

Figure 3: Portion of the IKONOS near infrared band.

Image Classification
Until recently, the best satellite imagery available for vegetation analysis was 20 and 30 meters multispectral SPOT and LANDSAT imagery. High-resolution IKONOS satellite imagery (4 meter multispectral) is now available and is better suited for the type of detailed analysis need for this complex, rugged landscape (Figure 3). In some instances, individual trees can be identified for ground control points, which was not possible using the SPOT or LANDSAT images.

Currently, this project is using an unsupervised classification scheme on an IKONOS four-meter, multispectral (blue, green, red, nir) satellite image. ERDAS 8.4 was used to create seven classes of land cover. Unsupervised classification was performed on an image created by the stack command in ERDAS consisting of all four bands.

RESULTS
Of the 26.06 square kilometers in the study area 7.05 km^2 is classified as xeric vegetation, 5.21 km^2 is mesic vegetation, 1.68 km^2 is wet forest or pasture, 0.88 km^2 is either developed land or light clouds, 0.54 km2 is covered by dense clouds, 10.58 km^2 is shadows, and 0.11 km^2 is unclassified as a result of zero reflectance (Table 1).

DISCUSSION
Satellite Image Classification
There are three important factors to consider when interpreting this data:
1. Denser/wetter vegetation is assumed to occur on the soil-covered landforms.
2. The unsupervised image classification has not yet been assessed for accuracy.
3. The landscape involves a high degree of heterogeneity.

Table 1: Results of unsupervised image classification.

Class Names	km^2
Xeric Vegetation	7.05
Mesic Vegetation	5.21
Wet Forest, Pasture	1.68
Developed, Light Clouds	0.88
Dense Clouds	0.54
Shadow	10.58
Unclassified (zeros)	0.11
Total	**26.06**

Because this project is in its infancy, it has not yet been possible to create or acquire a digital elevation model (DEM), nor to create a detailed geomorphic map of the study area, so the results and conclusions thus far must be regarded as preliminary. Our initial interpretation of the results, however, suggests that talus slopes and rocky hill summits occupy a significant proportion of the cockpit karst landscape. The xeric and mesic vegetation categories are strongly associated with talus slopes and rocky summits, and this may be the explanation for the relatively large areas that are categorized in these classes.

Wet forest and pasture occupy the third largest vegetation class. Pasture and some commercial agricultural land in the northern part of the image have the same spectral signature as the wet forest of the soil covered areas in the cockpits. Glades in the northern part of the image occupy a small portion (about 5%) of the study area. Cockpits, corridors, saddles, saddle slopes and flat hill summits are soil covered and probably account for the majority of the wet forest/pasture class. Corridors, saddles and saddle slopes are narrow and restricted landforms and the soil cover on flat hill summits is often restricted to a small depression in the hill top center. Shadows occupy such a large portion of the image classification because the image was acquired three hours after solar noon (about 3pm). Pixels that did not have spectral signatures registered zero and were categorized as unclassified.

Digital Elevation Model

We intend next to acquire or create a high resolution DEM, which will allow the positional accuracy of the satellite image to be improved significantly through applying terrain correction techniques. Development and application of an accurate high resolution DEM is critical to the future progress of this project. The DEM will be used to create several other data products for comparison with those derived from the satellite imagery. DEM derived products will include a geomorphic map of the karst landforms described above, a solar illumination model, a flow accumulation model, an accessibility map, and models of slope and aspect. The geomorphic map will be developed using cartographic modeling techniques described by Tomlin (1990). The solar illumination model may be useful in deciphering vegetation patterns that might not otherwise be apparent from the geomorphic map.

CONCLUSIONS

This project is still in its initial stages, and the results so far can only be regarded as preliminary. With respect to initial impressions, the area classified as wet forest is considerably less than anticipated and the area classified as xeric vegetation is considerably greater than expected. Although initially it appears visually that the Cockpit Country is composed dominantly of extensive cockpit depressions, the image classification suggests that talus slopes and rocky hill summits occupy a significant spatial proportion of the study area.

Additional field investigation is needed to clarify further the vegetation patterns within the Cockpit Country. A more sophisticated GPS receiver with an external antenna and post-processing capability will provide greater location accuracy.

Once this pilot project is completed the intent is to apply the same techniques to the entire Cockpit Country region. A high-resolution satellite image and a DEM can be used in conjunction to quantify and delineate the complex karst landforms and to assess vegetation patterns by reference to those landforms. A suitably designed GIS will permit the integration of additional data from previous and future scientific studies. Such data integration will permit more sophisticated analysis of the Cockpit Country's biotic and abiotic landscape, and will facilitate implementation of an appropriate management strategy.

ACKNOWLEDGEMENTS

We would like to thank Susan Konig and Mike Swartz of the Windsor Research Station. Their hospitality and company were greatly appreciated. Also, special thanks to the people of Windsor, who managed to lead us into and out of the cockpits each day.

REFERENCES

Asprey, G.F., and Robbins, R.G., 1953, The Vegetation of Jamaica. Ecological Monographs, vol. 23, issue 4, Oct. 1953, p. 359-412.

Aub, C.F.T., 1969a, The nature of cockpits and other depressions in the karst of Jamaica: 5th Int. Speleol. Cong., Paper M15, Stuttgart, 7 p.

Aub, C.F.T., 1969b, Some observations on the karst morphology of Jamaica: 5th Int. Speleol. Cong., Paper M16, Stuttgart, 7 p.

Bauer, B.O., 1996, Geomorphology, Geography, and Science. In: Rhoads, L. and Thorn, C. E., The Scientific Nature of Geomorphology, Wiley, New York, p. 381-413.

Day, M.J., 1986, Slope form and process in cockpit karst in Belize. In: Paterson, K. and Sweeting, M.M., New Directions in Karst, GeoBooks, Norwich, UK, p. 363-382.

Day, M.J., 1987, Slope form, erosion and hydrology in some Belizean karst depressions. Earth Surface Processes and Landforms 12 (5), p. 497-505.

Giles, P.T., 1998, Geomorphological signatures: classification of aggregated slope unit objects from digital elevation and remote sensing data. Earth Surface Processes and Landforms, 23 (7), p. 581-594.

Harvey, L.E., Davis, F.W., Gale, N., 1988, The analysis of class dispersion patterns using matrix comparisons. Ecology 69, no. 2 (1988) p. 537-542.

Kirby, M., 1995, Modelling the links between vegetation and landforms. In: Biogeomorphology, Terrestrial and Freshwater Systems, editors Hupp, C.R., Osterkamp, W.R., Howard, A.D., Elsevier, New York, p. 319-335.

Pickup, G., 1990, Remote sensing of landscape processes. In: Hobbs, R.J. and Mooney, H.A. (eds.), Functioning, Springer-Verlag, New York, p. 221-247.

Pregill, G.K., Davis, F.W., Hilgartner, W.B., Crombie, R.I., Steadman, D.W., Gordon, L.K., 1991, Living and late Holocene fossil vertebrates, and the vegetation of the Cockpit Country, Jamaica, Atoll Research Bulletin 348-354, no. 353.

Proctor, G.R., 1983, Cockpit Country and its Vegetation. In: Thompson, D.A., Bretting, P.K., Humphreys, M., Forests of Jamaica. Institute of Jamaican Publications Ltd., Kingston, Jamaica, p. 43-47.

Tinkler, K.J., 1985, A Short History of Geomorphology. Barnes and Noble Books, Totowa, p. 317

Tomlin, C.D., 1990, Geographic Information Systems and Cartographic Modeling. Prentice Hall, Englewood Cliffs, N.J., p. 249

Young, R.W., and White, K.L.,1994, Satellite imagery analysis of landforms: Illustrations from southeastern Australia, Geocarto International, 2, p. 33–44.

Geotechnical and Environmental Applications of Karst Geology and Hydrology, Beck & Herring (eds)
© *2001 Taylor & Francis, ISBN 90 5809 190 2*

The purposes of the main sinkhole project in the Latium Region of central Italy

ANTONIO COLOMBI, EUGENIO DI LORETO & FRANCESCO NOLASCO   Dipartimento Ambiente e Protezione
Civile, Regione Lazio, 00145 Roma, Italia, geolazio.utvra@usa.net

GIUSEPPE CAPELLI & ROBERTO SALVATI   Dipartimento di Scienze Geologiche, Università di Roma Tre,
00146 Roma, Italia, capelli@uniroma3.it

ABSTRACT

Some the territories of the Latium Region are exposed to catastrophic subsidence phenomena due to geological features as well as the coexistence of several other issues such as active tectonics, main groundwater circulation, peculiar geomorphological conditions, etc. So far very few studies have focused on sinkholes in Italy, and those mostly refer to the description of obvious phenomena. However, land-use planning and management necessities require more detailed studies focused on triggering issues, evolution modes, and temporal and spatial cyclicity.

The Geological Survey of the Latium Regional Government, with the scientific collaboration of the Departments of Geological Science of the two Universities of Rome, during the next two years is conducting a study of sinkhole areas of Latium. The main purposes of this study are:

- To find and critically analyze the widest references possible;
- To make a preliminary classification of sinkhole phenomena in general;
- To conduct a census of sensitive areas in the Latium Region and of sinkholes by means of both airphoto interpretation and field surveys;
- To establish a definitive sinkholes phenomena classification system relative to Latium.
- To develop an applied method, including an application for sampling areas (particularly a piedmont area in central-southern Latium close to the Lepini Mountains karst ridge, 200 km^2 wide and the San Vittorino plane, in the Apennines intra-mountains basin) for reconstructing the geological and structural setting with historical seismicity, former phenomena occurrence, active tectonic indicators, and gas emissions by conducting:
  a) Hydrogeological field surveys
  b) Geochemical and hydrochemical field surveys
  c) Geophysical campaigns: gravity and microgravity prospecting, geoelectrical tomography using the Schlumberger method, and analysis of seismic surveys using explosions in quarries
  d) Geotechnical campaigns (samples and laboratory analysis)
  e) Chemical and physical monitoring system network implementation
- To define modern zones where sinkholes are hazardous during the preparation of a Master Plan in sinkhole areas.

Presented here are data, ideas, methodologies and the purposes of the studies around the sinkhole area of the Latium Region. Also discussed is an Internet website for sinkholes which has been created to present the structure of the project, the phenomena occurring in our Region, and a page of site links useful for people interested in this problem.

GENERAL FEATURES

Sinkhole phenomena or "catastrophic subsidences" are present diffusely in the Latium Region and are found in different geological settings. The Latium Region contains the highest number of sinkhole events in Italy. This phenomenology becomes very hazardous where land use changes the landscape's natural pattern in an incisive way. In Latium the results of catastrophic subsidence are more than evident. The occurrence of catastrophic subsidence phenomena, as with any geological process, is neither random nor extemporaneous. Preliminary analysis of location, morphometry and evolution typologies, and their temporal recurrence clearly shows that sinkhole occurrence is preferential under certain conditions.

In consideration of the frequency and wide distribution of sinkhole phenomena in Latium, there is really an insufficient knowledge of sinkhole triggering issues, the ways they develop and evolve, and the relationships among the different geological factors. In particular, the relationships between geological-structural setting, tectonic activity, groundwater circulation and human impact that are generally accepted as the main triggering issues of such phenomena are completely absent from the wider international references.

Two main areas in the Latium Region are being studied because they represent two important situations where anthropic policy and natural hazards are strictly connected:

1. The Pianura Pontina area, a piedmont area in central-southern Latium close to the Lepini Mountains karst ridge, 200 km^2 wide
2. The San Vittorino plane about 80 km North-eastward the town of Rome, in the Apennines intra-mountains basin

The first area will be tested to develop a new numeric model and a conceptual matrix for sinkhole genesis (cfr. Salvati et al. in same volume) which will allow us to understand and to define a new sinkhole hazard classification. These products will hopefully offer new concepts for the administrative Master Plan..

In the second area, a project has been ongoing for several years studying these phenomena in clear risk situations where anthropic infrastructures (railway, highway, main river, drinkable springs, villages, quarries, thermal springs and agricultural activities) are commonly located close to more than fifty known sinkholes.

Through these experiences, our effort is to define and to map all the Latium Region's sinkhole phenomena, either actually known or potential sinkhole hazards. Figure 1 shows phenomena that have occurred in Latium (dark gray) and the areas to be studied in the next few years (light gray).

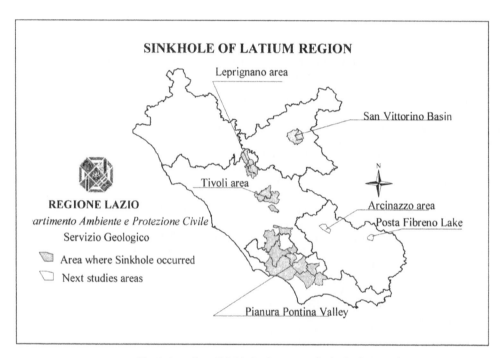

Fig. 1: Location of sinkhole phenomena in the Latium Region

ONGOING PROJECTS AND FUTURE

Some of the Latium territory is exposed to subsidence phenomena due to the geological features as well as to the coexistence of issues such as active tectonics, main groundwater circulation, and peculiar geomorphological conditions, etc. So far very few studies have focused on sinkholes in Italy, and they are mostly refer to the description of obvious phenomena while the land-use planning and management necessities require more detailed studies focused on triggering issues, evolution modes, and temporal and spatial cyclicity.

Following is the program which will be undertaken within the next four years:

- Step A - Knowledge
  1) Finding and critically analyzing the most references possible;
  2) Preliminary classification of sinkhole phenomena in general;
  3) Census of sensitive areas in the Latium Region and of actual sinkholes by means of both airphoto interpretation and field surveys
  4) Definitive sinkholes phenomena classification related to Latium evidences

- Step B – Methodology application
  Application to the sample areas, a piedmont area in central-southern Latium close to the Lepini Mountains karst ridge, 200 km^2 wide and an intramountain basin, of proposed methodology along with these elements:
  1) Geological and structural setting reconstruction through the historical seismicity, former phenomena occurrence, active tectonic indicators and gas emissions

2) Hydrogeological field surveys
3) Geochemical and hydrochemical field surveys
4) Geophysical campaigns
5) Geotechnical campaigns
6) Minero-petrographic analyses of clastic sediments
7) Chemical and physical monitoring network system implementation

- Step C – Sinkhole Policy
The third and last phase will be dedicated to defining the guidelines to mitigate the sinkhole risk in all areas of known hazard; these guidelines should be an important help to local administrations in matter of land-use planning.

In Italy the majority of sinkhole studies are limited to describing events that have occurred, but due to land-use planning and management imperatives, we have a strong need for more detailed investigations to develop a better understanding of the processes of catastrophic subsidence. It is important to learn more about event beginnings and factors. It is important to verify the geological and hydrogeological conditions of each particular area.

We have observed that not only in the Latium Region, but also in other Italian areas, such as Tuscany, Campania, the general geological conditions where a sinkhole occurred were:
- a piedmont area where the karst bedrock is 100-200 meters deep,
- the presence of strong, thick, clayey-sandy-gravel layers, with travertine intercalated,
- important hydraulic gradients,
- the presence of gas emission in the area with thermal springs or with anomalous concentrations of gas - radon, CO_2, etc.

The dimensions of the sinkholes vary as to their depth, but the constant, as mentioned, is a carbonate bedrock not less than 100 meters deep. The area of San Vittorino basin, the most studied area in Latium, has more than fifty sinkholes in a small but aligned area. In this area events occur or are discovered with good frequency while many old sinkhole phenomena have disappeared because of the agricultural activity.

In the Pianura Pontina area there are more than ten large sinkholes. One of them, Doganella di Ninfa's Sinkhole, has been monitored both topographically and geophysically.

SHORT POSTER DESCRIPTION AND INTERNET WEBSITE
Poster

The Poster is divided in two main areas:
- An upper sector which shows all the sinkholes that have occurred, the ongoing projects and the starting projects. Also shown are the geological features of the different sinkholes in the Latium Region.
- The lower sector presents the methodologies, applications and investigations used to defined and to map sinkhole phenomena and the different means of sinkhole propagation in these areas.

Internet

Due to interest in utilizing the Internet, the authors have created an Internet website, presented in both Italian and English, about the sinkholes in the Latium Region and the project to study them. Some of the pages on the website offer different geological and geomorphological situations of sinkholes of Latium Region, as well as the deadlines of the project. The website address is <http://www.geocities.com/sinkhole_prj2000>.

Along with the website, the first international sinkhole newsgroup has been created with the aim to facilitate scientific exchange between all researchers and parties interested in this phenomena. The newsgroup is hosted by eGroups.com, and the archive of messages can be viewed at <http://www.egroups.com/group/sinkholes>. One can subscribe by emailing <sinkholes-subscribe@egroups.com>. The list owner can be contacted by email at <sinkholes-owner@egroups.com>.

REFERENCES

Boyer L., Fiers S. & Monbaron M. (1998) - Geomorphological heritage evaluation in karstic terrains: a methodological approach based on multicriteria analysis. Suppl. Geografia Fisica e Din. Quaternaria 3/98 t.4

Capelli, G., Petitta M. & Salvati, R. (2000) - Relationships between catastrophic subsidence hazards and groundwater in the Velino Valley (Central Italy). SISOLS 2000, Ravenna-Italy, Sept. 2000

Capelli, G., Salvati, R. & Colombi A (2001) - Catastrophic subsidence risk assessment conceptual matrix for sinkhole genesis. 8th Conference of Sinkhole and Karst..., Louisville-USA, Apr. 2001 (this volume)

Colombi A., Salvati R, Capelli, G., Sericola A., Colasanto F., Crescenzi R., Mazza R., Meloni F. & Orazi A. (1999) - Problematiche da sprofondamento catastrofico nelle aree di pianura della Regione Lazio – Progetto Sinkhole del Lazio, "Convegno le aree di pianura" – Ferrara 6/9 Novembre 1999 Poster.

Faccenna C., Florindo F., Funiciello R. & Lombardi S. (1993) - Tectonic setting and sinkhole features: case histories from Western Central Italy. Quaternary Proceedings 3

Goldsworthy R. (1996) - The sinkhole project: technology, integration, and the environment. Learning and Leading with Technology 23

Kemmerly P.R. (1993) - Sinkhole hazards and risk assessment in a Planning Context. Journal of American Planning Association 59/2

Tharp T. M. (1999) - Mechanics of upward propagation of cover-collapse sinkholes. Engineering Geology

Geotechnical and Environmental Applications of Karst Geology and Hydrology, Beck & Herring (eds)
© 2001 Taylor & Francis, ISBN 90 5809 190 2

Karst development in the southern English chalk

ANDREW R.FARRANT   British Geological Survey, Keyworth, Nottingham, NG 12 5GG, England, arf@bgs.ac.uk

ABSTRACT

Recent geological mapping of the Upper Cretaceous Chalk of southern England has demonstrated that karst features are more widespread than previously recognised. Karst affects the engineering properties and the hydrogeological behaviour of the Chalk. As the Chalk occurs in the densely populated south-eastern half of the country, a good understanding of the geohazards and risks associated with karst features is essential to prevent costly engineering problems and to accurately model aquifer behaviour. The development of a new high resolution Chalk stratigraphy (Mortimore, 1986, Bristow et al., 1997) and its application by the British Geological Survey permits a more detailed assessment of the style, distribution and nature of karst features. Furthermore, improved mapping of the superficial Quaternary deposits that are a major influence on the density and distribution of surface karst landforms allows better prediction of where dissolution features may occur. This improved mapping, coupled with a developing GIS database of karst features, will permit delineation of karst geohazard domains for use by planners, developers, engineering geologists and hydrogeologists.

INTRODUCTION

The Upper Cretaceous Chalk is a very pure limestone consisting almost entirely of microscopic calcareous coccoliths with a small proportion of clay minerals. It forms Britain's major aquifer, occurring beneath much of southern and eastern England (Figure 1). Dissolution processes affect the morphology, engineering properties and the hydrogeological behaviour of the Chalk. Karst features are distributed across the outcrop, especially close to the contact with the overlying Palaeogene strata and Quaternary deposits. Because the Chalk occurs in the densely populated south-eastern part of the country, a good understanding of the potential geohazards and risks associated with karst features is essential to prevent costly engineering problems and to model aquifer behaviour accurately. Planning policy guidance and legislation in Britain now requires developers and local authorities to be aware of potential hazards when developing geologically unstable land.

HYDROGEOLOGICAL AND ENGINEERING IMPLICATIONS OF CHALK KARST

The Chalk is Britain's major aquifer and its protection and management are becoming increasingly important. Increased rates of groundwater abstraction and recent droughts have caused concern about water-supply and caused streams to dry up, endangering important wetland habitats, and creating problems with saline intrusion in coastal areas. Severe flooding has affected some areas, notably Chichester, possibly exacerbated by the reactivation of relict karstic conduits (Posford-Duvivier, 1994). Much of the Chalk outcrop is close to major urban centres or important transport corridors, with their attendant risks of pollution and groundwater derogation. Elsewhere rural areas are subject to intensive agricultural practices. Identifying rapid karstic groundwater flow-routes both in the saturated and unsaturated zones is crucial. Concerns about aquifer management and European legislation to protect wetland habitats have required water companies and environmental protection agencies to implement better models of aquifer behaviour. These models should incorporate the fundamental elements of karst hydrology and Chalk stratigraphy.

Chalk dissolution also has a significant impact for engineering projects (Mortimore, 1993). Dissolutional processes increase the number, density and width of fractures, and reduce the intact dry density of the surrounding chalk, commonly producing unstructured chalk putty. This diminishes the rock mass quality, softening the chalk and reducing chalk strength and load bearing capacity. This can lead to problems in the suitability of chalk as fill for embankments and earthworks, reducing the amount of suitable chalk available and creating disposal problems for the waste material. Formation of dissolution pipes at the soil-rock interface can create an irregular rockhead. The clay and gravel fills within these pipes are susceptible to increased mass compressibility, which can cause subsidence or excessive settlement. Similarly, flushing of the unconsolidated fill by water following heavy rain or from broken water pipes, can cause catastrophic subsidence. Large amounts of fill may cause disposal problems for contractors.

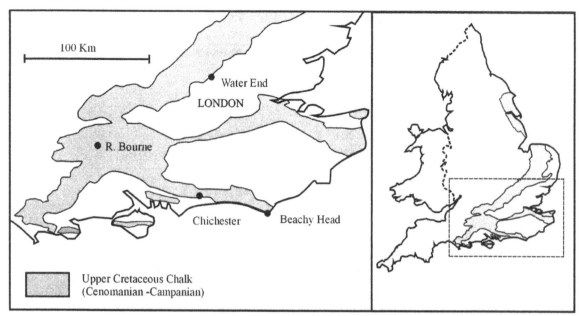

Figure 1. The distribution of Chalk in England.

KARST LANDFORMS

Although not normally regarded as a classic karst landscape, Chalk areas exhibit a wide range of karstic landforms. The density of dissolution features, stream sinks and resurgences in parts of the Chalk outcrop are comparable to the classic Carboniferous Limestone karst areas elsewhere in Britain (Waltham et al, 1997). The most abundant karst feature are dissolution pipes, which occur as conical sediment filled cavities developed from the surface extending down into the chalk. These can be several tens of metres deep, but most have little or no surface expression. They are best developed beneath a thin cover of superficial material where densities of over 250 ha[-1] have been recorded during road construction (Lamont-Black and Mortimore 1999). Fill type is dependent on the locally available superficial deposits and weathering products, and include clay-with-flints, sandy clay and silty clayey loam, Palaeogene sands and clays, and coarse flint gravels derived from river terraces.

Subsidence dolines (sinkholes) are commonly developed in the overlying superficial material. The best known examples occur in Dorset (Sperling et al. 1980; Waltham et al, 1997), but many others are known throughout the chalk outcrop. Edmunds (1983) provided an estimate of the density of dissolution features on the Chalk of southern England, based mainly on literature searches. For much of central southern England, densities of 5-10 dissolution features per 100 km[2] were identified. Recent geological mapping indicates that this is both an over simplification and a gross under-estimate. In reality, the spatial distribution is highly variable, and dependent largely on the presence or absence of a superficial cover. Where these cover deposits occur, sinkhole densities of 5-10 per km[2] are more realistic, but probably still underestimate the original density. Intensive land use, changes in agricultural practices, quarrying for agricultural lime and the infilling of dolines with refuse and waste material have obliterated many dolines, although some can still be recognised from air photographs.

Stream sinks are locally common, the vast majority occurring at the contact between the Chalk and the overlying Palaeogene sands and clays. Some, such as those at Water End, Hertfordshire are well known, but many are not recorded. Across the Chalk outcrop south of the River Thames, over 130 discrete stream sinks can be identified from topographical maps, yet less than 5% have been tracer tested. Those tests that have been conducted (for a review, see MacDonald et al, 1998) indicate that rapid turbulent karstic (conduit flow) occurs, with transport velocities between 2-6 km per day. Large springs and resurgences, many of which are utilised for public supply, occur throughout the Chalk outcrop, including one of the largest springs in England at Bedhampton, near Chichester, Hampshire, with a combined flow of around 104 000 m[3] per day.

Intermittent streams, known as 'winter-bournes' are common throughout the Chalk outcrop in southern England. During the summer months the lower part of the streams are fed from perennial springs. During the wet winter months, discrete springs known locally as 'bourne holes', become active farther up valley. Some streams, such as the Bourne River in Hampshire, have both influent and effluent sections, with the middle portion of the river remaining dry for much of the year.

Development of surface karst landforms

The abundance of surface karst features, notably dissolution pipes is related to the maturity of the contact between the Chalk and any overlying superficial deposits. This is influenced by the degree of point recharge, length of time during which dissolution processes have been operating, structural setting, and the type of chalk and superficial lithologies present. Chalk dissolution features are most common at the contact with the overlying Palaeogene strata, where extensive point recharge is concentrated. Elsewhere away from the Palaeogene margin, extensive outcrops of clay-with-flints occur. This is a thin (generally <10 m) *remanié* deposit formed by modification of the original Palaeogene cover and dissolution of the underlying chalk; it rests on a major sub-Palaeogene erosion surface that varies in altitude and stratigraphical level. Around the margins of the clay-with-flints, recharge and active dissolution have been concentrated for much of the Quaternary, at least during interglacial periods. Where the clay-with-flints has been eroded (and also depending on the level of dissection), remnants of earlier generation dissolution pipes may still be preserved below ground surface. Elsewhere, more recent 'Head' deposits resulting from periglacial solifluction of the clay-with flints mantle many valley sides and valley floors. Here less well developed dissolution features occur. Similar features are known from river terrace and raised beach deposits which occur throughout southern England, especially along the south coast and in the Thames Valley. The degree of karst development is largely related to the antiquity of the deposits, but many of these areas are prone to dissolution, pipe formation and collapse. Urban areas are especially prone to collapse due to modification of surface drainage and anthropogenic activity.

CHALK STRATIGRAPHY AND KARST DEVELOPMENT

The style of karst development is strongly influenced by the chalk lithology and its mass fracture characteristics; both of which are interdependent. Until recently, the Chalk of southern England was divided into three units, Lower, Middle and Upper Chalk, the latter Formation distinguished by the incoming of flints. This scheme was of little use in characterising the Chalk in terms of its susceptibility to karst development, hydrogeological or engineering properties. Mortimore (1986) devised a new lithological classification of the Chalk, amended by Bristow et al. (1997). The Chalk Group is now divided informally into White and Grey Subgroups and nine Formations (Table 1) that form the basis of the lithostratigraphical mapping currently used by the BGS. Furthermore, work by Mortimore and Pomerol (1991) demonstrates that intra-Cretaceous tectonism influenced sediment thickness, chalk lithology, fracture style; factors that are important for karst development.

Table 1. Chalk stratigraphy as used by the British Geological Survey. For details, see Bristow et al., (1997).

Chalk Group	Formation	Lithology	Preferential flow horizons
White Chalk (Subgroup)	Portsdown Chalk	Massive soft white flinty chalk	Bedding parallel joints, flint bands
	Culver Chalk	Massive soft-medium hard white chalk with major flint bands and marl seams	Bedding parallel joints, flint bands and marl seams
	Newhaven Chalk	Massive soft white chalk with marl seams.	Bedding parallel joints, marl seam - conjugate joint sets.
	Seaford Chalk	Massive soft-medium hard white chalk with major tabular and sheet flints.	Tabular and sheet flints, bedding parallel joints.
	Lewes Nodular Chalk	Hard nodular chalk with tabular and sheet flints, marl seams and hardgrounds	Sheet flints, open joints, faults, marl seams, bedding parallel joints.
	New Pit Chalk	Massive soft white chalk, marl seams and minor flint bands.	Marl seam – conjugate joint sets, bedding parallel joints.
	Holywell Nodular Chalk	Hard nodular chalk and marl seams	Open joints, faults, marl seams
Grey Chalk (Subgroup)	Zig Zag Chalk	Marly massive chalk, interbedded with limestone near base.	Perched flow above marly beds in limestone bands and on the 'Cast bed'
	West Melbury Marly Chalk	Marly chalk and limestone bands.	Perched flow above marl-rich beds.

Chalk lithology and mass fracture characteristics

The Grey Chalk Subgroup consists of two Formations, both of which contain rhythmic sequences of clay-rich marls and thin limestone beds, without any flint bands. Because of their high marl content, both the West Melbury Marly Chalk and the Zig Zag Chalk Formations have limited karstic development.

The White Chalk Subgroup consists of soft blocky white chalk interbedded with harder nodular chalk. Flints occur in the upper part of the sequence (from the New Pit Chalk upwards), and are abundant at certain stratigraphic levels, as are sponge beds. Some formations are characterised by abundant thin (<20 cm) marl seams, either of ash fall or detrital origin. They thin towards the basin margins and over structural highs. The soft white massive bedded chalk represent basinal deposition at times of high relative sea-level. The New Pit Chalk and the Newhaven Chalk consist of soft blocky white chalk with marl seams, but with generally few small flint seams. Both units are intensely fractured by conjugate joint sets that commonly dissipate along the marl seams. The more homogeneous soft to medium hard white Seaford and Culver Chalk Formations contain bands of large nodular and semi-tabular and sheet flints, but generally few marl seams. Both contain regular orthogonal and bedding parallel joint sets.

Nodular chalk forms in areas of reduced sedimentation, either during periods of low sea-level stand or at the margins of the basin or over structural highs. Two hard nodular chalk units occur in southern England; the Holywell Nodular Chalk and the Lewes Nodular Chalk. Being relatively hard and brittle, both are extensively fractured by open steeply inclined conjugate joint sets (Mortimore, 1993). The Lewes Chalk consists of medium- to high-density hard nodular chalk with marl seams and abundant sheet and nodular flint bands. Hardgrounds representing breaks in sedimentation and cementation of the sea-floor are locally common.

Lithological and mass fracture control on aquifer properties.

The aquifer properties of each lithology are different (Mortimore, 1993), depending on sedimentological variations and fracture style (orientation, trace length spacing, aperture, volume, connectivity and degree of infilling). In the West Melbury Marly Chalk, perched flow occurs above some of the more plastic marl seams along stress release induced open joints in the more brittle limestone beds above. Similarly, near the base of the nodular Holywell Nodular Chalk, at the junction with the Plenus Marls, stress release induced open fractures that are subsequently enlarged by dissolution are common. Flow in the more massive New Pit and Newhaven Chalk Formations is mainly controlled by conjugate joint sets which dissipate along marl seams. Water draining down the joints is commonly held up along these marls creating perched conduits. More pervasive conduit development takes place where these marl seams are at or close to the water table. In the Seaford and Culver Chalk Formations, bedding parallel joints, sheet and tabular flints form the preferred flow horizons. The largest conduits (for example Beachy Head Cave) occur on tabular or sheet flint bands such as the Seven Sisters Flint, but many occur along bedding parallel joints. The harder nodular Holywell and Lewes Chalk Formations support many open steeply inclined conjugate joint sets that are widened by dissolution. Conduit development is favoured where these intersect sheet flints and marl seams. These variations influence the pattern and style of conduit development and groundwater flow throughout the Chalk aquifer.

Conduit development in the Chalk.

The Chalk is an unusual aquifer in that it has several components of both porosity and permeability (Price, 1987). Although it possesses a high primary porosity of around 40%, hydraulic conductivities for bulk samples are very low (10^{-3} to 10^{-2} m/day), even though measurements of transmissivity commonly exceed 1000 m^2 day^{-1}. To account for this disparity, most authors favour the development of dissolutionally enhanced secondary permeability. Plenty of evidence from pump test data, tracer tests, tunnelling and geomorphology show that rapid groundwater flow occurs (MacDonald et al, 1998), although there is almost certainly an interchange throughout the aquifer between conduit and diffuse flow. Yet the karstic aspects of Chalk hydrogeology have been neglected. Examining aquifer behaviour using models of cave development (for example Ford and Ewers, 1978; Palmer 1991; Lowe, 1992), coupled with standard hydro-geological techniques may permit a better understanding of the extent and distribution of conduit flow.

Penetrable chalk caves are rare and generally small compared to those in more classic Carboniferous Limestone karst areas. Open cave entrances are scarce due to periglacial mass movement of unconsolidated Quaternary material and the susceptibility of chalk to weathering. The best known example of a chalk cave is Beachy Head Cave near Eastbourne in Sussex. It is a well developed relict phreatic tube about a metre in diameter and 354 m long, (Waltham et al. 1997). Large caves are rare, despite the capacity of the chalk to support significant roof spans. Diffuse recharge over much of the outcrop is commonly cited as the reason for the lack of sizeable cave development. Percolation water will rapidly loses its aggressivity within a few metres of the surface. Caves can still form at depth, however, by mixing of waters with differing chemistry. Mixing of highly saturated water from diffuse flow through the porous chalk with less saturated water flowing through open fractures and bedding planes will generate zones of enhanced dissolution, producing a sponge-work pattern of voids along bedding partings. However, mixing corrosion is unlikely to be major factor in conduit enlargement because the low rate of diffuse inflow restricts the rate of dissolution (Palmer 1991), but it plays an important role in conduit inception (Lowe, 1992), generation and determining patterns of conduit flow. Clearly, in areas of concentrated recharge, larger conduits will develop along the network of pre-existing and mixing generated proto-conduits and inception horizons.

Applying models of cave development may permit an understanding of the style of conduit development within the saturated zone. Ford and Ewers (1978) basic concept is that the optimum groundwater flow route will be chosen from all available fractures, inception horizons, bedding planes, joints and faults. The flow pathway utilised depends on the frequency and distribution of available fractures. Low fracture densities favour deep bathymetric systems. Conversely, higher fracture densities favour low gradient water-table conduits, where all parts of the aquifer are in good hydraulic communication. The Chalk is a well fractured aquifer with fracture spacing ranging between 0.1 and 2 m (see Bloomfield, 1996), although this varies significantly according to the tectonic province and lithology. The high density of fractures, lithological discontinuities and inception horizons provides many alternative flow pathways through the rock mass.

According to this model, the high fissure densities observed in the Chalk should favour the development of conduit systems at or just below the water-table. In the gently dipping Chalk, these will be developed mainly as strike or dip tubes along lithological and bedding-plane discontinuities. The discontinuities used will vary according to the water-table elevation and the chalk lithology as discussed above. Thus conduit flow is dispersed along a anastomotic mesh of interconnected smaller discrete conduits at or close to the water table, rather than a single 'main drain'. Generally, conduits follow horizontal discontinuities or 'inception horizons', along routes opened up by mixing corrosion, but where necessary, follow joints and faults to gain or lose stratigraphic elevation to reach the springs. These small conduits are prone to blockage by sediment washed in, causing locally increased hydraulic head, flow diversion and the creation of new conduits. Ultimately, the result is a network of largely choked small bore conduits.

Direct observation of cliffs, quarries and boreholes confirms the small-bore anastomotic nature of conduit development. Flow logging of boreholes in Chalk areas indicates that the bulk of water inflows into many boreholes come from a few specific horizons (Jones and Robins, 1999, p. 46-49). Furthermore, down-hole CCTV surveys demonstrate that conduits typically occur along sub-horizontal bedding parallel joints and to a lesser extent sub-vertical tectonic fractures (Waters and Banks, 1997). Open conduits observed in cliffs and quarries have a similar morphology. Tracer tests (Macdonald et al., 1998) also demonstrate flow separation, with dye reappearing at several springs as water migrates along the conduit mesh. Thus, many Chalk streams originate at a group of discrete springs rather than a single source.

Effects of Quaternary valley incision and sea-level fluctuation

Base-level changes may cause abandonment or reactivation of conduit systems. Given the history of periglacial valley incision and terrace development in many valleys in southern England during the Quaternary, several tiers of abandoned conduit systems may be present at any one locality, both above and below the present water-table. Elsewhere in the UK, similar tiers of cave development occur widely in the Carboniferous Limestone aquifer (Waltham et al., 1997). Relict conduits close to the present water-table may be become reactivated during wet periods when the water table rises, creating winter-bournes and potential flooding problems. This may have contributed to the severe flooding in Chichester in 1994, where flood levels dramatically increased once ground-water levels reached a certain elevation (Posford-Duvivier, 1994). Higher abandoned conduits may be utilised by water draining through the vadose zone, creating rapid flow pathways. Conduits formed during periods of low sea-level in coastal areas may still transmit substantial water flow at significant depths below the current water-table, as well as providing a potential rapid route for saline water ingress into coastal aquifers.

CHALK KARST DOMAINS

The idea of classifying Chalk outcrops into geomorphological 'domains' is not new (Mortimore, 1990, Lamont Black and Mortimore, 1999). However, the recent high resolution stratigraphical mapping across much of southern England has improved our understanding of the relationships between dissolution features, Chalk stratigraphy and Quaternary deposits. This has permitted the subdivision of the White Chalk Subgroup into karstic geomorphological domains (Table 2), expanding on the classification of Mortimore (1990).

Table 2. Geomorphological (karst) domains used to categorise karst geohazard risk, applicable for the White Chalk Subgroup.

Geomorphological (karst) Domain		Superficial Deposits	Potential hazards	Karst features and engineering hazards (co-dependent on chalk lithology and structure)
Hilltop and gentle dip-slope domains.	Mantled	Covered by thick, commonly highly disturbed clay-with-flints or other deposits (loess, till)	High risk	High density of dissolution pipes and dolines. Low quality soft and putty chalk, clay lined fractures and conduits. Prone to sinkhole collapse.
	Weathered	Thin or absent cover, no clay-with flints, but close to sub-Palaeogene or other erosion surface.	Moderate-high risk	Remnant dissolution pipes, pockets of low-quality weathered and softened chalk and pipe fills. Clay lined fractures and relict conduit systems
	Un-weathered	Bare chalk, little or no cover material present.	Low risk	Good quality chalk beneath thin weathered chalk rendzina.
Steep hill-slope domains	Mantled	Thin highly disturbed layer of clay-with-flints, 'head' or gravel.	Moderate risk	Undulating low amplitude dissolution pipes. Rare dolines. Thin zone of soft unstructured chalk.
	Un-weathered	Bare chalk, no cover material. Thin soils.	Low risk	Good quality chalk beneath thin weathered chalk rendzina.
Chalk valley domains	Dry valleys	Valley floor covered by variable thickness of disturbed gravelly sandy 'head' deposits	High risk	Bourne holes and ephemeral streams in the lower reaches. Much periglacial weathering and softening of chalk. Clay in fractures.
	River valleys	Valley floor covered by variable thickness of alluvium, peat, tufa and gravel.	High risk	Springs, bourne holes, influent and ephemeral streams. Much periglacial weathering and softening/puttying of chalk, clay in fractures.
Terrace and raised beach domains		Covered by permeable sands and gravels. Older terraces may be highly disturbed and soliflucted.	Moderate-high risk	Older (higher) terraces and beaches show greatest density of pipes. Open pipes and soft weathered chalk present. Prone to sinkhole collapse
Impermeable margin domains		Contact zone between chalk and adjacent impermeable lithologies, usually Palaeogene sand and clay.	Very high risk	Stream sinks, dolines and dissolution pipes common. Much weathered unstructured chalk. Clay filled fractures and open conduits present.

CONCLUSIONS

Recent geological mapping of the Chalk areas of southern England has demonstrated that the density of karst features is much greater than previously thought. Dolines, stream sinks, springs and dissolution pipes are widespread in many parts of the Chalk outcrop Their occurrence are mainly dependant on the presence of superficial deposits or adjacent impermeable strata, whereas groundwater flow through the Chalk is strongly influenced by chalk lithology and structure. The development of high resolution Chalk stratigraphy, combined with more traditional karst hydrology concepts provides a much better framework for understanding of the nature of conduit flow in the Chalk aquifer. Appreciation of these features, coupled with an appraisal of local topography and Quaternary history permits the classification of Chalk outcrops into karst 'domains'. All this information is being gathered for utilisation in a Geographic Information System (GIS). Eventually, this will enable hydrogeologists, engineering geologists and planners to evaluate, and minimise the risk from geohazards associated with Chalk karst effectively.

ACKNOWLEDGMENTS

Thanks must go to Dave Lowe, Tony Cooper and Pete Hopson for comments on an earlier draft of this paper. This paper is published with the permission of the Director, British Geological Survey (N.E.R.C).

REFERENCES

Bristow, R., Mortimore, R.N. and Wood., C.J. 1997, Lithostratigraphy for mapping the Chalk of southern England. Proceedings of the Geologist's Association, 109, 293-315.

Bloomfield, J. 1996. Characterisation of hydrogeologically significant fracture distributions in the Chalk: an example from the Upper Chalk of southern England. Journal of Hydrology, 184, 355-379.

Edmunds, C.N. 1983, Towards the prediction of subsidence risk upon the Chalk outcrop. Quarterly Journal of Engineering Geology (London), 16, 261-266.

Ford, D.C. and Ewers, R.O. 1978, The development of limestone cave systems in the dimensions of length and depth. Canadian Journal of Earth Science, 15, 1783-1798.

Jones, H.K. and Robins, N.S. 1999. The Chalk aquifer of the South Downs. British Geological Survey. Hydrological Report Series, Keyworth, Nottingham.

Lamont-Black, J., and Mortimore, R.N. 1999, Predicting the distribution of dissolution pipes in the Chalk of southern England using high-resolution stratigraphy and geomorphological domain classification. Hydrogeology and Engineering Geology of Sinkholes and Karst, Ed Pettit and Herring, Balkema, Rotterdam.

Lowe, D.J. 1992. The origin of limestone caverns: An inception horizon hypothesis. Unpublished Ph.D Thesis, Manchester Metropolitan University.

MacDonald, A.M., Brewerton, L.J. and Allen, D.J. 1998, Evidence for rapid groundwater flow and karst type behaviour in the Chalk of southern England. In: N.S.Robins (Ed), 1998, Groundwater pollution, aquifer recharge and vulnerability: Geological Society of London (Special Publication no.130). 95-106.

Mortimore, R. N. 1986; Stratigraphy of the Upper Cretaceous White Chalk of Sussex. Proceedings of the Geologists' Association, 97, 97-139.

Mortimore, R. N. 1990, Chalk or chalk? In: Chalk, Proceedings of the International Chalk symposium, Brighton, 1989, Ed Burland J.B. et al. Thomas Telford, London, 15-45.

Mortimore, R. N. 1993, Chalk water and engineering geology. In: The Hydrogeology of the Chalk of North-west Europe, Ed Downing, R.A., Price, M. and Jones., G.P. Clarendon Press, Oxford.

Mortimore R.N. and Pomerol, B. 1991, Upper Cretaceous tectonic disruptions in a placid Chalk sequence in the Anglo-Paris basin. Journal of Geological Society of London, 148, 391-404.

Palmer, A.N. 1991. Origin and morphology of limestone caves. Bulletin of the Geological Society of America, 103, 1-20.

Posford-Duvivier, 1994. River Lavant Flood Investigation. Haywards Heath: Posford Duvivier. (Consultants report)

Price, M., 1987, Fluid flow in the Chalk of England. In Goff, J.C. and Williams, B.P.J. Fluid flow in Sedimentary Basins and Aquifers. Geological Society Special Publications No 43, pp. 141-156.

Sperling, C.H.B., Goudie, A.S., Stoddart, D.R. and Poole, G.G. 1980, Dolines of the Dorset Chalklands and other areas in southern Britain. Transactions of the Institute of British Geographers, 2, 205-223.

Waltham, A.C., Simms, M.J., Farrant, A.R. and Goldie, H. 1997. Karst and Caves of Great Britain. Geological Conservation Review Series. Chapman and Hall, London.

Waters A. and Banks, D. The chalk as a karstified aquifer: closed circuit television images of macrobiota. Quarterly Journal of Engineering Geology (London), 30, 143-146.

Geotechnical and Environmental Applications of Karst Geology and Hydrology, Beck & Herring (eds)
© 2001 Taylor & Francis, ISBN 90 5809 190 2

Application of GIS technology to study karst features of southeastern Minnesota

YONGLI GAO & E.CALVIN ALEXANDER, JR.   Dept. of Geology & Geophysics, University of Minnesota, Minneapolis, MN 55455, USA, gaox0011@tc.umn.edu, alexa001@tc.umn.edu

ROBERT TIPPING   Minnesota Geological Survey, St. Paul, MN 55114, USA, tippi001@umn.edu

ABSTRACT

Several 2D and 3D maps have been developed to visualize karst feature distributions of southeastern Minnesota. ArcView and ArcInfo are used to sort locations of sinkholes according to the uppermost bedrock unit and depth to bedrock. A nearest-neighbor analysis of the sorted sinkholes reveals that the distributions of distances between sinkholes are highly skewed toward short distances (i.e., the sinkholes are strongly clustered). More detailed geostatistical methods such as histograms, probability estimation, correlation and regression have been used to study the spatial distributions of some of the mapped karst features of southeastern Minnesota. The addition of shaded relief maps from Digital Elevation Methods (DEMs) allows visualization of sinkhole locations on the surface topography of southeastern Minnesota. A sinkhole probability map for Goodhue County has been constructed based on sinkhole distribution, bedrock geology, depth to bedrock, GIS buffer analysis and nearest-neighbor analysis. A series of karst features for Winona County including sinkholes, springs, outcrops and dry valleys has been mapped and entered into the Karst Feature Database of Southeastern Minnesota. The Karst Feature Database of Winona County is being expanded to include all the mapped karst features of southeastern Minnesota.

The main goals of this research are to visualize large-scale patterns in the sinkhole distribution, to conduct statistical tests of hypotheses about the formation of sinkholes, and to create tools for land-use managers.

INTRODUCTION

Southeastern Minnesota is part of the Upper Mississippi Valley Karst that includes southwestern Wisconsin and northeastern Iowa. Karst lands in Minnesota are developed in Paleozoic carbonate and sandstone bedrock. Most karst features such as sinkholes are found only in those areas with less than fifty feet of sedimentary cover over bedrock surface. A significant sandstone karst has developed in Pine County (Shade and others, 2000). Much of the scientific karst literature (Davies and Legrand, 1972; Troester and Moore, 1989; and Dougherty and others, 1998) has focused on other parts of the country and world. Nevertheless, the karst lands of southeastern Minnesota present an ongoing challenge to environmental planners and researchers and have been the focus of a series of research projects and studies by researchers for over 30 years.

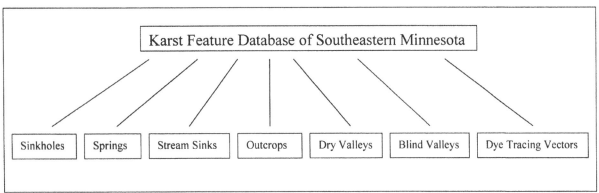

Figure 1. Database Structure of the Karst Feature Database of Southeastern Minnesota

Since the early 1980s, the Minnesota Geological Survey has been mapping karst features and publishing various versions of their results in the form of 1:100,000 scale County Geologic Atlases. In the mid 1990s, the Minnesota Department of Natural Resources was assigned responsibility for the hydrogeology portions of the County Atlases and are now responsible for the karst mapping. Dalgleish and Alexander (1984), Alexander and Maki (1988), and Witthuhn and Alexander (1995) published sinkhole distribution maps for Winona, Olmsted, and Fillmore Counties, respectively. Published Atlases of Washington, Dakota, and the counties of the

Twin Cities Metro area contain limited information on sinkhole occurrences. It has long been a desire of researchers and planners at the Minnesota Geologic Survey (MGS), the Minnesota Department of Natural Resources (MNDNR), and the Department of Geology and Geophysics at the University of Minnesota to combine the Geologic Atlases for the counties into a single map depicting the karst lands of southeastern Minnesota.

By working with research fellows in the Department of Geology and Geophysics and GIS, computer science and hydrology experts at the University of Minnesota, we have created a series of maps depicting the karst lands of southeastern Minnesota. This paper summarizes the current state of our research on karst feature distribution of southeastern Minnesota and discusses how to extend nearest-neighbor analysis and probability estimation to study sinkhole development. The main goals of this research are to look for large-scale patterns in sinkhole distribution, to conduct statistical tests of hypotheses about the formation of sinkholes, and to create management tools for land-use managers.

METHODOLOGY

Existing county and sub-county sinkhole databases have been assembled into a large GIS database capable of analyzing the entire data set. The basic structure of the Karst Feature Database of Southeastern Minnesota is shown in Figure 1. Several of the earlier maps had been produced by classic drafting techniques and had to be first digitized from the original stable base maps. Using ArcInfo and ArcView, we combined the sinkhole database with other geological data sets and plotted a series of 2D and 3D maps showing the relationships between sinkhole distribution and bedrock geology, depth to bedrock, surficial geology, and surface topography. One rendering of this combined data set is shown in Figure 2. Other renderings showing the correlations of sinkhole distributions across county boundaries with bedrock stratigraphy, surficial geology, and surface topography can not be adequately reproduced here in black and white.

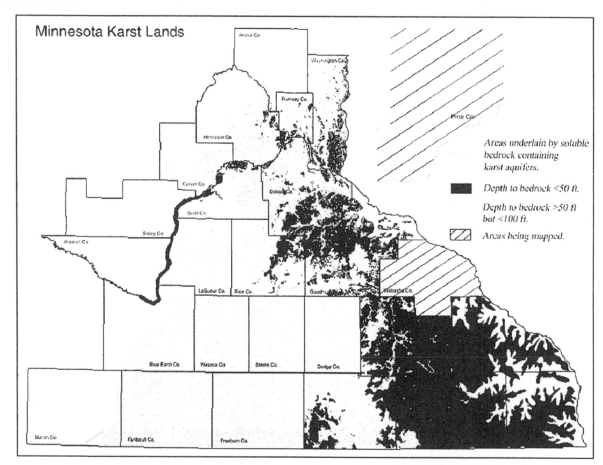

Figure 2. Minnesota Karst Lands. This map overlays the areas with < 50 feet and 50 to 100 feet of surficial cover over the area underlain by carbonate bedrock (and sandstone in Pine County.) Featureless counties in the southwestern portion of this map have not yet been mapped with sufficient accuracy to show the depth to bedrock information. The hatched areas are currently being mapped. Pine County is in east central Minnesota (about 100 miles north of Twin Cities).

Sinkhole probability map of Goodhue County

The construction of the sinkhole probability map was guided by similar efforts in Winona County (Dalgleish and Alexander, 1984a, b; Magdalene and Alexander) and Olmsted County (Alexander and Maki, 1988). The primary controls on sinkhole development in Goodhue County are stratigraphic position or bedrock geology and depth of surficial cover over bedrock surface. Major secondary controls appear to be structural geology, such as joints and position in the landscape. The division of the county into areas of varying sinkhole probability should be viewed as approximate. The boundaries are not sharply defined.

Nearest-neighbor analysis

A nearest-neighbor program written by Professor Randal Barnes of the Civil Engineering Department at the University of Minnesota was used to conduct the analysis. The program uses the UTM coordinates of sinkhole locations to calculate the direction and distance to each sinkhole's nearest neighbor.

The sinkholes in the Karst Feature Database of Southeastern Minnesota were categorized into three karst groups: 146 sinkholes in Cedar Valley Karst (Middle Devonian), 6940 sinkholes in Galena/Spillville Karst (Upper Ordovician/Middle Devonian), and 1161 sinkholes in Prairie du Chien Karst (Lower Ordovician). Nearest-neighbor analyses were conducted for sinkholes in each karst group. We also tested nearest-neighbor analysis on sinkholes in different counties, over different bedrock units, and in different probability areas. All the sinkholes in the sinkhole plains (highest sinkhole probability area) of Fillmore County were selected for an extended nearest-neighbor analysis and compared with Poisson distribution.

Winona County karst feature database and sinkhole development

A series of karst features for Winona County including sinkholes, springs, outcrops and dry valleys has been mapped and entered into the Karst Feature Database of Southeastern Minnesota. The Winona County sinkhole database was started by Dalgleish and Alexander (1984 a, b) and then updated by Magdalene and Alexander (1995). Magdalene recorded an age attribute for most of the sinkholes which can be used to visualize the sinkhole development in this area. IRIS ExplorerTM was used to visualize the sinkhole distribution and surface topography. A 3D USGS Digital Elevation Method (DEM) lattice file for SE Minnesota was registered and manipulated in Explorer to show the surface topography of Winona County. The resolution of this topography is one kilometer.

RESULTS AND DISCUSSION

The distribution of bedrock geology, depth to bedrock, and karst features form three bands of karst development that are arranged parallel to the Mississippi River. These bands of karst features extend into and are continuations of similar distributions in northeastern Iowa. These three bands comprise the Prairie du Chien, Galena/Spillville and Cedar Valley Karsts. The combinations of various sinkhole attributes and data sets are effectively displayed in color. Such color maps have proven to be effective tools for scientists, educators, and resource managers.

Sinkhole probability of Goodhue County

A sinkhole probability map for Goodhue County is currently in production. A draft map has been created based on sinkhole distribution, bedrock geology, depth to bedrock, surface topography, GIS buffer analysis, and nearest-neighbor analysis.

The only places in Goodhue County where sinkholes cannot form are in the northeastern portion of the county, where the Jordan Sandstone and St. Lawrence Formation are the first bedrock. In these deep river valleys erosion has removed all of the carbonate bedrock. The rest of the county was categorized into 5 sinkhole probability areas having some potential for sinkhole development. About half of Goodhue County is covered with irregular patches of surficial cover >50 feet thick. The sinkhole mapping and depth to bedrock map construction were conducted independently and then GIS techniques were used to compare the two data sets. The mapped sinkholes are found in the areas mapped as <50 ft to bedrock with encouraging fidelity.

Nearest-neighbor analysis

Figure 3 displays the histograms and cumulative fractions of the nearest-neighbor distances of the three sinkhole groups. As can be seen in Figure 3, the median distance to the nearest neighbor is significantly less than the mean distance to the nearest neighbor of all the sinkhole groups. Our nearest-neighbor analyses on other sinkhole data sets all showed a highly skewed distribution. All our nearest-neighbor analyses testify that sinkholes in southeastern are not evenly distributed in this area i.e., they tend to be clustered. This result confirms and expands Magdalene and Alexander's (1995) conclusions to the entire Minnesota data set. Figure 3 also reveals that the sinkholes in the Prairie du Chien Karst are spaced about three to five times further apart than are the sinkholes in the Cedar Valley and Galena/Spillville Karst. This implies that more isolated sinkholes occur in Prairie du Chien Karst.

Figure 4 shows that the nearest-neighbor distance distribution is distinctly different from the Poisson distribution within the sinkhole plains of Fillmore County. Since the Poisson Process describes randomly distributed data, this implies that sinkholes within the sinkhole plains of Fillmore County are not randomly distributed. The Poisson process does not adequately model the sinkhole distribution of southeastern Minnesota and should not be used to predict sinkhole occurrences.

Winona County karst feature database and sinkhole development

All the mapped karst features for Winona County are being entered into the Karst Feature Database of Southeastern Minnesota. This database is designed to be a relational database linked to GIS and hydrological models. The sinkhole attributes for Winona County include information on when the sinkholes formed.

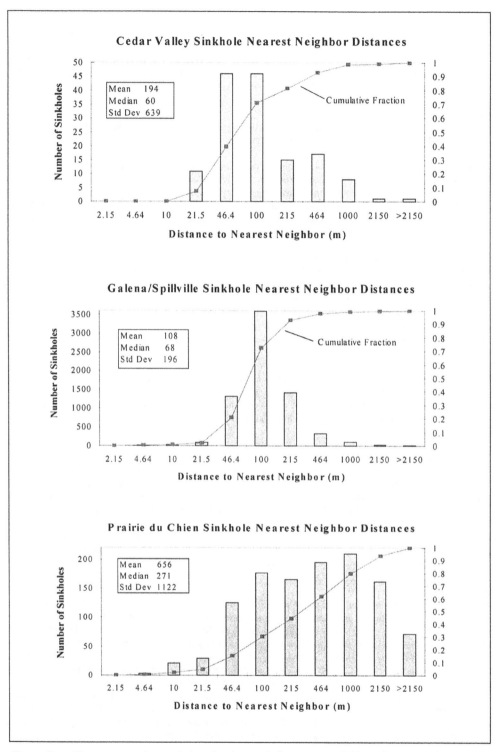

Figure 3. Histograms and cumulative fractions of the nearest-neighbor distances of the three stratigrapically distinct populations of sinkholes in southeastern Minnesota. Note the radically different numbers of sinkholes on the vertical axes of the three graphs. There are 146 Cedar Valley sinkholes, 6,940 Galena/Spillville sinkholes, and 1,161 Prairie du Chien sinkholes in these statistical analyses.

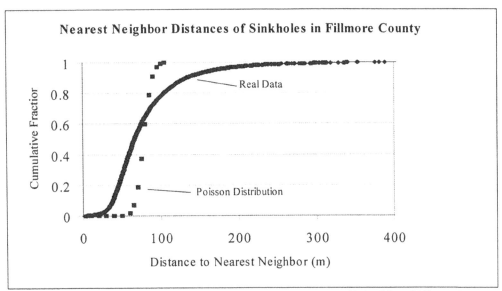

Figure 4. Nearest-neighbor and Poisson distribution of sinkholes in the sinkhole plains of Fillmore County.

Winona County karst feature database and sinkhole development

All the mapped karst features for Winona County are being entered into the Karst Feature Database of Southeastern Minnesota. This database is designed to be a relational database linked to GIS and hydrological models. The sinkhole attributes for Winona County include information on when the sinkholes formed.

The sinkholes in Winona County are divided into 4 age groups: before 1958, 1959 - 1977, 1978 - 1983, and 1984 - 1992. When displayed sequentially on DEM data sets, the age information allows the pattern of sinkhole development to be visualized. The most recently formed sinkholes are not randomly distributed among older sinkholes.

Future Work

We hope to integrate GIS, hydrogeological models, and visualization tools to investigate the hydrogeologic controls on the sinkhole distribution in southeastern Minnesota and to study groundwater contamination in Minnesota karst lands. Water quality and age and residence time data will be added to the investigation. Our initial efforts at 3D visualizations have yielded promising results. We hope to continue investigations using 3D visualizations of more attributes of sinkholes and other karst features with the growing variety of GIS compatible information. The steady improvements in both software and hardware enable significant increases in the size and complexity of the data sets and correlations that can be examined with available resources. We hope to add additional statistical analyses such as cluster analysis, orientation of nearest neighbor sinkholes to study sinkhole development and make sinkhole probability map more defensible in scientific, public and legal arenas. Finally, we plan to develop a web accessible version of the Karst Feature Database in Southeastern Minnesota. The goal is a user-friendly web site where a visitor can make on-line data queries and visualization.

SUMMARY

A karst database has been constructed that allows sinkhole and other karst feature distributions to be displayed and analyzed across existing county boundaries in a GIS environment. The resulting maps allow regional trends to be visualized and extend count-scale trends to larger scales. Southeastern Minnesota contains three distinctive bands of karst features that are continuations of similar trends seen in northeastern Iowa. These patterns simplify the determination of probability boundaries near county lines. Color maps, which combine the sinkhole distributions with other data sets and attributes, have proven to be useful tools for scientists, educators and resource managers.

A nearest-neighbor analysis indicates that the average distance to the next sinkhole in the Prairie du Chien Karst, the stratigraphically lowest of the three karsts, is 3 to 5 times greater than found in the Galena/Spillville and Cedar Valley Karsts. The results of the nearest-neighbor analyses were used to help create a sinkhole probability map for Goodhue County. The mapped distribution of sinkholes correlates well with <50 ft of surficial cover.

GIS, digital topography models and age information for sinkholes in Winona County have been combined to animate the history of sinkhole development in Winona County.

ACKNOWLEDGMENTS

This research was supported with funding from the Minnesota Department of Health. The databases used in this research have been built over the past 20+ years with support from The Legislative Commission on Minnesota Resources, Minnesota Department of Natural Resources and several counties. The sinkhole locating efforts of David Berner in Goodhue County are gratefully acknowledged. We thank Prof. Randall Barnes for the loan of his nearest-neighbor program and for many stimulating discussions of geostatistics.

REFERENCES

Alexander E.C., Jr., and Maki, G.L., 1988, Sinkholes and Sinkhole Probability: Olmsted County Atlas Series C-3, Plate 7, Minnesota Geological Survey.

Dalgleish, J.D., and Alexander, E.C., Jr., 1984a, Sinkholes and sinkhole probability: *in* Balaban, N. H., and Olsen, B. M., eds., Geologic Atlas of Winona County: County Atlas Series C-2, Plate 5, Minnesota Geological Survey.

Dalgleish, J.D., and Alexander, E.C., Jr., 1984b, Sinkhole distribution in Winona County, Minnesota: *in* Beck, B. F., ed., Sinkholes: Their geology, engineering and environmental impact: Boston, A. A. Balkema, p. 79-85.

Davies, W.E., and Legrand, H.E., 1972, Karst of the United States: *in* Herak, M., and Stringfield, V.T., eds., Karst, important karst regions of the northern hemisphere: Elsevier Pub Co., Amsterdam, p. 466-505.

Dougherty, P. H., Jameson, R.A., Worthington, S. R.H., Huppert, G.N., Wheeler, B., J., and Hess, J. W., 1998, Karst Regions of the Eastern United States with special emphasis on the Friars Hole Cave System: *in* Yuan D., and Liu, Z., eds., Global karst correlation: Science Press, Beijing, p. 137-155.

Magdalene, S., and Alexander, E.C., Jr., 1995, Sinkhole distribution in Winona County, Minnesota revisited: *in*: Beck, Barry F. and Person, Felicity M. eds., Karst Geohazards, Proceedings of the Fifth Multidisciplinary Conference on Sinkholes and the Engineering and Environmental Impact of Karst: Gatlinburg, Tenn., 2-5 April, 1995, A.A. Balkema, Rotterdam, p. 43-51.

Shade, B.L., Alexander, S.C., Alexander, E.C., Jr., and Truong, H., 2000, Solutional Processes in Silicate Terranes: True Karst vs. Psuedokarst with emphasis on Pine County, Minnesota (abstract), Abstracts with Programs, v. 32, n. 7, 2000 GSA Annual Meeting, p. A-27.

Troester, J. W., and Moore, J. E., 1989, Karst hydrogeology in the United States of American: Episodes, Vol. 12, no. 3, P.172-178

Witthuhn M.K. and Alexander, E. C., Jr., 1995, Sinkholes and Sinkhole Probability: Fillmore County Atlas Series C-8, Plate 8, Minnesota Geological Survey.

Geotechnical and Environmental Applications of Karst Geology and Hydrology, Beck & Herring (eds)
© 2001 Taylor & Francis, ISBN 90 5809 190 2

Karst unit mapping using Geographic Information System technology, Mower County, Minnesota, USA

JEFFREY A.GREEN & WILLIAM J.MARKEN   Minnesota Department of Natural Resources, Rochester, MN 55906, USA, jeff.green@dnr.state.mn.us, bill.marken@dnr.state.mn.us

E.CALVIN ALEXANDER, JR., & SCOTT C.ALEXANDER   Dept. of Geology & Geophysics, University of Minnesota, Minneapolis, MN 55455, USA, alexa001@tc.umn.edu, alexa017@tc.umn.edu

ABSTRACT

Mower County is in southeastern Minnesota in the area underlain by the sedimentary bedrock of the Hollandale Embayment. These rocks are Middle Devonian and Middle Ordovician karstic limestone and dolomite. As part of the Minnesota Department of Natural Resources County Geologic Atlas mapping project, the karst features of the county are being inventoried. These features include sinkholes, disappearing streams, caves, dry valleys, and springs. Previous karst mapping efforts for other county atlases in Minnesota have produced sinkhole probability maps and a springshed map. In Mower County, we are trying to develop a new type of karst map using a Geographic Information System to produce a karst unit map. Karst units are discrete three-dimensional bodies in which solution of the bedrock has resulted in the integration of surface water and groundwater. While the field mapping and hydrologic investigations were done with conventional methods, the karst unit delineation was done using a Geographic Information System. Using this technology allowed us to examine the county's karst using two-dimensional and three-dimensional views. Many different overlays of the karst elements were combined to better understand the landscape dynamics. Ultimately, it was the overlay of the karst features, hydrologic information, and depth to bedrock mapping on a shaded relief base map that allowed us to best delineate the individual karst units.

INTRODUCTION

In southeastern Minnesota, karst feature and karst flow mapping have been going on for nearly thirty years (Giammona, C.P., 1973, Wopat, M., 1974). As part of the state's county geologic atlas program, several generations of sinkhole maps have been produced (Dalgleish and Alexander, 1984, Alexander and Maki, 1988, Witthuhn and Alexander, 1996, Green and others, 1997). These maps were produced by a combination of field survey of selected areas, air photo interpretation, information from local residents, and reviews of the available soil surveys and United States Geological Survey (USGS) topographic maps. The maps show the known sinkholes and stream sinks and gave a probability ranking for future sinkhole development. The classes ranged from no sinkhole development to sinkhole plains.

These maps have proven to be popular with state and local regulatory agencies. While you can say with certainty that sinkholes are likely to occur and that conduit flow dominates the bedrock groundwater flow system in the sinkhole plains, it is more difficult to do so in the lower probability areas. Since much of southeastern Minnesota has been under agricultural cultivation for over a hundred years, many sinkholes have been filled. The presence or absence of these features on a probability map can change the ranking and the interpretation of a site and its suitability for development. Also, the maps reflect the present hydrology; changing the hydrology at a site can result in sinkholes forming and conduits reopening. This approach also does not include the areas between the sinkhole plains and the springs which serve as the discharge points for their conduit systems. Typically, these areas have few surface karst features but can have extensive subsurface cave systems. Based on a sinkhole probability map, they would be deemed to be less susceptible to groundwater contamination, even though they have caves and active conduits below them.

METHODOLOGY

We decided to try a new approach in mapping the karst of Mower County. The goal was to provide a map for citizens and government officials that would identify located karst features and provide an interpretation of the nature of the karst flow systems. Our method was to delineate karst units, discrete three-dimensional bodies in which solution of the bedrock has resulted in the integration of surface water and groundwater. They are delineated based on bedrock type, depth to bedrock, topography, surface and subsurface hydrology, spring chemistry, and the distribution, landscape position, and type of surface and subsurface karst features. These elements combine to form a series of discrete units that make up the landscape of Mower County. Water and material enter this system, move through the carbonate bedrock, and discharge either into streams or into lower bedrock aquifers. This has resulted in the integration of the county's ground and surface waters at a variety of time scales. In the shallow bedrock units that time scale is hours to months. In the deep karst units it is on the order of decades to centuries.

The inspiration for this approach came from karst and water quality studies done in the Devonian karst of Iowa (Libra and others, 1984). Based on the presence of sinkholes and their type, they were able to describe four distinct karst units. They also found that all areas with carbonate bedrock covered by less than fifty feet of surficial material had post 1953 water, based on tritium content, and elevated levels

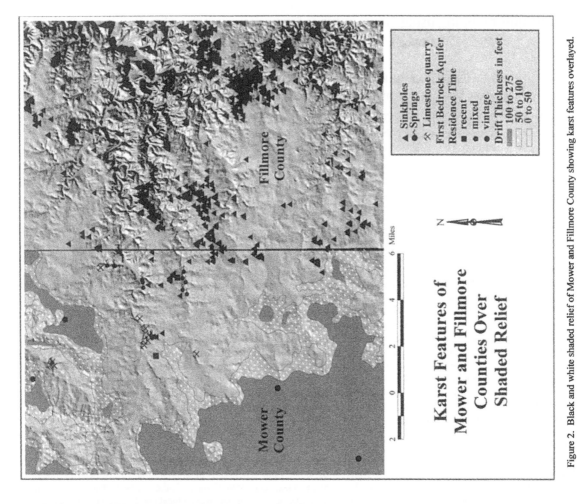

Figure 2. Black and white shaded relief of Mower and Fillmore County showing karst features overlayed.

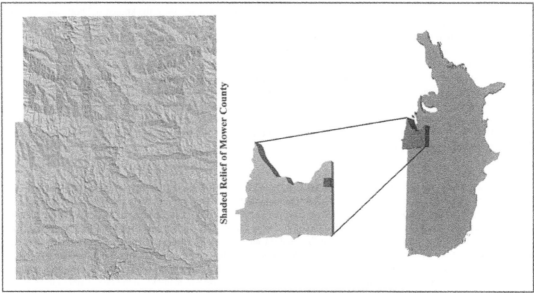

Figure 1. Mower County is located in southeast Minnesota which is located in the north central United States.

of nitrates. Using this as our basis, the first approximation of the karst units was done by examining the bedrock geology and depth to bedrock maps for the county (Mossler, 1998). Since the entire county is underlain by karst carbonate bedrock, the initial emphasis was on those areas with less than fifty feet of surficial material.

In order to delineate the karst units, a variety of data needed to be gathered. A karst feature inventory was created by field surveying as much of the county as possible. Features located included sinkholes, stream sinks, stream sieves, dry valleys, springs, and caves. While the field survey was as complete as resources allowed, it was not possible to investigate the entire area with less than fifty feet of surficial material. Additional information on sinkhole and spring locations was provided by local government sources and by county residents. They were of particular value for providing the locations of filled sinkholes. Aerial photographs were used but proved to be of limited utility. Quarries were also inventoried as karst features since every one we investigated had visible solution features present.

A limited number of dye traces were carried out, primarily in the southeastern corner of the county. Time of travel from stream sinks and sinkholes to springs was on the order of one-half to one mile per day. A similar time of travel was found in the northeastern part of the county from a stream sink to springs discharging in a large quarry. This trace was in a different bedrock unit than the traces in the southeastern part of the county. Fifteen springs across the county were also sampled for tritium. The results were categorized into pre 1953 water (vintage), post 1953 water (recent), and a mixture of pre and post 1953 water (mixed). As the springs were located it was noted if they were discharging directly from carbonate bedrock or if reducing waters were present. In our previous work in southeastern Minnesota we had found that springs discharging directly from carbonate bedrock had post 1953 water while reducing waters from springs were consistently a mix of pre and post 1953 water.

Department of Natural Resources (DNR) and local government staff had an on-going program to sample a variety of aquifers in the county. Thirty-one of those samples were from the first bedrock aquifer. These samples were broken down into pre 1953 water, post 1953 water, and a mixture of pre and post 1953 water. The samples that came from those areas with less than fifty feet of surficial material all were post 1953 water, indicating relatively recent recharge. The samples from those areas with more than fifty feet of surficial material were either all pre 1953 waters or a mixture of pre and post 1953 water. Generally, the samples that were a mixture were found in those areas that had less than one hundred feet of surficial material or were down gradient of areas with less than fifty feet of surficial material.

All of the information on karst feature locations, dye traces, spring water chemistry results, and groundwater samples were put into the ARCVIEW GIS package from Environmental Systems Research Institute Inc. (ESRI). When we converted that information into shape file coverages, we were then able to create a series of overlays. Our first step was to overlay the data assemblages on both the depth to bedrock and bedrock geology maps. This verified that sinkholes, the most common surface karst feature, were found only in those areas with less than seventy-five feet of surficial material. Then, we began a series of overlays at a variety of scales on shaded relief with a transparent depth to bedrock overlay. This allowed us to focus on those areas with less than seventy-five feet of surficial material and to still be able to look at the location of karst features in relation to landscape morphology. We were also able to join sinkhole and shaded relief maps from the county to the east providing additional insights into the dimensions of the karst units (Fig. 2).

In addition, we used ESRI's 3D ANALYST software to further our understanding of the county's karst. This package was used to create a three-dimensional view of the bedrock topography. From there, we could overlay the coverages of karst features, spring chemistry, quarry locations, groundwater chemistry, and depth to bedrock coverages. This process provided valuable insights into the distribution and occurrence of karst features. Additionally, a three-dimensional view of the karst features, shaded relief, and bedrock topography was created (Fig. 3). This view of the surface and subsurface validated our decision to describe the shallow drift areas "over" the bedrock highs as distinct karst units bounded by both transition karst (twenty-five to seventy-five feet of material over karst carbonate bedrock and groundwater residence time ranging from recent to mixed) and deep karst units (areas with more seventy-five feet of material over karst carbonate bedrock and groundwater residence time ranging from mixed to vintage).

RESULTS

A portion of the karst unit map for the county is shown in Figure 4. It covers approximately the same area of the county as shown in Figure 3. Based on our model, we have delineated a number of distinct karst units. If we had followed the existing sinkhole mapping protocol, much of this part of the county would have been mapped as either low probability (areas underlain by carbonate bedrock with greater than fifty feet of surficial material) or low to moderate probability (areas with isolated sinkhole clusters with an average of one sinkhole per square mile). The clusters of sinkholes would then have been separated out as higher probability areas. This approach is constrained by the mapper's ability to locate sinkholes and other surface karst features. Using our karst unit approach we have divided the landscape into named karst units. These units have distinct characteristics. For example, the Cedar River karst (Fig. 4) is a relatively flat plain dissected by the Cedar River and its tributaries. Surface features include sinkholes (both shallow bowls and collapse features), springs, exposed bedrock solution features and quarries. Potentiometric surface measurements have shown that groundwater flows to the Cedar River or its tributaries where it discharges from springs. Groundwater residence time at springs and wells range from mixed to recent. Vintage water likely flows from the adjoining *deep karst* units where it is then impacted by recent waters from this karst. Depths to bedrock range from zero to seventy-five feet.

Under this approach, newly formed or discovered sinkholes will not change an area's probability ranking. As new features are located, or more dye tracing is done, the description of the unit can be further refined. This model also recognizes the fact that groundwater recharge in those karst units occurs both at points (sinkhole sand stream sinks) and in areas that have shallow depths of surface material over carbonate rock. Additionally, you can account for differences between karst areas. In Mower county, the western part of the county has karst features that occur at greater depths to bedrock than those in the northeastern or southeastern parts of the county. Under this model, those differences can be used to delineate the units and be documented in their descriptions.

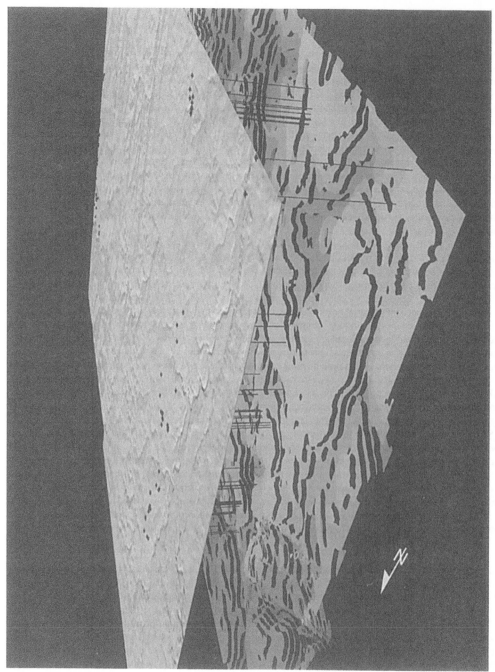

Figure 3. Three dimensional view showing a black and white shaded relief with sinkholes overlayed and projected down to a bedrock topography layer. The view is from the southwest looking towards the northeast. Bedrock topography is shown in 25 foot intervals.

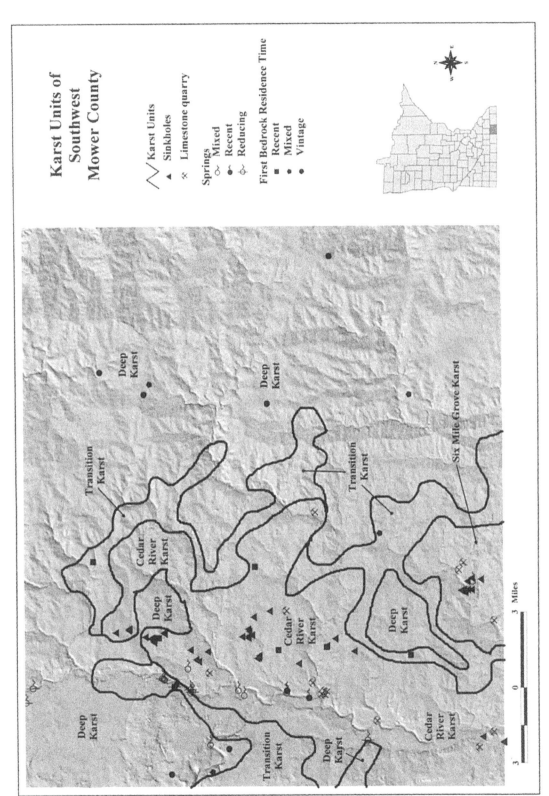

Figure 4. Karst units and karst features displayed over black and white shaded relief for southwest Mower county.

The map for the entire county will be published at a 1:100,000 scale. It will include known karst features, dye traces, and groundwater residence time information. Each karst unit's characteristics will be described. Since the map and the data coverages will be in ARCVIEW, it will be easy to keep the information up-to-date.

SUMMARY

Combining field mapping with GIS technology has allowed us to create a karst unit map. These units were delineated based on a variety of factors including karst feature locations, dye tracing, landscape morphology, depth to bedrock, and bedrock type. Converting this information into GIS coverages allowed us to more easily determine the nature and extent of the karst units.

ACKNOWLEDGMENTS

This project was a cooperative effort between the Minnesota Department of Natural Resources, the University of Minnesota Department of Geology and Geophysics, and Mower County. It would not have been possible to complete without assistance from county and other local government staff. Robert Libra of the Iowa Geological Survey provided valuable insights into the nature of the karst in this area. Dale Rossow of Dexter, MN provided us with the locations of a number of sinkholes and springs.

REFERENCES

Alexander, E. C., Jr., and Maki, G. L., 1988. Sinkholes and sinkhole probability, Plate 7 in Geologic Atlas of Olmsted County: Minnesota Geological Survey County Atlas Series C-3.

Dalgleish, J. D., and Alexander, E. C., Jr., 1984a. Sinkholes and sinkhole probability, Plate 5 in Balaban, N. H., and Olsen, B. M., eds., Geologic Atlas of Winona County: Minnesota Geological Survey County Atlas Series C-2.

Giammona, C. P., 1973, Fluorescent dye determination of groundwater movement and contamination in permeable rock strata. International Journal of Speleology, Vol. 5, pg 201-208.

Green, J. A., Mossler, J. H., Alexander, S. C., and Alexander, E. C., Jr., 1997. Karst hydrogeology of Le Roy Township, Mower County, Minnesota: Minnesota Geological Survey Open-File Report 97-2.

Libra, R. D., Hallberg, G. R., Ressmeyer, G. G., Hoyer, B. E., 1984, Groundwater quality and hydrogeology of Devonian carbonate aquifers in Floyd and Mitchell Counties, Iowa: Iowa Geological Survey Open-File Report 84-2, Pt. 1.

Mossler, J. H., 1998, Bedrock geology, pl. 2 and Depth to Bedrock and Bedrock Topography pl. 5 in Mossler, J. H., ed., Geologic atlas of Mower County, Minnesota: Minnesota Geological Survey County Atlas Series C-11, Pt. A.

Witthuhn, K. M., and Alexander, E.C. Jr., 1996. Sinkholes and sinkhole probability, Plate 8 in Faltisek, J., ed, Geologic Atlas of Fillmore County: Minnesota Department of Natural Resources County Atlas Series C-8, Pt. B.

Wopat, M., 1974. The karst of southeastern Minnesota, The Wisconsin Speleologist, Vol. 13, No.1, pg 1-47.

Geotechnical and Environmental Applications of Karst Geology and Hydrology, Beck & Herring (eds)
© 2001 Taylor & Francis, ISBN 90 5809 190 2

Wisconsin interagency karst feature reporting form

ROBERT PEARSON   The Planning and Mapping Subcommittee of the Wisconsin Groundwater Coordinating Council, Madison, WI 53705, USA, robert.pearson@dot.state

ABSTRACT

The Planning and Mapping Subcommittee of the Wisconsin Groundwater Coordinating Council (GCC) announced the creation of a karst feature reporting form and website in November 2000. Wisconsin recognizes the need for a coordinated, statewide effort to map karst features to benefit state and local agency programs, resource protection, human health, geologic mapping and land use development. Promoting the voluntary use of a standard karst reporting form is the first step of a long-term strategy for mapping Wisconsin karst.

Thick layers of glacial deposits cover about half of Wisconsin's karst. Areas that have visible karst features include the Silurian Dolomites in eastern Wisconsin and the Ordovician Dolomites located in the southern, eastern and western portions of Wisconsin. The most notable karst regions are the non-glaciated area of southwestern Wisconsin and the glacial scoured Niagara Escarpment in eastern Wisconsin, particularly the Door Peninsula.

The "Wisconsin Interagency Karst Feature Reporting Form" is provided to share ideas with others. Hopefully, use of the form, and the concept of voluntarily reporting of karst features, will provide useful information for future mapping activities.

INTRODUCTION

This paper shares an idea that the Planning and Mapping Subcommittee of the Wisconsin Ground Water Coordinating Council (GCC) is pursuing for mapping karst features. The GCC is an interagency group directed by Wisconsin statute to assist State agencies in coordinating non-regulatory programs and information exchange related to groundwater programs and issues. The membership includes staff from the Wisconsin Departments of Natural Resources (DNR); Commerce (COMM); Agriculture, Trade and Consumer Protection (DATCP); Health and Family Services (DHFS); Transportation (DOT); and the University of Wisconsin System (UWS) including the Wisconsin Geological and Natural History Survey (WGNHS). The subcommittee is responsible for identifying and coordinating the collection, use, and display of groundwater quality and quantity data.

The idea is to create a simple karst feature reporting form that is standardized and easily accessible for voluntary use by the public, private and government sectors via the Internet. State agency staff and others (e.g., environmental consultants) are encouraged to complete the form when karst features are discovered during routine field activities or projects. There is a concerted effort by the Planning and Mapping Subcommittee to promote the use of the form. Paper forms are stored in a central, publicly accessible WGNHS repository for karst information. The Subcommittee's long-term goal is to convert information in this file into an electronic database, to allow easy querying of records and development of a geographic information system (GIS) data layer for environmental or geologic mapping.

The definition of "karst features" is broadly defined because the form is intended for the general public and multi-disciplined professionals (e.g., biologists, geologists, engineers, landscape architects, planners, soil scientists, etc.). Features of primary interest include: sinkholes; caves; enlarged joints (fractures); disappearing streams and drainage (swallets); karst fens; epikarst; karst ponds; and deep subterranean solution features. Historical mines and spring locations are also included in the inventory.

PREVIOUS KARST INVENTORIES AND MAPPING

Inventorying karst features is not a unique idea. In 1985 the Pennsylvania Geologic Survey began to systematically map karst features on a countywide basis (Kochanov, 1999), and now has an online database in which sinkhole information can be queried by county location <http://www.dcnr.state.pa.us/topogeohazards/hazards.htm>. The National Park Service inventoried karst features around Mammoth Cave National Park to develop groundwater hazard maps in case of catastrophic chemical spills from nearby interstate highways (Meiman et al., 1997). Other states are also actively mapping karst features (Crawford and Webster, 1986; Hubbard, 1993; Kochanov, 1989; Panno and Weibel, 1993; Gao and Alexander, 1999).

Perhaps the first record of karst mapping in Wisconsin dates back to pre-statehood. In 1846 Alfred Brown, a land surveyor for the Western Territory, noted sinkholes in portions of Vernon County (e.g., "Sink hole 20 ft deep") which were eventually documented on the 1851 plat map for the 4th Principal Meridian in Township 14 North, Range 5 West (State of Wisconsin Board of Commissions of Public Lands, 1988). Some of the more recent soil conservation survey maps indicate "depressions, not crossable with tillage implements" (Slota, 1969) which occasionally are sinkholes.

The WGNHS maintains a paper file on caves in Wisconsin, and the Wisconsin Speleological Society periodically performed cave inventory surveys (Cronon, 1971). Private well construction logs or boring logs for subsurface investigations (e.g., landfill feasibility studies) sometimes note evidence of other subterranean karst.

Karst feature mapping is common for priority watershed studies in northeastern Wisconsin (Stieglitz and Johnson, 1986; Stieglitz and Dueppen, 1994; Stieglitz and Dueppen, 1995). Some sinkholes are mapped in southwestern Wisconsin (Day and Reeder, 1989). There is a karst map and a closed depression map for Pierce and St. Croix counties, respectively (Baker et al., 1990; Baker et al., 1991). These counties are located on the western border of Wisconsin and represent the northwest edge of Wisconsin's karst region.

Recently the Wisconsin DOT began mapping karst features for select highway projects (Burkel and Pearson, 1999; Swartz and Pearson, 2000). Old highway construction plan sheets or field notes occasionally document sinkholes, caves or mines (Pearson, 2000).

Several University professors and students have authored papers or reports with excellent karst information and local mapping. The schools most active in karst related research are UW-Green Bay, UW-Milwaukee, UW-River Falls, and UW-Madison.

WISCONSIN KARST OVERVIEW

About half of Wisconsin's karst is masked by thick layers of glacial deposits, particularly in eastern portions of the state. Figure 1 shows areas with visible, shallow karst (<2 meters below ground surface). This karst potential map combines bedrock groups and formations of Silurian and Ordovician carbonate deposits (dolomites). Wisconsin has 72 counties, and 23 of them are expected to contain evidence of near surface karst. Listed from west to east these counties are: St Croix, Pierce, Pepin, Buffalo, Trempealeau, La Crosse, Vernon, Monroe, Crawford, Grant, Richland, Iowa, Lafayette, Sauk, Green, Dane, Dodge, Jefferson, Waukesha, Calumet, Brown, Kewaunee, and Door Counties.

The dominant karst development process in Wisconsin is not thoroughly understood, and the regional history of multiple glaciation further complicates matters. Presumably much of Wisconsin's paleokarst has been obliterated or affected by glacial scouring and loading, drainage of meltwater, and exposure to repeated periglacial environments. Post-glacial weathering, recent topographic drainage, and human activity continue to influence Wisconsin's karst.

Some karst features are dominated by structural controls. For example, numerous solution-enlarged joints are preferentially oriented along regional joint sets found on the Niagara Escarpment of the Door Peninsula. These vertical, solution-enlarged joints tend to be several meters long and some exceed 100 meters. Joint widths are usually <1 meter and tend to narrow with depth. Joint depths are often <2 meters, but can be several meters. Other karst features appear to be influenced by stratigraphic position, bedding planes and changes in depositional facies within distinct bedrock formations.

Sinkholes in southwest Wisconsin tend to be funnel-shaped whereas in the northeast they seem to be elongated. Wisconsin sinkholes generally range from 5 to 10 meters wide and 2 to 6 meters deep. Larger sinkholes certainly exist, and several sinkholes serve as entry points for deeper caves.

Wisconsin caves are smaller and less extensive than caves found in Kentucky or Tennessee. Any void space large enough to be entered by a human is classified as a cave. The number of caves in Wisconsin is unknown, and estimates vary from 200 to 400. Most caves are small, usually <150 meters long and none reported are greater than 600 meters long (Schultz, 1986). The counties with the greatest number of reported caves include Dane (43), Door (39), Iowa (34), and Richland (48) (Cronon, 1971). Wisconsin contains five commercial caves.

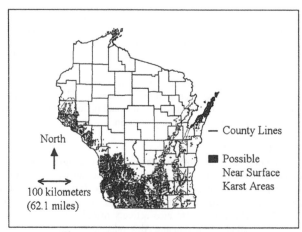

Figure 1. Wisconsin Karst Potential Map

WISCONSIN INTERAGENCY KARST FEATURE REPORTING FORM

This form is used to report the locations of "karst" features such as caves, sinkholes, enlarged fractures, disappearing streams or other surface drainage, and springs. Old/abandoned mine shafts are also included.

Please mail or FAX completed form to:
Karst Information File
WISCONSIN GEOLOGICAL & NATURAL HISTORY SURVEY
3817 MINERAL POINT RD
MADISON WI 53705-5100   fax: (608) 262-8086   phone: (608) 262-1705

WGNHS use only:
Form Received Date: _____
Database Entry Date: _____

(circle one)   Gov
County:_____ T. _____N R._____E or W _____¼ of _____¼ of Sec.__ Lot:_____

Fire #/Street Address:_____

City/Town/Village:_____ Zip Code: _____

Topographic Quad/Map Name:_____

Landscape Area (✓ *all that apply*)
☐ rural ☐ industrial
☐ urban ☐ highway
☐ other *(describe)*

Reporter Name (Last, First):_____ Reporter Phone:_____

Employer/Occupation:_____ Reporter Email:_____

Field Observation Date:_____ Reporting Date: _____

Property Owner Name:_____

Property Owner Address: _____ Owner Phone:_____

Feature Arrangement

(✓ all that apply)
☐ isolated features
☐ cluster of features

total # of features = _____

Feature Type (✓ *all that apply*)
☐ sinkhole
☐ enlarged fracture
☐ cave
☐ spring
☐ mine
☐ other *(describe)*

Concern (✓ *all that apply*)
☐ soil loss/erosion
☐ water quality
☐ collapse/safety
☐ endangered species
☐ flooding
☐ other *(describe)*

Karst Classification* *(enter one from list on back of this form):*_____

Shape (✓ *one*)
☐ circular
☐ elongate
☐ linear
☐ other *(describe)*

Size *(enter all that apply & circle unit)*
length: _____ feet / meter
width:_____ feet / meter
depth:_____ feet / meter
diameter:_____ feet / meter

Feature orientation *(compass)*: _____
Is feature open? ☐Y   ☐N
Is feature filled? ☐Y   ☐N
fill material:_____

Evidence of surface drainage into feature? ☐Y ☐N
Feature may receive polluted drainage? ☐Y ☐N

Drainage Area Size in Acres (✓ *one*)
☐ < 1   ☐ 1 – 10   ☐ > 10

Nearby Land Use & Estimated Distance (✓*all that apply, then enter distance & measurement unit*)
☐ high capacity well – municipal:_____
☐ high capacity well – agricultural:_____
☐ high capacity well – industrial:_____
☐ high capacity well – other:_____
☐ gasoline service station:_____
☐ animal waste lagoon:_____
☐ sanitary plant lagoon:_____
☐ storm water detention pond:_____
☐ chemical storage:_____

☐ house:_____
☐ building:_____
☐ septic field:_____
☐ quarry:_____
☐ gravel pit:_____
☐ potable well:_____
☐ salt storage shed:_____
☐ highway/street pavement:_____

☐ landfill:_____
☐ dam:_____
☐ farm field:_____
☐ parking lot:_____
☐ livestock pen:_____
☐ irrigation ditch:_____
☐ cemetery:_____
☐ other (describe):_____

Feature mapped as ... (✓ one)   X-coordinate   Y-coordinates
☐ point   ☐ line   ☐ area   *(e.g., Long., Easting)*   *(e.g., Lat., Northing)*

Referencing System:_____   _____   _____
(e.g., WTM, Lat/Long, State Plane, UTM, county system)   _____   _____

                                                   _____   _____
Datum (or Spheroid for Lat/Long):_____   _____
(e.g., datums = NAD91, NAD27; spheroids = WGS84, GRS80)   (see note* below)

***Attach another sheet or a diskette with an ASCII file of coordinates or ArcView shapefile, if more than four
x-y coordinates are collected for the karst feature(s) described on this form.***

Comments:_____
_____
_____
_____
_____
_____
_____
_____

Feature Drawing: *Plan view sketch should include: nearby landmarks (e.g., roads, fences, buildings),
approximate scale, north arrow, cross-section (if appropriate). Attach photos or other reference maps as needed.*

Karst Classification List:

Sinkhole: a topographic depression (unless filled) in which bedrock is dissolved or collapsed. Sinkholes may be open, covered, buried, or partially filled with soil, field stones, vegetation, weathered bedrock, water or other micellaneous debris. Sinkholes are usually circular, funnel-shaped, or elongated. Sinkhole dimensions vary by region. Wisconsin sinkholes generally range between 20 to 30 feet in diameter and 4 to 10 feet deep, although some can be wider and/or deeper.
Enlarged Fracture: solution enlarged or widened bedrock fracture that usually narrows with depth.
Pavement: extensive bare areas of exposed bedrock surfaces with many enlarged fractures or sinkhole features.
Fracture Trace: linear feature, including stream segment, vegetative trend and soil tonal alignment.
Spring/Seep: intermittent or permanent seepage of water from ground surface or bedrock outcrop or karst area.
Cave: natural cavity, large enough to be entered, which is connected to subsurface passages in bedrock.
Karst Pond: closed depression in a karst area containing standing water.
Swallet: a place where surface or storm water drainage disappears underground.
Karst Fen: marsh formed by plants overgrowing a karst lake or seepage area.
Mine Feature: a man-made shaft, tunnel, cave, hole, or other feature created for mining purposes.

KARST REPORTING FORM

A copy of the "Wisconsin Interagency Karst Feature Reporting Form" is provided before the reference section. The form captures general information on feature location, landscape area, feature arrangement, feature type, and possible issues of concern. Users can quickly review records for a local area (e.g., county or watershed), and get a relative sense of the karst setting. This information can be used for making general maps or identifying potential land use issues.

The latter sections of the form provide the opportunity to collect very specific information regarding karst characteristics, nearby land use, surveyed locations, and descriptive or visual documentation. These sections are valuable for detailed mapping purposes (e.g., watershed basin studies, highway design, site selection for waste storage, preservation plans for endangered resources, flood control plans, etc).

KARST WEBSITE

To support the use of the karst feature reporting form, the WGNHS maintains a karst information page on their website. This includes definitions and photographs of Wisconsin karst, and a preliminary karst potential map. The inventory form can be downloaded from the website as a pdf file (http://www.uwex.edu/wgnhs/karst.htm).

CLOSING REMARKS

Karst features can serve as direct and rapid conduits to valuable groundwater aquifers, wells, springs and streams. These features are very important when planning groundwater and surface water resource protection strategies. In addition to water quality concerns, karst landscapes can create unique construction, drainage, safety and liability issues. For example, foundation failures beneath buildings, roads, or sanitary lagoons occur periodically in karst settings.

The time has come to collect past, present, and future karst information into one central location for statewide mapping in Wisconsin. This mapping is needed for purposes such as watershed management, source water protection, storm water management, erosion control, highway and bridge construction, land use development, farm production, endangered resource protection, social history, and geologic research.

The Planning and Mapping Subcommittee of the Wisconsin Ground Water Coordinating Council solicits comments on their form and the idea of voluntarily reporting karst features. Comments, questions, and recommendations should be directed to Robert Pearson at (608) 266-7980 <robert.pearson@dot.state.wi.us>.

ACKNOWLEDGEMENTS

The Planning and Mapping Subcommittee of the Wisconsin Groundwater Coordinating Council includes: Lisa Morrison – chairperson (DNR), Tim Asplund (DNR), Steve Born (UWS), Cody Cook (DATCP), Ron Hennings (WGNHS), Leroy Jansky (Commerce), Michael Lemke (DNR), Robert Pearson (DOT), James VandenBrook (DATCP), and Chuck Warzecha (DHFS). Mindy James (WGNHS) is responsible for creating and maintaining the website which the subcommittee greatly appreciates.

REFERENCES

Alexander, E.C., Jr., 1980, Geology field trip: in Alexander, E.C., Jr., ed., An Introduction to Caves of Minnesota, Iowa, and Wisconsin – Guidebook for the 1980 National Speleological Society Convention, p. 143-190.

Baker, R.W., Bauer, E.J., Huffman, S.F., and Hass, E., 1990, Karst Map of Pierce County, Wisconsin. Wisconsin Geologic and Natural History Survey, Map Series 90-1.

Baker, R.W., Hughes, M, Huffman, S.F, and Nelson, A., 1991, Closed Depression Map of St. Croix County, Wisconsin. Wisconsin Geologic and Natural History Survey, Map Series 91-93.

Burkel, R. S., and Pearson, R.E., 1999, Karst feature mapping for Wisconsin state highway 57 expansion project (northeast Wisconsin, Door Peninsula): Seventh Multidisciplinary Conference on Sinkholes and the Engineering and Environmental Impacts of Karst, Harrisburg, Pennsylvania, April 10-14, 1999, Poster Session.

Crawford, N.C., and Webster, J., 1986, Karst hazard assessment of Kentucky - sinkhole flooding and sinkhole collapse: map prepared by the Center for Cave and Karst Studies, Western Kentucky University, for U.S. Environmental Protection Agency, Region IV.

Cronon, W., 1971, Wisconsin Cave Survey: the view from here. Wisconsin Speleological Society, Volume 10, Number 3, p. 77-88.

Day, M.J., and Reeder, P.R., 1989, Sinkholes and Land Use in Southwestern Wisconsin: Proceedings of the Third Multidisciplinary Conference on Sinkholes and the Engineering and Environmental Impacts of Karst, St. Petersburg Beach Florida, 107 - 113. (Sponsored by the Florida Sinkhole Research Institute, University of Central Florida, Orlando).

Gao, Y., and Alexander, E.C., Jr., 1999, Sinkhole distribution in southeastern Minnesota: extending GIS-based analysis and operations to support resource management: 44th Annual Midwest Groundwater Conference, St. Paul, Minnesota, October 13-15, p. 40.

Hubbard, D.A., 1993, Status report on the Virginia karst mapping program: Proceedings of the Fourth Multidisciplinary Conference on Sinkholes and the Engineering and Environmental Impacts of Karst, Panama City, Florida, January 25-27, p.

Kochanov, W.E., 1989, Karst mapping and applications to regional land management practices in the Commonwealth of Pennsylvania: Beck, B.F., editor, Engineering and Environmental Impacts of Sinkholes and Karst, Proceedings of the Third Multidisciplinary Conference, St. Petersburg Beach, Florida, October 2-4, p. 363-368.

Kochanov, W.E., 1999, The integration of sinkhole data and geographic information systems: An application of using ArcView 3.1: Digital Mapping Techniques '99 - Workshop Proceedings, Madison, Wisconsin, May 19-22, p. 183. U.S. Geological Survey Open-File Report 99-386 1999.

Meiman J., et. al., 1997, Completion and distribution of the groundwater hazard map of the turnhole spring karst groundwater basin. GRIST, vol. 41, number 1, U.S. Department of the Interior, National Park Service.

Schultz, G.M., 1986, Wisconsin's Foundations: A Review of the State's Geology and it's Influence on Geography and Human Activity. Kendall/Hunt, 210 p.

Slota, R.W., 1969, Soil survey of Vernon County, Wisconsin, USDA Soil Conservation Service, 82 p.

State of Wisconsin Board of Commissions of Public Lands, 1988, Original Plat Map Viewer Software Publication.

Stieglitz, R.D., and Johnson, S., 1986, Mapping and Inventorying of Geologic Features in the Upper Door Priority Watershed: Technical Report prepared for Door County, Wisconsin, 44 p.

Stieglitz, R.D., and Dueppen, T.J., 1994, Mapping and Inventorying of Geologic Features in the Sturgeon Bay-Red River Priority Watershed: Technical Report for University of Wisconsin-Green Bay; 85 p.

Stieglitz, R.D., and Dueppen, T.J., 1995, Mapping and Inventorying of Geologic Features in the Branch River Priority Watershed: Technical Report prepared for Brown and Manitowoc Counties, Wisconsin, 64 p.

Swartz and Pearson, 2000, Wisconsin Department of Transportation Sinkhole Mapping for USH 14/61 Vernon County – in progress.

Panno, S.V., and Weibel, C.P., 1993, Mapping of karst areas in Illinois: Proceedings of the Third Annual Conference, Illinois Groundwater Consortium, March 31 - April 1, Makanda, p. 259-69

Pearson, R.E., 2000, Wisconsin Department of Transportation Karst Inventory Records and Contract Special Provision Language.

IV Evaluating the Risk of and Remediating the Damage from Sinkhole Collapse

Geotechnical and Environmental Applications of Karst Geology and Hydrology, Beck & Herring (eds)
© 2001 Taylor & Francis, ISBN 90 5809 190 2

Characterization of a highway sinkhole within the gypsum karst of Michigan

RICHARD C.BENSON & RONALD D.KAUFMANN   Technos, Inc., Miami, FL 33142, USA, info@technos-inc.com

ABSTRACT

A small, five-foot diameter sinkhole formed within the southbound lane of U.S. Highway 23 in Iosco County, Michigan. The highway is parallel to Lake Huron and is adjacent to a gypsum quarry operated by U.S. Gypsum. There are reports of three other sinkholes in the area surrounding the U.S. Gypsum Quarry.

Geologic conditions consist of sand and clay over gypsum bedrock occurring at a depth of approximately 40 feet. While sinkholes and other karst features commonly occur due to dissolution of limestone, they can also occur in gypsum, and have been documented within the Michigan Basin.

A two-phased karst investigation was carried out that included geologic observations, detailed surface geophysical measurements, borings and geophysical logging. Efforts to characterize subsurface conditions were focused along a 2000-foot stretch of U.S. Highway 23 extending 1000 feet south and north of the sinkhole.

Surface subsidence along U.S. Highway 23 is caused by piping of unconsolidated materials into deeper cavity zones within the gypsum. The data do not indicate large cavities, but suggest the presence of smaller cavity zones within dissolution-enlarged bedding planes and fractures. The soil piping at the sinkhole on U.S. Highway 23 probably extends eastward under the adjacent southbound lane but does not extend to the west of the road. Areas of subsidence risk along U.S. Highway 23 were also identified and recommendations for remediation and long term monitoring were made.

BACKGROUND

A small sinkhole occurred in the western-most southbound lane of U.S. Highway 23 in Iosco County, Michigan (Figure 1). Figure 2 is a photograph of the small sinkhole. There have also been reports of a recent sinkhole on a nearby road about 1,200 feet northwest of the sinkhole on U.S. Highway 23, and two older sinkholes approximately 3,500 feet north of the sinkhole on U.S. Highway 23. Figure 1 shows the site location along with the location of these four sinkholes.

A closed portion of the US Gypsum Quarry lies west of U.S. Highway 23 between 2,000 and 7,000 feet north of the sinkhole. The quarry began operation in 1870 but has not been in operation since 1950. However, lower water level elevations have been maintained within the quarry.

PURPOSE AND SCOPE

The primary purpose of this karst investigation was to determine the cause of the sinkhole problem. Detailed measurements were acquired along U.S. Highway 23 and extended 1000 feet south and north of the sinkhole. Reconnaissance measurements were made along a longer, 1.8 mile, stretch of U.S. Highway 23 to identify other potential problem areas near the quarry. The investigation was divided into two phases. Recommendations were provided to Michigan DOT for remediation and long-term monitoring based upon the results of this two-phased investigation.

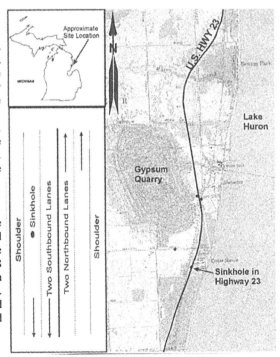

Figure 1. Site location map

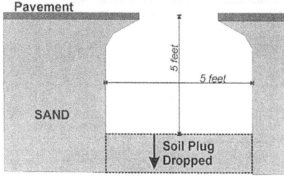

Figure 2. Photos of sinkhole and cross-section sketch

Phase I

The goal of Phase I was to characterize subsurface conditions and develop a basic conceptual model of karst conditions along U.S. Highway 23. A systematic approach was used in which regional geologic observations were integrated with existing data and geophysical measurements. Each component of Phase I was chosen to characterize regional and local geologic conditions and fulfill specific objectives (Table 1). Geophysical measurements, including ground penetrating radar, electromagnetic, and resistivity imaging were chosen to measure appropriate geologic properties to adequate depths of investigation (Figure 3).

Phase II

Additional radar and resistivity imaging measurements were made during Phase II along with microgravity measurements, borings and geophysical logging to better define the lateral and vertical boundaries of karst conditions (Table 1).

The locations of the borings and additional measurements were guided by the Phase I results, which allowed for an efficient and complete detailed investigation. The high- density data provided information on the size, depth, distribution and stability of cavities below and surrounding the roadway. Characteristic anomalies in the geophysical data over the known sinkhole allowed an optimum reconnaissance approach to be developed to identify other potential problem areas.

RESULTS
Geology
Regional Geologic Setting

Bedrock at the site consists of Michigan Formation gypsum deposits of late Mississippian Age, which were deposited within the Michigan Basin. These deposits outcrop below the glacial drift and consist of a series of gypsum beds with interbeded layers of sands and alternating layers of shale. The Michigan Formation dips gently at the estimated rate of 35 feet per mile in a southwesterly direction. This formation is mined at a number of locations within Iosco County and elsewhere in Michigan.

Regional Lineaments

A 1988 satellite photo was examined for evidence of regional lineaments that may be related to geologic structure and may infer regional fracture trends through the area. Two major lineaments were found to intersect within the sinkhole area on U.S. Highway 23.

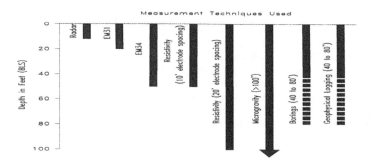

Figure 3. Depth of geophysical measurements

Table 1

Method	Parameter Measured	Depth of Measurements (feet)	Phase I Objective	Phase II Objective
Regional geologic literature	NA	NA	Background geologic data	---
Wells, Driller's Logs	Geologic strata	Few 100 feet	Local geologic data. Character and thickness of unconsolidated and depth to rock	---
Personal communication with U.S. Gypsum Staff	NA	NA	Local quarry history and details of gypsum	---
Satellite Imaging	Linear surface trends	Surface	Identification of lineaments (fracture trends) associated with geology	---
Observations	Regional	Surface	Karst surface conditions Burleight Township, 12 miles to the west	---
	Site specific	Surface	Karst surface conditions	---
Ground penetrating radar	Dielectric constant	10	Shallow high resolution data to identify soil piping or slumping	Expand spatial extent of data
EM31	Electrical conductivity	20	Lateral changes in unconsolidated material to 20 feet	---
EM34	Electrical conductivity	50	Lateral changes in unconsolidated material to 50 feet	---
2D Resistivity	Electrical resistivity	50	Details within unconsolidated material around the sinkhole	---
		100	Initial assessment of rock conditions	Additional assessment of unconsolidated material and rock conditions north and south of sinkhole
Gravity	Density	>100	---	Assess size and distribution of low density zones or open voids
Soil Borings	Samples of sand and clay strata Blow counts	40	---	Determine sand/clay strata Determine loose zones
Core Borings	Rock strata/RQD	80	---	Determine rock strata Determine weak rock and voids
Geophysical logging	Natural gamma	80	---	Clay and shale stratification
	Electrical conductivity	80	---	Clay and shale stratification
Water levels	Relative water level elevation	NA	---	Indication of groundwater flow direction

Regional Evidence For Karst

While sinkholes, cavities, and caves commonly occur due to dissolution of limestone, they can also occur in gypsum (Klimchouk, et al., 2000; Klimchouk, 2000, Ford and William, 1989). A cave was encountered during normal underground mining operations in the Grand Rapids Gypsum Company mine in July 1976 about 150 miles southwest of this site. The cave was approximately 100 meters long with sizeable rooms and passageways. The cave and mine are located in the same stratigraphic gypsum beds (Michigan Formation) in which the nearby US Gypsum mine is working (Elowski and Ostrander, 1977).

Other known gypsum related karst in the state that have been reported in literature are prominently exhibited in Burleight Township in the southwestern corner of Iosco County, about 12 miles west of the site. A linear string of collapse features consisting of several sinkholes and sink valleys, some dry and some water-filled, can be observed. The most recent collapse was reported on the White farm in early spring of 1968. Many other depressions, closed topographic lows, and internal drainage features indicate that this area is the locale for many other active karst systems (Elowski and Ostrander, 1977). These karst areas are also due to the solution of gypsum of the Michigan Formation.

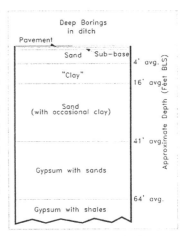

Figure 4. Simplified geologic cross-section (based upon boring logs)

Site Specific Geology

A generalized geologic section was developed from the boring logs obtained along U.S. Highway 23. Figure 4 shows a simplified geologic section constructed from the eight deeper borings made in the ditch just west of U.S. Highway 23.

The geologic strata encountered in each of the eight borings consists of:

- A layer of mostly sands to an average depth of 4 feet;
- A layer of mostly clays from 4 to 16 feet below grade;
- A layer of sand (with occasional clay) from 16 to 41 feet below grade;
- A layer of gypsum (with sand) from 41 to 64 feet below grade; and
- A layer of gypsum and shale from 64 feet to the maximum depth of the borings, 84 to 90 feet below grade.

The nearby US Gypsum mine (Figure 1) has 40 to 60 feet of overburden and excavates to depths to 70 to 75 feet. The upper zone of gypsum can be up to 20 feet thick. Mine operations' staff at US Gypsum have indicated that the lateral boundaries of mineable gypsum are often eroded and that widened fractures are often found filled with sands. They are not aware of any open cavities exposed during mining.

Ground Penetrating Radar Data

At this site, we are using radar to identify lateral changes in shallow soil conditions and to locate areas of soil piping. A GSSI SIR-2 ground penetrating radar system was used with a 100 MHz antenna pulled along the ground with an ATV. The data were digitally recorded with a time (depth) window set to 100 nS (approximately 12 feet deep).

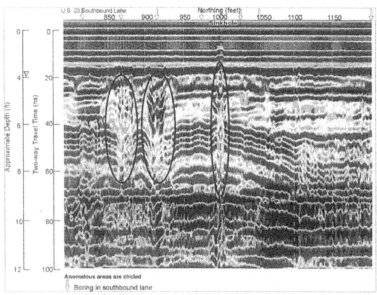

Figure 5. Example of radar data over sinkhole in southbound lane

The ground penetrating radar provided good quality data in the sandy soils to depths of up to 10 feet. Figure 5 shows an example of the ground penetrating radar data acquired in the westernmost southbound lane of U.S. Highway 23 over the sinkhole. Continuous layering of soils exist between Stations 700N and 1100N at a depth of 6 to 8 feet. Clear evidence of soil piping breaching the otherwise continuous layers exist at the location of the sinkhole (Station 1000N) and at Stations 875N and 915N (see circled features, Figure 5). The soil piping centered at the sinkhole extends to near the surface, while the other two larger soil piping features (Stations 875N and 915N) appear to end greater than 4 feet below the surface.

Geologic information from the borings in the southbound lane can be compared with the radar data in Figure 5. The borings show that the otherwise continuous clay layer has been breached in some areas. Loose zones are evident in the shallow sand layer (about 2 to 6 feet deep) in all the borings and in deeper clay layers within the boring at the sinkhole.

The radar data indicate that the soil piping features extend throughout the width of the westernmost southbound lane, and may continue to the east under the southbound lane. The soil piping associated with the sinkhole does not extend west to the southbound shoulder; however, the soil piping at Stations 875N and 915N may extend to the west of the road shoulder

Electromagnetic Data

Electromagnetic measurements were used to identify changes in conductivity related to changes in the unconsolidated materials. Electromagnetic data were acquired using a Geonics EM31 system, which measures electrical conductivity to a maximum depth of 20 feet, and a Geonics EM34 system which measures electrical conductivity to a depth of 50 feet. The EM34 system was mounted on a hand-carried boom so that measurements could be made continuously.

The EM31 and the EM34 data were acquired along the southbound and northbound shoulder. Data from the southbound shoulder are shown in Figure 6. Both the EM31 and the EM34 data show periodic changes which imply a relative change in clay concentration (lower values indicate less clay and higher values indicate more clay).

The spatial periodicity in these data ranges from 100 to 200 feet and averages about 100 feet. Both EM measurements also show generally increasing values north of Station 1400N.

The deeper EM34 data have slightly higher conductivity values than the shallower EM31, which suggests the average bulk conductivity (clay content) increases with deeper measurements. This is consistent with the geologic model (Figure 4).

The sinkhole along U.S. Highway 23 (Station 1000N) occurs in an area of lower conductivity (less clay - see Figure 6). Similarly, the two areas of soil piping identified in the ground penetrating radar data (Figure 5) at Stations 875N and 915N also occur in areas of lower conductivity values (less clay - see Figure 6). Both EM31 and EM34 values generally increase north of Station 1400N, which correlates to the trend observed along the southbound shoulder (See average trend line drawn through the data in Figure 6).

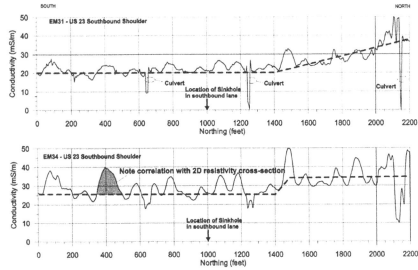

Figure 6. EM31 and EM34 data along U.S. Highway 23 southbound (Stations 0 to 2200)

The EM31 and EM34 data (Figure 6) acquired along the northbound shoulder have lower average conductivity values and show less variation than those along the southbound shoulder. This may suggest that the effect of dissolution and resulting subsidence may decrease to the east, further from the US Gypsum Quarry. The values generally increase north of Station 1400N, which correlates to the trend observed along the southbound shoulder.

2D Resistivity Data

Resistivity imaging measurements were used to characterize soil and rock stratigraphy to depths of up to 100 feet. An AGI Sting-Swift system with 28 electrodes was used with electrodes in a dipole-dipole array. The dipole-dipole array is very sensitive to horizontal changes in resistivity, and therefore is the best geometry to use for mapping vertical structures or isolated features such as fractures. Data were acquired along lines located in the ditches paralleling both sides of the highway to determine the lateral continuity of geologic conditions across the highway in the vicinity of the sinkhole.

The resistivity data contain values that are representative of the site geology. Generally, low resistivity values (less than 50 ohm-meters) are interpreted as clay and greater saturation. Higher resistivity values in the unconsolidated material are due to less clay content and higher sand content. The top-of-rock (gypsum) is defined by a gradient (100 to 150 ohm-meters) separating low resistivity unconsolidated material from the high resistivity unweathered gypsum.

Figure 7 shows the resistivity data acquired along the ditch adjacent to the southbound shoulder of U.S. Highway 23. The shallow, high-resolution data centered at the sinkhole show a shallow zone (less than 5 feet) of high resistivity material (dry sand) underlain by a much lower resistivity material between 5 and 10 feet deep (probably related to a combination of perched water and clay). A high degree of lateral and vertical variability exists between 10 and 40 feet, which is due to variations in the sand and clay content.

There are subtle indications of localized more sandy material (higher resistivity) centered at Stations 840N and 1000N, at a depth of 20 feet and at Station 920N at a depth of 40 feet. These sandy areas correlate with the location of the sinkhole, as well as the two soil piping features seen in the radar data (Figure 5).

The top of gypsum occurs at a depth of about 40 feet, where resistivity values start to increase. The upper zone of gypsum may be highly weathered with lower resistivity values (100 to 150 ohm-meters), while the deeper unweathered gypsum has much higher resistivity values (more than 300 ohm-meters).

Significant lateral changes in resistivity occur within the gypsum. Trends of lower resistivity values centered at Stations 420N, 870N, and 1180N are interpreted as fracture zones (dissolution-enlarged joints) within the gypsum in which cavities could form. Note that the sinkhole at Station 1000N is centered between two of these fracture zones. North of Station 1400N, the resistivity values become significantly lower, which probably indicates an area where gypsum has been eroded and filled with glacial till (similar to the gypsum deposit currently being mined by US Gypsum). This area also correlates with the trend of higher conductivity values observed in the EM data (Figure 6). Also note the correlation between the large fracture zone at Station 400 in the resistivity data (Figure 7) and the broad high conductivity identified in Figure 6.

Trends observed in the gypsum along the southbound cross-section appear to be continuous across to the northbound cross-section. Lower resistivity zones interpreted as fracture zones in the gypsum are evident at Stations 850N and 1170N in the northbound data. Also, a zone of deep, low resistivity values exists north of Station 1400N, correlating with the southbound resistivity data.

Microgravity Data

Microgravity data were obtained to determine if there are local areas of significantly lower density (possible cavities) beneath the roadway (i.e. are there larger localized cavities, or are the cavities smaller and more uniformly distributed over the site). Microgravity data were acquired at a 25-foot spacing along the southbound shoulder of U.S. Highway 23 between Stations 0N and 2200N. In addition to these measurements, closer spaced measurements (5 to 10 feet) were acquired directly over the sinkhole in the southbound lane. A Scintrex CG-3M gravimeter was used for this work. The raw data were reduced to Bouguer values using standard formulas and presented as profile lines.

Figure 7 shows the microgravity data acquired along the southbound shoulder of U.S. Highway 23. Two spatially broad zones of gravity lows exist in the data. These lows lie between Stations 200N and 675N and between Stations 900N and 1425N. At their

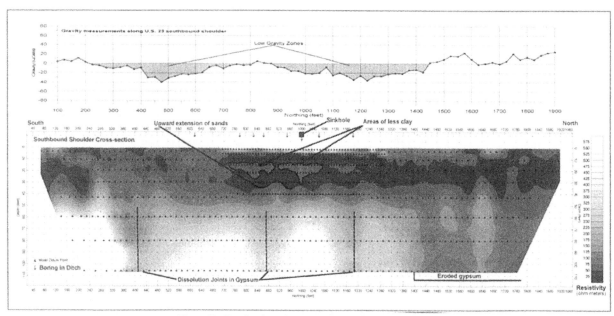

Figure 7. Gravity data shown with resistivity cross-section

center, the lows are approximately 40 μGals below the average value. The spatially broad nature of these anomalies indicates that they are either due to features within the bedrock, such as zones of greater weathering and fracturing, or due to broad shallow zones of low density.

North of Station 1425N, the gravity values gradually increase. This trend correlates with the change from gypsum to glacial till interpreted from the resistivity data at this location (Figure 7) and the EM data (Figure 6). It is likely that saturated clays in-filling eroded bedrock are slightly more dense than the weathered gypsum to the south.

To aid in resolving the cause of the gravity anomalies, forward models (Figure 8) were used to determine the size and shape of gravity anomalies caused by three possible geologic conditions beneath the highway:

1. A 5-foot wide zone of soil piping in the upper 40 feet of unconsolidated material (Figure 8a). This model results in a narrow (less than 10 feet) gravity anomaly of about 40 μGals.
2. A 50-foot wide fracture zone within the gypsum extending to a depth of 100 feet (Figure 8b). This model results in an 18 μGal anomaly with a full width of about 200 feet.
3. A 300-foot wide weathered zone in the upper portion of the gypsum (Figure 8c). This model results in a 30 μGal anomaly that extends over a wide, 400-foot area.

Note: These forward gravity models represent simplified subsurface conditions and average density values obtained from Telford et al., 1976.

Using the results of the forward gravity models, resistivity imaging cross-section, and borehole information as a guide, a gravity model was developed that provides a reasonable fit to the data obtained along the southbound shoulder of U.S. Highway 23 (Figure 9). The spatially-broad anomalies are modeled as a weathered zone in the upper portion of the gypsum with vertical fractures extending to a depth of 100 feet. North of Station 1425N, the trend of higher gravity is modeled as eroded gypsum filled with glacial till to a depth of 100 feet.

The main result of the microgravity survey is that there are no large cavities beneath the highway. The models suggest that the gravity anomalies are due to relatively small, low-density zones or cavities distributed within the epikarst zone and within vertical fractures.

Boring Logs

Boring locations were based upon the Phase I data and selected in both worst-case conditions (with loose sands and in fracture zones with dissolution) and "background" conditions (where soils were not piping and rock is massive with little dissolution). A total of sixteen (16) borings were drilled based on the results of Phase I work. Eight of these borings are located in the southbound lane of U.S. Highway 23 and eight are located in the ditch 10 feet west of the southbound shoulder. Figure 7 shows the eight borehole locations within the ditch (see arrows), which were used to provide detailed geologic data in this area.

All the borings in the southbound lane were drilled, geologically logged (by MDOT personnel), plugged and abandoned. The borings in the adjacent ditch were geologically logged (by MDOT personnel) and temporally cased with 3-inch PVC pipe within the unconsolidated materials. Four of these casings were open or had a slotted screen at the bottom for water level measurements. The drillers' logs and blow counts (SPT) for the borings located in the ditch west of the southbound shoulder are shown as cross-sections in Figures 10a and 10b.

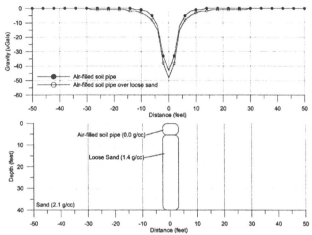

A. A vertical joint infilled with loose sand

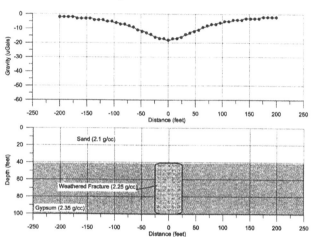

B: A vertical joint with weathered gypsum

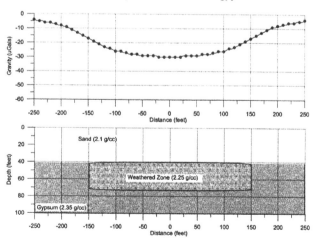

C: A weathered zone at the top of the gypsum

Figure 8. Three forward gravity models used to interpret the field gravity data.

The unconsolidated material consists of a shallow layer of sand over a layer of clay extending to an average depth of 16 feet. Fine sands with occasional clay occur between the clay layer (16 feet) and the top of gypsum (41 feet). The blow counts (Figure 10b) in the upper sand and clay layers are consistently below 20 bpf and increase from 20 to over 100 bpf in the deeper, fine sand layer. The SPT values generally indicate laterally uniform conditions. The boring in the ditch west of the sinkhole at Station 1000N shows relatively normal blow counts. This contrasts with the boring in the southbound lane over the sinkhole at Station 1000N, which shows very loose conditions to a depth of 50 feet. The contrasting blow counts indicate that the soil piping associated with the sinkhole does not extend west of the highway. The boring at Station 950N shows the lowest blow counts in the fine sand layer, averaging 30 bpf, and may be due to an extension of the soil piping observed at Station 915N in the radar data.

The depth to gypsum is at consistent average depth of 41 feet. Three of the borings [Stations 865N (Borehole #5), 1000N (Borehole #10), and 1185N (Borehole #15)] extend approximately 80 feet into the gypsum. These three deeper borings indicate that lenses of saturated sand are present within the gypsum to depths of 64 feet, below which lenses of shale are present. Small voids of 0.5 to 1.0 feet were encountered by the borings at Stations 865N and 1000N respectively.

Geophysical Logging

Geophysical logs provide further characterization of the geologic conditions as a function of depth within a borehole. Natural gamma and induction logs were obtained in the eight boreholes within the ditch west of the southbound lane.

The natural gamma logs (Figure 10c) indicate relatively uniform lateral conditions in the unconsolidated materials. Higher values between 4 and 16 feet correlate with the clay layer, while lower values above and below this layer correlate with the sands. Localized variations in the natural gamma logs indicate subtle variations in clay content. In the three deep borings, below 60 feet, higher gamma counts indicate layers of shale. The induction log shows similar results.

Water Level Measurements

Water levels were obtained to understand the direction of groundwater flow and its possible impact upon site conditions. Water levels were measured in the four screened or open borings. MDOT provided relative elevations of each of the boreholes, as well as water levels in Lake Huron, and two ponds within the nearby U.S. Gypsum Quarry.

Figure 11 shows the approximate locations of water level measurements along with a conceptual cross-section of groundwater flow at the site. Note that the portion of the quarry near U.S. Highway 23 was in operation between 1870 and 1950. The quarry has been shut down for nearly 50 years, but water levels are routinely pumped to maintain water levels in the quarry within 5 feet of a designated control level.

109

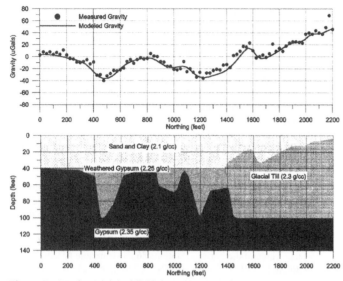

Figure 9. Gravity model of field data along southbound shoulder

The water level in the four boreholes (Figure 11) suggests that shallow groundwater flow will be from the road to Lake Huron, as would be expected. Within the vicinity of the quarry shallow perched groundwater will flow from the road to the quarry and deeper groundwater will flow from Lake Huron to the quarry.

CONCLUSIONS

The initial soil piping observed at the sinkhole was circular and less than 5 feet in diameter. This is a classic small diameter soil piping collapse where the soil material is slowly slumping downward into a deeper cavity. Since the gypsum bedrock lies about 40 feet below grade, the dissolution cavity must be greater than 40 feet deep. Note that such small diameter soil pipes (a few feet in diameter) are not uncommon as the result of initial subsidence within a thick unconsolidated material (Benson and Yuhr, 1987). However, if large open cavities are present, these small diameter soil pipes can enlarge at the surface over time becoming funnel shaped and leading to a much larger sinkhole at the surface.

The sinkhole was filled with compacted sand and covered with an asphalt patch. Within a period of about 40 days, it has settled 3 or 4 times and has been re-patched. This suggests the sinkhole may be still active or the subsidence may be due to the compacting the loose sands within the soil pipe.

The dissolution of gypsum has lead to small cavities within the weathered epikarst zone at the top of the gypsum and in enlarged bedding planes and vertical fractures within the gypsum. Many of these cavities may be filled with depositional materials or slumping sands and clays while some remain as open cavities.

While a large cave has been found in at least one underground gypsum mine near Grand Rapids, Michigan and many large sinkholes have developed in the gypsum in Burleight Township in the southwestern corner of Iosco County, about 12 miles from this site; subsidence has been relatively unknown in this part of Iosco County. However, cavities can exist for long periods of time without leading to surface subsidence. The trigger which causes subsidence is usually due to some change in surface or groundwater flow, change in groundwater level, or a combination of these factors. In this case, changes in surface and or groundwater flow could have been caused by one or more of the following:

- subtle changes in drainage associated with the addition of the passing lane;
- the long term dewatering of the nearby US Gypsum Mine since 1870;
- the lower levels of Lake Huron; and
- the installation and pumping of private water wells in the community just east of U.S. Highway 23 within 1,000 feet of the sinkhole.

FINAL CONCEPTUAL MODEL

The results from Phase I and II investigations were integrated into a refined geologic conceptual model shown as a cross-section in Figure 12.

The shallow radar data (Figure 5) clearly indicate soil piping is occurring within the shallow sands and clays at three locations, one of which has reached the surface. Based upon the radar data, the soil piping associated with the sinkhole extends throughout the width of the westernmost southbound lane, and may continue to the east under the easternmost southbound lane. The soil piping associated with the sinkhole does not extend west to the southbound shoulder; however, the soil piping occurring further south at Stations 875N and 915N may extend to the west of the road shoulder.

The linear nature of the soil piping indicated in the radar data suggests that it is caused by a linear geologic conditions (i.e., cavities aligned along a fracture). This is supported by the alignment of fractures within the 2D resistivity imaging data obtained along the southbound and northbound aprons of U.S. Highway 23.

The microgravity and boring data do not indicate the presence of large cavities, but suggest the presence of smaller cavity zones within dissolution-enlarged epikarst zone, bedding planes and fractures. The fact that no large diameter sinkholes have been reported or observed in this area further supports a localized (smaller) cavity system of limited extent.

Based upon the integrated data from Phase I and Phase II, we now know the general nature of the subsurface conditions and the cause of the sinkhole. However, the volume of cavity space and grout required for stabilizing conditions are as yet unknown. This will be resolved by the initial remediation grouting effort.

We have assumed that the long-term dewatering of the gypsum quarry is the dominant factor affecting groundwater flow and karst activity at the site. Furthermore, the concentrations of local residential water wells immediately east of the sinkhole could also be a trigger factor. Therefore it follows that subsidence activity should decrease with an increasing distance from the quarry and the local residential area. However, we do not know with certainty that this is the case.

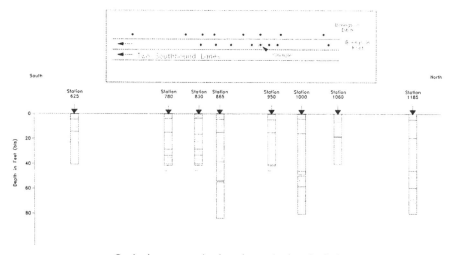

a. Geologic cross-section based upon borings in ditch

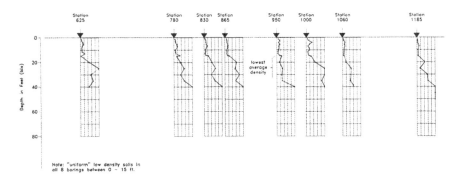

b. SPT (blow counts) cros-section based upon borings in ditch

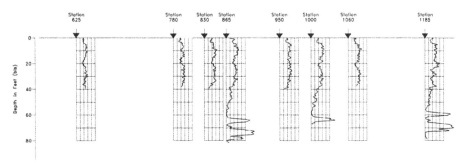

c. Natural gamma log cross-section based upon borings in ditch

Figure 10. Three forward gravity models used to interpret the field gravity data

REFERENCEs

Benson, R. C. and Yuhr, L. B., 1987. Assessment and long term monitoring of localized subsidence using ground penetrating radar. 2nd Multidisciplinary Conference on Sinkholes and the Environmental Impacts of karst, Orlando, February.

Elowski, R.C., and Ostrander, A. G., 1977. Gypsum karst and related features in the Michigan Basin. In: Guidebook of Alpena, Michigan. National Speleological Society, pp 31-38.

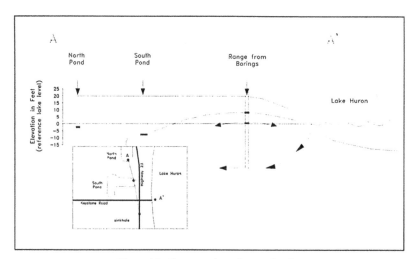

Figure 11. Cross-section of water levels

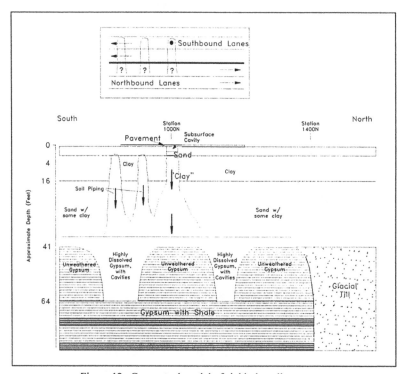

Figure 12. Conceptual model of sinkhole collapse area

Ford, D. C. and Williams, P. W., 1989. Karst Geomorphology and Hydrology. Chapman & Hall, New York, NY.

Klimchouk, Alexander, 2000. Speleogenesis in Gypsum. In: Speleogenesis, Evolution of Karst Aquifers. National Speleological Society, Inc., Huntsville, Alabama, pp 431-442.

Klimchouk, A. B.; Ford, D. C.; Palmer, A. N.; and Dreybrodt, W. (Eds), 2000. Speleogenesis, Evolution of Karst Aquifers. National Speleological Society, Inc., Huntsville, Alabama.

Telford, W. M., L. P. Geldart, R. E. Sheriff, and D. A. Keys, 1976. Applied Geophysics, Cambridge University Press, New York, New York, 860 p.

Geotechnical and Environmental Applications of Karst Geology and Hydrology, Beck & Herring (eds)
© 2001 Taylor & Francis, ISBN 90 5809 190 2

Evaluation of the risk of the formation of karst-related surface instability features in dolomite strata occurring in the Lebowakgomo area, Republic of South Africa

FREDERIK CALITZ   Southern Africa GeoConsultants (PTY) Ltd., 0700 Pietersburg, RSA, fcalitz@geocon.co.za

ABSTRACT

Recent engineering geological studies conducted in the vicinity of the important residential, legislative and industrial town of Lebowakgomo has revealed that the area may exhibit a potential for sinkholes and subsidence formation. A regional study is proposed to evaluate the risk of the formation of these features in an area underlain by dolomite and chert formation extending between the towns of Potgietersrus and the border between the Northern and Mpumalanga Provinces of the Republic of South Africa. Several mines, industries, agricultural development, residential settlements and a portion of an important highway are located in this area.

This document serves to define the scope, necessary project actions and proposed end-products of the study, and highlights the envisaged advantages of the implementation of the results. A preliminary study has shown that karst-related surface instability features are indeed present at localised positions throughout the study area, with some of these features already impacting on human activities. The results of a preliminary field survey, that included the evaluation of a number of existing sinkholes and a visit to caves located in the Makapans Valley near Potgietersrus, are discussed in detail, and the way forward is sketched.

INTRODUCTION

Evidence obtained from air percussion boreholes, open excavations, walk-over surveys, information on the location of known caves and reports by other authors clearly shows that interlinked subterranean caverns occur throughout the eastern portions of the northern Province of the Republic of South Africa that are underlain by dolomite and chert. This area includes the important residential, legislative and industrial town of Lebowakgomo, as well as portions of the industrial suburbs and the water supply wellfields of the town of Potgietersrus. Personal experience with civil engineering projects in this area has revealed that the local engineering fraternity lacks insight into the potential effects that a karst landscape may have on development.

A regional study is proposed to evaluate the risk of the formation of karst-related surface instability features in the southeastern portions of the Northern Province underlain by dolomite and chert. The area of interest extends roughly from the town of Potgietersrus in the west to the border between the Northern and Mpumalanga Provinces in the southeast (Figure 1), and covers a total area of approximately 21 900 Km2. Important platinum, diamond, gold and chrome mines, various light and heavy industries, agricultural development, rural and formalised residential settlements and a portion of an important highway with a toll booth are located in this area. Several extensive and well-known cave systems, for example: the Makapans and Echo caves (Figure 2), are located in this area.

GEOLOGICAL SETTING

The study area is underlain by dolomite, limestone, banded ironstone and chert of the Malmani Subgroup of the Chueniespoort Group, Transvaal Supergroup. The dolomitic strata are underlain by the Black Reef Formation (a siliciclastic unit deposited in a near shore-beach environment). The contact between the Black Reef Formation

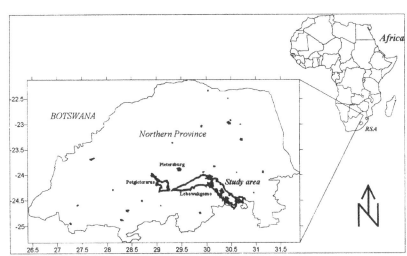

Figure 1: Map of the Northern Province showing the location of the study area

and the overlying dolomites is gradational. The dolomite succession is in turn subdivided into the Oaktree, Monte Christo, Lyttelton, Eccles and Frisco Formations. The Oaktree, Lyttelton and Frisco Formations exhibit a dark brown colour along weathered surfaces, and are characterised by the paucity

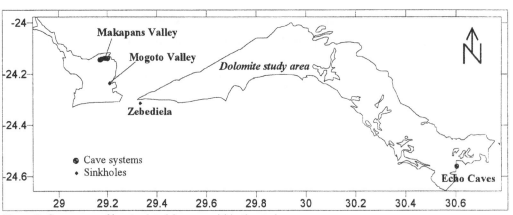

Figure 2: Occurrence of karst-related features within the study area

of chert layers and large megadomal stromatolitic structures typically associated with a shallow subtidal environment. The Monte Christo and Eccles Formations are light grey in colour along weathered surfaces, and are characterised by abundant chert layers and, as well as sedimentary and biogenic structures typically associated with an inter-to supratidal environment. The Monte Christo Formation may be locally subdivided into members, with the unit surfaces characterised by chert-in-shale breccias interpreted to be sequence stratigraphic boundaries.

The dolomitic strata in the Makapans Valley portion of the study area define a large fold structure with steep dips of between 40 and 65 to the east, southeast, southwest and west, depending on the position on the fold. The fold axis trends approximately northeast, and dips at a high angle to the southwest. Numerous strike-slip faults cut the stratigraphy, particularly in the western limb, while the stratigraphy in the eastern limb has strata duplicated by north to northwestwardly verging thrust faults. Igneous intrusions, in the form of dykes and sills, occur along the western limb, and may have serious implications as far as compartmentalization of the groundwater is concerned. The nose of the fold structure is overlain or covered by younger Karoo-age sediments and lavas.

These structural features hold significant implications for the development of karst structures at any scale, given the inferred relationship between structural weaknesses and the development of cave and sinkhole structures. The vertical and horizontal fault planes infer that dissolution of carbonates probably occurs along similar planes, which suggests that interlinked cave systems should be expected.

SCOPE OF THE STUDY

Objectives

The objectives of the proposed study are as follows:

- To define the precise extent of the dolomite strata in the eastern portion of the Northern Province
- To evaluate the general character of the dolomite strata in terms of stratigraphy, composition, surface expression, geohydrological setting and the effects of regional structural deformation
- To document and evaluate existing surface instability features occurring within the study area
- To determine the current extent of mining, industrial, commercial, agricultural and residential development, as well as transport infrastructure and important drainage features, within the study area
- To determine and evaluate the current risk of the formation of karst-related surface instability features within the study area on a regional basis, utilising different evaluation techniques
- To compile a regional karst risk map, highlighting developed areas where a possible high risk of the formation of sinkholes or subsidences exist
- To evaluate the effects of proposed future mining, industrial, commercial, agricultural development on the stability of the underlying dolomite strata by means of the regional karst risk map

Preliminary risk evaluation

The first phase of the study will entail the collation and evaluation of all available geological and physiographical information. The results of an ongoing regional geological mapping exercise being conducted by officials of the Pietersburg Regional Office of the South African Council for Geoscience will be of great importance to define the precise extent of the dolomite strata, as well as to delineate chert-poor and chert-rich strata, as well as occurrences of wad.

An attempt will be made to obtain detailed information on all known cave systems located within the study area, with special attention being given to its location, geological character, surface expression and subsurface extent. The more accessible caves will be subjected to a more detailed investigation, during which the surface expression, and geological and geohydrological character will be scrutinised. Specific attention will be given to any evidence of the upwardly migration of solution cavities in recent times.

All sinkholes and subsidences revealed by the available information will be investigated in the field and photographed. Details regarding the location, occurrence date (if known), the distance to the nearest human development, the type of feature, surface extent and depth, and inferred cause of each feature will be logged on a sinkhole/subsidence event record sheet.

Evidence of dolomite and possible karst zones occurring to a depth of 100 m in boreholes located within the study area and its immediate surrounds will be collated with the available information. This information will be useful to further refine the regional extent of the dolomite strata and to obtain point data on the occurrence of karst features within the dolomite strata that does not exhibit any surface expression.

Also included in this phase of the study will be the determination of the regional geohydrological character of the different portions of the study area through the evaluation of existing information held at the local offices of the Department of Water Affairs and Forestry. The current level at which the static water table occurs within the dolomite, as well as details regarding seasonal groundwater fluctuations are required to identify areas that may be undergoing dewatering, and to identify karst zones that are located above the static water table.

This phase will conclude with the compilation of a Regional Karst Databank utilising a Geographical Data Analyses System (GEODAS or GIS) to present the results graphically. It is envisaged that this databank will allow a user to obtain a detailed description of any karst-related feature, including a locality map, detailed logsheet and photographs.

One specific result of this phase of the study will be to prove, or disprove, the basic assumption on which most karst instability evaluation techniques are based, namely that the dolomite area in question is underlain by extensive networks of interconnected solution cavities at relatively shallow depth, thus defining the basic scenario to be utilised during the regional karst risk evaluation to be conducted during the latter stages of this study.

Current extent of human activity within the study area

The second phase of the study will focus on human activity within the study area. The extent of all urban areas, including rural, informal and structured development, will be determined and mapped. All industrial and commercial structures will be identified, and the location of active and dormant mines determined. The average population density for the villages and towns located within the study area will be obtained, based on the results of the 1999 population census.

The location of all minor and major dirt tracks, tarred roads, main roads and the N1 highway will be obtained, as well as an indication of the average vehicle count for each class of road.

Natural and man-made drainage features, including fountains, dams, pans, canals and wellfields, will be mapped to obtain a more accurate estimate of areas that may be undergoing dewatering. This information will be used to differentiate between areas for which a dewatering scenario will be used during the regional karst risk evaluation, and those for which a non-dewatering scenario will apply. Attention will also be given to the location and extent of water reticulation systems, including reservoirs, pipelines and irrigation schemes, in order to identify areas where relatively serious water leaks may occur. Areas where the localised re-introduction of water into the underlying dolomite aquifers by means of boreholes will also be noted.

The information obtained during this phase of the study will also be added to the Regional Karst Databank. These results will provide an estimate of the number of people, as well as the estimated value of buildings, that may be at risk in the event of the formation of sinkholes or subsidences.

Regional risk assessment

The risk of the formation of karst-related surface instability features can be assessed by means of the Method of Scenario Supposition (Buttrick, 1992) that relies on the evaluation of the probable impact that a specific development may have on a karst environment during the lifetime of the particular development. This method depends on the identification of factors that play a role in the process on sinkhole and subsidence formation, where the subsurface conditions in the dolomite profile that indicates potential susceptibility to karst processes must be determined. The susceptibility is generally expressed in terms of the risk of a certain event occurring.

As this method is ideally suited for the evaluation of the stability of site-specific developments, it will have to be adapted for application during regional risk assessment studies, due to regional variations in the character of the karst terrain. In this light it is envisaged that the study area will be divided into Karst Type Zones by grouping together areas that exhibit similar topography, geological setting, current land use and existing karst history. These zones will then form the basis of the regional risk assessment.

The resulting risk classification will be based on the evaluation of individual boreholes located within each Karst Type Zone, and will define a Regional Karst Risk Map. It must be noted that a degree of extrapolation will be necessary between the boreholes in areas where limited information is available, requiring that some sort of accuracy factor be added to the risk assessment.

Evaluation of risk to existing and future development

It must be noted that the Regional Karst Risk Map is not meant to replace site specific dolomite stability investigations, but is rather a tool available to the relevant decision makers and planners to ensure that those areas underlain by dolomite is developed in a sensible and safe manner, thus minimising the risk to inhabitants and users of these facilities.

It is envisaged that the Regional Karst Risk Map will be used on a strategic scale by local and Provincial Government Structures to identify existing development that falls within zones classified as exhibiting a potentially moderate to high risk of the formation of sinkholes or subsidences. This information will be of great importance to implement preventative and/or remedial measures in those high risk areas already developed to prevent loss of life or damage to existing structures. In the same light, the Regional Karst Risk Map can be used to identify areas with a low risk for the formation of karst-related surface instability features.

The results of this study will therefore facilitate sensible and cost effective planning of future development of the eastern portions of the Northern Province underlain by dolomite. The regional risk assessment will also define the implementation of precautionary measures, thus minimising the risk of severe structural damage to new developments and possibly saving lives.

Guidelines for the establishment and implementation of a community awareness programme with regard to the minimising of karst-related risk in order to educate the inhabitants of those villages underlain by dolomite will be drafted during this study.

PRELIMINARY RESULTS
Available information
A base map of the study area has been compiled by means of the ArcView 3.0a GIS software package. This map is based on 1 : 50 000 scale topographical maps, and contains the location of karst surface instability features and cave systems determined during a preliminary field visit on 25 November 2000. The map also shows the extent of existing townships and villages, as well as transport infrastructure. This base map is deemed to represent an initial version of the Regional Karst Databank.

Figure 3: Entrance to Peppercorn's Cave located in the Makapans Valley

A request for information on the known cave systems occurring within the boundaries of the study area was submitted to the South African Speleological Association. It was agreed that although information on all cave systems will be made available for use in this study, the location of only those caves deemed to be known to the public in general will be indicated on any maps and reports forthcoming from this study. This decision reflects the Association's viewpoint that the best way to protect an unspoilt cave is by not disclosing its location.

Caves located in Makapans Valley
A number of caves located in the Makapans Valley area between Potgietersrus and Zebediela (Figure 2) was visited on 25 November 2000. It must be noted that several of the caves in this area have been extensively disturbed through the mining of dolomite, limestone and travertine formations for use by the local cement industry. Early Man used these caves for shelter and protection, as ancient hearths, stone implements, pottery fragments and the remnants of cattle kraals can be seen near the openings of some of the caves.

The visits to the Peppercorn's (), Limeworks and Makapans caves clearly reveal that solution cavities, and thus cave formation, are generally restricted to possible bedding plane faults, fracture zones associated with near vertical faults, and radially arranged joints associated with anticlinal structures. Gradual enlargement of the caves have occurred through the collapse of sidewalls and unsupported cave roofs. The cave openings generally represent large elliptical sinkholes in the mountainside that are associated with extensive rubble cones scattered over the cave floors. The cave roofs are relatively smooth and reflect the topography of the base of the overlying rock layer. The cave sidewalls are generally composed of interlayered chert and dolomite breccia, travertine, and cavern infill material composed of semi-consolidated soil, rock and travertine fragments. Several fossil-rich layers, representing fossilised cave floors, were also noted.

Evidence of relatively recent roof and sidewall collapse can be observed throughout the caves (Figure 4), with the most recent collapse occurring around two months prior to the site visit. It is interesting to note that this rockfall uncovered some fossilised fauna remains occurring within cavern infill material. Frequent dark coloured deposits, representing the deposition of topsoil leached

Figure 4: Evidence of relatively recent collapse of a portion of the roof of Makapans Cave

through interconnected burrows and tunnels from Makondo structures on the surface, were encountered on cave floors throughout the area. Several medium to large sinkholes, associated with rubble cones, were noted in the roofs of several of the caves. One very large sinkhole, known as Katzenjammer Cave that is approximately 10 m across and 15 m deep, is deemed to represent the collapsed roof of large subterranean solution cavity.

It must be noted that an informal rural settlement comprising several "hut clusters" each housing up to 15 persons occurs directly to the west of the extensive Limeworks Cave. The community's water needs are addressed from a reservoir located nearby, but little or no surface water management is implemented and large ponds of "grey" water can be observed on a regular basis around the reservoir. Plans are currently afoot to relocate this community to a more suitable location.

Existing sinkhole in the vicinity of Lebowakgomo

An existing sinkhole, encountered during a recently completed borehole census, was investigated as part of the initial phase of this study (Figure 2). This feature is between 2 and 3 m wide, and approximately 1 m deep, and has been backfilled with gravel. Two production boreholes utilised for the abstraction of groundwater and three open boreholes occur within a distance of 100 m of the sinkhole. The sinkhole formed in an area located directly next to the junction between two minor dirt tracks in a citrus plantation. The area exhibits a gentle slope towards the area in which the sinkhole formed into which storm water that collects along the dirt tracks may drain after heavy precipitation. Two water supply pipes from the boreholes cut through this localised depression. In this light, it is assumed that the sinkhole formed as a result of localised dewatering, with the concentration of surface water in the lowest-lying area leading to the rupturing of one or both of the water pipes, finally triggering the formation of a sinkhole.

Two other sinkholes associated with production boreholes (and thus localised dewatering) are reported to occur in the Mogoto Valley (Figure 2), but could not be visited at this time, as it are located within the security area of the Klipspringer Diamond Mine.

THE WAY FORWARD

The results of the preliminary study clearly reveal that extensive cavernous conditions exist in localised portions of the study area, and that surface instability features associated with human activity has already occurred. In this light, it is evident that the dolomitic strata holds a definite risk to existing development, and that further development within this area may adversely affect the surface stability.

•   The following actions are deemed necessary to complete the preliminary risk evaluation:
•   To define the precise geological setting of the study area, including the delineation of all prominent faults and intrusions

- To define the precise characteristics that are required to adequately delineate the proposed Karst Type Zones
- To identify, log and evaluate all existing surface instability features
- To develop and refine the Regional Karst Databank

Project actions with regard to the determination and evaluation of existing human activity will follow the preliminary risk evaluation, and will conclude with the compilation of a first draft of the Regional Karst Risk Map.

ACKNOWLEDGEMENTS

The author would like to thank Frans and Hester Roodt of the Pietersburg Museum for their valuable inputs, Murray Obbes of the South African Council for Geoscience for sharing his knowledge on the local geology, and in particular the staff of GeoCon for their assistance and backing.

REFERENCES

Buttrick, D.B., 1992, Characterisation and appropriate development of sites on dolomite: Ph.D. thesis, Department of Geology, University of Pretoria, Pretoria, Republic of South Africa, 182 p.

Calitz, F., 1995, Applying the Method of Scenario Supposition to assess the stability of a site underlain by dolomitic rocks in South Africa: Land Subsidence Case Studies and Current Research: Proceedings of the Dr. Joseph F. Poland Symposium on Land Subsidence, AEG Special Publication No. 8, pp. 521-528.

Maguire, J.M., 1998, A guide to the palaeontological and archaeological sites of the Makapansgat Valley: Post Conference excursion, IV Congress of the International Association for the study of human palaeontology, International Association of Human Biologists, 90 p.

Maguire, J.M. (ed.), 1999, Makapansgat Valley Masterplan: Draft Masterplan Specialist Report, Cave Klapwijk and Associates.

Geotechnical and Environmental Applications of Karst Geology and Hydrology, Beck & Herring (eds)
© 2001 Taylor & Francis, ISBN 90 5809 190 2

Natural and induced halite karst geohazards in Great Britain

A.H.COOPER   British Geological Survey, Keyworth, Nottingham, NG 12 5GG, Great Britain, ahc@bgs.ac.uk

ABSTRACT

Halite or rock salt, in Great Britain occurs mainly within Permian and Triassic strata and is affected by natural and artificially induced dissolution, which has produced karstic landforms and subsidence. This natural and artificial salt karst associated with these deposits and the dissolution zone commonly extends down to a depth of between 30 and 130m, in the Triassic strata and up to 300-400m in the Permian rocks. In the Triassic rocks, this salt karst is buried by collapsed and foundered insoluble material that was interbedded with the salt or present in the overlying strata. In the Permian rocks a salt dissolution front is present and the overlying strata has collapsed into a monoclinal structure. In the Triassic areas, salt springs show that dissolution is active and they are associated with linear belts subsidence ponds. The occurrence of these salt springs in Britain is commonly indicated by the of place names ending in "wich", hence the naming of the towns of Northwich, Middlewich, Nantwich in the Cheshire saltfield and Droitwich in the English West Midlands. The salt deposits were exploited from pre-Roman times onwards utilising the brine from the natural springs ("wild" brine extraction). In Victorian times a flourishing salt and chemicals industry developed, but the techniques of uncontrolled wild brine extraction caused severe subsidence problems, the legacy of which is still present. Brine dissolution channels were developed extending many kilometres away from the abstraction points and an artificial salt karst was formed. The salt was also mined, but the mines commonly flooded and if the resultant brine was pumped out ("bastard" brine extraction) it caused catastrophic subsidence and the formation of large ponds and lakes. Legislation was introduced to deal with the problem and a levy is still charged for salt extraction to pay for any damage. Most modern salt extraction is by fairly deep dry mining or deep solution mining leaving large brine-filled cavities. The natural and induced salt karst, plus the mining problems have left a legacy of difficult ground for development. Some salt mines are still collapsing and the re-establishment of the hydrogeological regimes after brine extraction means that salt springs may flow again and cause further dissolution and collapse.

INTRODUCTION

In Britain, halite, or rock salt, occurs mainly within Permian and Triassic strata, mainly in the Cheshire Basin, but also in Lancashire, Worcestershire, Staffordshire, NE England and Northern Ireland (Figure 1 and Notholt and Highley, 1973). Where the saliferous Triassic rocks come to outcrop, most of the halite has dissolved and the overlying and interbedded strata have collapsed or foundered producing a buried salt karst. These areas commonly have saline springs, indicative of continuing salt dissolution and the active nature of the karst. The salt deposits were exploited from before early Roman times and salt springs or "wyches" were recorded in the Domesday Book (Sherlock, 1921). Consequently, British place names ending in "wich" indicate natural brine springs (except in most coastal areas where they indicate places where sea salt was produced). It is around such salt springs that the towns of Droitwich, Nantwich, Northwich and Middlewich developed in Cheshire and the West Midlands (Calvert, 1915). Halite dissolves rapidly and subsidence, both natural and mining-induced, has affected the main Triassic salt fields including Cheshire, Staffordshire, Worcestershire, coastal Lancashire and parts of Northern Ireland. Much of the mining has been by shallow brine extraction, a technique, the results of which intensify the natural karstification of the salt deposits.

Permian salt is present in Britain, at depth, beneath coastal Yorkshire and Teeside. Here the salt deposits and the karstification processes are much deeper than in the Triassic salt. The salt deposits are bounded up dip by a dissolution front and collapse monocline. Salt has also been won from these Permian rocks by dissolution mining and some ancient subsidence due to mining has occurred along the banks of the River Tees.

NATURAL AND INDUCED SALT KARST DEVELOPMENT IN THE TRIASSIC AND PERMIAN SALT DEPOSITS

The Triassic salt deposits in Britain were deposited in a semi-arid environment within mainly fault-controlled land-locked basins linked to the major depositional centres of the North and Irish Seas. The general Triassic sequence comprises thick sandstones and conglomerates overlain by red-brown siltstone and mudstone with halite and gypsum sequences. The Permian salt deposits form part of the Zechstein Group sequence that extends eastwards from Britain to Holland and Germany. Only the featheredge of the Zechstein Group is present on shore in Britain and the salt deposits are restricted to the coastal parts of northern England. Here the Permian salt is interbedded with thick dolomite, mudstone and anhydrite formations.

Where the Triassic saliferous rocks come to outcrop, salt karst has developed to depths of 30-130m. The development of this karst has been strongly influenced by groundwater flow and the modification of the groundwater regime during the last, and possibly earlier, glaciations. The last (Devensian) glaciation buried the north and central parts of Britain beneath a thick wet-based ice-sheet producing elevated groundwater heads linked to the hydrology of the ice-sheet (Howell and Jenkins, 1976). This elevated head is thought to have depressed and flushed out the saline waters forcing the dissolution surface of the salt to cut deeper into the sequence. When the ice-sheet melted, the groundwater regime changed and further dissolution of the salt occurred as new regional groundwater patterns formed. Because of the last glaciation, the depth of salt dissolution is commonly very deep. This mechanism may also help to explain why the dissolution extends down to about 130m in the Triassic salt of Cheshire, and to 3-400m in the Permian salt of northern England, where the cover of ice was probably much thicker. However, it must also be noted that the salt beds in the Permian probably would not come to outcrop since their facies is restricted to the deeper parts of the depositional basin (Smith, 1989).

The way in which the salt karst has developed is also dependent on the nature of the lithological sequence in which the salt beds were deposited. In the Triassic deposits, most of the salt is interbedded with units of siltstone and mudstone, commonly with gypsum. Where the groundwater has circulated and removed the salt, only the interbedded insoluble rocks remain as a breccia, concealed by the commonly less brecciated, but foundered cover sequence. The traditional term for the salt surface where the dissolution has occurred is "wet rock head" because it is from here that the original brine extraction took place. In contrast, the traditional name for the area where the salt has not dissolved because it is sealed beneath a cover of impermeable strata is "dry rock head" (Figure 2). The area of wet rock head encompasses the salt karst, but the depth of the karst surface generally lies some 30-130m below the ground surface and may even reach 180m in Cheshire (Howell, 1984).

The wet rock head area is the zone in which dissolution has occurred and the foundered ground above this zone commonly has numerous collapse features. These range from enclosed hollows 20-200m across to linear depressions up to several kilometres long. These depressions were called "brine runs" by the early brine miners and it was into these, and the brine springs that they sunk their extraction wells. This method of uncontrolled brine extraction was referred to as "wild" brining. It was an irresponsible way to abstract brine that caused the enlargement of the natural brine runs and resulted in widespread subsidence. The large-scale abstraction of brine in this way enhanced the development of the natural salt karst into an unnatural form. In many places the natural system was disturbed further by pillar and stall mining, commonly followed by brine extraction from the flooded workings. This method of working was called "bastard" brine pumping and produced large-scale collapses and a further perturbation of the salt karst system.

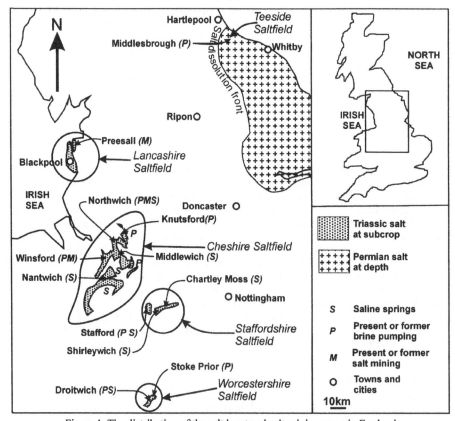

Figure 1. The distribution of the salt karst and salt mining areas in England

THE TRIASSIC SALTFIELDS OF THE UNITED KINGDOM
Cheshire

Two salt formations are present in the Triassic Mercia Mudstone Group in Cheshire. These are the Wilkesley Halite Formation (up to 300m thick) near the top and the Northwich Halite Formation (up to 200m thick, with mudstone partings) in the middle of the group (Earp and Taylor, 1986; Evans et al., 1993, Wilson, 1993). The sequence between and below the salt formations is dominantly red-brown gypsiferous and dolomitic mudstone and siltstone. The salt karst has developed below the collapsed and foundered strata over wet rock head. Brine escaping from this dissolution zone formerly emerged at the surface as brine springs, but subsequent brine extraction caused most of them to cease flowing. In Cheshire, brine springs occurred in many places including Northwich, Middlewich, Nantwich, Winsford. They were also present near Whitchurch (Shropshire) at Higher Wych and a place formerly called Dirtwich. Natural subsidence has occurred sporadically throughout the salt field, one of the earliest records being in 1533 at Combermere Abbey near Whitchurch (Sherlock, 1921). Many natural lakes such as Rostherne Mere near Knutsford were also formed by salt dissolution since the end of the Devensian ice-age and all the Cheshire saltfield is dotted with meres caused this way (Waltham et al., 1997).

The Cheshire saltfield has a long history of exploitation growing from its first pre-Roman development around the salt springs into a major industry. In the late 19th and early 20th Centuries the salt deposits were worked mainly by two methods: traditional mining and "wild" brine extraction. Most of the conventional mining was in shallow "pillar and stall" mines with networks of tunnels separated by narrow salt pillars. The "wild" brine pumping was commonly from wells sunk into natural brine runs. The method induced brine to flow towards the extraction boreholes aggravating the existing subsidence features or causing linear subsidence belts emanating along strike from the boreholes. In some situations, mine owners pumped the brine from flooded pillar and stall mines to the detriment of mine stability. This method of extraction was called "bastard" brine pumping a technique that was made illegal in 1930 (Calvert, 1915; Tony Waltham, pers. comm. 2000). Around Northwich and Middlewich, the resulting subsidence was catastrophic and widespread. New lakes (called "meres" or "flashes") appeared almost daily, many were hundreds of metres across (Earp and Taylor, 1986). These include significant lakes such as Moston Flash near Elworth (Waltham, 1989; Waltham et al., 1997). The subsidence in Cheshire was so severe that an Act of Parliament was passed that placed a levy on all locally extracted salt. This levy, which funded building reconstruction and compensation payments, is still paid to the "Cheshire Brine Subsidence Compensation Board", but at a lower rate to reflect the reduced risk from modern extraction (Collins, 1971). Modern salt extraction takes place mainly in 125-170m deep, dry, pillar and stall mines, or by controlled brine extraction. The latter leaves large deep underground chambers (at depths of between about 150-300m), that are left flooded and filled with saturated brine or used for waste disposal (Northolt and Highley, 1973).

The uncontrolled extraction of brine resulted in an artificial lowering of the brine/freshwater interface. This allied with groundwater abstraction from adjacent aquifers has disturbed the brine above the karst surface and introduced fresh water into some areas (Howell and Jenkins, 1976). Recent geochemical sampling by the British Geological Survey has shown that brine is still entering the rivers in the Cheshire saltfield. From river flow volumes, and the content of salt in the water it should be possible to calculate the amount of salt being removed annually from the Cheshire saltfield. However, not all the salt in the rivers is from the salt beds, since much of the rainfall includes salt derived by the weather coming across the Atlantic Ocean (Walling and Webb, 1981).

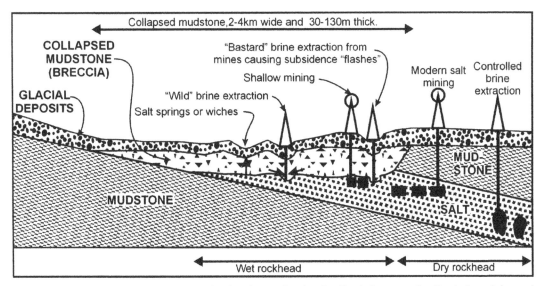

Figure 2. Cross-section through a typical Triassic salt area showing the dissolution zone, the dissolution mining and the present mining situation.

Staffordshire

The salt deposits under the town of Stafford lie in a synclinal structure faulted along its eastern side (Arup Geotechnics, 1990). The salt occurs interbedded with mudstone in a sequence 50-65m thick within the Mercia Mudstone Group. The near-surface deposits have dissolved to a depth of around 50m (wet rock head) and the salt has largely dissolved adjacent to the eastern fault zone. The natural salt springs around Stafford are not well documented, but the town is surrounded by the villages with names ending in wich, including, Baswich, Milwich, Gratwich, Colwich and Shirleywich (along with the village of Salt), suggesting the presence of brine springs. The springs at Shirleywich have a long history of exploitation from the 17th century to the late 18th century. The Shirleywich springs, and nearby ones at Weston appear to be fed from salt deposits lying to the north-east in a belt running through Chartley Moss and other small lakes. Lying 5km to the NE of Shirleywich, Chartley had the strongest saline springs in Staffordshire in the mid 16th century (Sherlock, 1921). Chartley Moss is a fresh-water pond and floating bog formed in a salt dissolution subsidence area. Localised collapse here causes minor seismic events that are felt on the surface of the bog, thus providing evidence of active salt dissolution and subsidence beneath the site (A. Brandon, pers comm. 2000). Brine was extracted from beneath the town of Stafford from about 1890 until a court case in 1970 over substantial subsidence damage caused the cessation of wild brine extraction from the area. Only about 10% of the volume of salt removed by this brine extraction has been accounted for by recorded subsidence and further subsidence may occur in the area. The main brine run trends NNE towards the extraction boreholes and about 2 square kilometres of land have been affected by subsidence. Since the 1940's about 20 properties have been demolished and 500 severely damaged (Arup Geotechnics, 1990).

Lancashire

The Lancashire coast beneath Blackpool and Preesall is underlain by a Triassic sequence with several salt units in the Mercia Mudstone Group. Two impersistent units the Rossal Halite and the Mythop Halite occur low in the sequence and the Preesall Halite high in the sequence (Wilson, 1990; Wilson and Evans, 1990). The Preesall Halite was formerly worked in salt mines at Preesall on the east side of the River Wyre. The east of the saltfield is marked by the Preesall Fault Zone. Adjacent to this the westerly dipping salt comes near to crop, but groundwater circulation has dissolved the salt to a depth of between 50 and 100m resulting in a collapse breccia down to wet rock head (Wilson and Evans, 1990). There are no confirmed saline springs in the area, their most likely position would be at the coast emerging beneath the sea. However, Sherlock (1921) notes the name of the village of Salwick near Kirkham just south of the district (however, because the coast is nearby, this name could equally relate to the production of sea salt). Probable subsidence areas, in which post-glacial peat deposits have formed, are also recorded in the district, especially around Mythop (Wilson and Evans, 1990). These subsidence areas range from 30 to 150m across and contain up to 10m of peat that formed over the last 12,000 years. They give an indication of the amount and rate of salt karstification that has taken place beneath the area.

Worcestershire

The Triassic saliferous rocks in Worcestershire occur in the Mercia Mudstone Group and include an upper and a middle group of salt beds approximately equivalent to the Cheshire salt beds further north. The saliferous sequence is about 90m thick of which 40% comprise siltstone and mudstone units. In Roman times, the town of Droitwich was called Salinae, a name indicative of the brine springs that rose in the valley of the River Salwarpe. In the 17th century shafts were sunk that increased the brine flow. In the 18th century deeper pits encountered artesian brine that could not be controlled and largely ran to waste (Poole and Williams, 1981). Some salt mining was undertaken in the district, but it encountered natural brine runs and the subsequent exploitation was by brine extraction that ceased in 1972. The brine abstraction caused the natural brine springs at the surface to dry up. It also caused a belt of subsidence and tilted buildings along, and spreading out from, the course of the original sinuous brine run. The brine run follows the wet rock head of the strata which dips to the south-east and it has a north-east trending course through Droitwich and Wychbold to Stoke Prior. Where the salt has dissolved, the zone of collapse over the wet rock head extends down to a depth of about 90m forming a belt 1-2km wide and 12km long. The route of the original brine run was the area in which the most subsidence and building damage occurred in the 19th century. In the 1980's after the cessation of brine extraction, the brine levels began to rise. This caused Poole and Williams (1981) to speculate that in the future brine would flow again from the sites of the original springs that were used by the Romans.

Northern Ireland

Triassic saliferous rocks are present in the area around Carrickfergus and Larne, bordering on Belfast Loch in Northern Ireland. The Triassic sequence includes up to 200m thickness of salt, proved in the Larne Borehole, but in the Carrickfergus salt field only about 40-50m thickness of salt has been proved. The salt deposits are largely protected by the overlying sequence and there is no record of wet rock head and extensive brecciation such as that found in Cheshire (Griffith and Wilson, 1982). A salt spring was known at Eden, where salt mining subsequently developed, and slight brecciation of the sequence was recorded in a few boreholes, but was not extensive. The most spectacular dissolution feature recorded in the area was the collapse of the Tennant Salt Mine (Griffith, 1991). This mine was a conventional pillar and stall mine, but subsequent owners removed substantial amounts of brine from the abandoned workings. This dissolved the pillars and produced a large collapse hole with concentric failure planes.

THE PERMIAN SALT AREAS OF THE UNITED KINGDOM

Teeside and Yorkshire

In the north of England, salt is present in the Permian strata of the Zechstein Group. This evaporite and carbonate sequence extends from northern England eastwards beneath the North Sea to Germany and beyond. Only the marginal part of the Zechstein basin encroaches onshore in England where it includes thick units of anhydrite, halite and potash. The salt and potash deposits are currently mined by pillar and stall workings near the coast at Whitby. Here they are up to about 80m thick and occur at depths of 1100-1250 metres (Woods, 1973). The halite was also been extracted by dissolution at Teeside in the early 20th Century (Tomlin,

1982) and its presence was largely responsible for the development of the chemical industry in the area. Anhydrite was also pillar and stall mined here at Billingham (Raymond, 1960). The salt karst features on Teeside are very deep-seated and may be palaeokarst or, as discussed earlier, may relate to deep dissolution during the last ice-age. At the coast the salt is present at a depth of about 500m. Westwards up dip there is a dissolution front in the salt (British Geological Survey 1987) and the overlying strata have collapsed and formed into a west-facing subsidence monocline (Figure 3). Further up dip the anhydrite units in the sequence pass into gypsum at depths of between 120 and 60 m and this too passes into another dissolution front with an associated subsidence monocline (Figure 3 and Cooper, 1998). Salt extraction by dissolution has largely ceased on Teeside, but some subsidence has occurred and it there may be some subsidence still to occur.

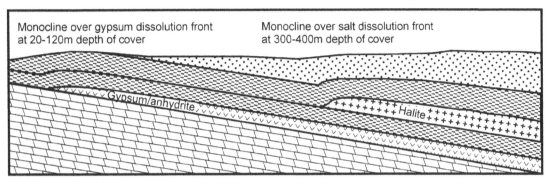

Figure 3. Simplified cross-section through the dissolution fronts of the Permian salt and gypsum/anhydrite in north-east England. In reality, there are several sequences of anhydrite and salt that can each produce their own monoclinal structures.

PLANNING FOR SOLUBLE ROCK GEOHAZARDS

Current planning procedures ensure that the modern exploitation lies largely outside of urban areas so that risks are considerably reduced. Furthermore, except for one natural brine extractor, the majority of the salt is mined either in deep dry mines, or by controlled brine extraction from depth. However, there is still a legacy of problems related to the salt deposits. These include old salt mines that have not collapsed, and compressible or unstable collapsed ground over former salt mines. In addition, natural salt dissolution at the rockhead interface, between the salt deposits and the overlying superficial deposits, can cause ground engineering problems and aggressive saline groundwater. With the ending of most near-surface mining and brine extraction, the hydrological system has, or is in the process of, re-balancing itself. It may be expected that natural groundwater flows will be re-established through the disrupted saltfields and further subsidence problems may occur. The accurate mapping of the rock salt and associated deposits, plus an understanding of their dissolution and collapse characteristics, can help development and planning in these subsidence-sensitive areas. These problems can then be reduced by careful planning and construction or by remediation of former mines. In some places, such as Teeside, the local planning guidelines for development in former salt dissolution extraction areas (Morris, 1975) place a zone, that is considered to be susceptible to subsidence, of between 150m and 300m radius around every former brine extraction borehole in the area. The size of this buffer zone appears to relate to the depth and amount of salt extracted from each hole. In these areas, it is recommended that avoidance or special precautions should be taken for certain types of construction. In Cheshire some former mines are causing subsidence and their remediation is being considered. In most of the salt karst areas, there is a legacy of difficult ground conditions produced by natural and man-made causes.

ACKNOWLEDGEMENTS

This paper has benefited from useful discussions with Dr Tony Waltham, Professor Martin Culshaw and Dr John Lamont-Black. Dr Andy Farrant and Alan Forster are thanked for critically appraising the manuscript. Published with the permission of the Director, British Geological Survey (NERC). The author acknowledges the contribution made towards this work by the ROSES (Risk of Subsidence due to Evaporite Solution) Project; ENV4-CT97-0603 and IC20-CT97-0042 funded by the EU Framework IV Programme.

REFERENCES

Arup Geotechnics 1990. Review of mining instability in Great Britain, Stafford brine pumping. Volume 3/vi. Arup Geotechnics for the Department of the Environment.

British Geological Survey 1987. Stockton Sheet 33. Solid and Drift.1:50 000 (Southampton: Ordnance Survey for the British Geological Survey).

Calvert, A.F. 1915. Salt in Cheshire. Spon Ltd, London, 1206 p.

Collins, J.F.N. 1971. Salt: a policy for the control of salt extraction in Cheshire. Cheshire County Council.

Cooper, A.H. 1998. Subsidence hazards caused by the dissolution of Permian gypsum in England: geology, investigation and remediation. In Maund, J.G. & Eddleston, M (eds.) Geohazards in Engineering Geology. Geological Society, London, Engineering Special Publications, 15, 265-275.

Earp, J.R. and Taylor, B.J. 1986. Geology of the country around Chester and Winsford. Memoir of the British Geological Survey, Sheet 109.119 p.

Evans, D.J., Rees, J.G. and Holloway, S. 1993. The Permian to Jurassic stratigraphy and structural evolution of the central Cheshire Basin, Journal of the Geological Society, London, Vol. 150, 857-870.

Griffith, A.E. 1991. 'Tennant's Ills'. Ground Engineering. Vol. 21, 18-21.

Griffith, A.E. and Wilson, H.E. 1982. Geology of the country around Carrickfergus and Bangor. Memoir of the Geological Survey of Northern Ireland, Belfast, HMSO. 118p.

Howell, F.T., 1984. Salt karst of the Cheshire Basin, England. 252-254 in Castany, G. Groba, E. and Romijn, E. Hydrogeology of karstic terranes. International contributions to hydrogeology, Vol. 1. International Association of Hydrogeologists.

Howell, F.T. and Jenkins, P.L. 1976. Some aspects of the subsidences in the rocksalt districts of Cheshire, England. Publication No 121 of the International Association of Hydrological Sciences, proceedings of the Anaheim Symposium, December 1976. 507-520.

Morris, C.H. 1975. Report on abandoned mineral workings and possible surface instability problems. County of Cleveland, Department of the County Surveyor and Engineer.63pp.

Notholt, A.J.G. and Highley, D.E. 1973. Salt. Mineral Dossier No. 7, Mineral Resources Consultative Committee, H.M.S.O. London, 36 p.

Poole, E.G. and Williams, B.J. 1981. 'The Keuper Saliferous Beds of the Droitwich area'. Report of the Institute of Geological Sciences, No. 81/2. 19 p.

Raymond, L.R. 1960. The pre-Permian floor beneath Billingham, County Durham, and structures in the overlying Permian sediments. Quarterly Journal of the Geological Society of London, Vol. 116, 297-315.

Sherlock, R.L. 1921. Rock-salt and brine. Special reports of the mineral resources of Great Britain, Memoir of the Geological Survey. H.M.S.O. London. 122 p.

Smith, D.B. 1989. The late Permian palaeogeography of north-east England. Proceedings of the Yorkshire Geological Society, Vol. 47, 285-312.

Tomlin, D.M. 1982. The salt industry of the River Tees. De Archeologische pers, Nederlands. 109pp.

Walling, D.E. and Webb, B.W. 1981. Water quality 126-169 in Lewin, J. (Ed.) British Rivers, Allen and Unwin.

Waltham, A.C. 1989. Ground subsidence. Blackie, Glasgow and London. 202p.

Watham, A.C., Simms, M.J., Farrant, A.R. and Goldie, H.S. 1997. Karst and caves of Great Britain. Joint Nature Conservation Committee. Chapman and Hall, London. 358pp.

Wilson, A.A. 1990. The Mercia Mudstone Group (Trias) of the East Irish Sea Basin. Proceedings of the Yorkshire Geological Society, Vol. 48, 1-22.

Wilson, A.A. 1993. The Mercia Mudstone Group (Trias) of the Cheshire Basin. Proceedings of the Yorkshire Geological Society, Vol. 49, 171-188.

Wilson, A.A. and Evans, W.B. 1990 Geology of the country around Blackpool. Memoir of the British Geological Survey, Sheet 66 (England and Wales), 82 p.

Woods, P.J.E. 1973 Potash exploration in Yorkshire: Boulby mine pilot borehole. Transactions of the Institution of Mining and Metallugry. Vol 82, B99-B106.

Geotechnical and Environmental Applications of Karst Geology and Hydrology, Beck & Herring (eds)
© *2001 Taylor & Francis, ISBN 90 5809 190 2*

The development of a national Geographic Information System (GIS) for British karst geohazards and risk assessment

A.H.COOPER, A.R.FARRANT, K.A.M.ADLAM & J.C.WALSBY   British Geological Survey, Keyworth, Nottingham, NG 12 5GG, Great Britain, ahc@bgs.ac.uk

ABSTRACT

Britain has four main types of karstic rocks, limestone, chalk, gypsum and salt, each with a different character and associated problems. Subsidence problems, difficult engineering and foundation conditions are widespread on these rocks. The triggering of subsidence by water abstraction and the enhancement of dissolution processes are relevant to some areas. Aquifer vulnerability and pollution tracing are concerns in most areas, especially the chalk, which is the major aquifer in southern Britain.

The British Geological Survey has embarked on a comprehensive digitisation scheme of the base 1:50,000 scale geological map information. In conjunction with this, the recording and assessment of geological hazards, including karst problems, are being undertaken to provide complementary digital information that enhances the basic map data. Digital map capture has been established at 1:10,000 scale using a customised interface with the ArcView GIS application. For the karst geohazards, this application is being extended to allow the digitisation of the karst features and the population of the associated Oracle database tables. For each type of karst feature, polygon or point attributes will be defined with suitable dictionaries for the appropriate morphological, stratigraphical or lithological entities. Linked features and databases for springs, stream sinks, dye tracing are being developed, tied to the existing hydrochemistry tables where appropriate. Cave survey and plan data derived from published sources may also be incorporated. From the factual data, hazard areas will be derived.

When the GIS is fully established and populated, the system will highlight the presence of karst features to non-specialists and allow the rapid interpretation of potentially hazardous karst areas by geological, hydrogeological and engineering geologists. With suitable links to polygon feature descriptions, basic geological reports derived semi-automatically are also feasible. The GIS will act as a desktop data capture facility with the intention that it can, ultimately, be extended to the capture in the field of information when suitable portable, robust computers become available.

INTRODUCTION

In Britain, many aspects of controlling the use of geologically unstable land, including karstic areas, have been written into Government planning policy. The main method of control is by the publication and implementation of planning policy guidance notes, the main one being "Planning policy guidance note 14: Development on unstable land" (Department of the Environment, 1990). This guidance information is currently being extended, and the extension (Annex 2; Department of the Environment, Transport and the Regions, 2000) is in the public domain for consultation. This guidance and the proposed additions put the emphasis on the developer to prove land is suitable for development, but the local authorities are also required to be aware of the problems in their areas and where possible to inform interested parties. The developers and the local government can only operate effectively if they know about the hazards that affect them and have access to suitable geological information. The British Geological Survey (BGS) is the main supplier geological and geohazard data.

The BGS has embarked on a major programme of map digitisation and modelling to provide 2D and 3D digital geological information for Britain. In tandem with this it has started a programme of capturing geological hazard information and the assessment of the level of the hazard present. Karst geological hazards in Britain are being considered as part of this geohazard programme. Traditionally, geological information for the solid and superficial rocks has been available as paper maps. The whole of the British 1:50,000-scale map set has been unified and is currently being digitised <http://www.bgs.ac.uk/digmap/home.html>. The stratigraphical descriptions and coded for the digitised geological units are defined in the British stratigraphical lexicon served via the internet <http://www.bgs.ac.uk/lexicon/lexicon_intro.html>. Similarly, the attribution of lithology is also indexed by tables available over the internet <http://www.bgs.ac.uk/bgsrcs/home.html>. The digitised map data is thus linked to database tables held in a standard way on Oracle servers. Views of index information and some geological information is currently available free over the internet from the BGS Geological Data Index (GDI) <http://www.bgs.ac.uk/geoindex/home.html>. These publicly available, datasets include views of solid and superficial geology at 1:625,000 scale (10-miles to one inch), borehole sites, geochemical and geophysical information.

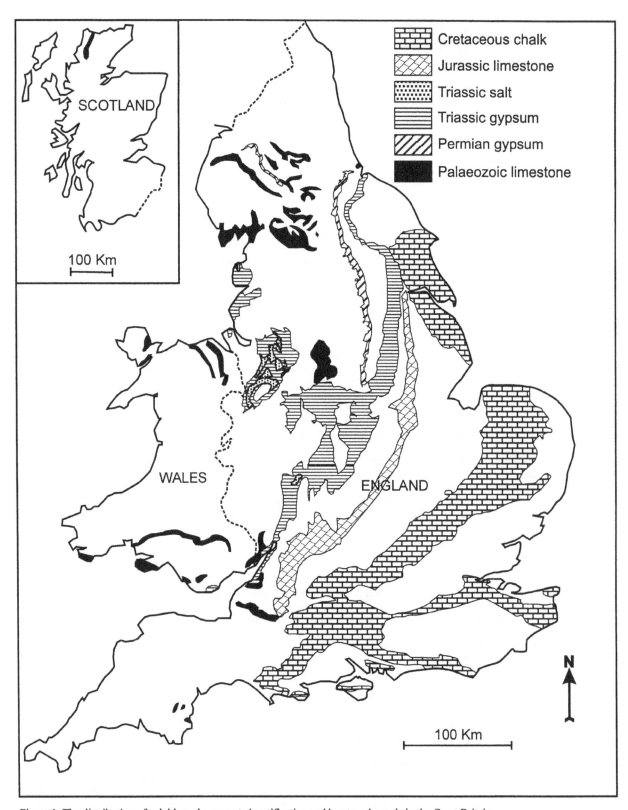

Figure 1. The distribution of soluble rocks prone to karstification and karst geohazards in the Great Britain

At the same time as these developments a Geographic Information System (GIS) interface utilising ArcView has been developed to allow the digitisation of geological mapping information directly by geologists. This system is constrained by live links to the Oracle lexicon tables that hold the definitions of all the approved polygon and data descriptors. This system is being extended to form the interface for the hazards data capture and the digitisation of karst hazard data.

The geological hazard information for Britain has never been produced as maps and most of the data exists as disparate sources including published papers, field maps, internal reports and some Government-sponsored studies, some with associated databases (Symonds Travers Morgan 1996; Applied Geology Ltd, 1993). Only a few specialists know their way around the information and if they are unavailable, or retire, the information can become difficult to retrieve and interpret. The philosophy of the British Geological Survey is to gather all relevant information and present it in a form that is usable by everyone. The system is currently being constructed around an ArcView interface linked to Oracle tables. This system forms the foundation for the eventual provision of data and interpretations on demand via the world-wide-web. Ultimately, the hazard layers can be linked to database tables and paragraphs of descriptive text to allow automated report generation. The BGS has started to build the system interfaces and databases that record and support the karst hazard information. This paper describes the structure and content of that programme as it relates to karst geohazards.

KARST GEOHAZARD AREAS AND TYPES IN BRITAIN

The karst areas of Britain include four major types and a few additional minor ones, their distribution is shown in Figure 1. The areas include limestones, chalk, gypsum and salt, each with different manifestations of the hazards related to them.

Limestones

The limestone karst of Britain was reviewed by Waltham et al (1997). The Carboniferous Limestone karst areas form much of the English Pennines and parts of Scotland, south-western England and South Wales. Areas of Jurassic limestones are also widespread in the east, centre and south of England. Small areas of Devonian limestones occur in Devon and Cambro-Ordovician limestone occurs in Scotland. The Carboniferous limestones of the UK host the largest and most extensive cave systems in the country. The major problems associated with the limestone karst areas relate mainly to water supply protection and some engineering conditions. The collapse of dolines occurs, but most of the limestone areas are in rural situations and the effects on property and infrastructure are slight. Dolines commonly form places where farm and other refuse or waste is illegally tipped, when this occurs it can cause rapid contamination of the karst groundwater and local drinking supplies fed from it. Items that are important for the GIS recording of the limestone areas include cave locations, stream sinks, springs, dolines and pollution sources.

Chalk

Chalk underlies much of eastern and southern Britain (Figure 1), it is the most widespread carbonate rock in the country and of immense importance for water supply. It is porous and includes less permeable calcareous mudstone beds and flints as nodules and beds. The rock is commonly strongly jointed with networks of microfractures that are karstified on a small, but extensive scale. Many of the karstic conduits are controlled by the positions of the flint and mudstone horizons, but the size of the conduits developed rarely exceeds one metre in diameter. The rock has traditionally been modelled as a large porous medium, but recent understanding is highlighting the karstic nature of the rock. It is therefore important to record the karstic features in the GIS including dolines, stream sinks, permanent and seasonal springs, and pollution sources. The seasonal streams are locally called winter-bournes (Waltham, et al., 1997) and the word bourne is commonly included in the place names of settlements on the Chalk. In addition to water supply, the Chalk also generates subsidence hazards and difficult conditions for construction and engineering projects (Edmonds, 1983). Surface and near-surface karstification of the rock, especially beneath and adjacent to superficial deposits, produces subsidence dolines, clay-filled pipes or fissures and uneven bedrock.

Gypsum

Gypsum karst is present mainly in a belt 3km wide and about 100km long in the Permian rocks of eastern and north-eastern England. It also locally occurs in the Triassic strata, but the effects of it are much less severe than those in the Permian rocks. The difference is mainly caused by the thickness of gypsum in the Permian sequence and the fact that it has interbedded dolomite aquifers. In contrast the Triassic gypsum is present mainly in weakly permeable mudstone sequences. In places where the major rivers have cut through the Permian sequence, the gypsum karst has formed phreatic cave systems. The rapid solubility rate of the gypsum means that the karst is evolving on a human time scale and active subsidence occurs in many places, especially around the town of Ripon (Cooper 1986, 1989, 1998). The active nature of the dissolution and the ongoing subsidence features, cause difficult conditions for planning and development (Symonds Travers Morgan, 1996; Paukstys et al., 1997; Cooper, 1998). The GIS entries for the gypsum karst include subsidence dolines (many with a known date of subsidence), sulphate-rich springs, stream sinks and the extent of the gypsum belt.

Salt

Salt in Great Britain occurs mainly in the Permian and Triassic strata of central and north-eastern England (Figure 1). Many towns on the Triassic strata have "wich" or "wych" in their names indicating that they are sited on former salt springs emanating from actively dissolving salt karst. These places became the focus for shallow mining and near-surface "wild" brine extraction, a technique that exacerbated the salt karstification (Arup Geotechnics, 1991; Calvert, 1915; Collins, 1971). Most extraction of natural brine has ceased and modern exploitation is mainly in dry mines or by deep controlled brine extraction leaving brine-filled cavities. Since the cessation of natural brine pumping, the saline ground water levels have returned towards their pre-pumping state. Brine springs are becoming re-established and natural karstification and subsidence may be expected to occur. The exact nature of the brine flow and how it might interact with mined and brined areas has yet to be studied. To understand these mechanisms it is important to collate the information and integrate it with data about the amount of salt in spring and river water. The BGS has a national geochemical database, which includes river water analyses for most of the saline areas. The generation of a karst hazards GIS will allow the rapid integration and analysis of the various datasets. The GIS data for the salt areas will include saline springs, subsidence dolines and areas of more widespread subsidence. Because the salt karst is strongly influenced by brine pumping and mining, there will also be entries into a similar mining database.

KARST HAZARDS DATABASE AND GIS

There is a wide range of karstified rock types in Britain, but they have many common characteristics including springs, sinks, subsidence features and distinctive groundwater chemistries. These are all features relevant to a British karst hazards database. In the past 20 years, attempts have been made to collate such karst information in Britain. An early study was of the Chalk (Edmonds, 1983) and this formed the starting point for the "Natural Underground Cavities" study for the then Department of the Environment (Applied Geology Ltd, 1993). This study produced a large complicated database of information held in Dbase4 architecture. However, it also included much information (such as geological map and formation data) that is now available as layers in a geographic information system (GIS). The practicalities of this database have been assessed to help define the BGS karst database structure; some data fields have been accepted, but the slightly populated or potentially GIS-derived data fields have been abandoned. Another study of karst hazards utilising GIS was done by Wadge et al (1993) using the data sets of Cooper (1986 and 1989). This study produced a prototype expert GIS system for the city of Ripon using ArcInfo running on Sun workstations. Later studies at Ripon have extended the karst hazard information and presented it in a suitable way for planning (Symonds Travers Morgan, 1996; Paukstys et al., 1997). It is now appropriate to extend these assessments beyond the small study areas to a national dataset. There is a need for good accessible geological information to allow the proper implementation of the planning guidance policy for unstable land (Department of the Environment 1990, Department of the Environment, Transport and the Regions, 2000). The BGS karst geohazards project and complementary projects on landslips and mining help BGS to supply these needs.

The fields that have been defined initially for the new British karst hazards database are shown in Tables 1-5, these will be reviewed in the early stages of database population. In designing this system numerous problematical areas have been found. The main problems are how to keep the system simple and how to record temporal information. Problems include how to link data sets such as tracer tests to sinks and springs where multiple links may exist and numerous dye tracer tests may have been undertaken. The definition of the way subsidence event dates are recorded is also difficult since some features may occur instantaneously, while others may have occurred over an extended time, or have only vague subsidence dates associated with them. As many of the geologists who will have to use the system may not be karst experts, it is important to keep the number of data fields to the essential minimum; options for more details will be held on secondary screens if the study requires them. It is envisaged that the proforma sheets required to record the karst data will eventually be displayed in the field on a portable computer, or hand-held data capture device, to facilitate direct database entry during fieldwork. In common with the BGS data architecture, each database entry will include the user code for the geologist responsible for its input, the National Grid Reference, the ground elevation and the observation date. These will either input manually or via a Global Positioning System during field data capture.

In addition to these databases, national databases already exist for borehole and water abstraction information. A separate exercise is producing national information about rockhead surfaces and the thickness of superficial deposits. These factors are important for the interpretation of mantled karst areas that are particularly prone to subsidence sinkhole formation. Used in conjunction with the karst information within a GIS environment, new interpretations of the relationships between the causes of the data can be readily made.

Table 1. The main database fields for sinkholes and subsidence features in the karst geohazards database

Item	Details/Type	Links
Dolines, subsidence hollows and pipes; point or polygon data Information such as stream sinks are linked to the sinkholes via the GIS	Size at surface	
	Shape (plan)	
	Shape (profile)	
	Type (if known)	
	Reliability of information	
	Date of subsidence	
	Fill deposits	Linked to the stratigraphical and lithological lexicons
	Property damage	Links to secondary field, recorded using the NCB scale of property damage 1-5 (NCB, 1975)
	Evidence of quarrying	
	Stream sink?	Links to stream sinks database
	Pollution	Linked to "lithological" lexicon defining waste types
	Source of information	
	Other data	

Table 2. The main database fields for stream sinks in the karst geohazards database

Item	Details/Type	Links
Stream sinks, point data	Name of stream sink	
	Size of stream sink under average conditions	
	Seasonality	
Linked to the sinkholes via the GIS	Type of sink	
	Reliability of data	
	Hydrochemistry data	If yes, link to hydrochemistry database and secondary data entry field
	Proven tracer tests	If yes link to hydrological links database
	Source of information	
	Other data	

Table 3. The main database fields for springs and resurgences in the karst geohazards database

Item	Details/Type	Links
Springs and resurgences; point data	Name of spring	
	Size of spring under average conditions	
	Seasonality	
	Type of spring	If borehole links to borehole database and secondary data entry sheet
	Reliability of data	
	Hydrochemistry data	If yes, link to hydrochemistry database and secondary data entry sheet
	Uses: Public water supply?	
	Proven tracer tests	If yes link to hydrological links database
	Source of information	
	Other data	

Table 4. The main database fields for caves in the karst geohazards database

Item	Details/Type	Links
Natural cavities; point, line and scanned data	Name of cave	
	Type of cavity	
	Size	
	Rock units	Link to stratigraphical lexicon
	Links to other caves	
	Hydrologically active	Link to hydrological links database
	Reliability of data	
	Source of information	
	Other data	

Table 5. The main database fields for hydrological links in the karst geohazards database

Item	Details/Type	Links
Hydrological links; attributed line data	Type of test	
	Link determinant	
	Operator	
	Tracer collection	
	Links to other caves	Links in database to other sinkholes and resurgences
	Flow time	
	Reliability of data	
	Source of information	
	Other data	

IMPLEMENTATION OF THE KARST HAZARDS GIS AND DATABASE

The karst hazard database is under construction and test areas are being populated with data to confirm the working practicalities of the system and the data fields. These test areas include Ripon for gypsum and part of Hampshire for the chalk. When the system is fully implemented, it will allow the karst geohazards data to be continuously revised as new information is acquired. Other entities will be derived from the database using GIS to interpret it. These will include the subsidence-prone areas or geohazard domains, which commonly cut across geological boundaries. They can only be defined by synthesis of the karst geohazard information outlined above, allied with geological map information. For example, the belt of subsidence caused by dissolution of the Permian gypsum in

Yorkshire affects three formations and the basal part of the overlying group. This area is about 3km wide forming a linear belt through the city of Ripon where special planning constraints are formally enforced (Symonds Travers Morgan, 1996; Paukstys et al., 1997). The karst hazards database and the GIS will allow this belt to be defined for the remainder of the gypsum karst area. Similarly, in the Chalk areas, the GIS synthesis will help with the identification of geohazard domains, which can then be added to the GIS as a layer to help with planning and groundwater protection.

The karst hazard database structure is designed to interact with numerous other databases held by the British Geological Survey. The database has a unified structure constrained by definitions served directly from the Oracle databases to prevent the erroneous entry of incorrect data. The database fields will each have a descriptive definition held in a linked table. This can then be used to produce automated reports of selected items or areas directly from the database. Because the database is served through a GIS interface, it allows the hazards information to be interrogated and linked with the underlying geological information, hydrological and hydrogeological datasets. The system also forms the foundation for gathering the information directly using computer technology in the field.

ACKNOWLEDGEMENTS
Richard Ellison and Alan Forster are thanked for critically reviewing the manuscript, Tony Waltham and Martin Culshaw are thanked for constructive comments. Published with permission of the Director, British Geological Survey (N.E.R.C.).

REFERENCES
Arup Geotechnics, 1991. Review of mining instability in Great Britain, section 3vi, Stafford brine pumping (Staffordshire). Arup Geotechnics, Newcastle upon Tyne.

Applied Geology Limited. 1993. Review of instability due to natural underground cavities in Great Britain. Royal Leamington Spa. Applied Geology Ltd.

Calvert, A.F. (1915) Salt in Cheshire. Spon Ltd, London, 1206 p.

Collins, J.F.N. 1971. Salt: a policy for the control of salt extraction in Cheshire. Cheshire County Council

Cooper, A H. 1986. Foundered strata and subsidence resulting from the dissolution of Permian gypsum in the Ripon and Bedale areas, North Yorkshire. 127-139 in Harwood, G M and Smith, D B (eds). The English Zechstein and related topics. Geological Society of London, Special Publication. No. 22.

Cooper, A H. 1989. Airborne multispectral scanning of subsidence caused by Permian gypsum dissolution at Ripon, North Yorkshire. Quarterly Journal of Engineering Geology (London), Vol. 22, 219-229.

Cooper, A.H. 1998. Subsidence hazards caused by the dissolution of Permian gypsum in England: geology, investigation and remediation. In Maund, J.G. & Eddleston, M (eds.) Geohazards in Engineering Geology. Geological Society, London, Engineering Special Publications, 15, 265-275.

Department of the Environment, 1990. Planning policy guidance note 14: Development on unstable land. London, HMSO.

Department of the Environment, Transport and the Regions, 2000. Planning policy guidance note 14. Development on unstable land. Annex2: Subsidence and planning. Consultation paper.<http://www.planning.detr.gov.uk/conindex.htm>

Edmonds, C.N. 1983. Towards the prediction of subsidence risk upon the Chalk outcrop. Quarterly Journal of Engineering Geology, London. Vol. 16, 261-166.

N.C.B. (1975) Subsidence Engineers' Handbook. National Coal Board Mining Department. UK, 111 pp.

Paukštys, B., Cooper, A.H. and Arustiene, J. (1997) 'Planning for gypsum geohazards in Lithuania and England'. 127-135 in Beck, F.B. and Stephenson, J.B (Editors) The Engineering Geology and Hydrogeology of Karst Terranes. Proceedings of the Sixth Multidisciplinary Conference on Sinkholes and the Engineering and Environmental Impacts of Karst Springfield/Missouri/6-9 April 1997. A.A.Balkema, Rotterdam.

Symonds Travers Morgan.1996. Assessment of subsidence arising from gypsum dissolution: Technical Report for the Department of the Environment. 228pp. Symonds Group Ltd, East Grinstead.

Wadge, G., Wislocki, A., Pearson, E.J. and Whittow, J.B. (1993). Mapping natural hazards with spatial modelling systems. In, Mather, P. (ed.) Geographical Information Handling - Research and Applications, Wiley, 239-250.

Watham, A.C., Simms, M.J., Farrant, A.R. and Goldie, H.S. 1997. Karst and caves of Great Britain. Joint Nature Conservation Committee. Chapman and Hall, London. 358pp.

Geotechnical and Environmental Applications of Karst Geology and Hydrology, Beck & Herring (eds)
© 2001 Taylor & Francis, ISBN 90 5809 190 2

Karstification below dam sites: a model of increasing leakage from reservoirs

W.DREYBRODT & D.ROMANOV   Karst Processes Research Group, Institute of Experimental Physics, University of Bremen, D-28334 Bremen, Germany, dreybrod@physik.uni-bremen.de

F.GABROVSEK   Karst Research Institute, Postojna, Slovenia, gabrovsek@zrc-sazu.si

ABSTRACT

Unnaturally steep hydraulic gradients below foundations or across abutments of dams may cause solutional widening of fractures in karstifiable rocks of carbonates or gypsum. This could cause increasing leakage which may endanger the performance of the construction. To investigate this problem recent models on natural karstification have been applied. We have performed numerical simulations of leakage below a model dam with a grouting curtain reaching down to 100 m below its impermeable foundation of 100 m width. Water is impounded to a depth of 100 m. The dam is located on a terrane of fractured rock dissected by two perpendicular sets of fractures with spacing of 5 m, and with a log-normal distribution of their initial aperture widths of about 0.02 cm. In the first state of karstification these fractures widen slowly, until a pathway of widened fractures below the grouting has reached the downstream side with exit widths of about 1 mm. This causes a dramatic increase of leakage and turbulent flow sets in. After this breakthrough at time T, in the second state of karstification, dissolution rates become even along these fractures and cause widening of about 0.1 cm/year for limestone, and at least of 1 cm/year for gypsum. This leads to an increase in leakage to unbearable rates within further 25 years for limestone, but only 5 years for gypsum. We have performed a sensitivity analysis of breakthrough time T for the various parameters, which determine the problem. The result shows breakthrough times in the order of several ten years for both limestone and gypsum. We have also modelled leakage to caves or karst channels 200 m below the bottom of the reservoir, which could induce the formation of sinkholes. The model can be extended to more realistic settings. In conclusion our results support the suspicion that increasing leakage at dam sites can be caused by recent karstification which is activated after filling the reservoir and can lead to serious problems within its lifetime.

INTRODUCTION

To understand the processes of the early state of conduit evolution in karstifying limestone one- and two-dimensional models have been developed that couple flow rates of calcite aggressive water to dissolutional widening of initial narrow fractures (Dreybrodt, 1990, 1996; Siemers and Dreybrodt, 1998; Dreybrodt and Gabrovsek, 2000; Dreybrodt and Siemers, 2000, Palmer, 1991, 2000; Groves and Howard, 1994). These models reveal a positive feedback loop coupling the rate of dissolutional widening to the flow rate of the aggressive water, driven by a constant head from an input of a percolating pathway of fractures to its output. This positive feedback loop causes an initially slow increase of the fracture widths and the flow rates, which suddenly is enhanced dramatically. The time when this happens is termed breakthrough time T. From then on constant head conditions of flow break down, because such high flow rates under most natural conditions can no longer be supported by the limited amount of water available at the input.

The cause for the breakthrough behaviour is the nonlinear dissolution kinetics of limestone (Dreybrodt and Eisenlohr, 2000, Eisenlohr et al, 1999) by which the dissolution rates are given as

$$F_1(c) = k_1(1 - c/c_{eq}) \ for \ c \leq c_s, \qquad F_n(c) = k_n(1 - c/c_{eq})^n \ for \ c > c_s \qquad (1)$$

c is the concentration of calcium in the calcite aggressive water, c_{eq} is its equilibrium concentration with respect to calcite, $c_s \approx 0.9 c_{eq}$ is the switch concentration, where the kinetics switches from a linear rate law to a nonlinear one with order $n \approx 4$. k_1 and k_n are rate constants in mol cm^{-2}s^{-1}. Recent research on gypsum rocks have revealed a similar rate law also for this mineral (Jeschke et al, 2000).

Using these rate laws in the models enables one to give analytic estimations of T, which reveals the parameters which determine early karstification. These are initial aperture width a_0 of the fracture, the distance L between input and output, the hydraulic head h, and the chemical parameters n, k_n, c_s and c_{eq}. From these parameters the breakthrough time T can be estimated for single fractures (Dreybrodt, 1996; Dreybrodt and Gabrovsek, 2000), and for percolation networks on a square lattice (Siemers and Dreybrodt, 1998; Dreybrodt and Siemers, 2000) by

$$T = (\frac{1}{a_0})^{\frac{2n+1}{n-1}} \left(\frac{L^2 \eta}{h c_{eq}} \right)^{\frac{n}{n-1}} \cdot (k_n)^{\frac{1}{n-1}} \cdot const \qquad (2)$$

where η is the dynamic viscosity of water, and the constant can be estimated for single conduits and simple networks.

After breakthrough the concentration c in the conduits drops close to zero and dissolutional widening is even along the conduit at about 1 mm/year for limestone (Buhmann and Dreybrodt, 1985) and at least 1 cm/year for gypsum (Jeschke et al, 2000; James, 1992).

Close to hydraulic structures in karst regions such as dam sites or artificial underground reservoirs, created by plugging karst channels (Milanovic, 2000), hydraulic heads are extremely high, and pathways of flow from the reservoir to base level are comparably short. Therefore extreme steep hydraulic gradients arise. This provokes the question, whether karstification below dams could be so fast, that within their lifetimes leakage increases to unbearable amounts. First models to answer this question have been developed on one-dimensional conduits by Palmer (1988) and Dreybrodt (1992, 1996). Both authors found comparable results, which demonstrated that hydraulic structures may be seriously affected within time spans of hundred years. Such single conduit models, however, are not realistic since no exchange of water between the evolving conduits and the surrounding fractures is considered. If water is allowed to flow from an evolving conduit into the net of fractures, more aggressive input water is driven along it and dissolution rates should be higher. This should accelerate its evolution. First approaches to this have been reported by Bauer et al. (1999).

In this work we deal with a two-dimensional cross section of a dam site, where the underlying rock is dissected by two perpendicular sets of fractures with a statistical distribution of their aperture widths. The spacing and the apertures of the fractures are chosen such that the hydraulic conductivity of the limestone is about 10^{-6} m/s, representative for moderately karstified rock.

MODEL STRUCTURE

The model consists of a two dimensional cross section of the soluble rock below the dam (see Fig. 1). The modelling domain is 500 m wide and 250 m deep. The fractures are represented by a square net with a spacing of 5 m by 5 m. This creates a network of 100 by 50 squares. The width of all fractures is 1 m. To account for the heterogeneity of the fracture system we assign a selected aperture width to each fracture within the net. Furthermore it is possible to consider different lithologies of the bedrocks by assigning to each fracture different values of the dissolution rate constants k_1, n, k_n and c_{eq}. By this way one is able to model also regions of insoluble rock or of highly soluble gypsum, located in the model domain. The impermeable dam with width W at its basis is also shown in Fig. 1. At its left-hand side water is impounded to depth h, supplying a constant head boundary condition. At its right-hand side, downstream, the hydraulic head is constant at zero. An impermeable grouting curtain below the dam extends to depth g. The lower domain boundary and the left-hand and right-hand borders are assumed as impermeable.

Water flow in the net is assumed to be laminar. The calculation of the head h_i for node i is based on mass conservation for each node, i.e. flow rate into the node is equal to that leaving it. The resistance of each fracture is calculated by the Hagen-Poiseuille law. The resulting set of linear equations is solved by preconditioned conjugate gradient method for sparse matrices (Steward and Leyk, 1994). Once all heads are known the flow rate in each fracture is calculated. Dissolutional widening F in each fracture is then calculated from the flow rates and the rate law of dissolution (eqn. 1). Details are given by Siemers and Dreybrodt (1998) and by Gabrovsek (2000). To obtain the aperture width profiles of the fractures and the flow rates as a function of time we use an iterative procedure. If $a(x,t)$ is the profile of fracture i along the coordinate x after a sufficiently short time step Δt, the new width profile is given by

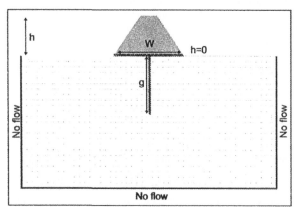

Figure 1: The model dam with fractured rocks below. Water is impounded to the left. The impermeable base and the grouting curtain are also shown.

$$a(x, t + \Delta t) = a(x,t) + 2\gamma F(x,t) \cdot \Delta t \qquad (3)$$

where $\gamma F(x,t)$ is retreat of bedrock in cm/year and Δt is a time step in years. After each time step laminar flow in all fractures is assured by calculating the Reynolds Number Re which remains well below 2000 until breakthrough occurs. The run is terminated, when Re exceeds 2000.

THE MODEL DAM ON LIMESTONE

Figs 2 a, b, c show a scenario, which is typical for a dam in limestone. The results for a dam in gypsum are also illustrated, but will be discussed later. The width of the dam is 100 m, grouting depth $g = 100$ m, and the depth h of the impounded water is 100 m. The fracture net consists of fractures one meter wide with a log-normal statistical distribution of their initial aperture widths with mean $a_0 = 0.02$ cm and $\sigma \approx 0.01$ cm. The chemical parameters have been taken $k_1 = 4 \cdot 10^{-11}$ mol cm^{-2}s^{-1}, $k_4 = 4 \cdot 10^{-6}$ mol cm^{-2}s, $c_{eq} = 2 \cdot 10^{-6}$ mol cm^{-3}, $c_s = 0.9\, c_{eq}$. These are typical for limestone (Dreybrodt and Eisenlohr, 2000). In this scenario we assume allogenic water with low calcium concentration impounded in the reservoir. Therefore the concentration of the water entering the fractures has been taken, $c_0 = 0$.

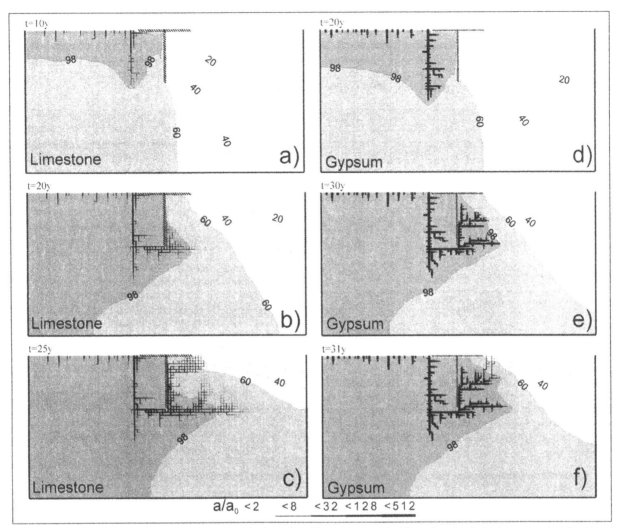

Figure 2: Evolution of karst channels below the model dam after filling the reservoir. Figs a, b and c depict limestone after 10, 20 and 25 years respectively. Breakthrough occurs at 25 years, when the fractures at the exit have widened to about 0.1 cm. Figs d, e and f show the evolution in gypsum for comparison at 20, 30 and 35 years respectively. The bars at the bottom indicate the widths of the fractures in multiples of the initial aperture width a_0.

Fig. 2a shows the situation ten years after filling the reservoir. First small channels started to grow. Close to the horizontal impermeabilization a main channel has penetrated to about 100 m downwards. The pressure isolines are also shown, to illustrate the distribution of hydraulic gradients. These are steepest just below the grouting curtain. Therefore flow is concentrated into this region. Consequently dissolutional widening is most active there, and after only 20 years a net of small channels with widths of about 1 mm have grown below the grouting. The steepest hydraulic gradient is now close to the right-hand side of the grouting wall and widening creates a channel directed upwards (Fig. 2b). After another 5 years this channel has reached the surface, and other channels follow the steep hydraulic gradients upwards (Fig. 2c). The fracture widths at those exits are about 0.1 cm. Fig. 3 shows the evolution of leakage Q in time. Three phases of behaviour can be observed:

a) A small increase of Q during the first 15 years;
b) After this time channels have approached closer to the surface and the total resistance to flow has decreased because only the region of steep gradients is utilized for flow. Therefore as the channels approach closer to the surface total flow increases.
c) Finally when the first channel has reached the surface breakthrough occurs and flow increases steeply. The run was terminated when turbulence sets in at a total leakage of 2000 $cm^3 s^{-1}$. For a dam with a width of 100 m this amounts to a leakage of 0.2 m^3/s. Shortly after breakthrough the calcium concentration becomes less than 0.1 c_{eq} everywhere in the leading flow path and dissolution rates are high and even, causing fracture widening of about 1 mm/year (Buhmann and Dreybrodt, 1985). This will not change during turbulent flow.

The further evolution of leakage can be estimated to a lower limit in the following way: The flow path of about 300 m length is approximated by one single straight channel with the hydraulic diameter of its exit, everywhere along the channel. For such a single channel we use the Darcy-Weisbach equation and the Colebrook-White formula for the friction coefficient to calculate the flow as a function of the fracture width of the channel (Dreybrodt, 1988). This estimation is also depicted in Fig. 3. Two limits are given. The upper curve is for a completely smooth channel, whereas the lower represents flow in a rough channel with roughness of 10% of the channel width. The flow rates increase in time t by a power law $Q \propto t^{3/2}$. About 25 years after breakthrough they have increased to about 100liter/s per meter of dam. For a dam, hundred meters wide, this amounts to $10 m^3 s^{-1}$. One should keep in mind that this is a lower limit. We will extend our program to turbulent flow to get more detailed information.

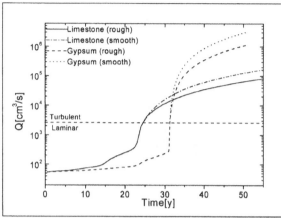

Figure 3: Evolution of leakage Q below the dam for limestone (full line) and gypsum (dotted line). The units of Q refer to the model section with 1 m widths of the fractures. The total leakage is found by multiplying with the width of the dam in m. The lower part of the figure depicts phase a) with laminar flow. At breakthrough turbulent flow sets in. The evolution of turbulent flow has been estimated for smooth and rough channels. See text. The limit of unbearable water loss is at 10^5 cm^3/s, which corresponds to about 10 m^3/s for a dam with a width of 100 m.

THE MODEL DAM IN GYPSUM

In this scenario we change the underlying rock from limestone to gypsum. This is achieved by changing the constants of dissolution to those of gypsum. These are $k_1 = 1.3 \cdot 10^{-7}$ mol $cm^{-2}s^{-1}$, $k_n = 3 \cdot 10^{-5}$ mol $cm^{-2}s^{-1}$, $n = 4.5$, and $c_{eq} = 15.4 \cdot 10^{-6}$ mol cm^{-3} (Jeschke et al., 2000).

In all calculations, in the same way as for limestone we have calculated the rates for pure diffusion control and compared them to the surface controlled rates. Then the smaller rates were used. Figs. 2 c, d, e show the results. They depict the channel patterns after 20 years, 30 years, and at breakthrough one year later. The behaviour is similar to that of Figs. 2 a, b, c, although owing to the much higher dissolution rates all channels are significantly wider. At breakthrough the exit widths are about 2 mm. The evolution of the flow rates is shown by Fig. 3. After breakthrough calcium concentrations are less than $0.1 c_{eq}$. Dissolution rates therefore are high, minimal 10 mm/year (James, 1992). Although with increasing flow rates the dissolutional widening may rise up to 10 cm/year we use the lower values to give a lower limit estimation. Fig. 3 shows also this estimation for turbulent flow. Only 5 years after breakthrough leakage rates have increased to 100 l/s.

The main difference between limestone and gypsum in this context is not the evolution of pathways before breakthrough. These are similar. It is the huge difference in the dissolution rates after breakthrough, which causes the much faster increase of leakage at gypsum sites. Although this gypsum scenario might not be very realistic, it has been described here for didactic reasons. More realistic scenarios, where gypsum occurs as layers in strata of clays, limestone, or sandstone can be handled by our model. This will be the target of future work.

SENSITIVITY ANALYSIS

To gain some estimation on the dependence of breakthrough times on the various parameters we used our model scenarios on limestone and on gypsum, and have varied one parameter, leaving all the others unchanged. However, when doing so, we have used different realizations of the statistical distribution of aperture widths. If one uses several different realizations leaving everything else unchanged breakthrough times vary by about ± 15%. This has been taken for the error bars in the following figures.

Fig. 4a depicts the dependence of breakthrough times on the average aperture widths a_0 of the fractures. Breakthrough times increase steeply with decreasing aperture width. They follow roughly a power law $T \propto a_0^{-3,8 \pm 0.2}$ which is close to the behaviour of breakthrough times for single conduits (cf. eqn. 2). Fig. 4b depicts the dependence of T on the head, h, which can be approximated by $T \propto h^{-1.3 \pm 0.3}$. With increasing depth g of grouting breakthrough times increase significantly for grouting depths larger than the impounded height h of the water (See Fig.4c).

An important parameter for dissolutional widening is the chemical composition of the water on the bottom of the lake, which determines both the input concentration c_0 to the fractures, and c_{eq} for limestone. Fig. 4d depicts the dependence of T on c_0. There is only little change for $c_0 < 0.5 c_{eq}$. If the water, however, becomes more saturated breakthrough times increase. After breakthrough the concentration drops to c_0 along the leading fractures and dissolution rates are accordingly lower. Thus the chemistry at the bottom of the lake is of utmost importance for limestone. In the case of gypsum, however, c_{eq} does not depend on the CO_2 concentration in the lake. Moreover impounded waters are usually not close to saturation with respect to gypsum. Therefore c_0 is much smaller than c_{eq}. Finally in Fig. 4e, we show the influence of c_{eq} to T for limestone, which again reminds to single channel behaviour with $T \propto c_{eq}^{-1.3 \pm 0.2}$. The boundary conditions of the impermeable right and left-hand sides of the domain may appear irrealistic since the soluble rocks may extend much further. We have therefore changed the domain width to 1000 m using a 200 by 50 net. No changes in breakthrough times were found.

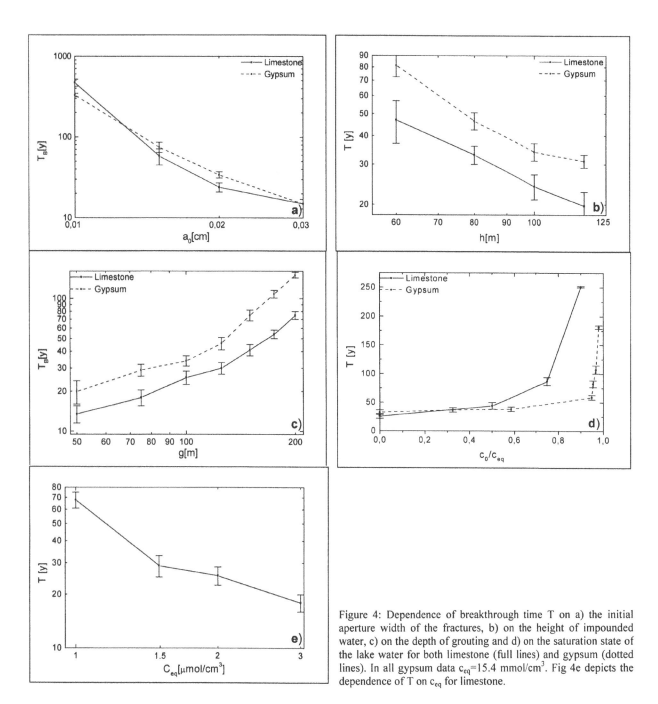

Figure 4: Dependence of breakthrough time T on a) the initial aperture width of the fractures, b) on the height of impounded water, c) on the depth of grouting and d) on the saturation state of the lake water for both limestone (full lines) and gypsum (dotted lines). In all gypsum data c_{eq}=15.4 mmol/cm^3. Fig 4e depicts the dependence of T on c_{eq} for limestone.

SINKHOLES

For a conference on sinkholes it is mandatory to deal with them also in this context. The formation of sinkholes on the floors of water reservoirs in karstic areas is a well known problem. If caves or karst conduits are located deep below the bottom of the lake a pathway of dissolutional widening could be directed from the floor of the reservoir to these conduits. We

have assumed that such conduits are located 200 m below. The location of the cave is marked by the black square. In the model we assign a hydraulic head of 10 m to all nodes comprising this area. Thus we simulate a large conduit completely filled with water which is directed towards a spring with an elevation of 10 m. In the case of a vadose cave the head which has to be chosen would be, $h = -200m$, the depth of location below ground. The network of fractures has been created with the standard log-normal distribution of aperture widths with the same distribution as in the previous cases. Fig. 5a shows the result at breakthrough for a limestone terrane. For limestone breakthrough is reached after only 11 years and the fractures at the cave have been widened to about 0.1 cm. After breakthrough the further evolution of leakage through this region is determined by turbulent flow and can be estimated by assuming a straight channel downwards as has been described above. Fig. 5b shows what happens when limestone is replaced by gypsum,

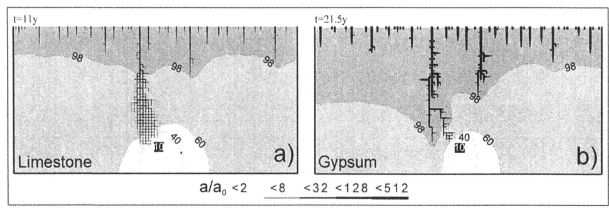

Figure 5: First stage of sinkhole evolution at breakthrough in limestone and gypsum respectively. The black square represents a karst channel at hydraulic head of 10 m, located 200 m below the reservoir bottom. Depth of the impounded lake is 100 m. The isolines of the head show the head contribution and illustrate the hydraulic gradient. The bars at the bottom indicate fracture widths in multiples of initial fracture width a_0

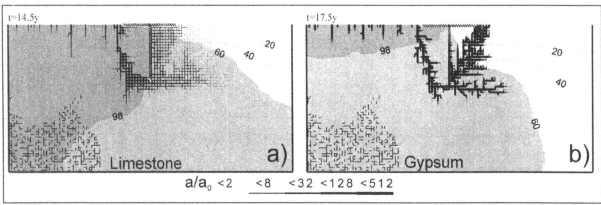

Figure 6: Conduit pattern and breakthrough for the scenarios of Fig. 2, but 30% of the fractions have a width of 0.035 cm. These fractures are evenly distributed on the entire net. A section is shown for illustration on the lower left-hand side of the net. In contrast to Fig. 2 maze like patterns are generated.

everything else left unchanged. Due to the different dissolution kinetics breakthrough is achieved after 21 years. At that time aperture widths close to the cave are about 0.2 cm. Such nets of small channels as shown by Fig. 5 can be regarded as initial states of sinkholes. To find out, how they develop requires extension of the programme to turbulent flow, which currently is in progress.

CONCLUSION AND FURTHER PERSPECTIVES

We have presented a computer model of karstification below dam sites and demonstrated that leakage from these hydraulic structures can be caused by dissolutional widening of initially narrow fractures of about 0.02 cm aperture widths. Such a geological situation, with such narrow fractures solely present, is highly idealistic. This is also seen from the initial leakage rates (cf. Fig. 3). For our standard dam with a width of 100 m initial leakage is only 6 litres/s, which would be ideal for each dam. In reality, even in moderately karstified regions there will be wider fractures and small channels distributed underground. This will greatly reduce breakthrough times. To obtain an idea about this we have assigned aperture widths of 0.035cm, to 30% of the fractures in our statistical standard. This was done by selecting each fracture and replacing its width by 0.035cm with an occupation probability of 0.3. Such a percolation net will not show percolating pathways connecting inputs and outputs by those wide fractures solely. (Siemers and Dreybrodt, 1998). Breakthrough times are reduced to half of those of the corresponding standard case. The conduit evolution at breakthrough is illustrated by Fig. 6. Although the changes in aperture widths seem to be moderate there is a striking difference of the conduit patterns compared to Fig. 2. The more dentitic pattern here is now replaced by a maze, with much more fractures contributing to the leakage from the reservoir. This shows that details in the structure of the rock below the dam site are of high relevance. Realistic geological settings require to incorporate into the programme wider fractures, fault zones, and already existing karst channels. This needs an intensive exchange of information between modellers and civil engineers. To extend the programme to such highly complex situations is tedious but should not present major problems.

In this paper we have intended to show on relatively simple cases that karstification below dam sites and close to other hydraulic structures may be induced by unnaturally steep hydraulic gradients acting on unnaturally short pathways of a few hundred meters. Although these models cannot claim that their predictions are highly reliable they give hints to what may occur during the lifetime of the dam, but probably after the lifespan of those who built it.

REFERENCES

Bauer, S., Birk, S., Liedl, R., and Sauter, M., 1999, Solutionally enhanced leakage rates of dams in karst regions. In: Palmer, A. N., Palmer, M. V., and Sasonsky, I. D., (Editors), Karst Modeling. Special Publication 5. Karst Waters Institute Inc., Charles Town, West Virginia, USA.

Buhmann, D., and Dreybrodt W., 1985, The kinetics of calcite dissolution and precipitation in geologically relevant situations of karst areas: 2. Closed system: Chem. Geol., 53, 109-124.

Eisenlohr L., Meteva, K., Gabrovsek, F., Dreybrodt, W., 1999, The inhibiting action of intrinsic impurities in natural calcium carbonate minerals to their dissolution kinetics in aqueous H_2O-CO_2 solutions: Geochimica et Cosmochimica Acta, 63, 989-1002.

Dreybrodt W., 1988. Processes in karst systems - Physics, Chemistry and Geology. Springer Series in Physical Environments 4, Springer, Berlin, New York, 188p.

Dreybrodt, W., 1990, The role of dissolution kinetics in the development of karstification in limestone: A model simulation of karst evolution: Journal of Geology, 98, 639-655.

Dreybrodt, W., 1992, Dynamics of karstification: A model applied to hydraulic structures in karst terranes. Appl. Hydrogeol., 1, 20-32.

Dreybrodt, W., 1996, Principles of early development of karst conduits under natural and man-made conditions revealed by mathematical analysis of numerical models: Water Resources Research, 32, 2923-2935.

Dreybrodt, W., and Eisenlohr, L., 2000, Limestone dissolution rates in karst environments: In: Klimchouk, A., Ford, D.C., Palmer, A.N., and Dreybrodt, W., (Editors), Speleogenesis: Evolution of karst aquifers. Nat. Speleol. Soc.,USA,136-148.

Dreybrodt, W., and Gabrovsek F., 2000, Dynamics of the evolution of a single karst conduit: In: Klimchouk, A., Ford, D.C., Palmer, A.N., and Dreybrodt, W., (Editors), Speleogenesis: Evolution of karst aquifers. Nat. Speleol. Soc.,USA,184-193.

Dreybrodt, W., and Siemers, J., 2000, Cave evolution on two-dimensional networks of primary fractures in limestone: In: Klimchouk, A., Ford, D.C., Palmer, A.N., and Dreybrodt, W., (Editors), Speleogenesis: Evolution of karst aquifers. Nat. Speleol. Soc.,USA, 201-211.

Gabrovsek, F., 2000, Evolution of early karst aquifers: From simple principles to complex models: Zalozba ZRC, Ljubljana, Slovenia, 150p.

Groves, C.G., and Howard, A.D., 1994, Early development of karst systems. 1. Preferential flow path enlargement under laminar flow: Water Resources Research, 30, 2837-2846.

James, A.N., 1992, Soluble Materials in Civil Engineering: Ellis Horwood Series in Civil Engineering: Ellis Horwood, Chichester, England, 434p.

Jeschke,A.A., Vosbeck, K., and Dreybrodt, W., 2000, Surface controlled dissolution rates in aqueous solutions exhibit nonlinear dissolution kinetics: Geochimica et Cosmochimica Acta, 65, 13-20.

Milanovic, P. T., 2000, Geological Engineering in Karst, Zebra Publishing, Ltd, Belgrade, Yugoslavia, 347 p.

Palmer,A.N., 1988, Solutional enlargement of openings in the vicinity of hydraulic structures in karst regions: In: 2nd Conference on Environmental Problems in Karst Terranes and Their Solutions. Assoc. of Groundwater Scientists and Engineers Proceedings, USA, 3-15

Palmer, A. N., 1991, The origin and morphology of limestone caves: Geol. Soc. of America Bull., 103, 1-21.

Palmer, A.N., 2000, Digital modeling of individual solution conduits: In: Klimchouk, A., Ford, D.C., Palmer, A.N., and Dreybrodt, W., (Editors), Speleogenesis: Evolution of karst aquifers. Nat. Speleol. Soc.,USA, 194-200.

Siemers, J., and Dreybrodt, W., 1998, Early development of karst aquifers on percolation networks of fractures in limestone: Water Resources Research, 34, 409-419.

Steward, D.E., Leyk, Z., 1994, Meschach:The matrix computation in C: Proceedings of the Centre For Mathematics And Its Applications, 32,The Australian National University, Canberra, Australia.

Geotechnical and Environmental Applications of Karst Geology and Hydrology, Beck & Herring (eds)
© 2001 Taylor & Francis, ISBN 90 5809 190 2

Remediation in karst – art, science, witchcraft, and economics

JOSEPH A.FISCHER, JOSEPH J.FISCHER, & RICHARD S.OTTOSON   Geoscience Services, Bernardsville, NJ 07924, USA, geoserv@hotmail.com

ABSTRACT

Subsurface remediation in karst terrain in the United States is by no means a new subject, although still a young one. As development pushes onto lands that were once too difficult or uneconomical to construct on in the past, more and more remediation work is becoming necessary. In the author's experience, remediating or quantifying karst hazards has gone from a sideline specialty to the front burner. Some of this increase is created by greater awareness, but that awareness came about as a result of increased construction over karst and subsequent failures.

Almost no amount of subsurface investigation would be able to characterize most karst sites with any accuracy. This is why more than science applies to developing a karst site at hazard levels on the order of a non-carbonate site. If the science is there, rarely is there enough money to utilize it fully. This is where art and witchcraft become necessary.

Remediating requires an understanding of the likely failure mechanisms to occur in a region for the conditions expected, then waving your magic wand, incanting the special words, performing an aerial reconnaissance on your broom, then going to work. Grouting in various forms (dental, slurry, compaction or a combination) appears to be the most prevalent form of remediation as it can be applied to a variety of conditions and failure mechanisms. Dynamic destruction (or dynamic compaction in conventional terranes) can be economical and useful depending on the planned construction. Brute force (either through deep or super-strong foundations) can also be utilized, though they can be quite expensive.

Thus, remediation requires a blend of engineering geologic understanding and extrapolation, geotechnical practice, and an appreciation for both the conservatism (or lack thereof) and the economics of the various concepts.

INTRODUCTION

Karst is an international problem, although there is only one location that truly typifies it - the province of Karst in what was formerly Yugoslavia. Areas of the United States and China certainly have true karst. However, in the United States, the term karst has been used for almost any area underlain by solutioned carbonate rocks. Obviously, there are differences in investigation, failure mechanism, and remediation concerns for areas underlain by:

- The geologically recent (Cenozoic) sediments of the Caribbean, Bahama Platform and Florida;

- Older (Mesozoic and Paleozoic) limestones and dolomites such as those found in the folded and faulted valleys and ridges of the eastern and western United States as well as the flat-lying carbonates typical of the mid-continent; and

- The occasional metamorphosed Proterozoic (solutioned) carbonates (marble).

Thus, developing appropriate remedial measures must include an understanding of the potential failure modes of the solutioned carbonates below a particular site, as well as the more common engineering parameters of building/facility type, loadings and potential remedial alternatives. Of course, a powerful player in any investigatory and/or remedial effort is economics. Not only should consideration be given to cost versus effectiveness, but the owner must also balance the overall expected costs against abandoning the facility.

FAILURE MODES

First, one must recognize that some carbonates are hard and some are soft. Some are old, faulted, folded and weathered. Some are old hard and flat-bedded. Some are recent, weak and flat. For purposes of this paper, we have considered failure in the structural sense (i.e. loss of foundation support), not in the environmental sense (i.e. contaminant flow through highly porous or solutioned carbonates). Such foundation failure can occur in a variety of ways, as noted below.

A.   Soft, recent karst is often characterized by "cover collapse" (solution) or "subsidence" sinkholes (e.g., Chen and Beck, 1989) where loose sandy materials fall or erode into generally near-vertical solution features within the underlying, relatively flat-lying "limerock". Foundation failures can essentially occur by the loss of support under all or part of a structure as indicated in Figure 1, below.

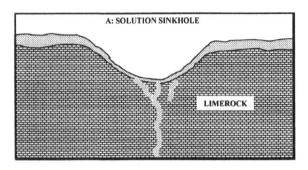
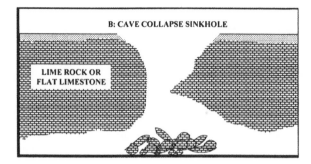
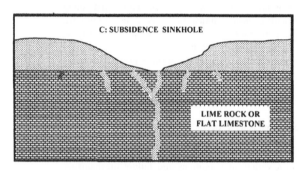
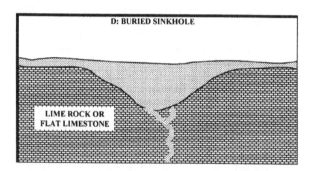

Figure 1 – Types of Sinkholes (Modified from Beck, 1991)

B.   Structural failure within the flat-lying, but hard Paleozoic rocks of the central United States generally occurs through "cave collapse" (e.g., Fischer, et al, 1996). The roof of a linear solution feature (cave) that has formed generally along a bedding plane, may collapse as a result of continued solutioning or increased loading.

C.   Conversely, a ravelling-type of failure generally occurs as a result of the erosion of overburden materials into solution features along inclined bedding or shear zones in, for example, the folded and faulted rocks of the Valley and Ridge Physiographic Province of the eastern United States (e.g., Fischer, et al, 1996). These sinkholes generally form from the rock surface upward and may not be evident until the soil arch (Figure 2A on the next page) collapses. Foundation concerns are obvious as the soil void can fail either from overloading or roof material removal.

D.   Solutioned cave features are often found in hard metamorphics of, for example, the western United States. Failure requires the overloading of a cavity roof or cutting into a cave (cave collapse sinkhole, see Figures 1B and 2B) rather than persistent dissolution or by erosion of the surficial soils into ancient cavities (Figures 1C or 2A).

    Thus, Beck, 1991, in characterizing sinkhole occurrences (Figure 1) used flat-lying rock. Revising this assumption and eliminating the limerock-type failure yields Figure 2 for folded crystalline rocks (Fischer, et al, 1996).
    These two-dimensional representations of failures yield a clue to what the geotechnical engineer planning the project must consider in three dimensions. Horizontal variations in the flat-lying sediments are primarily due to the existence of near-vertical solution channels ("cutters") usually found within these rocks (e.g., Figures 1A, 1C and 1D). Cavities (or even caves) can extend for several to hundreds of meters laterally. Erosion occurs along exposed bedding and often along beds at different elevations as a result of ground water level changes that may have occurred over long time periods. Often, different rates and/or quantities of precipitation will change the elevations of such a "conduit" flow phenomenon as well as the volume of flow within a conduit.
    For contrast, in folded rock terranes, significant strength and solubility variations can occur across bedding strike with, perhaps, down-dip continuity in physical properties. Of course, as fracturing can enhance permeability, variations of in situ stress magnitude and direction will further exacerbate the problem of trying to understand the variations in physical properties, performance and elevation across a folded and fractured, ancient Continental Margin area such as shown on Figure 3.
    Therefore, in developing site remediation or investigation protocol, an understanding of the nature of the subsurface and its likely failure mode (or modes), as well as the more conventional engineering concerns of loading (both structural and contaminant) and grading (cuts and fills) must be considered. Often, the folded rock terrane (Figure 2A) and limerock terrane (Figures 1A, 1C and 1D)

types of failures can be considered a <u>soil mechanics</u> constraint where a sinkhole or a zone of more compressible material results from the collapse of a soil void. In the more flat-lying, hard, crystalline rocks structural failures often result from the collapse of cavity roofs (more of a <u>rock mechanics</u> failure) or the softening of soils by moisture content increases and/or erosion of the fines. In addition, one must also recognize that <u>variations</u> in rock properties, including solutioning and weathering, will likely be more severe in a folded rock environment as a result of the numerous rock formations and/or beds that express themselves as the rock surface as one traverses the site (e.g., Figure 3).

The complexity shown on Figure 3 is not unusual in folded Appalachian terrane. The map shows a valley underlain by faulted Cambro-Ordovician-aged carbonates with the surrounding ridges and internal intrusives composed of Proterozoic-aged metamorphic rocks. The subsurface material variations along the cross-strike section A-A are enormous.

Thus in designing and estimating costs for a remediation program, it is necessary to have an understanding of the subsurface conditions and the likely failure mode(s). Sometimes, the exploration for a new site can be used to plan and estimate the costs for the remediation program. This is obviously the preferred route. However, if a structural failure occurs in a completed project without any suitable geotechnical information, it is necessary to use the

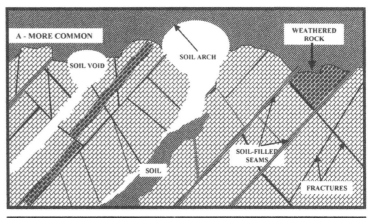
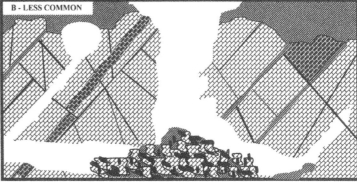

Figure 2 – Types of sinkholes (modified from Fischer, *et al.*, 1996).

readily available information for planning purposes. This process requires reviewing any state/federal/county/municipal data that might be available. These sources have ranged from helpful to useless in the authors' experiences. Even with a reasonable understanding of the site subsurface, cost estimates are often "S.W.A.G.s" (Scientific Wild-A__ Guesses) at best. With a site that has little to no subsurface information at hand, but with the existence of (or potential for) sinkhole formation, the scope and cost of a remedial program cannot be accurately estimated.

INVESTIGATION

Obviously, one would hope to have the data from a well-designed and properly executed field investigation in hand prior to planning a remedial program. Aerial photography, regional geologic information, test borings, test pits, and similar, will at least allow a general understanding of the problem (e.g., Fischer, et al, 1989). From a practical standpoint, there is never enough funding for a full investigation of a karst site even in its most simplistic manifestation.

REMEDIATION

Obviously, the size and nature of the sinkhole is important. Is it carbonate rock related or a debris pit? Is rock shallow or deep; flat-bedded, folded and/or faulted; soft or hard; sound or deeply weathered? What are the loads imposed upon the overburden soils or underlying rock? What are the importance and/or safety concerns for the facility/structure in peril? From a practical standpoint, can the owner afford the remediation?

Before the fact remediation can be accomplished by a variety of schemes:
1. Move the structure/facility or do not build.
2. Piles or caissons to sound rock.
3. Excavation to sound materials and filing with cement, rock, fly ash or a combination thereof.
4. Grouting (slurry and compaction grouting are the most common).
5. Dynamic compaction (or dynamic destruction).

Post construction remediation can usually be accomplished in several ways:
6. Excavation and backfill.
7. Underpinning.
8. Grouting (the most commonly used procedure).
9. Pouring concrete down the sinkhole.

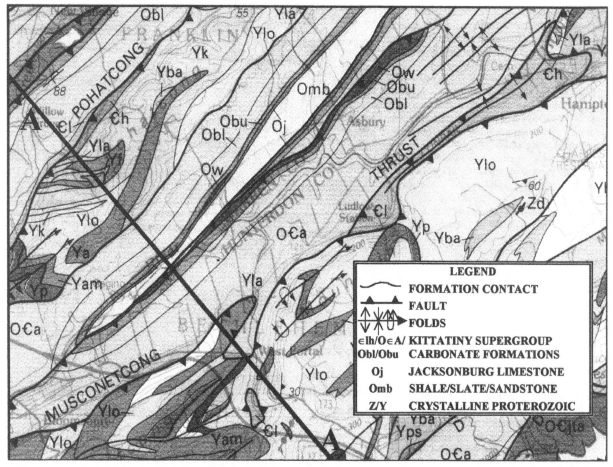

Figure 3 – Example of a "Continental-Margin" site (from Drake, *et al.*, 1996).

The following is a brief description of the remedial efforts described above using the same numbering system:

1. Moving to a better location on a large site or moving to another town has often been the positive response to karst concerns for major project sites, particularly prior to purchasing the property (while it is under contract).

2. Piles or caissons installed to (or generally into) sound rock is one of the more common foundation solutions for large and/or heavily loaded structures. Often, probe holes are planned for each column location. Pipe piles or caissons have the best chance of being founded on sound rock as a result of the ability to see "down the hole" before concrete is placed. Unfortunately, H-piles have been used as a result of their apparent cost advantages and flexibility in installation lengths. However, one has only to visualize the various lengths and configurations a flexible H-pile can achieve upon meeting with pinnacled limestone to understand the limitations and concern with using H-piles in karst.

3 & 6. Excavation to sound material can often be an economical procedure. However, without a knowledge of the depth to sound rock (which is usually highly variable) before starting excavation, the extent and ensuing costs can vary significantly. For example, a sinkhole area on a state highway in Maryland was first excavated to 6 or 7 meters and repaired by excavation and backfilling (Martin, 1995). After additional maintenance and investigation, the excavation continued to 22 meters before a final repair was effected. Often, this procedure is chosen for its probable speed in achieving a repair and for its apparent economics. Often, excavation to sound materials with some form of controlled backfill works well and effectively. However, time pressures on effecting a repair, for example on an existing roadway, can result in significant costs. In the Maryland highway case previously noted, repairs cost some $700,000.

Backfill can be impermeable (e.g., cohesive soils and concrete) or a graded filter that allows the passage of water into the subsurface. A combination of large rock fill and an impermeable cover has also been used. Again, the nature of the overlying construction, the degree of conservatism required and the local environmental concerns all must be considered.

4 & 8. Grouting is often the procedure of choice because of the ability to work close to structures and its relatively non-intrusive nature. Controlling grout flow by using thick mixes or quick-setting agents can maintain a limited area of repair. Compaction grouting can densify weak granular soils, fill voids or compress soft cohesive materials where appropriate. Larger voids can be filled with lower cost aggregate or transit mix grout. On-site batch plants can be used to place large quantities of a variable-mix grout where desirable, when grouting can be planned. Air- or hydro-track drilling of grout holes can provide useful subsurface data if drilled by knowledgeable technicians and monitored by an engineering geologist/geological engineer with experience at karstic sites.

5. Dynamic compaction, the dropping of large weights from great heights, can be most effective in remediating large areas such as stretches of roadway. The term "dynamic compaction" is likely still appropriate for southern limestone areas where the overburden and limerock is granular in nature. The term "dynamic destruction" seems to better fit areas underlain by cohesive residual soils that overlie older, harder (Proterozoic, Cambrian and Paleozoic) rocks. Dynamic destruction can collapse near-surface cavities or soils voids; seal underlying rock cavity openings with low permeability overburden soils, as well as densify both granular and cohesive soils. In addition, shallow rock voids may be exposed by these procedures and remediated by backfilling, usually with an appropriate cement slurry or lean grout. Of course, the area remediated looks like a World War I battlefield when completed. Backfilling and the use of borrow fill materials will likely be required to bring the area of concern back to grade.

7. The underpinning of structures can be used if the subsurface conditions are known prior to selecting the appropriate means (e.g., pin piles, caissons, pipe piles, etc.). The performance of any underpinning operation will require a relatively detailed knowledge of the subsurface, experienced designers, significant construction time, and likely will be quite expensive.

9. Placing transit mix grout or concrete in a sinkhole is often the first thought when an owner or constructor finds a sinkhole after a rainy night. Sometimes this works, but often the sinkhole is merely relocated a few feet away and discovered after the next heavy rain. Commonly, the wrong grout mix is used. The mix is too thick or coarse to flow through the existing subsurface channels; there is too little cement to bond the aggregate and reduce permeability and erosion; or it does not have a shrinkage-limiting agent added. While a quick fix, most "sinkhole mixes" (or alternately, flowable fills) do not work well. It is possible to improve the effectiveness of the quick fix by using the appropriate proportions of cement, water, fine aggregate (such as mason sand), combined with an appropriate anti-shrinkage agent. However, unless a definitive "throat" leading to the rock cavity can be sufficiently opened (usually by washing), even an appropriate flowable fill will merely fill the soil void, not the causative cavity. Even with the best of transit mix grouts, the chance of a sinkhole reopening in the future is fair to good in the authors' experience.

CONCLUSIONS

Appropriate remediation techniques at karst sites are a challenge. Many procedures are available, but all depends upon having or developing an understanding of the subsurface, the nature of the structure or facility of concern, the degree of conservatism desired, schedule requirements, and economics. One must be aware of the potential consequences of remediating one area and forcing the sinkhole to move laterally unless the whole subsurface cavity is filled or completely sealed.

Even with a detailed subsurface investigation (science), the variations existing below most karst sites required a great deal of extrapolation (art) and guesstimation (witchcraft). Then the reality of economics often necessitates changes and/or limitations to the best of investigatory and remedial programs.

REFERENCES

Beck, B.F., 1991, On calculating the risk of sinkhole collapse: Appalachian Karst, Proc. of the Appalachian Karst Symp., Nat. Speleological Soc. Redford, VA.

Chen, J. & B.F. Beck, 1989, Qualitative modeling of the cover-collapse process: Engineering & Environmental Impacts of Sinkholes & Karst, Proc. of 3rd Multidisciplinary Conf. on Sinkholes and the Engineering and Environmental Impacts of Karst, A.A. Balkema.

Drake, A.A. Jr., R.A. Volkert, D.H. Monteverde, G.C. Herman, H.F. Houghton, R.A. Parker, & R.F. Dalton, 1996, Bedrock geologic map of northern New Jersey: USGS Map I-2540-A.

Fischer, J.A., R.W. Greene, J.J. Fischer, R.W. Gregory, 1982, Exploration Grouting in Cambro-Ordovician Karst: Grouting, Soil Improvement and Geosynthetics, v. 1, ASCE Geotech. Publ. No. 30

Fischer, J.A., T.C. Graham, R.W. Greene, R.J. Canace, & J.J. Fischer, 1989, Practical concerns in Cambro-Ordovician karst sites: Engineering & Environmental Impacts of Sinkholes & Karst.

Fischer, J.A. and J.J. Fischer, 1995, Karst site remediation grouting: Karst Geohazards, A.A. Balkema.

Fischer, J.A., J.J. Fischer, R.F. Dalton, 1996. Karst site investigations: New Jersey and Pennsylvania sinkhole formation and its influence on site investigation: Karst Geology of New Jersey and Vicinity, Proc. of 13th Annual Mtg. of Geological Assoc. of NJ, Whippany, NJ.

Fischer, J.A., J.J. Fischer and R.J. Canace, 1997, Geotechnical constraints and remediation in karst terrane: Proc. of the 32nd Symposium on Engineering Geology and Geotechnical Engineering, Boise, ID.

Hannah, E.D., T.E. Pride, A.E. Ogden, and R. Paylor, 1989, Assessing ground water flow paths from pollution sources in the karst of Putnam County, Tennessee: Engineering & Environmental Impacts of Sinkholes & Karst.

Martin, A.D., 1995, Maryland Route 31 sinkhole: Karst Geohazards, A.A. Balkema.

Geotechnical and Environmental Applications of Karst Geology and Hydrology, Beck & Herring (eds)
© 2001 Taylor & Francis, ISBN 90 5809 190 2

New advances of karst collapse reseach in China

MINGTANG LEI, XIAOZHEN JIANG & LIYU Institute of Karst Geology, CAGS, Guilin, Guangxi 541004, China
mingtang@mailbox.gxnu.edu.cn

ABSTRACT

Covering 3.63 million km^2 of soluble rock, karst collapse (sinkhole) is the main geohazard in the karst regions of China. Before 2000, more than 1,446 collapse events and 45,037 sinkhole pits have been recorded. During the last decades, Chinese scientists have made great progress in the area of karst collapse research through a series of research projects. In this paper, the authors summarise some of these new advances. Firstly, the authors introduce the main achievements obtained in the 1980s. Secondly, the model experiment of karst collapse is discussed, which was established in Institute of Karst Geology CAGS and funded by the Foundation of Geological Science in 1993. Thirdly, the authors give a detailed description about the method of risk assessment of collapse hazards based on the Geographic Information System (GIS) through a case study in Liupanshui, Guizhou province. Finally, as an example of the integration of karst collapse data and GIS, the Sinkholes Information System of Liupanshui City, developed in 2000, is introduced.

INTRODUCTION

Covering 3.63 millions km^2 of soluble rock, China is one of the countries with most extensive distribution of karst collapse (Figure 1). According to incomplete statistics, there are more than 1,400 cases of karst collapses that are distributed over 22 provinces, with about 30,000 collapse pits formed. Karst collapses are concentrated in the southern part of China, including Guangxi, Guizhou, Jiangxi, Yunnan, Hunan, Hubei and Sichuan provinces; however, some northern provinces, such as Shandong, Liaoning, Heilongjiang and Hebei are also frequented by this geohazard (Table1). More than 38 mines have suffered from terrible karst collapses. For most collapses happening in developed regions with cities and mines, the damage usually is very large. Annual losses are more than 1,500 million Yuan (RMB).

In recent years, a series of collapse accidents have greatly impacted human life. For example, in April 6th, 2000, a critical sinkhole formed in Hongshan District, Wuhan, Hubei Province. There were 21 sinkhole pits formed, and the area of influence is more than 0.1km^2. The dimension of the largest pit is 54m, 33m and 7.8m in length, width and depth, respectively. Two buildings were swallowed and sixteen cracked. About 150 families suffered from the accident, and more than 900 residents are homeless.

On May 7, 1999, a serious sinkhole formed in Heshan, Guangxi province. The sinkhole was induced by waste storage of the Heshan Electricity Power Factory. There was only one pit formed, and it is 50 m in diameter and 12 m in depth. As a result, the waste (mainly ash) poured into the Heshan mine, putting the mine out of work, and the direct loss is more than 25 million Yuan (RMB).

On Nov. 11, 1997, a serious collapse, triggered by an explosion, happened in Zemu, a small town south of Guilin. Until 2000, 60 sinkhole pits were formed, and the influenced area is more than 0.2 km^2. As a result, about 100 resident houses were destroyed. Also in Guilin on June 23, 1999,

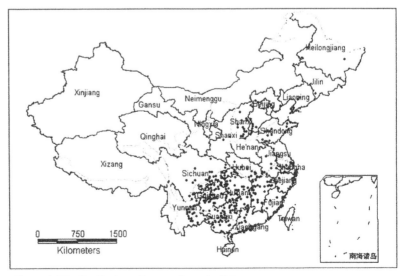

Figure 1: The distribution of karst collapse in China

Table 1: Collapses in each province

ID	Province	Induced collapse		Nature collapse		Uncertainty cause		Sum		1980-2000	
		Cases	pits	cases	pits	Cases	pits	Cases	pits	cases	pits
1	Anhui	93	274	2	5			95	279	3	11
2	Fujian	3	9					3	9	1	1
3	Guangdong	53	10066	3	9	7	8	63	10083	42	177
4	Guangxi	279	3474	173	2143	47	82	499	5699	232	1736
5	Guizhou	72	1881	42	55	1	5	115	1941	17	138
6	Hebei	27	346	5	251	2	8	34	605	3	19
7	Henan	3	24					3	24		
8	Heilongjiang	3	3					3	3		
9	Hubei	45	1041	18	61	7	32	70	1134	26	206
10	Hunan	90	18501	6	8	6	207	102	18716	33	364
11	Jilin	1	1					1	1		
12	Jiangsu	5	8	1	31	15	15	21	54	1	1
13	Jiangxi	47	1228	36	89	13	19	96	1336	7	124
14	Liaoning	6	57	1	108	1	20	8	185	2	21
15	Shandong	11	321	1	1	1	1	13	323	1	140
16	Shanxi	1	2	2	2	14	2633	17	2637		
17	Shaanxi	1	6	1	1			2	7		
18	Sichuan	4	8	12	17			16	25		
19	Tianjin	1	1					1	1		
20	Yunan	152	1710	7	22	36	36	195	1768	72	76
21	Zhejiang	45	45					45	45	11	11
22	Chongqing	23	132	20	29	1	1	44	162	2	2
	TOTAL	965	39138	330	2832	151	3067	1446	45037	450	3027

when a pupil was walking along the footway of Chuanshan road, a small sinkhole only 3 m in diameter and 4 m in depth occurred suddenly, and the child was swallowed and killed.

Among 1,400 collapse events, about 70% were induced by human beings. Additionally, groundwater drainage has become the most important factor triggering the formation of karst collapse, and the next is reservoir and mechanic loading. The data indicated groundwater pumping had caused more than 68.5% of induced collapses. According to statistics, only 900 collapse events have had the date recorded. Of more than 450 created after 1980, it is obvious, in China, with the urbanization of karst areas, that more and more collapse hazards will happen and influence our life.

Due to the extensive distribution and damage, the Chinese government and scientists have paid great attention to the geohazard of karst collapse. In the past couple of decades, a lot of work on the investigation, research, prevention and remedial engineering of karst collapses has been carried out in China, and a series of effective results have been produced.

ADVANCES OF KARST COLLAPSE RESEARCH
The national inventory of collapses

Since 1983, a series of special projects have been carried out by the Institute of Karst Geology. They are the first special and systematic researches on karst collapse in China. The purpose of these works is to research the regularity of collapse development in China, including the distribution, the mechanism, and evaluation.

Based on the survey, an inventory of karst collapses was established in 1989 using dBase software. About 800 collapse cases were documented. The main fields included the position, the date, the number of sinkhole pits, the origin, and the losses due to karst collapse. It became possible and convenient to get the latest information and statistics of karst collapses. In 1996, the inventory was integrated with GIS, and the Sinkhole Information System of China developed. From that point on, more data were added, and in 2000, there were about 1450 collapse cases recorded.

Model experiment of karst collapse

Supported by the Foundation of Geological Science a laboratory for a model experiment of karst collapse was established in the Institute of Karst Geology in 1993 (Figure 2). The main model is 3.0 m in height, and 2.0 m in both width and length. Consisting of two parts, the upper part, about 1.5 m in height, is filled with soil to simulate the sediments. The lower part consists of pipes simulating caves and openings in soluble rock. The pipe is connected with the soil box by an interface (Figure 3). The monitoring system consists of vibrating wire piezometers, vibrating wire pressure cells, a data logger, a water meter, a pressure gauge, a piezometer tube and a subsidence tube. The data measured includes the dynamics of karst water such as the flux, the velocity and the

Figure 2. Photo of model experiment

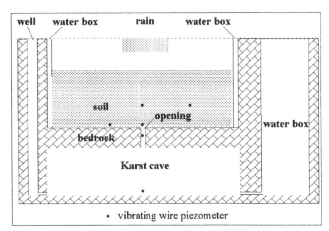

Figure 3. Sketch map of model

pressure in opening, the level of pore water, the stress and deformation of bottom soil, and the character of sediments dropping into opening.

There are six cities including Wuhan, Hubei province; Tangshan, Hebei province; and Xiangtan, Hunan province, that have taken part in the experiment. The purpose of the tests includes the determination of the critical condition triggering a sinkhole, the features of the properties and structure of sediments influencing the formation of collapse, and the mechanism of collapse induced by drainage. These experiments indicate:

1. The critical conditions triggering sinkhole development:
   - The change of pressure in karst opening:
     It has been shown that the stability of the overburden sediments mainly depends on the change of pressure in the karst opening, and the stress of the soil is inversely proportionate to the pressure. Table 2 is a list of the value in some places.
   - The critical velocity of karst water level decline:
     This is one of the most important results of the experiment of Lujiajie, Wuhan City, Hubei province. The critical value is about 0.65 cm/s and 0.146 cm/s corresponding to clay and sand respectively. Based on the velocity of karst level decline, it is possible to judge the failure of overburden soils.
   - The critical drop down of water level in well:
     The result of the experiment of Yulin had

Table 2: Critical change of pressure in karst opening

No	Critical Value	Description
1	40 kPa	Dunmu, Guilin city, Guangxi province
1	23 kPa	Xiangshuiba, Xiangtan city, Hunan province.
2	20 kPa	Zhongshanlu, Guilin city, Guangxi province.
3	10 kPa	Yucai, Yulin city, Guangxi province.
4	2 kPa	Xiaojie, Tongling city, Anhui province

shown that the drop down in the well closely relates to sinkhole formation, and it could be used as another critical condition of sinkholes. The test gives the critical value as about 300 cm. After the test, that value became the allowable drop down in wells in the whole of city.

A correlation exists among the change of pressure (dP), the velocity of the water level decline (dV), and the drop down in the well (dH). In Wuhan's experiment, the equation is: $dP = a*dH + b*dV + c$, where the unit of dP, dH and dV is kPa, cm and cm/s respectively. The coefficient of a, b and c depend on the properties of the soil and initial conditions, such as initial groundwater level, etc.

It is possible to monitor the change of pressure in a karst opening automatically by using the sensors and the data acquisition system; therefore, pressure change should be considered as the primary target for long-term monitoring.

2. The relationship between sinkhole formation and the influencing factors is confirmed as follows:
   - The feature of fluctuation in the groundwater level
     - If the water level is above and below the rock surface, it will be easier to trigger collapse.
     - If the water level stays under the rock surface, it will be more difficult for drainage to induce collapse, but leakage of surface water will become the major factor to induce collapse.
     - The period of easily induced collapse is when drainage begins and the karst water level falls just below the rock surface.
     - The decline of the karst water level forms negative pressure in karst openings, and, usually, that will speed up the seepage deformation of overburden soil.

- The character of sediments:
    - The structure and properties of the overburden soil, especially the properties of the bottom layer, are important for collapse formation. The higher the content of clay, the more difficult it is for collapse formation.
    - Compared to its structure and properties, the thickness of the overburden sediments isn't important for sinkhole development.

3. The mechanism of collapse:
    - The result of the model experiment confirms that the mechanism of soil failure caused by the decline of karst water is seepage deformation (piping).

As an extension of the model experiment, a project of sinkhole forecast, supported by the Ministry of Land and Resources, has been carried out since 2000 in Guilin, Guangxi province, by the Institute of Karst Geology. This project will be finished in 2002. The site of the project is about 0.2 km² in Zhemu, a small town south of Guilin, which has been suffering from sinkholes problem since November 11, 1997. More than 60 sinkhole pits have been formed there. According to the experiment's results, if the pressure change in karst openings can be monitored, it will be possible to forecasting the formation of collapse. In the project, the monitoring system contains vibrating wire piezometers, a data acquisition system, and ground penetrating radar (GPR). Twenty piezometers are kept in karst caves to record change of water pressure. To confirm an anomaly area in the soil, GPR will be used to scan the same section every 6 months.

Risk assessment of potential karst collapse

The key to evaluating the risk of potential collapse is to process and synthesize spatial data with different resources, such as karstification of bedrock, the structure and properties of the soil, the features of groundwater fluctuation, human activities, the types of landuse, infrastructures, and the condition of the social economy. It is difficult to combine these factors unaided. Aided by GIS, four projects concerning risk assessment of cities, including Tangshan, Xiangtan, Yulin, and Liupanshui, have been sponsored by the Institute of Karst Geology CAGS since 1993. IDRISI, the software of GIS developed by CLARK University, USA, is used. As an example, the risk assessment of Liupanshui City, Guizhou province, is summarized below.

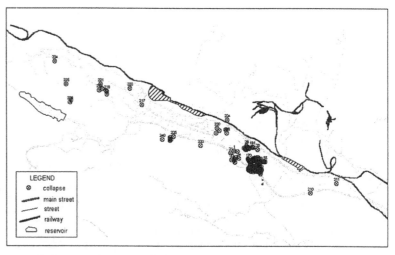

Figure 4: The map of sinkhole distribution

1. Background:
    Liupanshui is one of the main cities suffering from sinkhole hazards in China. More than 243 collapse cases have formed there since 1975 (Figure 4). All were induced by karst water drainage through 41 wells. The city is located in the Shuicheng Basin, and the urban area is about 26 km². The thickness of sediments is less than 31 m, and the level of groundwater is 0.8-12 m beneath ground. The bedrock is the Baizhuo group of Carboniferous, which consists of limestone.

2. Basic maps:
    Except geological maps which have a scale of 1:25,000, all thematic maps involved use 1:10,000.

3. Grid-cell definition:
    The grid-cell definition for analysis selected was at 10m×10m resolution so that all thematic maps contain 800 × 1600 = 1280000 cells.

4. Consideration of influencing factors:
    - Groundwater (F_w): water level in dry and rainy seasons in 1992.
    - Overburden soil (F_s): thickness of soil
    - Bedrock (F_r): the karstification of bedrock and the fault distance
    - Before evaluation, the value of all above factors are normalized to a range from 1 to 10 corresponding to the lowest and highest contribution to collapse formation, respectively.

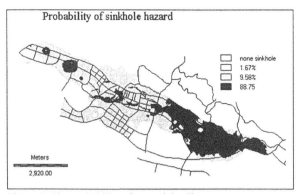

Probability of sinkhole hazard

☐ none sinkhole
☐ 1.67%
☐ 9.58%
■ 88.75

Meters
2,920.00

RISK MAP (In the unit of Yuan RMB)

☐ none sinkhole
☐ <50,000
☐ 50,000~100,000
☐ 100,000~200,000
☐ 200,000~500,000
■ 500,000~1,000,000
■ >1,000,000

Meters
2,920.00

Figure 5: The probability of potential collapse

Figure 6: Risk map of collapse hazard

5. Assessment collapse (H_coll):
   Use the following equation to assess sinkhole hazards: $H\_coll = (5*F\_w+3*F\_s+2*F\_r)/10$. This is basically a weighted average with the relative weight for each factor indicated by the numerical suffix of that factor. Combined with the distribution of collapses, the probability of collapse may be obtained (Figure 5).

6. Vulnerability assessment (V_soc):
   The influencing factors considered for vulnerability assessment include social susceptibility, vulnerability of structures, and economic vulnerability. All are defined based on landuse and lifeline types.

7. Risk assessment (R_coll) and zoning risk map:
   In the project, risk is computed as a function of sinkhole potential and vulnerability using the equation:
   $R\_coll= H\_coll*V\_soc$.
   As result of calculation, the whole study region is divided into 7 subareas with different risk levels. The risk map shows the results (Figure 6). On the map, the highest risk region is about 1.2 km^2. In this area, the loss from collapse hazards may be more than one million Yuan (RMB).

Management of karst collapses

A GIS gives scientists a powerful tool in manipulating information concerning geohazards. For many researchers in China, such as those working for governmental bodies, communities and urban planners, GIS and collapses are unfamiliar; therefore, it is difficult for them to use a commercial GIS application to manage so much information about collapse. Using GIS, the Institute of Karst Geology CAGS developed a series of special information systems in 1996 for such researchers. Some examples are the Sinkhole Information System of China and the Sinkhole Information Systems of Guilin, Yulin, and Liupanshui cities. In these systems, all layers are linked to pull-down menus which aid the user in selecting a desired display (Figure 7).

The introduction of the sinkhole information system of Liupanshui
1. Environment of the system
   • Hardware: IBM PC with PII CPU and more than 32M memory
   • Operation system: Windows 98
   • GIS applications: MapInfo Professional 5.0, Visual Basic 6.0 and Access 97

2. Data involved in the system
   In the system there are more than 40 thematic maps divided into four groups:
   • Sinkholes:
     – The distribution of 243 collapse accidents, and the fields include the date, position, number of pits, dimension of pits, causes, damage, and pictures.

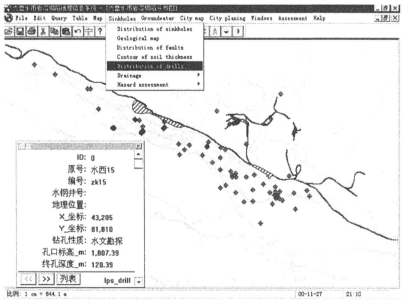

Figure 7: The windows of sinkhole information system of Liupanshui

149

- A geological map which contains fields for the stratum, lithology, and hydrogeological character.
- The distribution of 73 drills, and the fields include borehole logs, the character of the properties and thickness of soil, the bedrock, the number of caves, and the section map.
- The distribution of 41drainage wells, and the fields contain the position, the type of water, the owner, and the name of the aquifer.
- The contour of the sediments thickness
- A probability map of collapse and a risk map of collapse hazard, which are image files from risk assessment.
- Groundwater:
  - The distribution of 107 groundwater monitors, and the fields include the position, observations, and the statistical value of groundwater levels from 1991 to 1998.
  - A contour map of groundwater level in the dry and rainy seasons of 1992 and 1996.
- City maps:
  Include a series of maps such as a topographic map, a street map, a railway map, a map of streams, the distribution of main firms, and the type of land-use. All maps use a scale of 1:10,000.
- City planning:
  Consists of 10 maps of city planing, including maps of landuse, streets, railways, water supply, sewage, power supply, telecom, gas supply, and heating. All maps use a scale of 1:10,000.

3. Functions of the system:
   A commercial GIS has powerful functions such as controlling layers, general finding and SQL finding, redistricting, analysis of statistics, and thematic mapping. In addition, to meet the special needs of geohazard management, other functions are developed, such as displaying coupled layers, adding new points, and displaying pictures, etc.
   - Controlling layers is easier: Using the pull-down menu, if a user selects a layer, the others layers concerning the selected layer will be displayed together. It is unnecessary to look for separate files in a folder as it is when using a commercial GIS.
   - Adding data is simpler: Allowing a user to add new points of collapse, drilling, well, firm, and monitoring stations, using coordinates.
   - Displaying pictures: Using buttons on a toolbar, geotechnical borehole logs, geological cross sections from geophysical studies, and pictures of sinkholes, monitors and wells can be directly displayed.

Using this system, it is convenient to find detailed information about a certain area, such as the features of historical sinkholes, borehole logs, properties and thickness of sediments, the dynamics of the groundwater level, the drainage of wells and the karstification of bedrock. It is also possible to offer help to decision-makers and researchers dealing with the hazard when sinkholes happen.

CONCLUSION
1. During the last few decades, more and more collapse hazards have gained great attention from scientists and the government. The advances in karst collapse research in China mainly include research from a model experiment, risk assessment and geohazard management.

2. A large-scale model experiment has been constructed and has played an important role in collapse research in China. It is now possible to test and determine the mechanism of karst collapse, the relationship between collapse and its controlling factors, and the critical conditions for the occurrence karst collapse. It has shown that the pressure in a karst opening is the key parameter to inducing sinkholes.

3. The methodology of risk assessment has been established. By using GIS, it is convenient to process and combine the spatial data of the controlling factors of karst collapses and to determine risk zoning.

4. In order to meet the need of different users, a series of special information systems have been developed using software based on MapInfo Professional. It is considered helpful for governmental, research and engineering institutions in working with karst collapse problems.

ACKNOWLEDGEMENT
The Foundation of Geological Science of China, the Geological Survey of China, and the Foundation of Natural Science of Guangxi provided funding for various portions of this research. The authors wish to thank Prof. Yuan Daoxian, Prof. Xiang Shijun and Mr. Gao Yongli, a doctoral student of University of Minnesota, for their notable help.

REFERENCES

Benson, Richard C. , 1987, Assessment and long term monitoring of localized subsidence using ground penetrating radar: Proceedings of 2nd Mutidisciplinary conference on sinkholes and the Environmental Impacts of Karst, Orlando, 1987, pp161~170.

Jiang, Xiaozhen, *et al.*, 1994, The application of GIS to the evaluation of karst collapse: Proceedings of 7th International IAEG Congress, Lisboa, Portugal, 1994, pp4575~4579

Karnieli, Arnon, 1991, Stepwise overlay approach for utilizing a GIS with a soil moisture accounting Model: ITC Journal, 1991-1, pp11~18.

Lei, Mingtang, *et al.*, 1994, The model experiment on karst collapse: Proceedings of 7th International IAEG Congress, Lisboa, Portugal, 1994, pp1883~1889

Lei, Mingtang, *et al.*, 1996, The model experiment research of karst collapse in Xiangtan, Hunan province: research report of the Institute of Karst Geology, CAGS. 70p. (in Chinese)

Lei, Mingtang, *et al.*, , 1997a, Evaluation of karst collapse: Theory and Methodology: The Chinese Journal of Geological Hazard and Control, Vol. 8 supplement, pp38~42. (in Chinese, with English abstract)

Lei, Mingtang, *et al.*, 1997b, The sinkholes information system of Guilin, Guangxi, China: The Chinese Journal of Geological Hazard and Control, Vol. 8 supplement, pp.78~82. (in Chinese, with English abstract)

Lei, Mingtang, *et al.*, 1998, Review of sinkhole research in China during Last decades: The Chinese Journal of Geological Hazard and Control, Vol. 9, No.3, pp1~6. (in Chinese, with English abstract)

Lei, Mingtang, *et al.*, 2000, The risk assessment of karst collapse in Liupanshui city, Guizhou province: The Chinese Journal of Geological Hazard and Control, Vol.11, No.4. (in Chinese, with English abstract)

Mejia-Navarro, Mario and Garcia, Luis A., 1996, Natural Hazard and Risk Assessment Using Decision Support Systems, Application: Glenwood Springs, Colorado: Environmental &Engineering Geoscience, Vol.II, pp299~324.

Stangland, Herbert G., 1987, Use of ground penetrating radar techniques to aid in site selection for land application sites: Proceedings of 2nd Mutidisciplinary conference on sinkholes and the Environmental Impacts of Karst, Orlando, 1987, pp171~177.

Yuan, Daoxian, 1987, Environmental and engineering problems of karst geology in China: Proceedings of 2nd Mutidisciplinary conference on sinkholes and the Environmental Impacts of Karst, Orlando, 1987, pp1~11.

Geotechnical and Environmental Applications of Karst Geology and Hydrology, Beck & Herring (eds)
© *2001 Taylor & Francis, ISBN 90 5809 190 2*

A karst case history revisited

RICHARD S.OTTOSON, JOSEPH J.FISCHER, & JOSEPH A.FISCHER   Geoscience Services, Bernardsville, NJ 07924, USA, geoserv@hotmail.com

ABSTRACT

In a previous paper, *Wyndham Farms – A karst case history* (presented at the Sixth Multidisciplinary Conference on Sinkholes and the Engineering and Environmental Impacts of Karst in April, 1997), we described a cost effective site planning and geotechnical study program performed for a large residential subdivision overlying solution-prone carbonate rocks in northwestern New Jersey. The program consisted of a review of the available geologic data and aerial photography, a geotechnical reconnaissance and widely spaced test pits and test borings performed using special procedures and equipment as outlined in Fischer and Canace, 1989. These data were utilized to develop a "geologic model" of the site subsurface. This model resulted in dividing the site into six segments that attempted to anticipate the conditions in areas of similar subsurface hazard. Prior to the development of each of these segments, a more complete geotechnical investigation was to be performed in order to increase the understanding of each section's specific subsurface conditions and design/construction concerns. Central to the planned program was to be ongoing construction inspection of critical areas by geotechnical personnel with extensive experience with development in karst terranes. The construction process started in full accordance with these concepts.

Since the presentation of that paper, however, a combination of, perhaps, cost consciousness, the booming market for new single family homes within the northeastern United States, and the surprisingly rapid sellout of the 242 acre/240 lot site led the developer to reconsider the previously planned pre-construction subsurface investigation and construction inspection program. The market for these ±$300,000 to 400,000 homes dictated that this project be built within the <u>perceived</u> minimum amount of time. These economic pressures resulted in the use of construction procedures that were likely not economical for a site overlying solution-prone carbonate rocks.

This paper describes the extensive delays and costs that eventually resulted from the over-blasting of rock, indiscriminate site grading, and the elimination of the utility and building foundation excavation inspection program. The attempt to decrease construction time resulted in a lengthy sinkhole remediation program to ensure the safe construction of roads, utilities and houses.

The remediation program took in excess of 1½ years of continuous drilling and grouting work and resulted in placing some 3,000 cubic yards of grout and truck-mixed flowable cement grout fill after the initial, less frenzied Phase I construction. Overall, the remediation program was very likely more costly than the originally planned procedures and probably resulted in minimal or no timesavings.

INTRODUCTION

Wyndham Farms is a 242-acre, 240-lot residential subdivision located in one of northwestern New Jersey's carbonate valleys in Warren County near Phillipsburg, New Jersey. The region is mostly farmland currently under development pressures as a result of ongoing westward expansion in an attempt to find less expensive developable property. Wyndham Farms is the fourth major development to be built within the Township and the first to be approved under a recently adopted "limestone ordinance". The "limestone ordinance" required the developer to investigate the subsurface including a study of the site and regional geology as well as more traditional geotechnical foundation studies. Additionally, the "limestone ordinance" required that the developer agree to use construction procedures minimizing possible damage to the subsurface carbonates that could result in sinkhole formation and surface subsidence after project completion and the developer leaving the area. In our previous paper, *Wyndham Farms – A karst case history* (presented at the Sixth Multidisciplinary Conference on Sinkholes and the Engineering and Environmental Impacts of Karst in April 1997), we described a cost-effective site planning and geotechnical study program performed for this site. The program consisted of a review of the available geologic data and aerial photography, a geotechnical reconnaissance and widely spaced test pits and test borings performed using special procedures and equipment as outlined in Fischer and Canace, 1989. These data were utilized to develop a "geologic model" of the site subsurface. This model resulted in dividing the site into six segments that attempted to anticipate the conditions in areas of similar subsurface hazard. Prior to development in each of these segments, a more complete geotechnical investigation was to be performed in order to increase the understanding of each segment's specific subsurface conditions and design/construction concerns. Central to the planned program was to be ongoing construction inspection of critical areas by geotechnical personnel with extensive experience with development in karst terranes. The developer agreed with the Township to comply with this program and the construction procedures for karst terranes including minimal grade changes to avoid sinkholes, rock excavation by hydraulic hammer or controlled blasting for minimal vibration/disturbance and site grading with the drainage of trenches to prevent surface water ponding. The construction process started in full accordance with these concepts.

SUMMARY OF SITE CONDITIONS

Geology

Physiographically, the site lies in one of the intermontane valleys of the New Jersey Highlands, just to the south of the Kittatinny Valley (in the Valley and Ridge Physiographic Province). The rocks that underlie the site are part of the Kittatinny Supergroup and the overlying Jacksonburg Formation. These rocks were deposited in Early to Middle Ordovician time in a warm inland sea that covered the area. The formations existing below the site include the dolomites of the Allentown Formation and Beekmantown Group (or Epler and Rickenbach Formations) and the overlying limestone of the Jacksonburg Formation. The oldest formation is the Allentown Formation at the northern end of the site with the youngest, the Jacksonburg Formation, at the south end of the site. Beekmantown Group deposits underlie the central portion of the site. Most of the site lies upon an up-thrust block bounded by the Brass Castle Fault that outcrops within the Rickenbach Formation in the northerly portion of the site. The axis of an anticlinal fold is located within the Jacksonburg Formation close to and approximately parallel to the southern site boundary. A synclinal fold axis was found approximately 900 feet to the north of the axis of the anticline. The geologic formation contacts and the structural features described are shown on the Site Plans, Figures 1 and 2.

The carbonate rocks found underlying Wyndham Farms have been solutioned. A pinnacled rock surface was found underlying the overburden soils over most of the site. The overburden was mostly glacial tills overlying residual soils with minor amounts of topsoil and fill.

Surface

The surface topography of the site is gently rolling. The highest area of the site is near the southern boundary, coincident with the fold (anticline) in the underlying rocks, which then slopes gently downward to the area of the syncline to the north. Drainage from the higher, southerly areas of the site flows from the syncline into a low lying area (ghost lake) at the Jacksonburg/Epler contact that is drained by a low-gradient swale trending northwest near the western boundary of the site. This swale also drains higher areas to the east.

A number of sinkholes were found on the site prior to development. Two areas of clustered sinkholes were found just south of the contact of the Jacksonburg and Epler Formations and just north of the Brass Castle Fault along the western boundary of the site. Sinkhole activity was relatively common in the southern area of the site underlain by the Jacksonburg Formation, especially near the axes of the anticline and syncline. No pre-existing sinkholes were found in the areas underlain by the Beekmantown Group rocks with the exception of the area just to the north of the Brass Castle Fault described previously. Two pre-existing sinkholes were found in the northern end of the site underlain by the Allentown Formation.

Subsurface

The site was divided into six (6) segments upon the bases of the existing subsurface conditions and the presumed geotechnical constraints. As shown on Figures 1 and 2, the site is bisected by Greenwich Street. The segments were started at the south end of the project in the youngest rocks, Jacksonburg Formation (Limestone) and continued to the north through the increasingly older rock formations (Beekmantown and Allentown Formation dolomites). Segment 1 was located from the southern boundary to midway between the axes of the anticline and syncline shown on the Site Plan, Figure 1. Segment 2 extended across Greenwich Street to the northern boundary of the Jacksonburg Formation. Segment 3 extended to the contact between the Epler and Rickenbach members of the Beekmantown Formation. Segment 4 extended to the Brass Castle thrust fault in the Rickenbach member of the Beekmantown Formation. The area of the site to the north of Greenwich Street is shown on the Site Plan, Figure 2.

SITE DEVELOPMENT

The first section developed was in the southwest corner of the site and included a study of the subsurface, the use of karst friendly construction procedures and the foundation inspection/remediation program as described above. The work was planned and accomplished with only one house foundation requiring remediation of a sinkhole. However, the market for housing in western New Jersey was skyrocketing and the developer was able to sellout the homes faster than he could construct them. Given these sales conditions and the perceived need to complete and close on as many homes as possible before the hot market cooled, the developer decided to deviate from the previously agreed upon geotechnical investigations prior to construction of each section; the use of karst friendly construction procedures and foundation inspection/remediation prior to foundation construction. The Township required that the developer remediate all sinkholes encountered in the development in accordance with the requirements of the "Limestone Ordinance" and certified by a technically qualified professional with extensive, recognized experience in local sinkhole remediation.

The developer started work in future sections using uncontrolled rock blasting for utility and foundation installations. Trenches were not drained to prevent ponding of rainwater during utility installations. Backfill was poorly compacted resulting in utility "swales" rather than firm, level trench areas. Upon completion of the utility installation, roadway curbing was laid and roadways paved with an asphalt base course prior to house construction, the final placement of topsoil and grading of the lots. No drainage for housing foundation excavations was provided permitting water to pond in open foundation areas. The resulting lack of drainage during the home construction period permitted ponding of large amounts of water behind curbing and within foundation excavations. The uncontrolled blasting combined with the lack of drainage and the resulting ponding virtually guaranteed that sinkholes would form. The developer was advised continuously of the effects of his chosen construction procedures on the formation of sinkholes. The decision made was always to continue development using the procedures that appeared to speed construction and reduce construction costs with little apparent concern for sinkhole remediation costs and delays.

As expected, large numbers of sinkholes developed throughout the site during the construction. Sinkhole occurrence was directly related to the extent of blasting for utility and basement construction and the amount of rainfall during the construction period. The number of sinkholes was greatest in the eastern portion of Segments 1 and 2, along the Brass Castle Fault, the northernmost area of Segment 5 and most of Segment 6. The great number of sinkholes found in Segment 6 was greatly influenced by the incidence of Hurricane Floyd in the fall of 1999. At the time the hurricane hit with extensive rainfall and flooding, the utilities, curbing and roadway subgrade preparation had just been completed awaiting the placement of asphalt paving for the roads. The resulting ponding of water in the areas of shallow rock that required blasting produced a great number of large and small sinkholes in and around the roadways and home foundations.

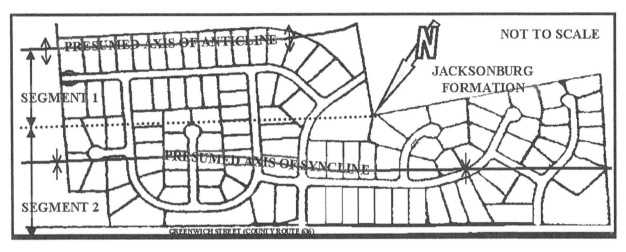

Figure 1 – Site Plan (southerly portion of site).

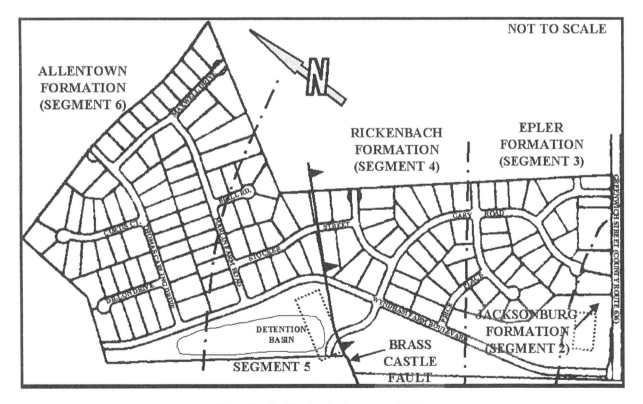

Figure 2 – Site Plan (northerly portion of site).

SINKHOLE REMEDIATION

Procedures

Most of the sinkholes that developed during the construction of Wyndham Farms were not overly large nor had open "throats" to the cavities within the rock. As a result, the most typical method of remediation including cleaning out loose soil, running water to open a "throat" and filling with transit-mix grout to seal off the rock opening, was not appropriate. Where large open "throats" were found this method was used, but for the most part transit-mix grout was only used to fill a sinkhole void to permit ongoing construction, such as paving, to continue prior to final remediation completion.

The Wyndham Farms sinkholes generally resulted from many small open fractures in the blasted rock, not large open voids in the rock. In many cases the fractures extended great distances laterally and vertically from the intended boundary of the rock excavation. To effectively seal off the multiple, generally small, openings into the rock, numerous probe holes were drilled surrounding the sinkhole to intercept as many of the fractures as possible. Plastic "tremie" pipes were placed in the probe holes and a site-mixed, very fluid cement/sand grout (slurry grouting method) was pumped to the bottom of the probe holes through the "tremie" pipes. Where very small sinkholes were found, plastic "tremie" pipes were jetted into the sinkhole void to open underlying fractures/voids and then grouted with site-mixed grout. This grouting method permits the maximum penetration of the grout into the small fractures and open joints in the rock to completely seal off the rock and prevent future sinkhole formation. The site-mixed grout permitted rapid variation in the grout mixture to match the "flowability" required to effectively seal the rock surface.

The drilled probes consisted of both rotary wash-drilled borings using only water as a drilling fluid and pneumatic percussion- (air track) drilled probes. The rotary-drilled borings were utilized at the start of remediation efforts in each segment to obtain soil samples and rock cores to better identify the subsurface conditions in the area. Air track probes were utilized for the majority of probe holes once the subsurface conditions and causes of sinkholes were better understood.

Grout Takes

Remediation of sinkholes began in the first section developed in August 1997 and continued sporadically through June 1998. Remediation work was performed almost continuously from mid 1998 through June 2000 after the initiation of site work in Segment 1 at the south end of the site and continuing through Segments 4, 5 and 6. A total of 561 drilled probes and 72-jetted probes were installed and grouted for the remediation of the entire site. A total of 2,880 cubic yards of grout was used to remediate the entire site. Table 1 presents a breakdown of the probes installed and the grout takes of site-mixed grout and plant-mixed grout for each of the defined subsurface segments shown on Figures 1 and 2.

TABLE 1 – Remediation Program Results					
SEGMENT	NO. OF PROBES		GROUT VOLUME		
	DRILLED	JETTED	SITE-MIXED	TRANSIT-MIXED	TOTAL
1 & 2	120	38	675.5	59.7	735.2
3	22	2	21.5	15.0	36.5
4	0	1	0.5	1.0	1.5
5	135	6	325.0	25.0	350.0
6	285	25	1665.0	92.0	1,757.0
TOTALS	561	72	2,687.5	192.7	2,880.0

As can be seen from Table 1, Segments 1 and 2 underlain by the Jacksonburg Formation, Segment 5 near the Brass Castle Fault and Segment 6 underlain by the Allentown Formation developed the most sinkholes and required the greatest number of probes and the most grout to complete the remediation. Segments 1, 2 and 6 had the smallest depths to the rock surface and therefore required the most blasting for installation of utilities and foundations. The sinkholes in Segment 5 that required the greatest number of probes and amount of grout were along a deep trench blasted in the rock for a sanitary sewer line. This trench was located virtually on the Brass Castle Fault trace within the northwest trending drainage swale.

As we felt at the outset of the development of the Wyndham Farm site, a reasonable amount of investigation of the subsurface conditions prior to the development of each section of housing and use of karst friendly construction procedures, where possible, would have dramatically reduced the number of sinkholes that developed during the actual construction. The high costs, bad publicity for the development and delays caused by sinkhole remediation could have been significantly less than that actually experienced.

At Wyndham Farms, the "Limestone Ordinance" gave the Township officials the authority to demand the developer remediate the sinkholes that occurred as a result of the developer's use of inappropriate construction procedures for site development in areas underlain by solutioned carbonate rock formations. A strong "Limestone Ordinance" with subsequent enforcement can minimize the risk of future sinkhole development with potentially adverse effects for homeowners, utilities and Township roadway maintenance budgets.

REFERENCES

Drake, A.A., Jr., R.A. Volkert, D.H. Monteverde. G.C. Herman, H.F. Houghton, R.A. Parker, & R.F. Dalton, 1995, Bedrock geologic map of northern New Jersey: USGS Map I-2540-A.

Fischer, J.A., R.W. Greene, J.J. Fischer, & R.W. Gregory, 1982, Exploration grouting in Cambro-Ordovician karst: Grouting, Soil Improvement and Geosynthetics, v. 1, ASCE Geotech. Publ. No. 30.

Fischer, J.A., T.C. Graham, R.W. Greene, R.J. Canace, & J.J. Fischer, 1989, Practical concerns Cambro-Ordovician karst sites: Engineering and Environmental Impacts of Sinkholes & Karst.

Fischer, J.A. & R.J. Canace, 1989, Foundation engineering constraints in karst terrane: Foundation Engineering: Current Principles and Practices, v.1. ASCE.

Fischer, J.A., & J.J. Fischer, 1995, Karst site remediation grouting: Karst Geohazards, A.A. Balkema.

Fischer, J.A., J.J. Fischer, & R.J. Canace, 1997, Geotechnical constraints and remediation in karst terrane: Proc. of 32nd Symp. on Engineering Geology and Geotechnical Engineering, Boise, ID.

Fischer, J.A. & J.J. Fischer, 1997, Wyndham Farms – A karst case history: The Engineering Geology and Hydrogeology of Karst Terranes, A.A. Balkema.

Geotechnical and Environmental Applications of Karst Geology and Hydrology, Beck & Herring (eds)
© 2001 Taylor & Francis, ISBN 90 5809 190 2

Catastrophic subsidence risk assessment: a conceptual matrix for sinkhole genesis

ROBERTO SALVATI & GIUSEPPE CAPELLI   Dipartimento di Science Geologiche, Università di Roma Tre,

00146 Rome, Italy, salvati@uniroma3.it, capelli@uniroma3.it

ANTONIO COLOMBI   Regione Lazio – Dip. U.T.V.R.A., Geological Survey, 00145 Rome, Italy,
a.colombi@flashnet.it

ABSTRACT

Cover-collapse sinkholes, generally referred to as catastrophic subsidences, are widely present in the Latium Region within different geological frameworks. These sinkholes show different hazard grades depending upon the hydro-geomorphologic context of each area, and they become a major geological risk wherever the land undergoes heavy alterations due to human activities. The peculiarities of Italian sinkholes and the major problems in the management of sinkhole-prone areas are represented by their location in groundwater discharge areas (Salvati et al., 2000) and by the thickness of the overburden that exceeds 150 meters (Tharp et al., 2001).

The authors present herewith guidelines for a conceptual model for evaluating the sinkhole hazard. This model contributes to and offers a different approach to the topic, compatible with commonly accepted geomorphological models.

The Latium Regional Government and the Geological Sciences Department of Rome TRE University, through a biennial research program, will study these sinkholes. The goals will be to: catalog sinkholes, delineate their territorial evidences, and classify their different genetic forms. In addition, one aim of the project is to give local administrations a methodology to follow to identify the areas at risk in the territorial planning phase.

The wide range of parameters and variables in sinkhole genesis led the authors to seek a conceptual model that could define different potential scenarios and, after their calibration upon Latium Region territory, to export them to well known geomorphological models.

CONCEPTUAL MATRIX MODEL

The majority of international references about sinkholes are primarily Anglosaxon and North American. The majority of published Italian papers regarding catastrophic subsidence are, indeed, limited to the general aspects of sinkholes and only obvious sinkholes whose geomorphological description has considered the genetic-evolution features (Bono, 1995; Facenna et al, 1993).

Sinkholes are located in different areas of the Latium Region, and it is pretty clear that these phenomena are present in nearly every geological context. They range from the alluvial plains that border karst ridges - as in S.Vittorino Plain (Capelli et al, 2000), Doganella di Ninfa and Sprofondi of Pontina Plain (Capelli et al., 1999) or the Posta-Fibreno area - to the Plio-Quaternary depositional basins such as the Puzzo lakes in the Roman countryside or, eventually, to a volcanic rocks context such as Giulianello Lake in the Alban Hill Volcanic Complex.

A common feature to all the sinkhole-prone areas is their localization in areas of groundwater discharge (Salvati et al, 2000). "Sinkholes occurred in groundwater discharge areas near the base of karst ridges, in zones with major groundwater circulation, thick overburden, and proximity to deep-source gas or thermal upwelling" (Salvati et al, 2000). Such a complex framework for sinkhole development is difficult to reconcile with the traditional conceptual model, even for the upward propagation of the deformation (Tharp et al., 2001).

In the last few years, the increasing necessity for local administrations to adopt new and more functional territorial planning and management tools has collided with the definite lack of work standards regarding the relationships between sinkholes and urbanization. It is necessary to perform detailed and in-depth studies and investigations to achieve, if possible, a preventive management of the sinkhole hazard. It is, therefore, important to gain a deep insight of sinkhole inducing issues: their triggering and evolution modes and, for each different area, verification of the historical events in order to delineate the phenomena's main features.

In sinkhole prone areas, particularly where risk conditions are already clear, experimental methodologies become more and more important and significant. After validation of these experimental methodologies, they will help in comprehending the phenomenon of sinkholes and prevention wherever hazard conditions exist.

Presently, our experience in combination with the international state-of-practice enables us to suggest a research program that will give regional administrations a census and typology of sinkhole events in sinkhole-prone areas. Furthermore, the regional government will achieve an operational methodology that results in better planning and management solutions, in conjunction with the contributions of smaller local administrations.

It is, therefore, fundamental to acquire an in-depth knowledge about the triggering issues and the relationships existing between the different local geological, hydrogeological and geological-structural frameworks. This may be obtained by delineating a conceptual matrix model that validates the main points as well as the relationships among the different geological-environmental situations existing in an area and the developments that this situation induced.

The conceptual matrix model can be represented as a flow diagram with bounded options, which lead to the final hazard grade definition. The matrix has five cognitive elements represented respectively by: Geology, Hydrogeology, Tectonics, Seismicity, and Anthropization. Each of these is further divided into classes that take into account the different features that characterise them, i.e., Geology is divided into the classes of bedrock origin, bedrock depth and cover type. Each class is defined by its presence/absence and then the eventual relationships that it should have with the other classes.

The final result is a variable-hazard-grade definition of sinkholes. Each grade is subsequently associated with the vulnerability derived from the analyses of local anthropic aspects, as well as the seismic hazard. The hazard scenarios thus become fundamental for cognition of triggering issues, and they are strictly local being influenced by the interactions among local aspects.

Basic knowledge will be derived from a two-year work schedule consisting of various phases. During the first phase, the census of sinkhole-prone areas, sinkhole events will be catalogued and classified. In a second phase, detailed studies on a test area will be undertaken; during this phase, hydrogeological, geochemical, geotechnical, geophysical, mineralogical and geological-structural surveys and field sampling will be conducted. At the same time, a monitoring system will be implemented by drilling boreholes and equipping them with chemical, physico-chemical and physical water parameters probes.

This conceptual matrix will consequently be integrated and refined "in progress" by means of an iterative process using collected and historical data allowing even more detailed and representative hazard scenarios. The third and last phase of the Project will be devoted to the so-called "lifelines" of prevention/intervention delineation. Those lifelines will be differentiated for each area at risk, and they will enable the regional administrations to give real estate answers to smaller local administrations and to professionals concerning the attention and care to take when conducting Territorial Planning studies.

FINAL REMARKS

The conceptual model presented here is designed to be a new suggested methodology aimed at a complete knowledge of sinkhole triggering issues, and the understanding of relationships among these issues, that could induce hazardous situations in different geological and morphological frameworks. Such a matrix model is not at all antithetic to commonly accepted geomorphologic models for sinkhole description, but it aims to be a complement to them. Different iterative tracks that analyze factors, local and surrounding geological conditions and their interactions, determine hazard scenarios.

Taking into account the increasing interest and concern that catastrophic subsidence phenomena are gaining both in Italy and over all the world, and considering that so far in Italy little or no attention has been given to sinkhole deriving risk, the "Sinkholes in the Latium Region" work group has created a website. The website address is <http://www.geocities.com/sinkhole_prj2000/index.html>, and it is the tool by which we would like to share our knowledge and suggestion with anybody interested in the specific project or more generally with anybody interested in sinkholes and their environmental impact effects. The website has been created to allow the best dissemination of the results achieved and activities undertaken by the Sinkhole Project.

Coupled with the website, the authors have initiated a newsgroup at the URL <http://www.egroups.com/group/sinkhole>. This is the first international newsgroup for scientists and researchers throughout the world to communicate and share experiences, suggestions and news about sinkholes and any related topic.

REFERENCES

Bono, P. 1995. The sinkhole of Doganella (Pontina Plain, Central Italy): *Environmental Geology*, v. 26, p. 48-52.

Boyer L., Fiers S. & Monbaron M. (1998) - Geomorphological heritage evaluation in karstic terrains: a methodological approach based on multicriteria analysis. Suppl. Geografia Fisica e Din. Quaternaria 3/98 t.4

Buttrick D., van Schalkwyk A. (1998) - Hazard and risk assessment for sinkhole formation on dolomite land in South Africa. Environmental Geology 36/1-2

Capelli, G., Petitta M. & Salvati, R. (2000) Relationships between catastrophic subsidence hazards and groundwater in the Velino Valley (Central Italy). Proceedings Sixth International Symposium on Land Subsidence (SISOLS 2000), Ravenna-Italy, Sept. 2000

Faccenna C., Florindo F., Funiciello R. & Lombardi S. (1993) - Tectonic setting and sinkhole features: case histories from Western Central Italy. Quaternary Proceedings 3

Goldsworthy R. (1996) - The sinkhole project: technology, integration, and the environment. Learning and Leading with Technology 23

Kemmerly P.R. (1993) - Sinkhole hazards and risk assessment in a Planning Context. Journal of American Planning Association 59/2

Roth M.J.S., Ruggles R., Berrier N., Doyle S. & Fish D. (1998) - Using metadata sources to assess risk in areas prone to sinkholes formation. 1 Int. Conf. on Geospat. Info in Agri and Forestry 2

Tharp T. M. (1999) - Mechanics of upward propagation of cover-collapse sinkholes. Engineering Geology 52

Nearing, M.A., Parker, S. C. Bradford, J. M. and Elliot, W. J., 1991. Tensile strength of thirty-three saturated repacked soils. Soil Sci. Soc. Am. V., 55: 1546-1551.

Salvati, R., Sasowsky, I.D. & Capelli, G. (2000) – Conceptual model for cover collapse sinkholes in areas of groundwater discharge (Central Italy). Geological Society of America Meeting 2000, Reno (NV).

Tharp, T.M., 2001. Conceptual model for geotechnical evaluation of sinkhole risk in the Latium region. Proc. 8th Multidisciplinary Conf. on Sinkholes and the Engineering and Environmental Impacts of Karst (this volume)

Geotechnical and Environmental Applications of Karst Geology and Hydrology, Beck & Herring (eds)
© 2001 Taylor & Francis, ISBN 90 5809 190 2

Conceptual model for geotechnical evaluation of sinkhole risk in the Latium Region

ROBERTO SALVATI   Dipartimento di Science Geologiche, Università di Roma Tre, 00146 Rome, Italy, salvati@uniroma3.it

THOMAS M.THARP   Dept. of Earth & Atmospheric Science, Purdue University, West Lafayette, IN 47907, USA, ttharp1@purdue.edu

GIUSEPPE CAPELLI   Dipartimento di Science Geologiche, Università di Roma Tre, 00146 Rome, Italy, capelli@uniroma3.it

ABSTRACT

This paper discusses the preliminary application of a poroelastic algorithm (Tharp, 1999) to cover-collapse sinkhole events in Central Italy. These events occur in the presence of a thick sediment cover, ranging from ten to two hundred meters, overlying carbonate bedrock. The sediment cover is a locally complex continental sequence of fluvial, alluvial and lacustrine sediments. The hydrogeological setting is characterized by a major confined circulation in the karst aquifer and several smaller and shallower groundwater circuits in the confined and unconfined aquifers within the overburden. The sinkhole prone areas are located in areas of shallow water table and groundwater discharge. Formation of new cover-collapse sinkholes in this environment is believed to result from piezometric surface drawdown caused by a combination of seasonal variation in recharge, pumping for irrigation, and perhaps even changes in stress state or hydrology resulting from fault displacement.

We suggest that under these circumstances hydraulic fractures occur parallel and near to the walls of soil voids at the soil-karst interface as a result of water pressure drops in the karst below. This results in sloughing of the walls of the void. Once initiated this progressive sloughing may propagate to the ground surface. A standard comprehensive geotechnical investigation will allow calculation of the poroelastic parameters necessary to analyze pore pressure transients around the soil void. Tensile strength of the soil will also be measured to allow evaluation of the critical effective stress state for hydraulic fracturing. This investigative program allows a quantitative assessment of the risk of sinkhole formation resulting from changes in the groundwater pressure regime.

INTRODUCTION

Sinkhole occurrences in areas of groundwater discharge are relatively uncommon (Capelli *et al.*, 2000; Salvati *et al.*, 2000) and so far very poorly investigated. In the Latium Region seven sites have been identified as sinkhole prone areas. More than 70 open sinkholes are present and those filled up in the last century may be numbered in the hundreds. Furthermore, in these areas the great thickness of overburden, up to 200 meters, and the coexistence of several causative factors including human presence, mostly in the form of groundwater exploitation, poses a major problem for the identification of a geotechnical model that might explain the observed occurrences.

One of several goals of the Sinkhole Project in the Latium Region (Colombi *et al.*, 1999; Colombi *et al.*, 2001) is to define a possible numerical model that might describe sinkhole genesis and evolution in such complex areas. This paper represents the first step toward that geotechnical model implementation using as a starting point the model suggested by Tharp (1999, 2001). The following discussion will be focused on the Pontina Plain sinkhole prone area only.

GEOLOGICAL AND HYDROGEOLOGICAL FRAMEWORK

The Pontina Plain is the southernmost portion of an elongate downthrown area that developed between the first Apennine belt ridges and the shoreline starting in the Lower Pliocene. This depression is located among the Ausoni and Lepini Mounts (carbonate platform facies) to the east and the current Tyrrenian margin (where calcareous-silici-marl deposits form the Meso-Cenozoic buried bedrock) to the west (see Figure 1). The Pontina Plain structural setting is furthermore complicated by the existence of secondary tectonic elements joined to the general geological-structural evolution of this sector.

Interpretation of deep boreholes and geophysical investigation data shows that the carbonate ridges extend below the Quaternary sediments with a tiered geometry, due to the presence of normal fault sets that cut the carbonate structures. The Pontina Plain sedimentary succession developed from a marine depositional system to a transitional fluvial-coastal system early on, and to a fluvial-continental depositional system later (all of this Pliocene-Pleistocene). Therefore, the entire system is characterized by both vertical and lateral variability.

The hydrogeological setting, as usual in Italian scenarios, is strongly constrained by the geological-structural setting and its setting below the Plain is complex. Within the overburden an unconfined aquifer is present near the surface. Multiple confined and semi-confined aquifers of limited extent underlie this, within the overburden. Below this, at some depth, there is a regional confined aquifer

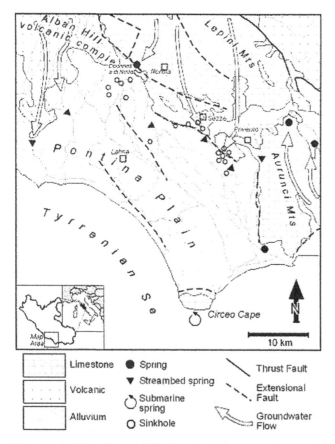

Fig.1. Map of the Pontina Plain showing bedrock geology, tectonic features, plain deposits, springs, groundwater flow, and sinkholes. Redrawn from (Boni *et al.*, 1988).

Legend:
- Limestone
- Volcanic
- Alluvium
- Spring (●)
- Streambed spring (▼)
- Submarine spring
- Sinkhole (○)
- Thrust Fault
- Extensional Fault
- Groundwater Flow

10 km

that is recharged by circulation from the karst ridge. In the Lepini carbonate ridge, this unit behaves as an unconfined aquifer. In addition to these two groundwater circuits, a hydrothermal circuit exists in the buried karst bedrock of the Plain, with main circulation along the fault network (Boni *et al.*, 1980). In this area of groundwater discharge, a condition not generally conducive to dissolution, the dissolution of carbonate rock that is ultimately responsible for sinkhole potential may result significantly from the pressure of deep-seated gaseous sources (carbonate or sulfuric).

The river network in the Plain is mainly fed by the several springs located along the Lepini ridge western border and the total discharge amount can be estimated at 15 m^3/(530 ft/s). Springs that drain the karst ridge are mainly discrete orifices, though some stream-bottom springs are found along the Uffente River where it flows close to the ridge border. The spring waters show highly variable physico-chemical characteristics (Boni *et al.*, 1980; Salvati *et al.*, 2000)

Sinkhole occurrence is limited to a narrow strip between the Lepini Mountains western border and the SS.7 Appia freeway. Although their density and distribution is not high they do represent a potential hazard as shown by the two latest occurrences. The "Ninfa sinkhole" (1989) occurred along a country road, breaking it up and barring subsequent use of the road (Bono, 1995). The "Pettinicchio sinkhole" (1995) occurred a few meters away from the SS.7 Appia freeway and close to an electric power station.

Sinkhole distribution seems to be influenced by tectonic pattern, as they mostly occur in trends oriented in directions very similar to the Lepini west border fault pattern. Other karst features such as dolines and small uvalas located inside the Lepini Mounts ridge are also thus oriented.

FUTURE PARAMETER COLLECTION AND GEOTECHNICAL ANALYSES

The unique local characteristics of the Italian sinkholes require a new approach in mathematical modelling. The principal difficulties lie in formulation of a model that will cope with the stratigraphic complexity in simulating upward propagation through an overburden of several hundreds of meters with many horizons characterized by different geomechanical properties and behaviours (Fig. 2). To collect the data needed for the first modelling attempt, several field surveys and investigations will be undertaken. In particular, three boreholes will be drilled near the "Pettinicchio Sinkhole" in order to investigate the geological and hydrogeological framework five years after it collapsed. The "Pettinicchio Sinkhole" is one of the test sites chosen for this study and the drill holes will assist the definition of the stability condition after a collapse, and the collection of data needed to implement the mathematical model cited above. In particular, data will be collected for characterization of geomechanical properties of rocks for every significant horizon (uniaxial, triaxial, shearbox, and tensile strength tests), hydraulic properties of every significant horizon (both my means of insitu tests and lab tests), grain size analyses and pressure meter tests. In addition to the geomechanical tests and sampling, a series of geophysical surveys will be undertaken. Detailed electrical resistivity profiles and electrical downhole tomography will be carried on in order to achieve the best resolution and possibly to allow extension of the geotechnical parameterization to the entire sinkhole area.

MECHANICAL ANALYSIS OF SINKHOLE FORMATION

One of several goals in this study is to measure the material parameters necessary to evaluate the conditions for sinkhole formation. In common with other regions with a high water table and substantial depth to the karstic formation, sinkhole formation in the Latium region, particularly in the Pontina Plain which is the test site for this study, seems to be associated also with drops in the piezometric surface. These may result from unusually severe seasonal declines, pumping, or rapid changes resulting from nearby earthquakes.

Formation of a surface sinkhole is preceded by formation of a soil void that forms by movement of soil into an opening in the rock below. Tensile failure will allow the roof of the soil void to assume a no-tension domal shape. This stable configuration can occur only in cohesive soils, generally clay rich. Our geotechnical investigation will characterize cohesive strata that allow formation of domal voids. It will also identify non-cohesive or other weak strata such as uncemented sands that will flow freely downward into the karst opening if breeched by an upward propagating soil void. Such flow would continue until the next cohesive layer above is exposed.

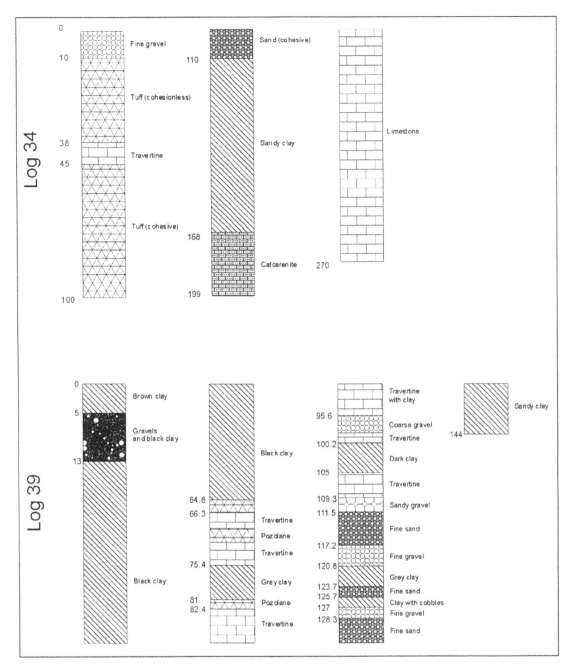

Fig. 2. Borehole logs from Pontina Plain. The two boreholes, located less that 1 km (0.6 miles) apart, show the major variability of the stratigraphic sequence within the Plain. It should be noted that in borehole 39 the downthrown carbonate bedrock lies more that 140 (460 ft) below the surface while in borehole 34 it is present at a depth of 168 (551 ft). Redrawn from (Cassa del Mezzogiorno-P.S. 29, 1982)

In the geohydrologic environment of the Pontina Plain the failure mechanism leading to upward propagation of a soil void is probably hydraulic fracturing. The failure process outlined here is explained in more detail in Tharp (2001). In the Pontina Plain the soil void at the karst surface is submerged, with water in the void in communication with the karst aquifer below. When the piezometric surface in the karst aquifer is drawn down the water pressure on the wall of the soil void is immediately reduced, reducing the radial compressive stress in the soil. The pore pressure in the soil around the opening exhibits a more gradual drop, with the result that high pore pressure gradients may occur. If pore pressure at some radial distance into the wall of the soil void is greater than the small radial compressive stress plus tensile strength of the soil, hydraulic fracturing will occur. This will cause the innermost layer of soil around the soil void to slough off, exposing soil with even higher pore pressure to the same boundary conditions. Failure at this second stage will be immediate, resulting in sloughing of a second layer of soil into the void. Sloughing of successive layers will proceed rapidly, allowing the soil void to propagate upward until it reaches a stratum of high tensile strength, or the region of low pore pressure near the surface. Because the soil void widens in its upward propagation, failure to the ground surface is common.

Because the soil matrix deforms easily, a change in pore pressure causes deformation and changes stress in the soil. Deformation of the soil matrix also generally affects pore pressure. These interactions require a poroelastic analysis. For simplicity and to accommodate the required very small finite difference grid spacing near the wall of the soil void, the soil void and boundary conditions are approximated as spherically symmetric. This allows a one-dimensional analysis. The derivation of the poroelastic equations and methods for determining the necessary material parameters are discussed by Tharp (2001).

Figure 3 illustrates the influence of two parameters that are important in assessing sinkhole risk and will be measured in our investigation. The results of the analysis for a soil void 2m (6 ft) in diameter at 30m (98 ft) depth show total head drop to cause hydraulic fracturing (failure) versus rate of head drop. Figure 3A shows the effect of hydraulic conductivities spanning the range for a typical residual soil (Tharp, 2001). For a given total head drop the order of magnitude difference in hydraulic conductivity can change the critical drawdown rate by an order of magnitude. Figure 3B shows failure conditions in the same soil for different tensile strengths. The lower strength is the mean of measured values for four clay soils (Nearing *et al.*, 1991), the greater is set arbitrarily an order of magnitude higher. This variation in tensile strength also produces an order of magnitude variation in critical drawdown rate. Sinkholes may form in response to drawdown rates ranging from meters per hour to meters per week, but within this range the conditions for formation are very sensitive to such parameters as hydraulic conductivity and tensile strength.

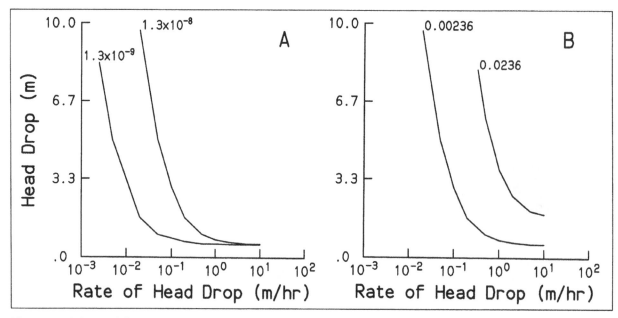

Fig. 3. Head drop at failure (hydraulic fracturing) versus rate of head drop within the soil void. A) Failure curves for hydraulic conductivities of 1.3×10^{-9} m/sec (4.3×10^{-9} ft/sec) and 1.3×10^{-8} m/sec (4.3×10^{-8} ft/sec). B) Failure curves for tensile strength of 0.00236 MPa (0.34 psi) and 0.0236 MPa (3.4 psi).

CONCLUSIONS

The complexity of sinkholes and their formation process has often frustrated efforts to come to an intellectually satisfactory or practical understanding of the phenomenon. We have embarked on a multidisciplinary geophysical, geologic and geotechnical investigation to carefully characterize several important sinkholes. In addition to characterizing the three-dimensional geometry of the sinkholes, we are collecting both standard geotechnical data and data that are particularly relevant to analyzing sinkhole formation. With the results of this field and laboratory investigation we will be able to evaluate in a quantitative and practical way the conditions likely to cause new sinkholes in the Latium region.

REFERENCES

Boni. C, Bono, P., Calderoni, G., Lombardi, S., and Turi, B. 1980. Hydrogeological and geochemical analyses on the relationships between karst circulation and hydrothermal circuit in the Pontina Plain (Southern Latium). (Indagine idrogeologica e geochimica sui rapporti tra ciclo cirsico e circuito idrotermale nella pianura Pontina (Lazio Meridionale)): *Geologia Applicata e Idrogeologia*, 15: 203-245.

Bono, P. 1995. The sinkhole of Doganella (Pontina Plain, Central Italy): *Environmental Geol.*, 26: 48-52.

Cassa per il Mezzogiorno. 1982. Progetto Speciale 29. Roma

Capelli, G., Petitta, M. & Salvati, R. 2000. Relationships between catastophic subsidence and groundwater in the Velino Valley (Central Italy). Proceedings of Sixth International Symposium on Land Subsidence (SISOLS 2000), Ravenna Italy.

Colombi, A., Capelli, G., Colasanto, F., Crescenzi, R., Mazza, R., Meloni, F., Orazi, A., Salvati, R. & Sericola, A. 1999. Probelmatiche da sprofondamento catastrofico nelle aree di pianura della Regione Lazio – Il "Progetto Sinkhole del Lazio". Proceedings Conference "Le Pianure", Ferrara, Italy.

Colombi, A., Salvati, R. Capelli, G. 2001. Catastrophic subsidence risk assessment. A conceptual matrix for sinkhole genesis in complex areas. Proc. 8th Multidisciplinary Conf. on Sinkholes and the Engineering and Environmental Impacts of Karst (this volume)

Nearing, M.A., Parker, S. C. Bradford, J. M. and Elliot, W. J., 1991. Tensile strength of thirty-three saturated repacked soils. *Soil Sci. Soc. Am.*, 55: 1546-1551.

Salvati, R., Sasowsky, I.D. & Capelli, G. 2000. Conceptual model for cover collase sinkholes in areas of groundwater discharge (Central Italy). Geol. Soc. of Am. Meeting 2000, Reno, NV.

Tharp, T.M., 1999. Mechanics of upward propagation of cover-collapse sinkholes. *Engr. Geol.*, 52: 23-33.

Tharp, T.M., 2001. Conceptual model for geotechnical evaluation of sinkhole risk in the Latium region. Proc. 8th Multidisciplinary Conf. on Sinkholes and the Engineering and Environmental Impacts of Karst (this volume)

Geotechnical and Environmental Applications of Karst Geology and Hydrology, Beck & Herring (eds)
© 2001 Taylor & Francis, ISBN 90 5809 190 2

Geotechnical characterization and modeling of a shallow karst bedrock site

TIMOTHY C.SIEGEL   S&ME, Inc., Mount Pleasant, SC 29464, USA, tsiegel@smeinc.com

DANIEL W.MCCRACKIN   CH2M Hill, Inc., Oak Ridge, TN 37830, USA, dmccrack@ch2m.com

ABSTRACT

A site with shallow karst bedrock is characterized for engineering purposes by examination and testing of rock samples and by classification of the rock mass according to the RMR system. On the basis of six measurable parameters (intact rock strength, rock quality, joint spacings, joint conditions, groundwater conditions and joint orientations), RMR values were assigned to the rock mass. For analysis purposes, cave roof conditions were considered separately, as well as, incorporated with samples from other site areas. For all of the rock samples, an RMR of 65 was assigned. For the cave roof conditions, an RMR of 50 was assigned. On the basis of the RMR values, a failure criterion (or model) was established using the Hoek-Brown failure criterion. The authors believe that this is the first published application of this approach for evaluation of naturally occurring openings in rock (i.e., caves).

INTRODUCTION

The site of a proposed landfill is located within Valley and Ridge Physiographic Province of Tennessee, and is underlain by hard limestone and dolostone of the Knox Group and the Middle Ordovician Lenior Limestone. The site is rare in that the ground surface is mainly comprised of rock outcrops and, in some areas, a relatively thin soil layer underlain by rock. Naturally occurring caves and sinkholes are present. From the initial review of the site conditions, it was apparent that stability of the rock surface was critical to the permitting and design of the proposed landfill. To allow cave stability analysis, it was necessary to characterize the condition of the rock mass in engineering terms. Extensive examinations and tests were performed on rock core samples to classify the rock conditions according to the Rock Mass Rating (RMR) system (Bieniawski, 1973 and 1989). Failure criterion was developed for the site based on an empirical relationship between rock mass condition and rock mass strength proposed by Hoek and Brown (1980a; 1980b; 1988; and 1997).

SITE GEOLOGY AND HYDROLOGY

Structural elements of the site reflect the typical patterns of deformation found throughout the Valley and Ridge Province of East Tennessee. Bedrock strikes and the trends of fault traces and fold axes are northeast-southwest. Bedrock strike at the site varies between N 30° E and N 70° E. At least six joint sets were recognized by field survey. Neither faults nor joints are considered active, showing no obvious signs of recent crustal movements. They are primarily products of deformation associated with Paleozoic tectonic events hundreds of millions of years ago. Folds within the bedrock are integrally related to the faults, with fold axes oriented northeast-southwest. However, the style of folding varies across the site with northwest fold limbs tending to be steeper (almost vertical in places).

Cave orientations are consistent with bedrock rock strike (i.e., N 30° E and N 70° E), with beds dipping northwest and southeast depending on the location of the bedrock with respect to the folding. Cave orientations are also consistent with the six dominant joint sets measured in rock outcrops and cores.

The site exhibits karst development characterized by (1) numerous coalescing sinkholes, (2) closed depressions, (3) cave development, (4) sinking streams, (5) carbonate rock outcrops, and (6) conduit-fed spring discharges. Geologic structure imposes significant control on groundwater flow at the site, with solution channels (i.e., conduits) developed mostly in the direction of the bedding strike and rock mass permeability significantly higher in the northeast-southwest direction, as opposed to perpendicular to bedding. The larger dissolution features and caves occur within the vadose zone closer to the ground surface, and the intensity and frequency of the larger dissolutional features decreases with depth.

ENGINEERING CLASSIFICATION OF ROCK

Classifications systems allow site-specific rock conditions to be compared to a database of observed rock behavior (e.g., strength, stability). Early in the development of geotechnical engineering, Terzaghi (1946) recommended steel support for tunnel excavations based on empirical data by means of visual classification using terms such as "intact", "blocky", and "seamy". Since then, more sophisticated classification systems have been developed to characterize rock-masses for engineering purposes including Rock Mass Rating (RMR) system (Bieniawski, 1973 and 1989), the Q-system (Barton, et al., 1974), and the recently introduced Geologic Strength Index (GSI) (Hoek, 1998).

For the subject site, the near-surface rock was classified according to the RMR system based on examinations and testing performed on rock samples collected from 13 test borings broadly distributed over the site, including one test boring (designated PAC-7) drilled through the roof of an existing cave. The size designation of the rock cores were NQ with the exception of boring PAC-7 that was HQ size. The following six parameters are used to classify the rock mass according to the RMR system: (1) intact rock strength, (2) rock quality, (3) joint spacings, (4) joint conditions, (5) groundwater conditions, and (6) joint orientations.

Table 1: Results of Uniaxial Compressive Tests

Boring	Depth (m)	γ (kN/m^3)	q_u (MPa)	E/10^3 (MPa)	ν
GB-3	6.0	26.6	115.2	51.7	.30
GB-4	1.8	27.3	112.6	83.1	.32
GB-4	3.0	27.3	-	50.3	.33
GB-6	1.8	26.5	67.0	69.0	.30
GB-9	6.0	26.5	67.0	-	-
GB-10	12.0	26.4	59.7	20.0	.32
EB-3	17.0	27.2	73.5	87.9	.30

γ – unit weight, E – elastic modulus, ν – Poisson's ratio

Intact Rock Strength

The strength of intact rock core samples was estimated by uniaxial compressive tests (ASTM D3148). For this test, select samples were cut to a length-to-diameter ratio of approximately two, and their ends were machined to an acceptable flatness. To allow strain measurements during the application of the compressive load, strain gages were epoxied to the rock core samples. Incremental compressive loads were applied along the axis of the rock core sample until failure (i.e., fracture). The average uniaxial compressive strength (q_u) from the uniaxial compressive tests is 82 MPa (11,900 psi). The results of the uniaxial compressive tests on intact rock core samples are summarized in Table 1.

Rock Quality Designation (RQD)

Rock Quality Designation (RQD) is defined as the total length of pieces of sound rock core with length greater than 10 cm (4 inches) divided by the length of the core run (Deere, 1968). The average RQD for all of the rock core samples, weighted by the run length, is 83%. If the upper most core run is not considered (since the upper rock is typically irregular), the average RQD becomes 92%. According to Deere and Deere (1989), an RQD above 90% classifies the bedrock as "excellent" with respect to the stability of underground openings. Although the average RQD is representative of the general rock conditions, cave roof conditions may be better represented by the conditions at PAC-7. Test boring PAC-7 was drilled through the longest spanning cave roof and the resulting average RQD is approximately 36%. Space limitations preclude a complete presentation of the RQD measurements in this paper.

Table 2: Summary of Joint Spacings

Location	Joint Spacing Categories (Bienawski, 1989)				
	< 60mm	60-200mm	200-600mm	0.6-2m	>2m
PAC-7	14%	19%	28%	34%	5%
Site	13%	23%	30%	27%	7%

Joint Spacings

Joint spacings were quantified by measuring the distance between similar joint types (e.g., near-vertical joints, along-bedding joints, and cross-bedding joints) in the rock core samples supplemented by observations of rock outcrops. The majority of the joint spacings were relatively evenly distributed from 0.01 m to 1 m (0.2 ft to 6.6 ft). In comparison to the joint spacings from all of the borings, the joint spacings observed in the rock samples from PAC-7 are generally consistent with the rest of the site. Table 2 summarizes the results of the joint spacing measurements performed on rock core samples.

Table 3: Summary of Joint Roughness and Weathering

Location	Joint Condition Designation						
	Healed	S,U	S,M	S,H	R,U	R,M	R,H
PAC-7	27%	1%	8%	4%	-	2%	56%
Site	53%	5%	7%	<1%	1.4%	13%	20%

Healed – A bonded joint with filling (e.g., calcite)
S,U – smooth, unweathered   R,U – rough, unweathered
S,M – smooth, moderately weathered R,M – rough, moderately weathered
S,H – smooth, highly weathered   R,H – rough, highly weathered

Joint Conditions

The condition of joints includes roughness, filling, weathering and separation. Examination of rock core samples indicates that slightly more than one-half of the joints (approximately 53%) were healed with calcite filling. The remaining joints exhibited a range of roughness, degree of weathering and separation. An indication of the high strength of the healed joints was that, in many cases, laboratory testing indicated that mechanical breaks in the core occurred through continuous rock rather than along nearby healed fractures. Comparison between PAC-7 and the results of all of the borings indicates that PAC-7 has a significantly higher percentage of weathered discontinuities. The joint conditions based on examination of the rock samples are summarized in Table 3 and Table 4.

Groundwater Conditions

Considering the extensive groundwater and geo-chemistry studies performed at this site, a thorough discussion of the groundwater flow at this site will be published at a later date. In summary, groundwater flow within the karst aquifer occurs below the depth pertinent to this evaluation (i.e., the depth of the caves). Furthermore, stormwater control measures will be implemented during

Table 4: Summary of Joint Separation

Location	No Separation	< 1mm Separation	>1mm Separation
PAC-7	24%	20%	56%
Site	52%	24%	24%

construction of the proposed landfill, and, once in place, the landfill bottom liner will preclude the downward flow of surface water. For these reasons, the cave conditions are expected to be moist, but not inundated.

Joint Orientations

The RMR classification includes an adjustment for the orientation of joints. This adjustment considers steeper dip angles as less favorable. It also considers an orientation of the underground opening parallel to strike as less favorable. The latter adjustment is more applicable to man-made openings (e.g., tunnels) in which the orientation of the opening is less variable than naturally occurring caves. For purposes of determining a failure criterion, a rating of "fair" was conservatively assigned for the orientation of the joints.

Table 5: Site-Specific Rock Mass Rating

RMR Parameter	PAC-7	Site
Strength of Intact Rock	8 (assumed)	8
RQD	7	18
Joint Spacings	9	9
Joint Conditions	16	20
Groundwater	15	15
Joint Orientation	-5	-5
RMR	50	65

Rock Mass Rating

On the basis of the six parameters (i.e., intact rock strength, rock quality, joint spacings, joint conditions, groundwater conditions, and joint orientations), the RMR values shown in Table 5 were assigned to the rock mass at the subject site. It is useful to consider the results for PAC-7 separately (as a conservative estimate of rock mass strength) as well as incorporated with the other rock samples. For all rock samples, an RMR of 65 was assigned. For PAC-7, an RMR of 50 was assigned.

HOEK-BROWN FAILURE CRITERION

Using the RMR values determined for the subject site, the failure criterion shown in Figure 1 was established for use in cave stability evaluations. The Hoek-Brown failure criterion is an empirical approximation of the strength of jointed rock masses, and is defined, in terms of the major and minor principal stresses, by the following equation:

$$\sigma_1 = \sigma_3 + (mq_u\sigma_3 + sq_u{}^2)^{1/2}$$

where σ_1 and σ_3 are the major and minor principal stresses at failure, respectively, q_u is the uniaxial compressive strength of the intact rock, and m and s are empirical constants which depend on rock type and condition of the rock mass. As suggested by Hoek and Brown (1980), the dimensionless empirical strength constants m and s were estimated using the following correlations with the RMR:

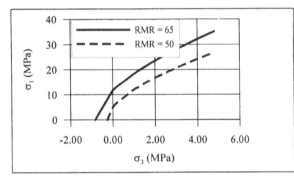

Figure 1: Site-Specific Hoek-Brown Failure Criterion

$$s = \exp\left(\frac{RMR - 100}{9}\right) \qquad m = m_i\exp\left(\frac{RMR - 100}{28}\right)$$

where the dimensionless constant m_i is determined by triaxial testing. From statistical analysis of published triaxial strength data, Brady and Brown (1993) presented the values of m_i as a function of rock type. For the carbonate rock types present at this site, a value of 7 was assigned to m_i.

CONCLUSIONS

A site with shallow karst bedrock is characterized by examination and testing of rock samples, and by classification of the rock mass according to the RMR system. Using the RMR, a failure model was established using the Hoek-Brown failure criterion. The authors believe that this is the first publish application of this approach for the evaluation of naturally occurring caves.

ACKNOWLEDGEMENTS

The authors gratefully acknowledge Janet Carwile for preparation of the final manuscript and Dr. Matthew Mauldon of the Virginia Polytechnical Institute (formerly of the University of Tennessee) for his valuable contributions to this paper.

REFERENCES

ASTM (American Society of Testing and Materials). (2000). "Standard test method for elastic moduli of intact rock core specimens in uniaxial compression." *Annual Book of ASTM Standards*, Vol. 4.08, Section Four, 330-334.

Barton, N., Lien, R., and Lunde, J. (1974). "Engineering classification of rock masses for the design of tunnel support." *Rock Mechanics*, Vol.6, 183-236.

Bieniawski, Z. T. (1989). *Engineering rock mass classifications*. Wiley and Sons, New York, New York.

Bieniawski, Z. T. (1973). "Engineering classification of jointed rock masses*." Transactions of the South Africa Institute of Civil Engineering*, Vol. 15, 335-344.

Brady, B. H. G. and Brown, E. T. (1993). *Rock mechanics for underground mining*. Chapman and Hall, New York, New York.

Deere, D. U. (1968). "Geologic considerations." *Chapter 1, Rock mechanics in engineering practice*, eds. K. Stagg and O. Zienkiewicz, Wiley, New York, New York, 1-20.

Deere, D. U. and Deere, D. W. (1989). "Rock quality designation (RQD) after twenty years." *Contract Report GL-89-1*, US Army Corps of Engineers.

Hoek, E. and Brown, E. T. (1997). "Practical estimates of rock mass strength." *International Journal of Rock Mechanics and Mineral Science*, Vol.34 (8), 1165-1186.

Hoek, E. and Brown, E. T. (1988). "The Hoek-Brown failure criterion – a 1988 update*." Proceedings, 15th Canadian Rock Mechanics Symposium*, University of Toronto.

Hoek, E. and Brown, E. T. (1980a). *Underground excavations in rock*. Institution of Mining and Metallurgy, London, Engl.

Hoek, E. and Brown, E. T. (1980b). "Empirical strength criterion for rock masses." *Journal of Geotechnical Engineering*, ASCE 106 (GT6), 1013-1035.

Terzaghi, K. (1946). "Loads on tunnel supports." *Chapter 4, Rock tunneling with steel supports*, Commercial Shearing and Stamping Company, Youngstown, Ohio.

Geotechnical and Environmental Applications of Karst Geology and Hydrology, Beck & Herring (eds)
© 2001 Taylor & Francis, ISBN 90 5809 190 2

Geosynthetic reinforcement above sinkholes to protect landfill liners

TIMOTHY C.SIEGEL & P.ETHAN CARGILL, S&ME, Inc., Mt. Pleasant, SC 29464, USA, tsiegel@smeinc.com

DANIEL W.MCCRACKIN CH2M Hill, Inc., Oak Ridge, TN 37830, USA, dmccrack@ch2m.com

ABSTRACT

The two-dimensional finite difference code FLAC is used to model the interaction of a landfill bottom liner supported by geosynthetic-reinforced soil above a sinkhole. The soil is assumed to be a homogeneous, elastic-plastic material. FLAC grids were prepared to represent a range of sinkhole throat (i.e., rock opening) sizes and liner elevations above the sinkhole throat. The sinkhole throat was simulated by removing a portion of the FLAC grid beneath the reinforcing geosynthetic. The model behaves as expected in several respects. The stress and strain in the reinforcing geosynthetic increases with increasing sinkhole-throat width. Also, the strain in the liner decreases as the vertical distance between the liner and sinkhole throat increases. An increase in the number of geosynthetic layers reduces the strain in the liner, especially when the liner is immediately above the geosynthetic reinforcement.

Although the application of FLAC to this problem appears promising, several limitations are apparent when evaluating the results. One primary issue is that the tensile stresses developed in the reinforcing geosynthetic are significantly lower than expected based on the computed tensile strains. It is suggested that either the FLAC encounters a computation difficulty when computing the stress-strain relationship for the reinforcing geosynthetic or the parameters selected for the analysis do not effectively represent the actual behavior of the soil-geosynthetic interface.

Even though the model presented in this paper requires further refinement and improvement, FLAC modeling offers a promising approach for the evaluation of geosynthetic reinforcement above sinkholes where strains within the overlying materials is of concern, such as landfill liners in this case. The flexibility of FLAC is illustrated in this paper in application to static evaluation in homogenous soil. It is anticipated that FLAC models similar to the one presented in this paper can be applied to more complex conditions such as various sinkhole configurations (e.g., circular), layered soils, seepage, and/or seismic motions.

INTRODUCTION

Landfills constructed over karst terrain are exposed to risk associated with sinkholes. Development or enlargement of a sinkhole beneath a landfill can potentially allow the bottom liner to be unsupported, and thus render it susceptible to damage due to high tensile strain. To mitigate this risk, geosynthetics can be used during landfill construction to reinforce the soils above areas of sinkhole activity. Geosynthetic reinforcement is most effective at sites with relatively thin soil overburden, where: 1) the sinkhole throat can be identified and measured during subgrade preparations, and 2) the risk of sinkhole instability is greater as compared to sites with relatively thick soil overburden due to the effects of soil arching (Yang and Drumm, 1999). In this paper, the geosynthetic reinforcement over a sinkhole throat is modeled using the two-dimensional finite difference code FLAC (Itasca Consulting Group, 2000). The model assumes that the sinkhole throat can be identified and measured during landfill subgrade preparation, and that the geosynthetic reinforcement can be constructed over the sinkhole throat.

LITERATURE REVIEW

Early in the evolution of geosynthetics, Giroud (1982) developed tensioned-membrane theory for application to geotextile reinforcement of a geomembrane over a void. Bonaparte and Berg (1987) combined tensioned membrane theory with soil arching (Terzaghi, 1943; Handy, 1985) to estimate the geosynthetic strength requirements for a roadway over karst terrain. Giroud, et al. (1988) prepared simple charts for the combination of tensioned membrane theory and soil arching for design of geosynthetic-reinforced soil. Berg and Collin (1993) presented a variation to Giroud, et al. (1988) by including the contribution of the geomembrane to the stability of the landfill liner. Several published case histories (Paulson and Parker, 1993; Stelmack, et al., 1995; Alexiew, 1997) illustrate the application of the approach originally developed by Giroud and advanced by others to the design of landfills over sinkholes. One assumption common to all of the aforementioned papers is that soil arching develops fully. The applicability of this assumption depends on the amount of soil movement necessary to fully develop soil arching, and its compatibility with the mobilization of tensile resistance in the geosynthetic reinforcement. The model presented in this paper is perhaps more realistic as it achieves compatibility between soil arching and tensile resistance.

Gabr, et al. (1992) modeled a landfill liner over a sinkhole using finite elements and confirmed that the use of geosynthetic reinforcement can reduce the stresses and strains in the liner. The finite element model by Gabr, et al. (1992) assumes that the geosynthetic reinforcement is located immediately below and in contact with the liner. The model presented in this paper has the geosynthetic reinforcement immediately above the rock surface (and sinkhole opening), with the vertical distance between the liner and geosynthetic reinforcement varied to analyze the stress-strain conditions in the liner.

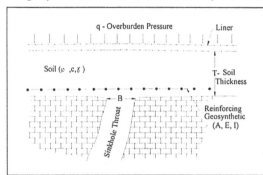

Figure 1: Conceptual Illustration of Sinkhole Model

SINKHOLE MODEL
Geometry
A conceptual illustration of the sinkhole model is shown in Figure 1, where T is the fill thickness between the top of the bedrock surface and the bottom liner, and B is the width of the sinkhole throat. The sinkhole throat is conservatively assumed to be a trench of infinite length, and therefore, plain-strain conditions are applied. The vertical pressure, q, is due to the soil below the critical height. The vertical pressure due to material above the critical height, as determined by numerical analysis, is transferred directly to the rock surface on each side of the sinkhole throat as the result of arching.

Soil properties
The soil behavior is defined by the Mohr-Coulomb plasticity model. The model parameters are defined as follows: ϕ is the friction angle, c is the cohesion intercept, γ is unit weight, K is the bulk modulus, and G is the shear modulus. The friction angle and cohesion intercept define the boundary between elastic and plastic soil behavior. The bulk and shear moduli determine the deformation of the soil at a stress level below yield. For stress levels at or above yield, the soil model behaves perfectly plastic. For simplicity, the soil is modeled as homogeneous with the parameters presented in Table 1.

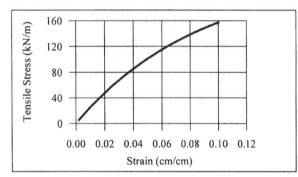

Figure 2: Stress-Strain Curve for Reinforcing Geosynthetic

Geosynthetic properties
The reinforcing geosynthetic is modeled as a beam-type structural element with nominal bending and compressive stiffness. The model parameters are defined as follows: A is cross-sectional area per unit width, E is elastic modulus, and I is the moment of inertia. As shown in Figure 2, the tensile stress-strain relationship of geosynthetics are non-linear. To account for this aspect, the FLAC analysis was iterated until compatibility was achieved between the stress and strain in the reinforcing geosynthetic.

Interface properties
A interface element was included in the FLAC grids to represent the contact of the reinforcing geosynthetic with the surrounding soil. The interface allows slip and/or separation along the contact surface. The model parameters, as listed in Table 1, are defined as follows: kn is the normal stiffness, ks is the shear stiffness, and β is the interface friction angle.

Table 1: Summary of Sinkhole Model Parameters

Soil Parameters		Reinforcing Geosynthetic Parameters		Interface Parameters	
ϕ	25 deg.	A	.007 m^2/m (0.0233 ft^2/ft) per layer	kn	157,000 kPa/m (1000 ksf/ft)
c	0	E	see Figure 2	ks	314 kPa/m (2 ksf/ft)
γ	19.5 kN/m^3 (121 pcf)	I	0	β	17 deg.
K	6700 kPa (140 ksf)	-	-	-	-
G	3070 kPa (64.2 ksf)	-	-	-	-

FLAC ANALYSIS AND RESULTS
Determination of the critical height
The initial step in the FLAC analysis was to determine the *critical height*. The critical height is the distance above the rock surface beyond which the surcharge load is transferred directly to the rock surface on each side of the sinkhole throat as a result of arching. To determine the critical height, FLAC grids with different heights were subjected to gravitational forces. To avoid the effects of the model boundary constraints, the width of the FLAC grids were at least five times the sinkhole throat width. This is consistent with numerical models prepared by others (DeBoarst and Vermeer, 1984). Once each of these grids achieved equilibrium,

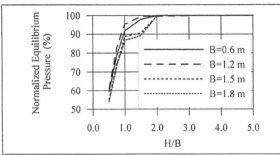

Figure 3: Determination of the Critical Height

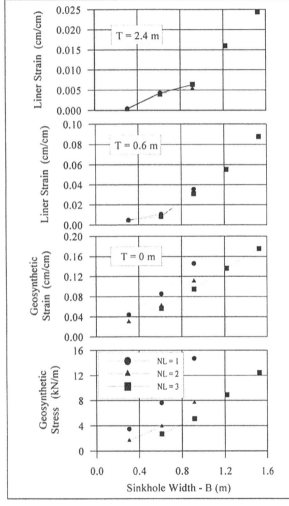

Figure 4: Summary of FLAC results

an opening or trapdoor was then created in the bottom. As the soil moved downward into the trapdoor, the principal stresses rotated and the vertical stresses immediately above the trapdoor dramatically reduced.

The minimum upward pressure (labeled hereafter as the *equilibrium pressure*) necessary to resist the downward movement of the soil through the trapdoor was determined using an iterative process. By incrementally increasing the grid height, the equilibrium pressure approached a constant value. The soil height where the equilibrium pressure became constant with increasing grid height is the critical height. Figure 3 is a plot of the normalized equilibrium pressure versus the ratio of soil height-to-trapdoor width (i.e., H/B), and indicates that the critical height occurs at a soil height-to-trapdoor width ratio of approximately 2. In Figure 3, the equilibrium pressure at different values of H/B is normalized to the equilibrium pressure at the critical height (i.e., $H/B=2$).

Evaluation of geosynthetic-reinforced sinkhole

Using the model parameters presented in Table 1, FLAC grids were prepared to represent a range of conditions. The sinkhole width was varied between 0.3 m and 1.5 m. As many as 3 layers (NL) of geosynthetic reinforcement were included in the FLAC grids. The distance from the rock surface to the liner was varied from zero (i.e., the liner is directly above the reinforcing geosynthetic) to 2.4 m. Reinforcement provided by the liner was conservatively ignored in this analysis; however, it would be quite easy to include such resistance.

For each FLAC grid, a sinkhole throat was simulated by removing a portion of the FLAC grid beneath the geosynthetic. Then FLAC solved for the forces within the grid, including the tensile stress developed within the geosynthetic reinforcement, in accordance with static equilibrium. Once an acceptable level of static equilibrium was achieved, the stress and strain conditions in the geosynthetic reinforcement and the strain condition in the liner were extracted from the FLAC output.

The results of the FLAC sinkhole model are graphically summarized in Figure 4. In general, the FLAC sinkhole modeled behave as expected in several respects. First, the stress and strain in the reinforcing geosynthetic increased with increasing sinkhole throat width (B). Second, the strain in the liner decreased as its distance (T) to the rock surface increased. And third, additional layers of reinforcing geosynthetic reduced the strain in the liner, especially when the liner is immediately above the geosynthetic.

For ease in evaluating the results, the liner was also modeled as a beam-type element; however, the properties were selected so that the behavior of the FLAC grids was not influenced by the presence of the liner.

Evaluation of the FLAC results also indicates that there are several limitations to the model presented in this paper. A primary limitation is evident by the observation that the tensile stresses developed in the reinforcing geosynthetic are significantly lower than expected based on the computed tensile strains. As a consequence, the use of multiple reinforcing layers has significantly less influence than expected. The reason(s) for this limitation was not apparent from the analysis performed for this paper. However, the authors believe that either FLAC encounters a computation difficulty when computing the stress-strain relationship in the geosynthetic reinforcement or the parameters selected for the analysis do not effectively represent the actual behavior of the soil-reinforcing geosynthetic interface. Further evaluation is necessary to identify to cause of this limitation and to improve the FLAC models so that the stress-strain behavior of the geosynthetic reinforcement is effectively represented.

CONCLUSIONS

FLAC modeling offers a promising approach for the evaluation of geosynthetic reinforcement of soil above sinkholes where strains within the overlying materials is of concern, such as landfill liners in this case. FLAC solves for static equilibrium, which inherently achieves compatibility between (1) development of soil arching, and (2) mobilization of tensile resistance in the reinforcing geosynthetic. The flexibility of this approach is illustrated by its application to a range of sinkhole sizes, layers of reinforcing geosynthetic, and liner elevations with respect to the sinkhole throat. In this paper, the soil is assumed to be a homogeneous, elastic-plastic material. It is anticipated that FLAC models similar to the one presented in this paper can be applied to more complex conditions such as various sinkhole configurations (e.g., circular), layered soils, seepage, and/or seismic motions. The analysis results include stresses and strains in the reinforcing geosynthetic and strains in the liner that may be compared to survivability criteria established during design. During design, the computed liner strains can be compared to laboratory determined stress-strain criteria established for the liner to determine if the resultant strain is within tolerable limits (i.e., typically within a strain of 5% to 10%).

ACKNOWLEDGEMENTS

The authors gratefully acknowledge Janet Carwile for preparation of the final manuscript and Kelvin Engram for preparation of the illustrations.

REFERENCES

Alexiew, D.A. (1997). "Bridging a sinkhole by high-strength high-modulus geogrids." *Proceedings of Geosynthetics '97,* Long Beach, California, 13-24.

Ben-Hassine, J., Booth, P.E. and Riggs, D.R. (1995). "Static and seismic stability of residual soils over a bedrock cavity." *Karst geohazards*, Balkema, Rotterdam, The Netherlands, 357-362.

Berg, R.R. and Collin, J.G. (1993). "Design of landfill liners over yielding foundations." *Proceedings of Geosynthetics '93,* Vancouver, Canada, 1439-1453.

Bonaparte, R. and Berg, R.R. (1987). "The use of geosynthetics to support roadways over sinkhole prone areas." *Proceedings of the 2nd multidisciplinary conference on sinkholes and the environmental impacts of karst*, Orlando, Florida, 437-445.

DeBorst, R. and Vermeer, P.A. (1984). "The possibilities and limitations of finite elements for limit analysis." *Geotechnique*, 34, No.2, 199-210.

Gabr, M.A., Hunter, T.J. and Collin, J.G. (1992). "Stability of geogrid-reinforced landfill liners over sinkholes." *Earth reinforcement practice*, Balkema, Rotterdam, The Netherlands, 595-600.

Giroud, J.P. (1982). "Design of geotextiles associated with geomembranes." *Proceedings of the 2nd international conference on geotextiles*, Las Vegas, Nevada, 37-42.

Giroud, J.P., Bonaparte, R., Beech, J.F. and Gross, B.A. (1988). "Load-carrying capacity of a soil layer supported by a geosynthetic overlying a void." *Proceedings of the international geotechnical symposium on theory and practice of earth reinforcement*, Balkema, Rotterdam, The Netherlands, 185-190.

Handy, R.L. (1985). "The arch in soil arching." *Journal of geotechnical engineering*, ASCE, Vol.111, No.3, 302-318.

Itasca Consulting Group (2000). *FLAC User's Guide*, Minneapolis, Minnesota.

Koutsabelouis, N.C. and Griffiths, D.V. (1989). "Numerical modelling of the trap door problem." *Geotechnique*, 39, No.1, 77-89.

Merry, S.M., Bray, J.D. and Bourdeau, P.L. (1995). "Stress-strain compatibility of geomembranes subjected to subsidence." *Proceedings of Geosynthetics '95*, Nashville, Tennessee, 799-811.

Paulson, J.N. and Parker, L.W. (1993). "Multiple geotextile layers used for geomembrane support in a landfill: the Marion County (Florida) landfill project." *Proceedings of Geosynthetics '93*, Vancouver, Canada, 1287-1300.

Stelmack, M.J., Sheridan, J.T. and Beck, B.F. (1995). "Cap reinforcement over a sinkhole as part of a landfill closure." *Karst geohazards*, Balkema, Rotterdam, The Netherlands, 389-395.

Terzaghi, K. (1948). *Theoretical soil mechanics*, J. Wiley and Sons, New York, New York.

Yang, M.Z. and Drumm, E.C. (1999). "Stability evaluation for the siting of municipal landfills in karst." *Hydrogeology and engineering geology of sinkholes and karst*, Balkema, Rotterdam, The Netherlands, 373-380.

Geotechnical and Environmental Applications of Karst Geology and Hydrology, Beck & Herring (eds)
© 2001 Taylor & Francis, ISBN 90 5809 190 2

Pin piles in karst topography

FRED S.TARQUINIO & SETH L.PEARLMAN   Nicholson Construction Co., Cuddy, PA 15031 USA,
info@nicholson-rodio.com

ABSTRACT

Karst topography is a problematic geology located throughout the United States. It is difficult because it is highly unpredictable and variable in nature (see Figure 1), since the mineral composition of the bedrock (limestone) is vulnerable to the formation of solution cavities and sinkholes. The upper layers of the limestone are typically polluted with voids, massive clay seams and artesian conditions. The overburden soils above the limestone are typically silts and clays that are susceptible to large settlements when heavy foundation loads are placed on them. For heavily loaded structures or structures that are sensitive to settlement, deep foundations are necessary. Typical deep foundations are drilled shafts or driven piles. These foundation types transfer the foundation loads to the refusal material or bedrock. However, in the karstic formations, the refusal material may be underlain by voids and clay seams. Therefore, foundations bearing above these problem zones may be unstable. The solution to this problem is the use of Pin Piles. Pin Piles are drilled and grouted piles typically ranging from 5 to 12 inches in diameter. Their capacity in soil and rock is derived entirely from friction. In karst areas where several strata of bedrock must be penetrated, the Pin Piles are readily advanced through the hard strata and socketed into competent bedrock. The piles are drilled with an eccentric down the hole hammer that advances the casing as the hole is being drilled. The hammer reacts to hard drilling in a way that is both visible and audible; therefore verifying the competency of the rock.

Figure 1 – Karst Uncovered

INTRODUCTION

Many cases exist where it is not practical or economical to install more traditional deep foundation elements such as driven piles or drilled shafts. In these cases, such as in ground where obstructions, boulders and solution cavities are present, Pin Piles have been selected. Pin Piles are drilled-in elements typically ranging from 5 to 12 inches in diameter, which consist of steel casing, steel reinforcement, and cement grout. They derive capacity in the ground from side friction and work equally well in both compression and tension.

These piles were developed in Europe in a simpler form, typically using only a central rebar core encased with cement grout placed into a small diameter drilled hole. These minipiles or micropiles were installed as individual elements or groups with cumulative benefit.

In the United States, particularly when considered for building applications, it was realized that traditional mini or micro piles were limited by their structural capacity. That is, the rebar core with virtually unconfined grout surrounding it was not very resistant to high compressive loads or any manner of lateral bending or eccentric loading.

Since these elements are most often installed using a steel drill casing, it was the innovation of the American contracting community to incorporate the steel casing into the pile designs. This ductile steel casing pile, Pin Pile, provides a high degree of structural resistance in the soft upper soils, and allows for the full optimization of the underlying competent geology. It is often the physical constraints that act as a trigger for use of Pin Pile technology. Such situations may be:

- subsurface obstructions or difficult ground,
- limited overhead clearance,
- vibration or noise sensitivity,
- settlement sensitivity,
- limited plan access,
- the need to install elements in close proximity to or through existing footings, columns, walls, or other structures.

Geotechnical complications may act as a single driving force for Pin Pile selection, or may complement the already difficult physical limitations of a project. The geotechnical situations other than what might be called "conventional" which Pin Piles may be conveniently installed are:

- karstic limestone geology (that includes voids or soil-filled solution cavities),
- bouldery ground or glacial till,
- variable and/or random fill,
- underlying existing foundations or man-made obstructions,
- rock formations with variable weathering,
- or soils under a high water table.

GEOTECHNICAL ASPECTS

Pin Piles derive their load carrying capacity from side friction in suitable ground stratum, either soil or rock. Ground types that are capable include:

- stiff or hard non-plastic clays or silts,
- sands and gravels,
- rock formations,
- combination materials such as glacial tills or residual soil formations with variably weathered rock inclusions.

Since the primary load carrying capacity in the ground is derived from frictional bond in soil or rock, the Pin Piles can develop high capacity in both compression and tension. In compression, Pin Piles typically range in working load from 50 to 200 tons. In tension, their capacity is nearly identical geotechnically, and is primarily limited by the amount and detailing of the core reinforcement and casing joints used in the elements.

When subjected to lateral loadings, the piles derive resistance from the horizontal response of the adjacent soils and can sustain significant lateral deflection within the available structural pile capacity. Seven-inch diameter Pin Piles have been laterally tested in variable urban fills, with a free head condition, to up to 19 kips with 0.75 inch deflection. On a site with stiff alluvial soils, in a free head condition, a seven-inch Pin Pile was tested to 24 kips with 0.3 inch deflection. Where significant bending capacity is desired due to lateral loads, a larger upper casing can be installed to provide a pile with a two-part cross-section.

INSTALLATION METHODS

The ability to install Pin Piles in the most difficult and problematic geotechnical situations is a major advantage in their use for building foundations. This capability is gained principally by the optimal selection of drilling and grouting techniques. Through proper selection and experienced execution, good results are obtained.

Pin Piles are installed using rotary drilling techniques similar to those used in the oil and gas industry. The piles develop their geotechnical capacity through grout to ground adhesion in the bond zone. In soils this bond is typically developed using pressure grouting, and in rock, tremie grouting. The principle type of drilling and grouting techniques used by the writers in karstic areas is described below.

Rotary Eccentric Percussive Duplex Drilling

This method uses an inner rod and an outer casing, with the spoils flushed inside the casing. The bit on the inner drill rod is equipped with a down-the-hole hammer. The hammer bit is specially designed to open up during drilling to a diameter slightly larger than the outside diameter of the drill casing (see Figure 2). This bit provides a slightly oversized hole through obstructions or rock and thereby allows the casing to simultaneously follow it down. Compressed air is used to drive the hammer and also acts as the drilling fluid to lift the cuttings. This drilling method is used in soils containing large amounts of obstructions such as cobbles, boulders or demolition waste and is also very effective in advancing a drill casing through highly fractured rock zones in karst. This method is most effective for installation in karst.

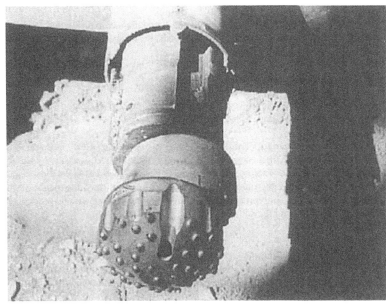

Figure 2 – Eccentric Hammer Inside the Casing

Tremie Grouting

This is a method used to place grout in a wet hole. A grout tube is lowered to the bottom of the drill casing or rock socket. Grout is pumped through the tube as it is slowly removed from the hole. As the grout fills the drill casing or hole, it displaces the drilling fluid. Tremie grouting is primarily used where the Pin Pile bond zone is founded in rock. When working in highly broken or fractured rock or in voided, karstic situations, grout loss is possible and may warrant testing for a sealed bond zone. When this is done it is typical to perform water testing and seal grouting as required, then re-drill, and test again. This potentially repetitive process requires commercial compensation using unit prices for these variable and unpredictable quantities.

Installation Sequence

The response of the down the hole hammer indicates whether or not the rock of sufficient quality is penetrated. Once a competent bond zone is established, the casing is withdrawn to the top of the bond zone and the pile is filled internally with grout. Once the grout level has stabilized in the bond zone, the centralized reinforcing steel is placed. A typical cross-section is shown in Figure 3.

DESIGN
Materials

Pin Piles are typically constructed using steel casing with special machined flush jointed threads. The casing meets the physical properties of ASTM A-252 Grade 3, except that the minimum yield strength is typically 80 ksi. This material is most often mill secondary API drill casing. As such, material certifications are not available. The physical properties are confirmed by cut coupons from representative pieces of casing. The core reinforcing steel is Grade 60, 75 or 80 reinforcing bar (ASTM A615, A616 or A617) or Grade 150 prestressing bar (ASTM A722)

The grout typically consists of neat cement and water mixed with a high shear colloidal type mixer. This grout has a fluid consistency, a water/cement ratio of about 0.45, and a typical minimum unconfined compressive strength (from cubes) of 4,000 psi in 28 days.

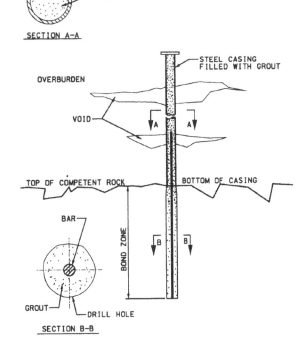

Figure 3: Typical cross-section

Structural Capacity

The structural design of Pin Piles for building foundations is typically not found in local building codes. However, based on our experience in conjunction with some applicable codes, the following equation is used.

$$P_{all} = (0.40 \text{ to } 0.50)F_y A_s + (0.35 \text{ to } 0.45)f'_c A_c$$

where: F_y = Yield Stress of casing or rebar
  A_s = Cross-sectional area of casing or rebar
  f'_c = Unconfined Compressive Strength of the grout at 28 days
  A_c = Cross-sectional area of the grout

The most applicable factors are selected upon the experience of the contractor, the experience of the structural engineer, and the observed performance of the test piles. The equation above is conservative as long as the pile is structurally configured to take advantage of confinement. That is, the grout needs to be confined inside the pipe casing, and in the rock bond zone, the grout receives lateral confinement from the ground. This effect has been confirmed many times through load test performance at ultimate loading, with stresses far in excess of typical values. More research and modeling is needed in this area to further quantify this benefit so that further economy of design is possible.

Geotechnical Capacity

The bond length is determined by experience and by previous load tests in the rock. A typical allowable bond stress in karstic bedrock with unconfined compressive strengths ranging from 15,000 to 20,000 psi is 100 psi. The bond zone capacity is calculated as a typical friction pile. Tip resistance is usually neglected. The following calculation is used.

$$P_{all} = \sigma \pi d L$$

where: σ = Allowable bond stress of competent rock in bond zone
  d = Diameter of bond zone
  L = Length of bond zone

Testing

All projects of any significance justify full scale testing (ASTM D1143) of at least one pile unless there is significant confidence and prior experience with the founding stratum. The purpose of the testing is to verify both the geotechnical capacity of the bond zone and the structural performance of the pile.

CASE HISTORIES
Exton Mall and Garage, Exton, Pennsylvania

Exton is a city located approximately 20 miles west of Philadelphia. The owner of the Exton Square Mall was constructing a second-story addition and a new parking garage. The initial foundation contractor encountered difficulty in drilling through the karstic limestone underlying the site. This contractor's method of installation resulted in several pile failures during the load test program. Nicholson was called in on short notice to take over the construction. The construction for the addition to the mall was completed in two areas: drilling from inside the mall itself, and drilling at close proximity to the perimeter of the building. The work on the inside of the mall took place during hours when the mall was non-operational. This meant construction crews would work night shift, ensuring the stores were cleaned and functional for normal business during the day. Special piping was used to remove the spoils from the drilling up through the roof of the mall, and down into refuse containers on the ground. Drilling conditions inside the mall involved tight access and limited headroom (12 feet). Crews moved merchandise and protected it where necessary.

Bedrock at the site consisted of karstic limestone with voids and clay seams. The top of competent bedrock ranged from 20 feet to 150 feet below the existing ground surface. The maximum design working load was 300 kips in compression. A total of 294 Pin Piles were installed in the interior of the mall and 111 Pin Piles were installed around the perimeter. The average pile lengths were approximately 34 feet, ranging from 20 to 150 feet below the existing slab elevation. A total of 355 piles were installed for the new parking garage, with pile lengths averaging 43 feet, ranging from 25 to 85 feet. The Pin Piles consisted of a 7 inch O.D. by 0.50 inch wall outer steel casing above the competent rock bond zone and two #18 Grade 75 all-thread bars from the bottom of the 10-foot bond zone overlapping 5 feet into the steel casing. The casing had physical properties equal to or exceeding ASTM A252, Grade 3, except that the minimum yield strength was equal to or greater than 80 ksi. Physical properties of the steel casing were determined by coupon tests on random samples. Pile grout consisted of Type I Portland cement grout (w/c = 0.45), with a nominal 28 day strength of 4,000 psi.

The piles outside the existing mall were installed with a large track-mounted Casagrande C12 drill rig, and the piles inside the facility were installed with both Davey Kent DK-50 or Klemm mini drill rigs. All drills utilized rotary eccentric percussive drilling techniques. For the outside piles, the casing was advanced in 20 foot, flush joint threaded sections to the bottom of the bond zone. Piles drilled inside were advanced with 5 foot, flush joint threaded casing due to overhead limitations. Once the bottom of the bond zone was established, the casing in all piles was retracted to the top of bond zone elevation. Consolidation grouting of the bond zone was then performed using a 1.5-inch diameter steel tremie tube.

Three production Pin Piles installed were tested in accordance with the Standard Test Method for Piles Under Static Axial Compressive Load (ASTM D1143-81, Reapproved 1994). In particular, the Quick Load Test Method for Individual Piles (Section 5.6) was used. The maximum test load applied was 600 kips equal to twice the design load.

Beaver Stadium Expansion, Penn State University, State College, Pennsylvania

The Beaver Stadium Expansion for Penn State University in State College, Pennsylvania is currently under construction. When finished, the expansion will add approximately 10,000 new seats, including new luxury boxes. Also, a new scoreboard was constructed. Since the expansion ties in to the existing stadium, the new structures are very sensitive to differential settlements.

The project site is underlain by karstic geology, consisting of very hard dolomitic limestone (unconfined compressive strengths over 20,000 psi). The hard bedrock is pinnacled, with almost vertical bedding planes. This geologic formation is prone to voids, clay seams, varying water levels and greatly varying overburden depths. The top of bedrock at the project site ranged from about 5 feet to over 75 feet below the floor slab elevation.

Two pile designs were used, supporting both new interior and exterior columns. The exterior piles were designed to support loads of 150 tons in compression and 75 tons in tension. The interior piles were designed to carry 50 tons in compression. All of the piles were installed vertically and consisted of three distinct components. The first was the outer casing that provided additional bending capacity and lateral stiffness. The second component was the centralized reinforcement bar that extended to the tip of the pile and overlapped inside the outer casing. The third component was the grout which was tremied into the pile.

Exterior piles consisted of a 7-inch O.D. x 0.5 inch wall thickness threaded flush joint steel drill casing with a minimum yield of 80 ksi. The interior piles consisted of 5-1/2 inch O.D. x 0.415 inch wall casing. The bond zone section was made up of either one #18 Grade 75 rebar (50 ton pile) or two #18 Grade 75 rebar (150 ton pile) in a 5 to 10 foot long tremie grouted bulb with the bar overlapping 5 feet into the casing. The neat cement grout had a minimum 28-day strength of 4,500 psi. The casing was made from mill secondary API drill casing. The piles were installed using rotary eccentric percussive duplex drilling with air as the flushing medium.

Three compression and tension pile load tests were conducted on piles with 150 ton compression capacity and 75 ton tension capacity to verify the adequacy of the pile design and installation methods. To be acceptable, a test pile had to carry a test load equal to two times the design load and less than 0.5 inches deflection. The Beaver Stadium Expansion project involved installing over 600 Pin Piles. The Pin Piles were successfully installed despite drilling in the very difficult karst geology that included pinnacled limestone, clayseams, voids and solution cavities.

ACKNOWLEDGMENTS

The authors would like to thank the numerous owners and design consultants who provided the forward thinking by allowing the use of this technology in their projects. We also wish to thank other employees of Nicholson Construction for their time.

REFERENCES

Kenny, J. R., "Behavior and Strength of Composite High Strength Steel Tubular Columns", Thesis for Master of Science, University of Pittsburgh, 1992

Groneck, P.B., Bruce, D.A., Greenman, J.H., and Bingham, G., "Pin Piles Save Silos", *Civil Engineering Magazine*, September 1993

Munfakh, G.A., Soliman, N.N., "Back on Track at Coney Island", *Civil Engineering Magazine*, December 1987

Morschauser, G.B., Davis, J.E.B., "Replacing an Urban Foundation", *Civil Engineering Magazine*, December 1990

Pearlman, S.L., and Wolosick, J.R., "Pin Piles for Bridge Foundations", Proceedings of the 9th Annual International Bridge Conference, June 1992, pp. 247 - 254, #IBC-92-40.

Pearlman, S.L., Richards, T.D., Wise, J.D., and Vodde, W.F., "Pin Piles for Bridge Foundations: A Five Year Update", Proceedings of the 14th Annual International Bridge Conference, June 1997, pp. 472-480, #IBC-97-53

Pearlman, S.L., "Pin Piles for Structural Underpinning", Proceedings of the 25th Annual Deep Foundations Institute Meeting, October 2000

Tarquinio, F.S. and Pearlman, S.L., "Pin Piles for Building Foundations", Proceedings of the 7th Annual Great Lakes Geotechnical and Geoenvironmental Conference, May 1999

Geotechnical and Environmental Applications of Karst Geology and Hydrology, Beck & Herring (eds)
© 2001 Taylor & Francis, ISBN 90 5809 190 2

The reflection of karst in normative building documents of Russia

V.V.TOLMACHEV & M.V.LEONENKO   State Venture "Antikarst and Shore Protection", 606023 Dzherzhinsk, Russia,
karst@kis.ru

ABSTRACT

Twenty percent of Russia is karst terrain. The karst process has a detrimental affect on many types of economic activity. The nature of karst hazards is comprised of several types of activity that are reflected in normative documents. Normative documents which are now in force regarding karst are being examined at the federal, provincial and departmental levels. The main strengths and weaknesses of these documents have been noted, with the main weakness being poor coordination. Discussions are being held on suggestions to change the number of karst hazard categories. Under conditions of covered karst, it becoming apparent that special normative requirements are needed for the maximum use of geophysical methods to predict spatial and temporal karst hazards.

According to a map of karst hazards in Russia, prepared by the Institute of Geoecology of the Russian Academy of Sciences in 1993 under the direction of V.I. Osipov and V. M. Kutepov, and a schematic map of karst occurrence in the USSR, presented in Building Standards and Rules (SNiP) 2.01.01-82 , about 20 percent of the country is karst terrain which influences economic activity in one way or another. As a rule, karst processes have a detrimental effect on both the economy and the ecology. According to SNiP 22-01-95 "Geophysics of Hazardous Natural Impacts" (Supplement 6), karst is assigned the classifications of "most hazardous," "moderately hazardous", and "hazardous" dependent upon indicators which characterize only karst formations on the soil surface.

These classifications are important only for some, but not all, surface structures. It is unlikely that this assignment of importance is correct. Karst hazards are affected differently by human activities. Table 1 lists different activities with the differing harmful results of each activity.

There is a need to always consider the probable impacts of karst processes when making any predictions of karst hazards. It is appropriate to classify these impacts as shown in Table 2.

Table 1. Types of Karst Hazards

Human Activity	Karst Hazard
Operation on the surface of civil, industrial and transportation structures	Damage to structure as a consequence of bedding deformation
Construction and operation of hydrotechnical structures	Water leaks from reservoirs and channels
Waste disposal operations	Intensive pollution of the geological environment and the activation of karst processes
Construction and operation of underground structures	Sudden breaching of water into an adit when drifting. Deformation of structures due to increasing pressure (as on the timbering in mines)
Mining	Reduction in balance stocks of mining materials. Decrease in stability of underground mines and quarries. Sudden flow of water into adits and quarries.
Water-supply and sewage systems	Decrease in productivity of water-well pumps of underground water due to karst activation and piping in adjacent areas. Damage to water-supply communications due to karst piping, causing activation of the karst-piping process in adjacent areas.
Agricultural production	Decrease in arable land, complications in reclamation works, death/damage to livestock by falling into sinkholes, large degree of pollution in the geological environment from mineral fertilizers.

Table 2. Classifications of social and economic consequences of karst hazards

Economic consequences	Social Consequences – the threat to human life and safety		
	1. In great number	2. In single cases	3. Practically as an exception
A. Extraordinarily detrimental	1A	2A	3A
B. Considerably detrimental	1B	2B	3B
C. Small detrimental	1C	2C	3C

In most cases, the precision of karst risk evaluation depends upon engineering geological investigations which classify Type 3C consequences as having the least probable consequences and reach the maximum probable consequences in Type 1A.

At the present time, there is no specific Russian document, either a SNiP or Code of Practice (Svod Pravil or SP) which encompasses a variety of questions that should be considered regarding economic development in karst terrains, i.e., questions concerning engineering investigations, design, construction and operation of structures, and possibly liquidation of structures (waste-disposals). To our mind, it is unlikely a document such as this is worthwhile. This is due to the fact that in Russia the man-made and natural factors which determine karst formation and development are too varied. A detailed standardization of the activities of various specialists, as applied to technical natural, man-made, engineering/construction and operational conditions is not useful. However, the establishment of definite standards and rules concerning works in karst terrains within the framework of a Federation (province) contained in a "Provincial Normative Document" (Territorialniye Stroitelniye Normi – TSN), or within the framework of a town, a large enterprise, or a combination of both, a "Company Standard" (Standarti Predpriyatiy – STP) would be very worthwhile. Moreover, the specific type of design and operation of some structures, i.e., hydrotechnical or underground, need a special approach in karst terrains; therefore, it would be wise to create special departmental standards for this class of structures.

Presently a number of SNiPs and SPs treat the problems of engineering investigations and design in karst terrains with different amounts of focus. A brief analysis of them is shown in Table 3.

Some SNiPs have only a basic requirement that it is essential to work out individual projects for structures in karst terrains, e.g., SNiP 2.05.02-85 "Motor Roads". An analysis of a number of projects for such structures has shown that the given requirement is not sufficient and that individual projects are most often not being worked out. It is worth noting that there are designs for which it is necessary to take into account the peculiarity of a building and/or operation of a structure in karst conditions. However, existing SNiPs on these designs have no specific requirements whatsoever. Among these SNiPs are 2.05.03-84 "Bridges and Pipes", 2.06.03-85 "Structures of Land-Reclamation Systems", 32-03-96 "Airports", 2.04.084* "Water Supply – Outside Networks and Structures", 2.02.03-85 "Pile Foundations".

As may be seen from Table 3, some SNiPs are unrelated to each other and/or in contradiction to each other. Some do not take recent research into account, nor accepted practices of development in karst terrains in both Russia and other countries. This is primarily true in SNiPs on sectoral design objects.

The main concern of building development in karst terrains is evaluation of the karst hazards. In SNiP 1.02.07-87 "Engineering survey for construction. Basic principles" and in the draft of SP 11-105-97, Part II "Engineering/geological investigations in regions of hazardous geological engineering/geological process", the evaluation of karst hazards is based in the zoning of the territory by the intensity of collapse formations and the average diameters of collapses. In regard to the intensity of collapse formations, a five-magnitude system has been properly seen in relation to the frequency of karst collapses in a year per 1 km^2 (λ). In this system, the most hazardous category ($\lambda > 1,0$) is Category I, and the least hazardous ($\lambda < 0,01$) is Category V. However, there is also a Category VI, which is when karst collapses are impossible, as in karst-free terrains.

This classification of terrains has been taken from "Recommendations on engineering/geological investigations and evaluation of areas for industrial and civil building in karst terrains of the USSR" (PNIIIS, 1967). Before this SNiP 1.02.07-87 was the only official document which regulated the evaluation of karst hazards, although such classifications were offered in earlier (1947-1962) published scientific works by L.A. Makeev, G.A. Maximovich and I.A. Savarensky. A rather detailed analysis of these works is given in the work of V.V. Tolmachev, et al., 1986.

To a large degree, both "Recommendations on engineering/geological investigations and evaluation of areas for industrial and civil building in karst terrains of the USSR" (PNIIIS, 1967) and "Recommendations on design of structures in karst terrains of the USSR" (PNIIIS and others, 1967) have helped ease problems in building in karst terrains and have great import for the economics of karst terrains. However, it is felt that in the draft of SP 11-105-97, Part II, there should be a more logical numbering system of categories, with Category 0 - most safe, Category 1 – least hazardous, Category 2 – more hazardous, etc. Professor G.A. Maximovich, one of the founders of karstology suggested such a system in 1961. Such a system will not limit the categories to only six and will make the accepted quantitative indicators of karst and correspond more logically to the categories. For example, such an approach to numbering is inherent in the building standards for karst terrains in Germany in which Category 0 territory is absolutely safe and 7 is the most hazardous territory in which building in not recommended (Buechner, 1991). The exact same system of numbering is used for describing the majority of hazardous natural processes, e.g., seismic processes. It is felt that the only warranted objection to this system of numbering will be the probable misunderstanding between specialists in investigation and design at the first use of it, since it has not been their habit since 1967.

Table 3. The problems of building development of karst terrains reflected in normative documents

SNiP, SP	Main Karst Concepts	Notes and Commentaries
SNiP 2.01.07-85: "Loads and Impacts", 1987	Impacts determined by bed deformations accompanied by bed subsidence in karst terrains refer to peculiar impacts	In karst terrains, bed deformations are connected not only with subsidences but also with collapses. There is a lack of coordination of terminology accepted by SNiP 2.02.01-83* and SP 11-105-97.
SNiP 2.01.15-90: "Engineering protection of buildings and structures against hazardous geological processes. Main design concepts", 1991	Antikarst measures – a list of them and their requirements	There is a need to point out that the character and amount of antikarst protection is determined in many cases by the number of engineering investigations
SNiP 11-02-96: "Engineering survey construction. Basic principles", 1997	The report on engineering investigations must have a prediction of karst processes and the risks from them	It would be well to point out that the prediction of hazard from natural processes should be of a spatial nature as well as temporal with regard to scheduling and deadlines for utilization of a structure
11-105-97, Part I: "Engineering geological site investigations for construction", 1997	If necessary research organizations should be drafted in making the technical task and program on engineering/geological investigations under complex natural conditions	The demand are actual for karst terrains
SP 11-105-97, Part II: "Engineering/geological investigations in regions of hazardous geological processes", 2000 (the draft).	Methods of investigating in karst terrains for all stages of design are completely presented The classification of karst formations on the soil surface (collapses, local subsidences, common subsidence) is given The categories of stability	The interplay between investigations and the peculiarities of designed and operational structures is shown to an insufficient extent. It would be better to use the term "karst hazard" instead of the term "stability" and to change the numbers of categories.
SNiP 2.02.01-83* "Foundations of buildings and structures", 1995	The requirements for the initial data for designing the foundations of buildings and structures in karst terrains are given and include the use of probable characteristics of karst deformations	The methods of determining the rated span of a karst collapse under a foundation have been stated according to these requirements
SNiP 2.05.06-85: "Arterial pipelines", 1988	It is well to provide measures to strengthen pipelines with a availability of collapse close to the route of the pipeline as collapse can influence the safe operation of the pipeline	There is danger from all types of surface karst formations, especially those where large subsidences are predicted, to pipelines. We must deal with the probability of collapse, not with the availability
SNiP 2.01.28-85 "Waste disposal", 1985	It is not permissible to locate waste disposal facilities in zones of "active" karst	The term "active" is indefinite

185

The probability of a more differentiated evaluation of karst hazard according to the intensity of collapse formation is also determined by a wider application of various methods of soil section investigation. As a rule, karst processes are discrete processes, i.e., cavities, rarefied zones in covered soils, sinkholes and local subsidences. That is why geophysical methods can be the most effective methods of investigation, e.g., the detection of anomalies and observing their development over time, and this must be reflected in the appropriate regulations. It is felt that it is advisable to show the following points in such regulations:

- The variety of geological/geophysical situations requires an individual approach for the choice of effective geophysical methods in concrete conditions. In the majority of cases, the development of karst-piping processes is characterized by a complex geological/geophysical environment; however, the majority of geophysical methods are effective for comparatively simple geological/geophysical environments.
- It would be well to formulate more concrete requirements for the results of geophysical researches necessary for the evaluation of the degree of karst hazard.
- Methods of geophysics do not disrupt the geological environment so, consequently, do not provoke the activation of karst-piping processes during investigations, thus allowing the site to be monitored over time, and this is extremely important to the economic objectives of karst monitoring.
- The list of approved methods of geophysical investigations in karst terrains should contain definite conditions for their application, under which they will usually prove themselves precise in the prediction of the degree of karst hazard.
- Normative documents should contain regulations regarding the futility of using unscientific methods, e.g., extrasensory, in the prediction of the degree of karst hazard and antikarst protection.

In Russia and the Ukraine, unscientific methods such as extrasensory perception were used frequently in the late 1990s and most often involved private "one-day" firms or unique individuals. An analysis of the efficiency of such investigations in the Nizhny Novgorod region has proven them totally unsound. In some countries there is a prohibition on the official use of psychics in the practice of engineering/geological investigations. An analogous requirement is contained in TSN 22-308-98 NN of the Nizhny Novgorod region (TSN, 1998). Unfortunately, there have been occasions when psychics were used for the operational prediction of karst hazards, even in terrains where there are ecological dangers.

In spite of the weaknesses in the general normative documents of Russia, it should be noted that the Russian baseline for the development of karst terrains was more completely developed than in some other countries. Nevertheless, rather serious mistakes are being made in investigations, design and operations which are causing much economic damage and accidents. Analysis of the causes of damage to structures in the Nizhny Novgorod region shows that the main cause of such damage is not karst "chaos" but is the human factor, i.e., insufficiently qualified investigators and designers, mistakes in understanding and simplification of karst processes, preponderance of departmental or industrial self-interest over national economic interests, insufficient judicial responsibility for accidents, and gaps between regulations in general Russian documents and concrete situations in object design. The State Venture for Antikarst and Shore Protection considers making territorial and territorial/departmental documents regarding karst will significantly reduce the impact of these factors. The Venture has presented the following documents:

- "Temporal Instructions on Control Over the Condition of Railway Beds in Karst Sites of the Gorky Railway", 1995. This document has been confirmed by the director of the Gorky Railway.
- "Basic Track Maintenance Instructions for Hazardous Terrains", 1997. This document has been confirmed by the Ministry of Communication Ways of Russia.
- "Engineering Investigations, Design, Building and the Operation of Structures in Karst Terrains in the Nizhny Novgorod Region" 1999 (TSN 22-308-98 NN), a provincial normative document regarding construction. This document was presented to the Order of Architecture and the Town Building Committee of the Nizhny Novgorod Region Administration and has been confirmed by the governor. After two to three years of approval of the TSN, it is suggested that a second issue of the document be prepared to supplement it with proposals of design and investigating organizations of Russian, and the experience of foreign countries, primarily Germany, the United States of America and the Republic of South Africa, in the field of engineering karstology.
- "Recommendations on Organizing and Performing Karst Monitoring" is now being presented to the Order of the Ministry of Natural Resources of Russia.

ACKNOWLEDGMENT: The authors wish to express thanks to Gayle Herring for her editorial skills.

REFERENCES

Tolmachev, V.V., Troitsky, G.M., Khomenko, V.P., 1986, "Engineering/Building Development of Karst Terrains", Moscow, Stroyisdat, 177 p, (Russian).

Maximovich, G.A., 1961, "Solidity of Karst Sinkholes and the Stability of Karst Terrains", Geology and Surveying, No. 7, pp. 17-32, (Russian).

Buechner, K.H. 1991, "Die Gefaerdung von Bauwerken durch Erdfaelle im Vorland des Westharzes", Geologisches Yahrbuch. Reihe C., Heft 59, Hannover, 40 p, (German).

TSN 22-308-98 NN, 1999, "Engineering Investigations, Design, Building and the Operation of Structures in Karst Terrains in the Nizhny Novgorod Region", 71 p, (Russian).

Geotechnical and Environmental Applications of Karst Geology and Hydrology, Beck & Herring (eds)
© *2001 Taylor & Francis, ISBN 90 5809 190 2*

Application of a standard method of sinkhole detection in the Tampa, Florida, area

E.D.ZISMAN   BTL Engineering Services, Inc., Tampa, FL 33684, USA, ezisman@ozline.net

ABSTRACT

This manuscript discusses the application of the Standard Method of Sinkhole Detection. The method is based on evaluating eleven sinkhole characteristics that may be found during a subsurface investigation. Application of the method is shown with data obtained in the Tampa, Florida area during the past two years. The standard method addresses investigative characteristics such as the density of soil material at different depths, drilling properties, stratigraphic conditions and ground water levels. Once the total numbers of sinkhole characteristics are identified the Sinkhole Score for the site is determined. The Sinkhole Score is the number of sinkhole characteristics found compared to the number of sinkhole characteristics possible. The manuscript also discusses the importance of considering other factors that can cause sinkhole-like distress, discusses methods of detection and minimum investigative requirements.

INTRODUCTION

The sinkhole detection method has been presented (see Zisman, 2000) as a method for detection of sinkholes based on various site conditions found in a sinkhole investigation. The purpose of this manuscript is to discuss the application of the method and to provide insight into the complexity of sinkhole formation and the variety of sinkhole-like conditions that should be considered in sinkhole evaluations.

The occurrence of sinkholes is a very complex process. It is subject to the almost random nature in which sinkholes occur and the complex factors that control their occurrence. Further compounding the problem is methods used to evaluate sinkholes are limited in the extent of the subsurface that can be "viewed" and accurately analyzed. In essence, our knowledge of the subsurface is limited to the narrow column of soil and rock that is sampled by the boring and supplemented by the broad-based shallow data obtained by geophysical methods such as GPR. These limitations, in most instances, restrict our ability to accurately predict and analyze sinkhole occurrence.

Because of the complexity in the formation and prediction of sinkholes, and in many instances the great length of time involved in their formation, investigators have developed different sets of criteria for determining the presence of a sinkhole. This manuscript presents the results of site investigations in which the Standard Method for Sinkhole Detection was used during a two-year period in the Tampa, Florida area. Increased use of the Standard Method of Sinkhole Detection in sinkhole investigations would result in more consistent classification of sinkhole prone areas and would provide for uniformity and fairness in the evaluation of sinkhole insurance claims (Florida is one of the few States that has mandatory sinkhole insurance coverage).

The author has reviewed reports from some investigators who have refused to acknowledge sinkhole presence if evidence of raveling is not found during the investigation. In contrast, other investigators have declared sinkhole presence based on very few confirming sinkhole features. To standardize the wide diversity in the identification of a sinkhole, this manuscript provides a review of the Standard Method of Sinkhole Detection that can be used in sinkhole investigations to provide uniformity in sinkhole evaluation. The reader is referred to Zisman, 2000, which provides a more detailed discussion of the standard method and provides additional considerations in its application.

SINKHOLE DETECTION

Complex Subsurface Conditions

Because of the complex and essentially random nature of the factors controlling sinkhole development, sinkhole detection is difficult and in some instances, a sinkhole can go undetected although structural damage is present. Complicating the problem of detection is that only a small area of the ground surface may be affected by the sinkhole making sinkhole detection more difficult. And, for those investigators dependent on finding the small raveling zone in their investigation, sinkhole detection becomes exceedingly difficult. (Raveling is the downward erosion of soil material carried by water flowing through a pervious zone or erosion pipe that extends to a cavity in the limestone (Frank & Beck, 1991)). Randazzo, 1997 has recognized the problem stating that borings will likely miss the small conduits associated with solution sinks.

The small raveling zone may propagate upward to the ground surface while effecting only a relatively small zone of soil material. The disturbance of only a small zone of surface soil is typical of most raveling and cover subsidence sinkholes. Moreover, it is possible that the sinkhole will not be aligned over the limestone cavity; the raveling zone may, in fact, "travel" in its movement upward considerable distances laterally as a result of varying conditions in the composition of overlying impervious or pervious materials.

Overview of the Various Methods

Because only the SPT method provides a soil sample, in the author's opinion, other methods are at a disadvantage to SPT methods. GPR methods provide a broad view of the subsurface but are hampered because of interference with clay zones and limitation in the depth of penetration of the GPR signal. In many projects, it is advantageous to supplement SPT borings with GPR surveys and the use of CPT methods to further refine subsurface interpretations. CPT methods can provide quick, continuous measurements of soil properties and can furnish relatively rapid and accurate water table readings that are not generally possible with SPT borings.

Resistivity methods can be useful in subsurface investigations particularly where clay material is found close to the ground surface. However, for sites that do not contain clay, GPR methods are generally more cost effective. Seismic methods are not well suited for sinkhole investigations because of the difficulty in discerning the small-scale features that are significant in a sinkhole investigation.

Probability of Finding a Raveling Zone

During a sinkhole investigation where there is no surficial expression of a sinkhole, the probability is very low that a boring will actually encounter the raveling zone in the overburden or the area directly above a sinkhole. It has been estimated that on a one-acre site, to establish a 90 % confidence level of finding a cavity 23 meters (75 feet) in diameter, 10 borings would be required. And, 1,000 borings would be required to find a cavity 2.3 meters (7.5 feet) in diameter (Benson & La Fountain 1984). Although the probability of finding evidence of a sinkhole is significantly increased where an investigation is directed to finding sinkhole caused building distress, it is still difficult to find direct evidence of the sinkhole mechanisms.

Minimum Investigation Requirements

In view of the low probability in finding direct sinkhole features and given the complexity in the formation of sinkholes, a discussion of the minimum requirements for a sinkhole investigation will help in decreasing the odds of finding sinkhole features and provide recommended minimum standards among investigators. Following is a list of the minimum investigative requirements based on experience with other sinkhole investigations.

1. A minimum of three borings should be drilled as close to the structure as possible and on different sides of the structure with one boring near the area of distress.
2. All borings should be drilled at least five feet into well-indurated rock (N>50).
3. A minimum of three hand augers should be drilled on each side of the structure to sample for clay and organic material.
4. The footing should be excavated in at least two locations around the structure to check dimensions, quality of concrete, exposure of reinforcement etc.
5. All cracks should be plotted on a sketch of each wall of the house in an effort to determine patterns and trends in the cracking. In addition, location of cracking should be shown on the boring location plan.
6. Photographs should be taken that clearly show damage to the structure.
7. The history and condition of surrounding structures should be noted.
8. Reference should be made to old topography and air photos
9. If possible, a GPR survey should be performed before borings are drilled.

Non-Sinkhole Causes of Structural Distress

A sinkhole investigation is not complete without considering other potential causes of building distress other than those resulting from sinkhole activity.

Four potential causes of building distress are (1) organic material, (2) shrink/swell clay, (3) poor construction and (4) unusual conditions. In evaluating these potential causes, it should be considered that more than one may be present at the site. Care should be taken to fully measure and evaluate the potential impact of the cause. For example, if organic material is present, is it sufficiently close to the structure and in sufficient quantity to impact the structure? Similarly, if clay is present is it expansive and sufficiently close to impact the structure?

STANDARD METHOD OF SINKHOLE DETECTION

Eleven Conditions

The following is a general review of Standard Method of Sinkhole Detection. The method consists of evaluating the characteristics generally associated with sinkhole activity as tabulated in Table 1. This table consists of eleven sinkhole characteristics that are based on consistent sinkhole features found in hundreds of site investigations completed at many sites in the west central area of Florida. Note that with the exception of possibly Condition 10, none of the other characteristics are by themselves conclusive evidence of a sinkhole.

Table 1. Eleven Characteristics Generally Associated with Sinkhole Activity

No.	Condition	Cause	Site Condition	
1	$N_{av.} < 7$, in depth interval from 0 to 4.6 meters (15 feet) below ground surface in granular material.	The presence of very loose to loose material within about 4.6 meters (15 feet) of the ground surface indicates raveling has possibly effected the surface soils within a zone of significant influence of the building foundations.		Yes / No
2	$N_{av.} < 15$ in depth interval from 4.6 to 12.2 meters (15 to 40 feet) in granular material.	The presence of loose material within a depth range of about 4.6 to 12.2 meters (15 to 40 feet) is an indicator of possible raveling in a zone that can, in time, effect the integrity of soil supporting the building.		Yes / No
3	Loss of drill fluid.	Losses of drilling fluid are significant in that they often indicate that loose and permeable zones are present. Sandy soils can readily move through these areas creating weak or raveling zones.		Yes / No
4	Variation in depth to water table in short distance	Variation in depth to water table, in the borings, may indicate the presence of soil raveling or steep groundwater gradients indicative of sinkhole development.		Yes / No
5	Caving of drill hole or a drop of drilling tools.	Caving of the drill hole as the boring is being advanced or weight of rod (WOR) conditions are indicators of voids or very soft subsurface conditions that may result from karstic erosion.		Yes / No
6	Substantial decrease in SPT values with increasing depth.	An abrupt decrease in SPT values with increasing depth can be indicative of loose raveling zones.		Yes / No
7	Loose or soft material overlies highly fractured rock	An indication that piping of soil material into voids in the rock may have occurred causing a loosening of the surrounding soil.		Yes / No
8	Absence of clay confining layer over limestone bedrock	Presence of clay confining layer prevents piping of soil material into voids in limestone rock.		Yes / No
9	Highly variable limestone surface over short distance	An indication of highly solutioned limestone rock that likely contains voids.		Yes / No
10	Settlement (near or adjoining structure).	Depressions of the surrounding ground surface particularly any that form as the boring is being advanced or evidence of recent settlement of ground around structure could indicate an active sinkhole. Presence of this condition alone may be sufficient to conclude sinkhole damage is present.		Yes / No
11	Soil strata not in normal sequence or uniform thickness	Absence of continuity in the stratification between borings could indicate karst features. Presence of muck at depth may be due to in-filling of an ancient sinkhole or cavity. Presence of disturbed strata above the rock surface. Evidence of sagging of the overlying strata.		Yes / No
	Total number of positive indicators (Yes)			

Sinkhole Score

The ratio of the number of sinkhole conditions found in the investigation to the total number of sinkhole conditions possible is calculated to determine the Sinkhole Score as follows:

Sinkhole Score =   No. of Sinkhole Conditions Found in Investigation x 100
                    Total No. of Sinkhole Conditions Possible (*)

(*) Total No. of sinkhole conditions possible may vary depending on the scope of the investigation.

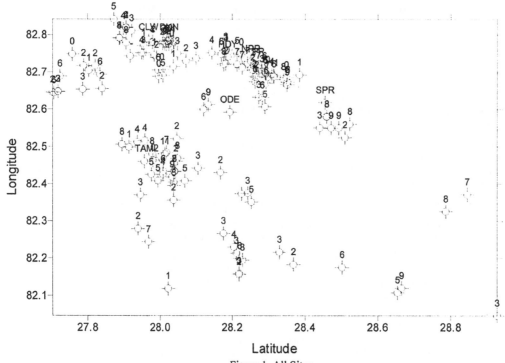

Figure 1. All Sites

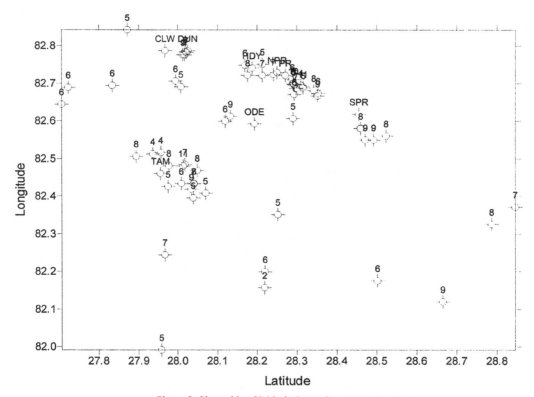

Figure 2. Sites with a Sinkhole Score Greater or Equal to 4

A sinkhole score of four is a critical value in the predication of sinkhole presence. This score represents a gray area in which one can conclude either for or against the presence of a sinkhole. In many instances, we have concluded that a score of 4 indicates that a sinkhole is present. This conclusion requires collaborative data from other site features. For example in Table 1, Item 10 (settlement) or Item 4 (variation in depth of water table) or Item 5 (caving of the drill hole or rod drop); these conditions would sway the results in favor of a sinkhole.

Relative Weight
In Table 2, the relative weight of Table 1 conditions is shown. A value from one to twelve has been assigned to the Table 1 conditions with twelve being the most important and one the least. The table provides a general indication of the relative importance of the various factors. The factors can be assigned to the Table 1 conditions to develop a Weighted Sinkhole Score.

Table 2. Relative Weight of Table 1 Conditions

No.	Condition	Weight
1	$N_{av.}$ < 7, in depth interval from 0 to 4.6 meters (15 feet) below ground surface in granular material.	5
2	$N_{av.}$ < 15 in depth interval from 4.6 to 12.2 meters (15 to 40 feet) in granular material.	3
3	Loss of drill fluid.	7
4	Variation in depth to water table in short distance	11
5	Caving of drill hole or a drop of drilling tools.	7
6	Substantial decrease in SPT values with increasing depth.	10
7	Loose or soft material overlies highly fractured rock	11
8	Absence of clay confining layer over limestone bedrock	12
9	Highly variable limestone surface over short distance	11
10	Settlement (near or adjoining structure).	12
11	Soil strata not in normal sequence or uniform thickness	11
	Total	100

APPLICATION OF SINKHOLE DETECTION METHOD
Comparison of Data
The plots shown in Figures 1 and 2 provide the results of investigations completed in the Tampa, Florida area in which the Standard Method of Sinkhole Detection was used. Figure 1 shows the location of all investigations with the Sinkhole Score indicated above the site. In Figure 2, the data has been restricted to show only those sites that have a Sinkhole Score equal or greater than 4. It is apparent from the figures that the application of the method tends to reduce the scatter in the location of sinkhole sites while accentuating the areas of known sinkhole activity.

Other investigators are encouraged to adapt this method and contribute their data to the database. The inclusion of large amounts of data into Figures 1 and 2 can result in development of an additional condition in the Weighted Sinkhole Score. That is the development of a Regional Susceptibility Factor. The Weighted Sinkhole Score can be applied to marginal sites having a Sinkhole Score equal to 4 where additional data may be required.

CONCLUSIONS
A method has been presented for establishing sinkhole potential based on determining eleven diagnostic sinkhole indicators during a site investigation using normal SPT or CPT methods. Using these methods, sites can be classified by the Standard Method of Sinkhole Detection, which permits standardizing a site's sinkhole potential based on a site's sinkhole score.

Presently, investigators use their own methods of analyzing sinkhole data with no uniformity in classifying the sinkhole potential of a site. This manuscript has provided a method of standardizing sinkhole investigations using SPT and to some extent CPT data to establish the likelihood of a sinkhole occurring. The use of geophysical data such as GPR surveys can serve to increase the accuracy and broaden the coverage of an investigation.

The methods discussed herein have provided techniques whereby sinkhole determination can be related to common SPT methods of site investigation. The method can provide for uniformity in the identification of sinkholes and can lead to standardizing the techniques used to establish the presence of a sinkhole.

ACKNOWLEDGEMENTS

The author acknowledges the help of Messrs. Larry Brown and Larry Gordon of BTL Engineering Services, Inc. for their support, review and help in the preparation of this manuscript.

REFERENCES

Zisman, E.D., 2000, A Standard Method of Sinkhole Detection in the Tampa, Florida Area, Association of Engineering Geologists, In press.

Benson, R.C. & Fountain, L. K., 1984, Evaluation of subsidence or collapse potential due to subsurface cavities, Proceedings of the First Multidisciplinary Conference on Sinkholes/Orlando/Florida, p. 201-215.

Frank, E.F. & Beck, B.F., 1991, An Analysis of the Cause of Subsidence Damage in the Dunedin, Florida Area 1990/1991, Florida Sinkhole Research Institute, University of Central Florida Orlando, Florida

Randazzo, A.F. & Jones, D.J.; Editors, The Geology of Florida; University Press of Florida, 1997, ISBN 0-8130-1496-4.

V Groundwater Flow and Contaminant Movement in Karst

Geotechnical and Environmental Applications of Karst Geology and Hydrology, Beck & Herring (eds)
© *2001 Taylor & Francis, ISBN 90 5809 190 2*

Development of a coupled surface-groundwater-pipe network model for the sustainable management of karstic groundwater

RUSSELL ADAMS & GEOFF PARKIN   Water Resources Systems Research Laboratory, Dept. of Civil Engineering, University of Newcastle, Newcastle Upon Tyne, NEI 7RU, UK, r.adams@ncl.ac.uk

ABSTRACT

This paper considers the hydrogeological simulation of groundwater movement in karstic regions using a hydrological modelling system (SHETRAN), which has been adapted for modelling flow in karstic aquifers. The new model has been developed and used within the STALAGMITE (Sustainable Management of Groundwater in Karstic Environments) project, funded by the European Commission. The project has wide-ranging objectives covering: groundwater pollution and land-use management in karstic areas; assessing the efficiency of management policies for the sustainable use of karstic groundwater in Eastern Europe; and the development of a Decision Support Tool (DST) to implement these policies.

The SHETRAN model is physically-based in so far as most of the parameters have some physical meaning. The modifications made to SHETRAN to simulate karstic aquifers are: (1) the coupling of a pipe network model to a variably-saturated 3-d groundwater component (the VSS-NET component), to simulate flow under pressure in saturated conduits; (2) the coupling of surface water features (e.g. sinking streams or "ponors", and spring discharges) to the conduit system; (3) the addition of a preferential "bypass" flow mechanism to represent vertical infiltration through a high conductivity epikarst zone. Lastly, a forward particle tracking routine has been developed to trace the path of hypothetical particles with matrix and pipe flow to springs or other discharge points. This component allows the definition of groundwater protection zones around a source for areas of the catchment (watershed) which are vulnerable to pollution from non-point sources (agriculture and forestry).

INTRODUCTION

The STALAGMITE (Sustainable Management of Groundwater in Karstic Environments) project has several objectives. The most important of these are concerned with the development of decision support tools and procedures for the holistic management of karstic groundwater in a sustainable manner. The project focusses on the eastern European countries of Bulgaria, Slovenia and Slovakia where karst waters account for a significant proportion of the water supply. At the centre of the project is a predictive groundwater modelling capability based around the development of an existing model (SHETRAN) to simulate groundwater flow in karstic aquifers. The model will be used to predict the effects of different catchment (watershed) management strategies on water supplies, and also the effects of climate change on water resources by linking the model with Europe-wide climate data sets.

THE SHETRAN MODELLING SYSTEM

SHETRAN is a physically-based distributed hydrological modelling system. Currently it is capable of simulating water flow, contaminant transport and sediment transport (Ewen *et al.* 2000). A modular system of interconnected components representing the hydrological cycle has been adopted for the structure of SHETRAN. This structure allows existing components to be upgraded or new components to be added to the model without serious alterations to the basic structure of the system. Each component is based around the finite-difference equations for flow and transport which are solved by robust and accurate numerical methods. A Variably-Saturated Subsurface (VSS) component has been recently added to represent three-dimensional variably-saturated flow in heterogeneous porous media (Parkin, 1996). VSS allows the simulation of multiple confined and unconfined aquifers, perched aquifers and stream-aquifer interactions. Figure 1 shows a schematic diagram of the flow processes represented in the SHETRAN model.

To allow the simulation of groundwater flow in abandoned deep mine systems, the VSS-NET component has been developed (Adams & Younger, 1997, 2000). VSS-NET has been used to model mine water rebound - a term used to describe the flooding up of abandoned mines to the ground surface. The rebound process usually gives rise to long-term discharges of highly polluted water. VSS-NET comprises a turbulent flow component (NET) which is based on a water-supply distribution network model, coupled to the existing laminar-flow (VSS) component, to enable flows in the shafts and access tunnels (roadways) found in abandoned mines to be simulated. These shafts and roadways are of a similar size to karst shafts and conduits and were usually engineered for permanence.

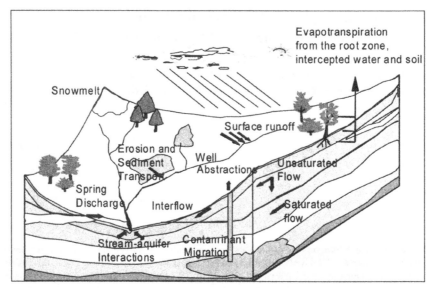

Figure 1 Schematic Diagram of Flow Processes Represented in the VSS Component

A KARST HYDROGEOLOGICAL MODEL FOR SHETRAN

The combined SHETRAN and VSS-NET models from the basis for an extended model for flow in karstic aquifers. A similar modelling approach has previously been used, by adding conduits to an existing porous media groundwater flow model (MODFLOW) to simulate the formation of a cave system through calcite dissolution (Clemens *et.al.* 1996). As far as possible the new model is physically-based insofar as the model parameters have some physical meaning. Where direct parameter measurements are not available, various sources of data to parameterise the model (for example spring hydrograph analysis) are being explored (e.g.. Felton, 1994). The following additional flow processes to those represented in the basic SHETRAN model have been identified as important in karst aquifers and are shown schematically in Figure 2.

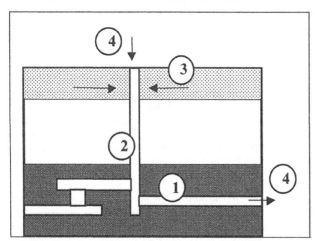

Figure 2 Karstic Flow Processes Modelled in SHETRAN

1) *Flow in a sub-horizontal cave network in the saturated zone*

The existing VSS-NET component can directly simulate flow through a network of interconnected conduits in the saturated zone. The exchange flows between the conduits and the matrix cells are calculated using Darcy's law, therefore a coefficient of proportionality is required for each pipe. The flows in the conduits are calculated using the Darcy-Weisbach or Hazen-Williams formulae for turbulent flow. The model requires values of the conduit diameter (D) and effective roughness (k) to define the headloss - flow relationship in addition to the length of each pipe. Gale (1984) tabulated some values for the ratio k/D for karstic conduits and Jeannin & Maréchal (1995) used a value of 0.25 for a cave system in Switzerland when modelling the system using a discrete conduit network model. A boundary condition with a fixed head (e.g., a spring discharge where the water surface elevation is measured) is required for each network of conduits. Each cave section can be represented by a straight length of pipe (a conduit), and a network consisting of looped and/or branched caves can be simulated by the model. Cave systems are rarely mapped with the same resolution or certainty as mine tunnels (White, 1999), so this component of the model is suitable for systems where the inflows and outflows to the cave network are well-known, for example through tracer tests.

2) Shafts

These represent sub-vertical features connecting the cave networks in the saturated and unsaturated zones and sometimes extending to the ground surface. In the new model shafts may receive rainfall or flow from sinking streams in addition to inflow from the conduit network and connected aquifer. The water level in shafts can also define fixed head boundaries in the saturated cave network. The new model simulates only major, clearly-mapped shafts which connect the ground surface to the cave network.

3) Epikarst bypass flow generation

The epikarst region near the ground surface absorbs and transmit infiltration rapidly into the subsurface region. Infiltrating water flows laterally towards vertical shafts which form conduits to the saturated zone (Jeannin & Grasso, 1995). Kiraly *et.al.* 1995 assessed the role of epikarst in a three-dimensional finite-element model and concluded that usually the representation of an epikarst layer was necessary to reproduce the dynamics of a karst aquifer. Therefore, a semi-empirical model for SHETRAN has been developed which allows the instantaneous transfer of water between an upper epikarst layer and the lower aquifer layer. The transfer flux is calculated as a function of the depth of water stored in the epikarst layer above a critical threshold. The model therefore requires two parameters: the threshold water depth and the constant of proportionality.

4) Interactions with Surface flows.

The new model allows interactions between the VSS-NET and OC (Overland/Channel flow) components of SHETRAN in both directions. Stream inflows into either shafts or conduits can be simulated to represent the sinking streams (ponors) found in karst regions. Spring discharges from conduits into streams (which are modelled by the OC component) can also be simulated.

IMPLEMENTATION OF A FORWARD PARTICLE TRACKING MODEL IN SHETRAN

Defining the area of ground over which the groundwater catchment is vulnerable to surface pollution is a major objective of the STALAGMITE project. In karst regions, actions at the ground surface can rapidly create impacts at large distances away. COST ACTION 65 (1995) produced recommendations for groundwater protection in 16 European countries with karst aquifers. In non-karstic aquifers particle tracking is often used to define protection zones around groundwater abstraction boreholes.

To identify vulnerable areas of karst aquifers and the catchment areas supplying sources, a forward particle tracking model has been developed for SHETRAN. This method allows the particles to be released at the start of the simulation from a number of locations in the catchment (usually an aquifer grid cell) which the user can specify. Particles can be tracked in both the VSS component (in the saturated and unsaturated zones, through both the rock matrix and epikarst) and the NET component (through conduits) to an outlet source, usually a spring or borehole, and the travel time to the source can be calculated. The user can then draw isochrones of travel time around the source or pathlines showing the origin of each particle (thus delineating the catchment area of the source).

MODEL VERIFICATION AND APPLICATION

The new model described above has recently been applied to steady-state simulations on a simple hillslope catchment drained by a single conduit spring connected to a single channel representing a small stream. These simulations have verified the components of the flow model described above. Figure 3 shows a plan view of the catchment with a series of particle pathlines depicted by the solid lines. The catchment comprises 64 100m square elements, and a single particle has been released from the centre of each element at the start of the simulation. The spring is located near to the centre on the southern boundary of the catchment. It is connected to three 100m long pipes orientated in the north-to-south direction, which drain the catchment. The upstream (northern) boundary condition is a fixed-head. The pathlines show the movement of particles towards the pipes and then directly to the spring.

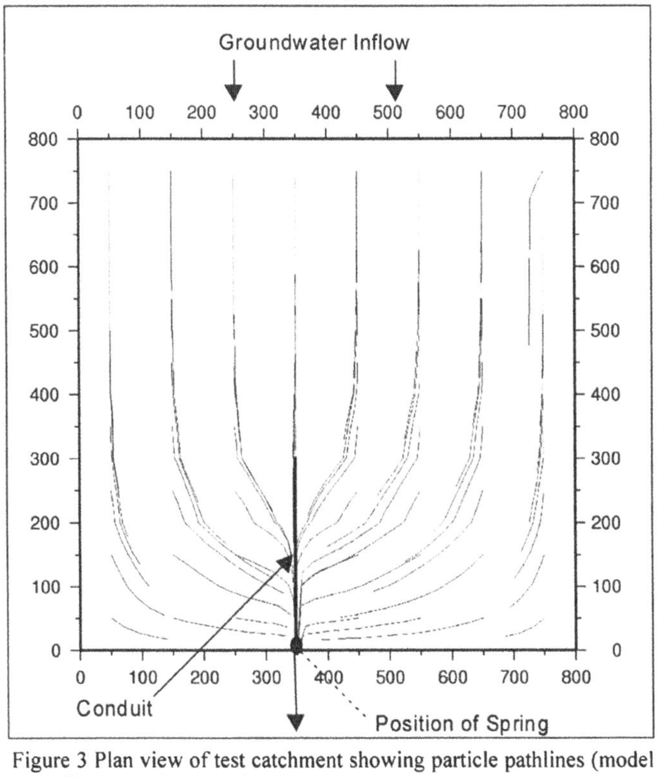

Figure 4 shows the discharge from the conduit spring for a time period of three months. Steady-state conditions have been achieved after around one month (744 hours) when the discharge falls to approximately 5 l/s.

Following initial testing, the SHETRAN karst hydrogeological model will be applied to karst aquifers in Slovenia, Slovakia and Bulgaria. The model will then be transferred to the relevant institutions in the project countries for future modelling studies. Simulations of the Nanos karst massif in Slovenia are currently underway. This massif comprises two sinking streams (Belscica and Lokva ponors), with a mean discharge of around 0.1-0.2 m^3/s which drain an area of impermeable flysch (clay), connected by a karst conduit over 12 km long to a group of major springs (Vipava), from which the total daily mean flow is around 6 m^3/s (KRI 1997). Tracer tests have proven the connection between the ponors and Vipava springs and the total recharge area of the springs has been estimated at 149 km^2. The study area is shown in Figure 5, the area of flysch being located in the south-east corner. Simulations will compare several versions of the conceptual model with and without the pipe network included, and calibration will take

Figure 3 Plan view of test catchment showing particle pathlines (model co-ordinates are in metres)

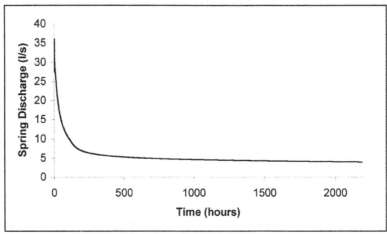

Figure 4 Time-Series Plot of Discharge from Conduit Spring

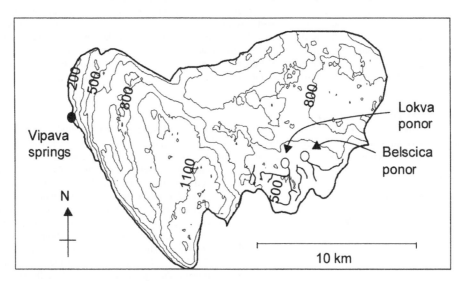

Figure 5. Map of Nanos karst massive (Slovenia), showing elevation, locations of ponors and
Vipava Springs. (Elevations are in metres above sea level).

place using the observed spring discharge. The model showing the best fit will then be used to compare different scenarios of land use change in the Nanos massif including deforestation or reforestation.

CONCLUSIONS

The physically-based distributed modelling system SHETRAN has been enhanced by the addition of a new component capable of simulating the major flow processes in karst aquifers, allowing both surface and subsurface processes to be modelled. A forward particle tracking routine has been added to enable the delineation of catchment areas around vulnerable karst water supply sources and to calculate the travel times of contaminants to the sources. The model will be used to assess the efficiency of land management policies for the sustainable use of karstic groundwater in Eastern Europe and also the possible impact of climate change on water quantity.

ACKNOWLEDGEMENTS

The SHETRAN model development and testing described in this paper was carried out under EC Project IC15-CT98-0113.

REFERENCES

Adams, R. and P.L. Younger, 1997, Simulation of Groundwater Rebound in Abandoned Mines Using a Physically Based Modelling Approach. In Minewater and the Environment: Proceedings of the 6th International Minewater Association Congress, Bled, Slovenia 8-12 September 1997. Vol. 2, 353-362.

Adams, R. and P.L. Younger, 2000, A Strategy For Modelling Ground Water Rebound In Abandoned Deep Mine Systems. Paper accepted by Ground Water.

Clemens, T., Huckinghaus, D., Sauter, M., Liedl, R. and Teutsch, G., 1996, A combined continuum and discrete network reactive transport model for the simulation of karst development. Calibration and Reliability in Groundwater Modelling (Proc. ModelCARE 96 Conf., Golden, Colorado, Sept. 1996). IAHS Publ. No. 237.

COST ACTION 65 1995, Hydrogeological aspects of groundwater protection in karst areas: guidelines. Bulletin d'Hydrogéologie, No. 14, Centre d'Hydrogéologie, Université de Neuchâtel, Switz. 35-48

Ewen, J., Parkin, G. and O'Connell, P.E. 2000, SHETRAN: a coupled surface/subsurface modelling system for 3D water flow and sediment and solute transport in river basins. ASCE J. Hydrologic Eng., 5(3) 250-258

Felton, G.K. 1994, Hydrologic response of a karst watershed. Trans. of the ASAE, 37, 143-150.

Gale, S.J. 1984, The Hydraulics of Conduit Flow in Carbonate Aquifers. J.Hydrol.,70, 309-327.

Jeannin, P-Y. and Grasso, D.A., 1995, Recharge respective des volumes de roche peu perméable et des conduits karstiques: role de l'épikarst. (in French, abstract in English). Bulletin d'Hydrogéologie, No. 14, Centre d'Hydrogéologie, Université de Neuchâtel, 95-111.

Jeannin, P-Y and Maréchal J-Ch., 1995, Lois de Pertes de Charge dans les Conduits Karstiques (in French, abstract in English). Bulletin d'Hydrogéologie 14. Centre d'Hydrogéologie, Université de Neuchâtel, 148-176.

Kiraly, L., Perrochet, P. and Rossier, Y. 1995, Effect of epikarst on the hydrograph of karst springs: a numerical approach. Bulletin d'Hydrogéologie, No. 14, Centre d'Hydrogéologie, Université de Neuchâtel, 199-220.

KRI 1997, Karst Hydrogeological Investigations in South-Western Slovenia. Karst Research Institute, Postojna, Slovenia KRI 1997

Parkin, G. 1996, A Three-Dimensional Variably-Saturated Subsurface Modelling System for River Basins. PhD Thesis, Dept. of Civil Engineering, University of Newcastle upon Tyne, UK.

White, W.B. 1999, Groundwater Flow in Karstic Aquifers. In: The Handbook of Groundwater Engineering [J.W. Delleur, Ed.], Chapter 18, CRC Press.

Geotechnical and Environmental Applications of Karst Geology and Hydrology, Beck & Herring (eds)
© 2001 Taylor & Francis, ISBN 90 5809 190 2

Groundwater dye tracing: an effective tool to use during the highway development process to avoid or minimize impacts to karst groundwater resources

DAVID BEDNAR, JR.   Michael Baker Jr., Inc., Shreveport, LA 71107, USA, dmbednar@mbakercorp.com

THOMAS ALEY   Ozark Underground Laboratory, Protem, MO 65733, USA, oul@tri-lakes.net

ABSTRACT

Detailed hydrologic studies must be employed when building highways through karst terrain. Underlying karst aquifers are highly vulnerable due to the rapid flow of surface water into the groundwater flow system through discrete recharge zones. These waters ultimately reach springs and sometimes wells, when they may adversely impact water quality. Groundwater dye tracing studies provide essential hydrogeologic information on the functioning of the karst aquifers and on the likely impacts from highway construction and operation. Expanding the use of high quality dye tracing studies in karst terrains during the highway planning and development process will help avoid or minimize adverse impacts on karst groundwater resources.

INTRODUCTION

The construction of new highways throughout the United States has been a continual process for the last 50 years. Many of these highways have been built through karst areas. The unique geologic features and complex hydrologic characteristics of these areas need assessment approaches not normally required in other landscapes. Groundwater dye tracing is an effective tool for identifying and characterizing groundwater flow routes. It is the authors' view that groundwater tracing investigations are practical and cost-effective, and that they yield valuable data which should routinely be developed for highway corridors proposed in karst areas.

This paper identifies and discusses groundwater dye tracing methods generally appropriate to highway corridor studies. Groundwater dye tracing is deceptively simple, and there are many critical decisions which must be made relative to the design and conduct of such investigations.

Four groundwater dye trace studies designed to assess highway corridor impacts on karst groundwater resources are discussed. The selected studies illustrate some of the diverse ways in which tracing studies can be used. The dye tracing studies were associated with highway projects in West Virginia, Arkansas, and Tennessee.

GROUNDWATER TRACING METHODS

The selection of the dyes to be used, the location of dye introduction points, the manner in which the dyes are introduced, the sampling strategy employed, and the analytical approach used must be tailored to the hydrogeologic setting, the issues of concern, and the quality and credibility of the data needed for the study. The following paragraphs summarize these considerations relative to highway corridor studies.

The most useful tracer dyes for most highway corridor studies are those which can be adsorbed onto activated carbon samplers. The most appropriate dyes for corridor studies are fluorescein (Acid Yellow 73, Color Index Number 45380); eosine (Acid Red 87; Color Index Number 45350); rhodamine WT (Acid Red 388, no assigned Color Index Number); and sulforhodamine B (Acid Red 52, Color Index Number 45100). Other dyes exist and may be appropriate in some special situations. While two or three of these dyes can often be used concurrently, the concurrent use of too many dyes in corridor studies is likely to create confusion and unclear results rather than enlightenment and fully credible conclusions.

Selection of dye introduction points and the design of methods for introducing the dyes are often critical considerations. In many cases natural features such as sinkholes and losing streams (surface streams which sink into the groundwater system in localized areas) are appropriate dye introduction points for highway corridor studies. If information on very localized areas is needed, borings specifically constructed to facilitate dye and water introduction into the epikarstic zone are appropriate. Corridor studies are commonly conducted prior to final decisions on highway alignments, and access for equipment (such as water-hauling trucks) is often difficult or impractical. As a result, most dye introductions for highway corridor studies must utilize natural runoff waters. "Dry sets", which place dye where it will be introduced into surface flow derived from storms, can often be used.

The dye sampling approach used for most highway corridor studies places primary reliance upon activated carbon samplers with secondary reliance upon grab samples of water. Aley (1999) provides detailed information on these approaches. Activated carbon samplers adsorb and retain (and thereby concentrate) all four of the tracer dyes identified earlier. The samplers are placed in appropriate wells, springs, surface streams, and any other potentially relevant points to which the introduced tracer dye may flow. Samplers are periodically collected, and new samplers placed, at intervals appropriate to the particular study and the issues of concern. Weekly sampling is common. Background sampling prior to any dye introduction and sampling of some control stations during the study is essential for credible investigations.

Grab samples of water are typically collected each time a sampling station is visited. The water samples provide data on dye concentrations at particular points in time whereas the activated carbon samplers provide data on the total quantity of dye accumulated on the samplers. At most sampling stations one does not have continuous custody of activated carbon samplers; this may become an issue in subsequent public hearings or litigation. This limitation can be offset by analyzing the grab samples of water from stations where dyes are detected since an investigator can maintain custody of water samples and can determine whether or not the dye analysis results from activated carbon samplers and water samples are consistent with one another.

Dye analysis work should be conducted by a well equipped and experienced laboratory. The four identified dyes are all fluorescent, and quantification of the fluorescence is the best and most sensitive analytical approach. In most cases the analysis work should use a spectrofluorophotometer where the width of excitation and emission slits can be set and where samples can be synchronously scanned with a constant separation between the excitation and emission wavelength.

No uniform method for dye analysis work associated with water tracing has been developed by the limited number of entities which do this work, nor by organizations such as EPA or ASTM. Each of the better laboratories has its own standard approaches, and these are generally similar although there are important differences. The lack of a uniform method of analysis is justified by the fact that tracer dyes are used in many different ways and in many different settings; one size does not fit all. Prices charged for dye analysis vary among laboratories due in large part to differences in the extent of quality assurance and quality control steps and to the expertise of the personnel involved. As an illustration, large volumes of reagent water are routinely needed to prepare activated carbon samplers for analysis. Some laboratories use municipal tap water as their reagent water even though it contains a chlorine residual which can destroy some of the adsorbed dye. At the other end of the spectrum, another laboratory has its own unchlorinated deep groundwater supply and quality assurance tests which ensure that the reagent water is always free of dyes and compounds which might interfere with dye analysis work. Those who select a laboratory based upon the lowest analysis cost per sample may not get the quality of results they need, especially if the data must withstand technical or legal challenges such as those commonly encountered in highway corridor issues.

EXAMPLE 1: APPALACHIAN CORRIDOR H HIGHWAY PROJECT
Hydrogeologic Setting and Issues

The West Virginia Department of Transportation - Division of Highways and the Federal Highway Administration proposed to construct an approximately 161 kilometers (100 miles) long highway from Elkins, West Virginia eastward to just west of the West Virginia/Virginia state line on WV 55. The proposed Corridor H highway facility would provide a divided, four-lane highway with partial control of access. The highway would traverse through portions of the West Virginia counties of Randolph, Tucker, Grant, and Hardy (USDOT, 1994). Groundwater dye tracing investigations were used to assess potential highway impacts on Wardensville Spring, Capon Springs, the Lost River, and springs in Greenland Gap.

Wardensville Spring is the sole source of water for the town of Wardensville, which is located at the base of Anderson Ridge. The West Virginia Department of Health and Human Resources (1994) developed a preliminary delineation of the wellhead protection area for the Wardensville Spring. The recharge area for the spring was determined to be along Anderson Ridge. Anderson Ridge is a plunging anticline composed of the Oriskany Sandstone and the spring discharges near the contact between the Oriskany Sandstone and the Marcellus Shale that underlies the adjacent valley. The preferred alternative would cross the wellhead protection area designated for the spring. Dye tracing studies were conducted to determine if areas that would be impacted by the preferred alternative had the potential to contribute water through groundwater conduit systems to Wardensville Spring.

Capon Springs is located about 12 kilometers (7.5 miles) northeast of Wardensville in Hampshire County, West Virginia. It is a significant resort that first became prominent in the 1850's. Three warm springs, collectively called the Capon Warm Springs Complex for this project, discharge from a nearly vertical outcrop of Oriskany Sandstone at a gap on the northwestern limb of the Great North Mountain Anticlinorium. Based largely on tritium concentrations in the water from Warm Capon Spring, Lessing et al. (1991) concluded that the Oriskany Sandstone was the sole recharge unit for the springs. In contrast, flow and temperature measurements by McColloch (1986) indicated that the springs are more interconnected with surface conditions than Lessing et al. (1991) concluded. Carbonate units of the Tonoloway and Wills Creek Formations and the Helderberg Group lie adjacent to the Oriskany Sandstone; these units are known to contain caves in West Virginia. There was public concern that some of the flow of the Capon Warm Spring Complex might be derived from these carbonate units in areas crossed by the preferred alternative. Groundwater dye tracing assessed this possibility.

The Lost River sinks in an area about four miles west of Wardensville Spring. The sinking area is within the Helderberg Group which is mostly cherty limestone. The preferred alternative crosses the Lost River about 91 meters (300 feet) downstream of the sinks. As a result, highway runoff or hazardous material spills in this area could impact the groundwater system. Tracer dyes provided a surrogate for investigating this possibility.

Greenland Gap is located in Grant County, West Virginia slightly east of the Allegheny Front. Greenland Gap formed from the downcutting of Patterson Creek through the Willis Creek Anticline (locally known as New Creek Mountain). The Gap is owned by the West Virginia Chapter of the Nature Conservancy. The Nature Conservancy expressed concern that the Corridor H project might damage the unique geological and biological conditions in Greenland Gap. Patterson Creek, which flows through the gap, is a West Virginia high quality trout stream and receives recharge from a karst spring discharging upstream from the gap. This spring discharges at the base of Walkers Ridge and is located slightly west of New Creek Mountain; there are numerous sinkholes on top of this mountain. The preferred alternative for Corridor H crosses upslope of many of these sinkholes. As a result, contaminants or spills from the highway could enter these sinkholes. If these sinkholes contributed water to springs supplying water to Patterson Creek they could adversely impact that spring.

Dye Trace Studies

On May 19, 1994, dyes were injected for the Wardensville Spring, Lost River, and the Capon Warm Springs Complex dye traces. Activated carbon samplers were collected at sampling station on weekly intervals for six weeks. Activated carbon samplers were collected at the Wardensville Spring and Capon Springs for 104 days and 62 days respectively. On May 24, 1994, dyes were injected for the Northern, Middle, and Southern Greenland Gap Traces. Activated carbon samplers were collected at weekly intervals for six weeks.

Wardensville Spring Trace

Rhodamine WT dye was introduced into a stream of water that was intentionally discharged from the Wardensville water tank. The tank is on Anderson Ridge approximately 213 meters (700 feet) upgradient of the spring. This dye introduction location was selected because the Wardensville Water Superintendent had noticed that water discharged from the tank onto the ridge immediately sank into the subsurface within a few feet.

One dye injection site and ten sampling stations, including Wardensville Spring, were used for this trace and are shown on Figure 1. No dye was recovered at any sampling stations after approximately 42 days. No dye was recovered at Wardensville Spring after 104 days after dye introduction. The USDOT (1994) recommended that, if the highway is constructed upgradient of the spring, monitoring wells should be placed every 305 meters (1,000 feet) across the wellhead protection zone.

The Wardensville Spring Trace was a relatively simple trace that addressed concerns brought forth by local officials and residents that waters could rapidly enter the subsurface on Anderson Ridge and discharge at Wardensville Spring within a few hours or days and impact their sole source of drinking water. This trace demonstrated that rapid infiltration of water on Anderson Ridge does not necessarily indicate rapid water movement through the subsurface to Wardensville Spring.

Capon Warm Springs Complex Trace

Fluorescein dye was introduced into a sinking stream approximately 6.4 kilometers (4 miles) from the warm springs. The injection site and sampling stations are shown on Figure 1. This sinking stream flows across the stratigraphic units which might contribute recharge waters to the springs. No dye was recovered at the Capon Warm Springs Complex during the 62 days of sampling. One sampling station downgradient of the sinking stream was the only station where dye was detected. Large concentrations of dye at this station demonstrated that most of the dye reached this sampling station.

This simple dye trace addressed the concerns of the owners of the Capon Springs and Farm Resort concerning the effects that construction of Corridor H may have on their historic warm springs. The results from this trace, along with the study by Lessing, et al (1991), indicated that waters entering the groundwater flow system in the immediate vicinity of the preferred alternative would not ultimately reach the Capon Warm Springs Complex.

Lost River Trace

Eosine dye was introduced into a sinking portion of the Lost River on the east side of the river about 366 meters (1200 feet) upstream of the area where the preferred alternative crosses Lost River. At the time of dye introduction there was continuous flow along Lost River. The dye introduction was made into boulders on the river bank where water was clearly sinking into the subsurface. A sampling station on Lost River about 244 meters (800 feet) downstream of the dye introduction point demonstrated that no detectable quantities of dye returned to the surface along this portion of Lost River. The dye introduction and sampling stations are shown on Figure 1.

Dye was recovered at sampling station 5 located downstream of the point where the Lost River emerges and becomes the Cacapon River. No dye was detected at any other sampling station including the Wardensville Spring.

Results from this simple trace addressed two important issues. First, the possibility that the sinking point on the Lost River contributed appreciable groundwater recharge to Trout Run or the Wardensville Spring was demonstrated not to be the case. Second, that spills may enter the sinking point on Lost River and contaminate the groundwater flow system. Results demonstrated that waters that enter the sinking point on Lost River contributed recharge to the Cacapon River two miles downstream in about six days. This indicated that there is very little storage in the portion of the karst aquifer traversed by the dye and that hazardous substances would be carried through the groundwater flow system rapidly.

Northern, Middle, and Southern Greenland Gap Traces

Three dye injection sites (using three different dyes) and twenty-four sampling stations were used for this study in the Greenland Gap area. The tracing work was conducted to determine discharge points of waters entering sinkholes located downslope or downstream of the preferred alternative for Corridor H. These three traces were used to characterize groundwater flow directions which would be followed by water derived along the preferred alternative. Dye injection sites and sampling stations are shown on Figure 2.

The Northern Greenland Gap Trace was designed to determine the discharge points for water entering a sinkhole located on Walkers Ridge that drains a substantial amount of land traversed by the preferred alternative. The dye injection site receives water from a surface stream and the preferred alternative would cross this stream about 823 meters (2,700 feet) upstream of the sinkhole.

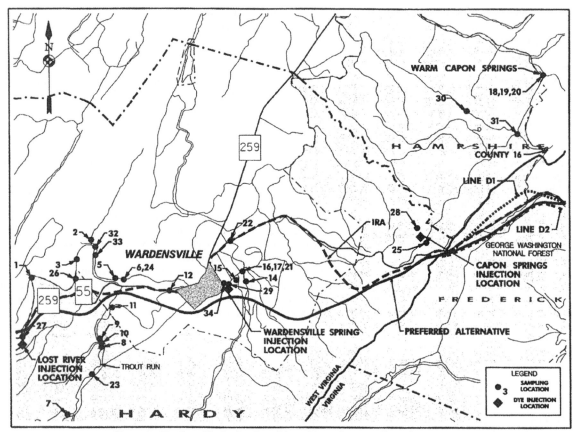

Figure 1. Wardensville Spring, Lost River, Warm Capon Springs dye traces.

The Middle and Southern Greenland Gap Traces were conducted in a valley between Walkers Ridge and New Creek Mountain. Two sinkholes were used as injection locations which received runoff from land traversed by the preferred alternative. There were two important reasons for these traces. First, to determine of the sinkholes into which dye injected recharged springs in the valley. Some of these springs are used as domestic water supplies. Second, to determine which springs in the Patterson Creek basin received recharge from the selected dye injection sites.

The three groundwater traces demonstrated that waters derived from this portion of the preferred alternative will subsequently discharge from two locations. The primary discharge point is Munsing Spring (Sampling Station 113), an important cold-water tributary to Patterson Creek. The second location is a spring or springs in or tributary to Patterson Creek in the stream segment between Sampling Stations 118 and 119. These results demonstrate that any stormwater runoff and or hazardous spills from the highway would enter the dye injection locations and into regional karst groundwater flow system and reach Munsing Spring and directly yield contaminants to Patterson Creek within seven days. Rapid groundwater travel rates and appreciable concentrations of dyes recovered during the tracing work indicated that the groundwater system is highly permeable and provides ineffective natural cleansing.

As the result of the dye traces performed in the Greenland Gap area, mitigation measures would be employed to help prevent groundwater contamination. These measures would include construction of detention ponds to contain spills and peat sand filters up gradient of the sinkholes to intercept and treat highway runoff before entering the groundwater flow system (USDOT, 1994).

EXAMPLE 2: HIGHWAY 71 RELOCATION
Hydrogeologic Setting and Issues

The Arkansas Highway and Transportation Department planned to construct a four-lane interstate highway on new location that would connect the Fayetteville Bypass with U.S.71 at Mckissick Creek at Belle Vista, Arkansas. The length of the project was approximately 40 kilometers (25 miles) long and would pass through portions of the Arkansas counties of Benton and Washington (USDOT, 1979). Two alternative alignments, Alternative A and Alternative B, were developed in the project area.

The project area is mostly underlain by the Boone Formation that consists of limestone and chert that varies in proportion horizontally and vertically. Extensive solution of the bedrock exists throughout most of the project area. This solution has created discrete recharge zones which lack surface expression. At these points, surface water enters the groundwater flow system and provides a supply of water to wells and springs. These springs drain areas where the proposed highway would pass through. In addition, the lower portion of the Boone Formation contains significant caves. Therefore, delineation of the recharge areas for these springs was necessary to access impacts to karst groundwater resources and associated cave fauna.

Dye Trace Studies

Five dye traces were conducted to identity recharge areas for major springs within the project area. Small quantities of fluorescent dye were injected at locations where surface water was seen disappearing into the groundwater flow system. Major springs identified during this investigation included Ozark Spring, Brush Spring, Cave Springs, Osage Springs, and Ford Spring.

Among other findings, this investigation demonstrated that initial alignment for Alternative A passed through 6 kilometers (4 miles) of the recharge area for Cave Springs and 9 kilometers (5.7 miles) through the recharge area of Ford Spring. Any hazardous spills would impact important cave fauna which includes the grey bat (*Myotis grisescens*) and the Ozark Cavefish (*Amblyopsis rosae*) at Cave Springs. Likewise, any contamination impacting Ford Spring would impact populations of the Oklahoma salamander in Little Sugar Creek which receives recharge water from Ford Spring. At the time of the study the grey bat was a federally listed endangered species. Subsequent to the

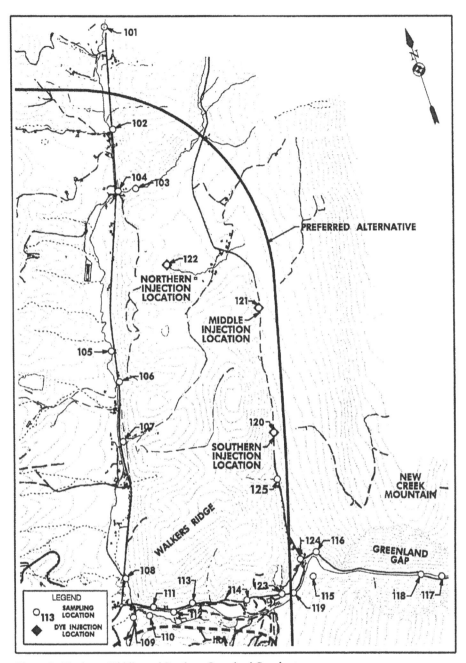

Figure 2. Northern, Middle, and Southern Greenland Gap dye traces.

study the Ozark cavefish has been federally listed as a threatened species. A principal recommendation from the dye tracing investigations was that Alternative A should be moved beyond the recharge area of Cave Springs.

After reviewing the results of the initial dye trace studies, the Arkansas Highway and Transportation Department authorized three additional groundwater traces to refine the eastern boundary of the Cave Springs recharge area. Based upon this tracing work, the highway alignment was shifted outside of the Cave Springs recharge area, and the highway was constructed along a new alignment.

This groundwater tracing investigation identified a potential problem (namely that the new highway passed through the recharge area for the cave which provided habitat for a federally listed endangered species, and for the Ozark Cavefish, which was subsequently federally listed as a threatened species). Tracing data identified a corridor route which would not impact this habitat, and that corridor was used. The Arkansas Highway and Transportation Department evaluated the possibility of building spill containment structures in the event the highway was constructed along the initial alignment, but this approach was determined to be less desirable than changing the location of the highway corridor.

EXAMPLE 3: U.S. HIGHWAY 65 IMPROVEMENT PROJECT

Hydrogeologic Setting and Issues

The Arkansas Highway and Transportation Department planned to improve U.S. 65 north of Harrison, Arkansas to a four-lane facility. The Department recognized that two spring systems, Bear Creek Springs and Smokehouse Spring, might be impacted by construction activities and operation of the highway. Both of these springs are integral assets of local businesses and warrant protection from adverse impacts. To address these concerns seven dye traces were conducted to delineate the recharge area for these springs and to characterize interactions between the two springs.

The project area is located in the Springfield Plateau subdivision of the Ozark Mountains Province of northern Arkansas. Bear Creek Springs and Smokehouse Spring discharge from the lower portion of the St. Joe Limestone Member of the Boone Formation. Bear Creek Springs discharges from a cave at the base of a bluff on the north side of Rolands Fork.

Smokehouse Spring only discharges water to the surface during storm events. The spring discharges from a cave passage in a sinkhole along the east side of Denning Creek. Smokehouse Spring is an estavella in that it will accept water during low flow conditions, but will discharge water during higher flow conditions. There are three additional estavallas in the channel adjacent to Smokehouse Spring. Estavella 1 is located along Denning Creek about 15 meters (50 feet) upstream of Smokehouse Spring. Estavella 2 is located along Denning Creek directly across the stream from Estavella 1. Estavella 1 and 2 are located along a limestone outcrop that extends completely across the stream. Estavella 3 is located approximately 6 meters (20 feet) downstream of Estavella 2.

Dye Trace Studies

Eight dye injection sites and eighteen sampling stations were used for this study. The tracing work resulted in the delineation of the recharge area for Bear Creek Springs, and the delineation of an area which contributed recharge to both Bear Creek Springs and Smokehouse Spring. An area was identified where highway construction and operation activities would be likely to not adversely impact one or both of the springs.

The result of the dye tracing studies was that the Arkansas Highway and Transportation Department made revisions to the design of the new alignment. The revisions included a bridge over part of the recharge area and a shift and reconfiguration of the interchange area to move it outside of the identified high vulnerability area. In addition, a geophysical survey was conducted to ensure that bridge footings would not impact underground caverns or flowing water that contribute recharge to the springs. The ultimate result of this work was that natural resources were protected and an improved highway was constructed.

EXAMPLE 4: FOOTHILLS PARKWAY

Hydrogeologic Setting and Issues

From June to December 1990, three groundwater dye trace studies were conducted along a portion of Section 8D1 for the Foothills Parkway project in the Great Smoky Mountains National Park. The purpose of the study was to access the potential impacts that highway construction might have on the hydrology of the area and to determine measures to minimize impacts to groundwater resources during construction and operation of the highway. A cave stream flows beneath right-of-way limits and discharges at a spring on private land which is used as a source of drinking water and for a large pond stocked with fish.

The study area is located in Sevier County Tennessee within a geologic structure known as the Wear Cove Window. Wear Cove is located within an anticline where Cambrian strata were thrust over Paleozoic strata. A karst terrain has developed on units of the Knox Group carbonates; caves, sinkholes, and springs are common in the area.

Dye Trace Studies

Four dye introductions were made and a total of twelve sampling stations were used. The tracing work demonstrated a complex groundwater flow system where dye introduced at one location would be recovered at multiple groundwater discharge points. Rapid groundwater flow occurred in some cases, yet in other cases appreciable water detainment occurred within the epikarstic zone. The studies concluded that very careful treatment of sinkhole areas would be necessary if the highway were constructed along the proposed alignment.

CONCLUSIONS

The four examples demonstrate use of dye tracing investigations for assessing potential highway impact upon groundwater resources. Investigations conducted for the U.S. 71 and U.S. Highway 65 projects provided convincing evidence that resulted in design changes to protect the karst groundwater resources and an endangered species. No design changes are expected along the preferred alternative of Corridor H in the vicinity of Greenland Gap, Lost River, Wardensville Spring or Capon Springs. The use of groundwater dye tracing during the highway development process is a valuable tool which should be used more routinely in the assessment of proposed highway projects.

ACKNOWLEDGEMENTS

The senior author wishes to express thanks to all those who made submission of this paper possible. I thank Mr. Lynn Malbrough of the Arkansas Highway and Transportation Department and Mr. Thomas Aley of the Ozark Underground Laboratory who provided examples that were discussed throughout this paper. I thank Mr. Randy Luketic of Michael Baker Jr., Inc. for the preparation of illustrations. Additionally, I thank Michael Baker Jr., Inc. for providing me with the financial support that resulted in the inclusion of this paper in the proceedings.

REFERENCES

Aley, Thomas. 1999. Ozark Underground Laboratory's Groundwater Tracing Handbook. Ozark Underground Laboratory. 35p.

Lessing, et al. 1991. Relations Between Springs and Geology Delineated by Side-looking Airborne Radar Imagery in Eastern West Virginia. U.S. Geological Survey, Water Resources Investigations Report 88-4096. pp. 20-28.

McColloch, Jane S. 1986. Springs of West Virginia. West Virginia Geologic and Economic Survey, Vol. V-6A. 493p.

U.S. Department of Transportation. 1994. Supplemental Draft Environmental Impact Statement - Appalachian Corridor H, Elkins, West Virginia to Interstate 81, Virginia. Federal Highway Administration. Charleston, West Virginia.

U.S. Department of Transportation. 1979. Final Environmental Impact Statement - Highway 71 Relocation, Fayetteville to McKissick Creek. Federal Highway Administration. Little Rock, Arkansas.

West Virginia Department of Health and Human Resources. 1994. Letter from Lewis Baker, Geologist, to Mr. John Bowman, Mayor of the Town of Wardensville dealing with the Wellhead Protection Area for Wardensville Spring. 1p. and map.

Geotechnical and Environmental Applications of Karst Geology and Hydrology, Beck & Herring (eds)
© 2001 Taylor & Francis, ISBN 90 5809 190 2

Changes in ground-water quality in a conduit-flow-dominated karst aquifer following BMP implementation

JAMES C.CURRENS Kentucky Geological Survey, University of Kentucky, Lexington, KY 40506, USA,
currens@kgs.mm.uky.edu

ABSTRACT

Water quality in the Pleasant Grove Spring karst ground-water basin, Logan County, Ky., was monitored to determine the effectiveness of best management practices (BMP's) in protecting karst aquifers. Ninety-two percent of the 4,069-hectare (10,054-acre) watershed is used for agriculture. Water-quality monitoring began in October 1993 and ended in November 1998. By the fall of 1995 approximately 72 percent of the watershed was enrolled in BMP's sponsored by the U.S. Department of Agriculture Water Quality Incentive Program (WQIP).

Pre-BMP nitrate-nitrogen concentration averaged 4.65 mg/L. The median total suspended solids concentration was 127 mg/L. The median triazine concentration measured by immunosorbent assay was 1.44 µg/L. Median bacteria counts were 418 colonies per 100 ml (col/100 ml) for fecal coliform and 540 col/100 ml for fecal streptococci. Post-BMP, the average nitrate-nitrogen concentration was 4.74 mg/L. The median total suspended solids concentration was 47.8 mg/L. The median triazine concentration for the post-BMP period was 1.48 µg/L. The median fecal coliform count increased to 432 col/100 ml after BMP implementation, but the median fecal streptococci count decreased to 441 col/100 ml.

The pre- and post-BMP water quality was statistically evaluated by comparing the annual mass flux, annual descriptive statistics, and population of analyses for the two periods. Nitrate-nitrogen concentration was unchanged. Increases in atrazine-equivalent flux and triazine geometric averages were not statistically significant. Total suspended solids concentration decreased slightly, while orthophosphate concentration increased slightly. Fecal streptococci counts were reduced.

The BMP's were only partially successful because the types available and the rules for participation resulted in less effective BMP's being chosen. Future BMP programs in karst areas should emphasize buffer strips around sinkholes, excluding livestock from streams and karst windows, and withdrawing land from production.

INTRODUCTION

Karst aquifers in Kentucky provide ground water to uncountable wells and springs used by individual households. Large springs are the water source for many public supply systems. Furthermore, the flow of streams is maintained during the dry months by discharge from karst springs. Many cities obtain their water from spring-fed streams. Because replacing these water sources would be impractical if not impossible, protecting the quality of ground water is vital for human health and economic development in Kentucky.

Reducing agriculturally derived nonpoint-source pollution of karst aquifers is important because some of the most productive agricultural lands in Kentucky occur in the karst regions. The 35 counties that are predominantly karst terrane produce over 50 percent of the annual agricultural receipts in Kentucky (Kentucky Agricultural Statistics Service, 1998). Logan County is one of these counties, and typically is among the top 10 agriculture producing counties in the state. The Pleasant Grove Spring karst ground-water basin was selected to test whether an economic incentive program intended to encourage farmers to adopt best management practices (BMP's) as part of the U.S. Department of Agriculture (USDA) Water Quality Incentive Program (WQIP) designed to protect ground water would reduce ground-water pollution in a karst aquifer.

PREVIOUS RESEARCH

Pesticides in ground water became a matter of public concern in the 1980's (U.S. Environmental Protection Agency, 1986; Gish and others, 1990) and had been found in ground water in 40 states by 1988 (Williams and others, 1988). The presence of agricultural pollutants in karst aquifers also began to be studied in the 1980's. The Big Spring karst ground-water basin in the Galena aquifer in Iowa was studied by Hallberg and others (1983, 1984), Libra and others (1984, 1986), and Libra (1987). The Galena aquifer is overlain by Pleistocene glacial outwash and till, and has significant diffuse flow characterized by fewer direct inflows and long residence time compared to other karst aquifers. Nitrate-nitrogen concentrations were typically under 45 mg/L, but occasionally exceeded

70 mg/L. Concentrations of atrazine seldom exceeded 0.85 µg/L, but peaked at 5.1 µg/L during one spring season storm. Hippe and others (1994) studied two karst springs in Pennsylvania and found only low concentrations of herbicides and nitrate, no spring season pesticide pulse, and little difference in atrazine or nitrate between springs overlain by agricultural and residential land uses. Their sampling schedule, however, included only single grab samples during a few high-flow events. Work in southeastern West Virginia by Boyer and Pasquarell (1994, 1996) and Pasquarell and Boyer (1995, 1996) found a strong positive linear relationship between nitrate concentrations and percentage of the ground-water basin area used for agriculture. Furthermore, the occurrence of atrazine coincided with application season, and fecal bacteria counts increased with percentage of land in agricultural production. Boyer and Pasquarell (1994) found average fecal coliform counts of less than 1 colony per 100 ml (col/100 ml) in a karst spring draining a pristine basin. Panno and others (1998) studied the Fogelpole Cave ground-water basin in southwestern Illinois. They found that triazine and sediment increased in response to spring season storms during the application period and that nitrate-nitrogen became elevated in the winter and peaked during the spring.

In Kentucky, Felton (1991) monitored Garretts Spring, which drains the Sinking Creek basin of northwestern Jessamine County in the Inner Blue Grass Region. Felton found that nitrate concentrations varied seasonally and were highest during wet, winter months, but pesticide concentrations only occurred in low concentrations throughout the remainder of the year. Ryan and Meiman (1996) identified discrete periods of high-flow events when waters containing increased suspended sediment and bacteria counts arrived from agricultural lands bordering Mammoth Cave National Park. Quantitative dye traces from the suspected source areas were conducted simultaneously with the storm to identify the water from the source of the sediment and bacteria.

Currens (1999) reported for 1990 through October 1994 that significant contaminants in the Pleasant Grove Spring basin are herbicides (specifically atrazine), bacteria, and sediment. Nitrate-nitrogen was the most widespread contaminant in the basin, but concentrations averaged 5.2 mg/L and generally did not exceed U.S. Environmental Protection Agency (EPA) maximum contaminant levels (MCL's). Atrazine had been consistently detected in low concentrations, and other pesticides were occasionally detected. Concentrations of triazine (including atrazine) and alachlor had exceeded drinking-water MCL's during spring season high flow. Maximum concentrations of triazine, carbofuran, metolachlor, and alachlor in samples from Pleasant Grove Spring were 44.0, 7.4, 9.6, and 6.1 µg/L, respectively. Bacteria counts were always significant and occasionally exceeded standards for drinking-water sources. A biological assessment showed an adverse impact on aquatic biota downstream, probably because of sedimentation. Several other sites in the basin were sampled with similar results.

STUDY AREA

Geographic location

The Pleasant Grove Spring karst ground-water basin is located in Logan County, southwestern Kentucky (Fig. 1). The spring discharges from a 4,069-hectare (10,054-acre) karst ground-water basin as determined with dye-tracing techniques. The topography is mature karst developed on Mississippian carbonates mantled with residuum and minor loess. The spring discharges to Pleasant Grove Creek, which flow 2.4 km (1.5 mi) to its confluence with the Red River, a major tributary of the Cumberland River.

Geology and hydrogeology

Only the St. Louis Limestone and the overlying Ste. Genevieve Limestone are mapped at the surface in the basin (Shawe, 1966; Miller, 1968). Reconnaissance reports of the area hydrogeology are by Brown and Lambert (1962) and Van Couvering (1962). A prominent bedded chert at the top of the St. Louis Limestone has a significant influence on karst development. This chert is probably the equivalent of the Lost River Chert (Garland R. Dever Jr., Kentucky Geological Survey, oral commun., 1994). The strata dip gently to the northwest at 11 meters per kilometer (60 ft/mi) into the Illinois structural basin. There are no mapped faults within the basin.

The Pleasant Grove Spring ground-water basin is a shallow, unconfined, carbonate aquifer (Currens, 1994). Sinkholes and sinking streams dominate the gently rolling landscape. The northern, headwaters half of the basin is characterized by a slow-flow regime, whereas the southern, downstream half is predominantly fast flow. Although two flow regimes are recognized, the preponderance of ground-water flow in both regimes is turbulent and concentrated in a trellised system of tributary conduits and caves. An unquantified, but probably significant, percentage of the water storage in the basin is in the epikarst.

Water flowing in Upper Pleasant Grove Creek persists at the surface through most of the year, gradually ceasing headward as drought periods lengthen. Perennial surface-flowing streams occur only in the headwaters of the basin. Flow in the southern end of the basin is wholly underground except during extreme high flow. The intake capacity of George Delaney swallow hole is exceeded during floods, and water flows south in a normally dry channel to Johnson swallow hole. Under exceptional flood conditions, discharge also exceeds the intake capacity of Johnson swallow hole and flows overland to a confluence with Pleasant Grove Creek, downstream of Pleasant Grove Spring, and out of the ground-water basin. Thus, Pleasant Grove Spring is an alluviated, underflow spring (Worthington, 1991) discharging from a continuously submerged cave.

Soils

Soils in the basin are silt loams derived from loess and limestone residuum, which are classified in the Pembroke-Crider association (Dye and others, 1975). The soils are moderately permeable, well drained, and have deep root zones with loamy or clayey subsoils. The soils are deep and have a thickness as great as 2 m (76 in.) to the base of the subsoil. Six auger holes drilled by the Kentucky Geological Survey in the central part of the watershed ranged in depth from 1.5 to 5.8 m (5 to 19 ft) and did not hit rock.

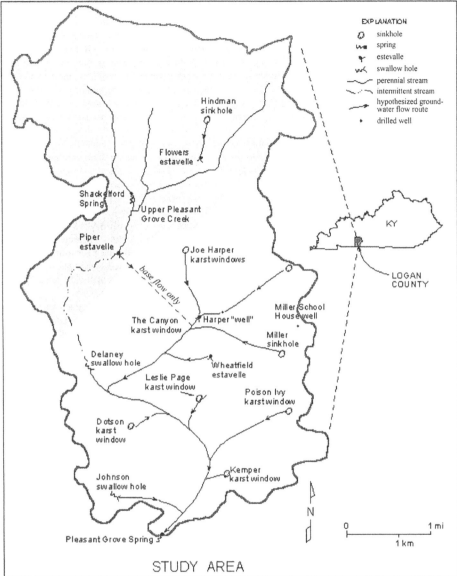

EXPLANATION
- ⦾ sinkhole
- ᴖ⛿ spring
- ⦙ estevalle
- ᴗᴗ swallow hole
- ⌒⌒ perennial stream
- ⌒⌒ intermittent stream
- → hypothesized ground-water flow route
- ⦁ drilled well

Hindman
sinkhole

Flowers
estavelle

Shackelford
Spring

Upper Pleasant
Grove Creek

Piper
estavelle

Joe Harper
karstwindows

base flow only

The Canyon
karst window

Harper "well"

Miller School
House well

Miller
sinkhole

Delaney
swallow hole

Wheatfield
estavelle

Leslie Page
karst window

Poison Ivy
karstwindow

Dotson
karst
window

Johnson
swallow hole

Kemper
karst window

Pleasant Grove Spring

KY

LOGAN
COUNTY

N

0   1 mi

1 km

STUDY AREA

Figure 1. Location of the Pleasant Grove Spring Basin, showing important field sites and hypothesized groundwater flow routes within the basin.

Land use

Most of the area of the basin (92 percent) is used for agriculture. About 70 percent of the basin area is row crop, largely in the northern two-thirds of the study area (Currens, 1999). Another 20 percent is hay fields and pasture. The predominant production system is no-till corn on a 2-year rotation with winter wheat and soybeans. Large fields of wheat, oats, rye, soybean, alfalfa, corn, milo, hay, and tobacco are grown. Livestock production is mostly beef and dairy cattle with some swine. Almost all row crops are cultivated using conservation tillage, although some conventional tillage was still practiced at the beginning of the project. The only nonagricultural business in the watershed is a privately owned tractor and automobile repair garage. The largest residential community is Oakville in the northeastern quadrant of the study area, which consists of about 30 scattered houses and mobile homes. Other residential development is limited to farmsteads.

ASSUMPTIONS

The assumptions made were that farm chemical usage could be estimated from area of crops grown and recommended application rates, that the contaminant load could be determined as it leaves the basin, and that the effect of BMP's on water quality would be measurable. Although monitoring results vindicated the latter assumptions, the chemical usage estimate was inaccurate. The WQIP chemical usage inventory showed considerable annual variability and generally less average usage of atrazine than recommended.

METHODOLOGY

This research compared water quality before and after BMP implementation to judge the effectiveness of the BMP program. The initial reconnaissance of the hydrogeology and water quality of the basin began in August 1990. Pleasant Grove Spring was instrumented for continuous monitoring and intensively sampled beginning in October 1992, prior to BMP implementation. A major grant was awarded by the USDA to the Logan County office of the Natural Resources Conservation Service (NRCS) to implement a Water Quality Incentive Program in the ground-water basin. BMP's sponsored by the WQIP were mostly in place by the fall of 1995, and continuous monitoring continued through the end of the project in October 1998.

Tasks not part of the monitoring program were an inventory of chemical applications, a livestock census, and routine domestic water-well monitoring. They were omitted because of budgetary, logistical, and manpower limitations but also because it was thought they would both have a poor response and create an atmosphere of distrust. As a condition of receiving WQIP money, however, farmers were required to report chemical usage, acres of crops, and head of livestock to the NRCS for calendar years 1996, 1997, and 1998.

The details of the ground-water dye-tracing methods, sampling protocols, sampling frequency, and analytical techniques used can be found in Currens (1994, 1999). Procedures followed U.S. Geological Survey (1982) or EPA (1983) standard methods. Results from nonstandard sampling or analytical techniques were related through linear regression to data obtained from standardized methods. For example, duplicate samples were analyzed for atrazine and metolachlor determined by gas chromatograph (GC) and for triazine and metolachlor determined by enzyme linked immunosorbent assay (ELISA). The ELISA determined concentrations were then correlated with concentrations determined by GC. Samples at Pleasant Grove Spring initially were collected monthly, supplemented by storm samples. The sampling frequency was increased to biweekly during the winter months and bi-daily supplemented with storm samples during the planting season. Continuous discharge was estimated from stage data using rating curves developed from direct discharge measurements. The stage/discharge rating curve at Pleasant Grove Spring takes into account the occasional overflow at Johnson swallow hole.

The monitoring strategy focused on the principal discharge point and contaminants while conducting less intensive monitoring at upgradient sites. Four contaminants of likely agricultural origin were identified during reconnaissance of the basin: atrazine, nitrate, suspended sediment, and bacteria from animal waste. Orthophosphate was later identified as a pollutant. The strategy compared results of mass-flux calculations and concentration statistics for pre- and post-BMP periods. Mass flux, or flow loading, is the quantity of a constituent moving past a water-quality monitoring station during a specific interval. For this project, flux was estimated by multiplying the discharge during each 10-min interval by the most recently determined concentration for that constituent. The annual mass flux is calculated by summing the increments over the year. Flux was calculated for GC atrazine equivalent of ELISA triazine, GC metolachlor equivalent of ELISA metolachlor (a minor contaminant), nitrate-nitrogen, and total suspended solids. Arithmetic averages, geometric averages, medians, and total flux were calculated for each year of monitoring at Pleasant Grove Spring for triazine, nitrate, and suspended solids. Arithmetic averages, geometric averages, and medians were calculated for bacteria and orthophosphate. The population of pre-BMP annual statistics and other data were then compared to post-BMP values using various statistical tests. Time-series analysis was not used because logistical constraints prevented the collection of samples at uniform intervals.

Qualitative ground-water dye tracing was used to delineate the ground-water basin boundary and quantitative tracing was used to determine travel times from George Delaney swallow hole and Leslie Page karst window to Pleasant Grove Spring. In addition, samples were collected for nitrogen isotope ratio analysis ($^{15}N/^{14}N$) at Leslie Page karst window and Pleasant Grove Spring. The channel reach immediately downstream of Pleasant Grove Spring was inventoried for biological diversity by Kentucky Division of Water personnel before BMP implementation, and at the end of the project. Land use (crop type) was determined from aerial photographs provided by the USDA Farm Services Administration and crop areas were measured by the NRCS.

In addition to Pleasant Grove Spring, upstream sites were monitored to identify the general source-area pollutants. Water gaging stations were placed at four locations. The first two were on Upper Pleasant Grove Creek at George Delaney swallow hole and at a box culvert where Johnson-Young Road crosses the creek. Other gaging stations were placed at Spring Valley karst window and Leslie Page karst window. An additional site, Miller School House well, was added in 1994. This abandoned, drilled domestic well is just inside the eastern mapped drainage boundary of Pleasant Grove Spring. It was used as a control because the well is upgradient of most agricultural activity.

RESULTS

Best management practices

In the spring of 1995, BMP implementation under the WQIP began in a significant percentage of the basin and was essentially completed by that fall. Approximately 2,919 hectares (7,213-acres) of farmland, or 72 percent of the watershed, was ultimately enrolled. The most popular practices sponsored by the program and applied to the study area were conservation crop rotation and crop residue use. Significant areas were also converted to pasture, hay land, and conservation cover. BMP's used in smaller areas included exclusion of cattle from streams and karst windows, and use of or extension of grass filter strips.

Miscellaneous investigations

The nitrogen isotope analysis suggested that nitrate-nitrogen at Leslie Page karst window is primarily derived from synthetic fertilizer, whereas nitrate-nitrogen discharging from Pleasant Grove Spring is mixed biogenic and synthetic fertilizer nitrate. These findings are consistent with the dominance of row crops at Leslie Page karst window and the mixed land use of the basin as a whole. The biological inventories indicate mixed results. Pollution-sensitive aquatic fauna did not reappear, but the population of macroinvertebrates increased after the BMP's were implemented (McMurray, 1999). The most significant finding of the land-use measurements is the steady loss of forested areas in the basin. Wooded areas declined during the project from 15.8 percent to 9.7 percent, whereas land in row-crop production increased from 68.5 percent to 77.9 percent of the basin area.

Monitoring

Of the contaminants found in the Pleasant Grove Spring Basin, nitrate-nitrogen was most widespread and persistent. Nitrate-nitrogen concentrations, unlike pesticides, remained nearly constant throughout the year (Table 1) and usually did not exceed the EPA's maximum contaminant level for drinking water of 10.0 mg/L. The maximum nitrate-nitrogen concentration measured in the basin between 1990 and 1998 was 13.10 mg/L (at Leslie Page karst window) and the average concentration was 5.12 mg/L. The maximum orthophosphate concentration was 1.397 mg/L (at Pleasant Grove Spring) and the median for the basin was 0.05 mg/L. The total suspended solids maximum was 3,267 mg/L and the median was 46 mg/L. The median fecal coliform bacteria count was 220 col/100 ml and the median fecal streptococci count was 600 col/100 ml for all sites. Maximum bacteria counts were 200,000

212

col/100 ml for fecal coliform and 810,000 col/100 ml for fecal streptococci; these samples were collected at Pleasant Grove Spring and George Delaney swallow hole, respectively, during high-flow events.

Table 1: Annual statistics for contaminants discharged in significant concentrations from Pleasant Grove Spring calculated from the complete sample set, except those for bacteria, which are from monthly samples.

Water - year	Nitrate-nitrogen, average (mg/L)	Triazine, median (µg/L)	Total suspended solids, median (mg/L)	Orthophosphate, median (mg/L)	Fecal coliform, median (col/100 ml)	Fecal streptococci, median (col/100 ml)
1991–92	4.20	1.40	NA	0.036	100	350
1992–93	4.98	0.83	45	0.045	NA	NA
1993–94	4.86	1.30	65	0.053	300	200
1994–95	4.57	1.62	189	0.051	1,100	1,700
1995–96*	4.88	1.05	46.5	NA	1,000	1,400
1996–97*	4.56	1.17	33	0.254	1,100	700
1997–98*	4.77	2.21	64	0.153	200	250

NA: Not available   *: BMP period

The maximum concentrations for three of the four pesticides routinely analyzed by ELISA during the project were 7.4, 12.0, and 29.6 µg/L for carbofuran, alachlor, and metolachlor, respectively. Median concentrations of these pesticides, however, are near detection limits and below their MCL's. Concentrations of alachlor and atrazine (represented by triazine as measured by ELISA) exceeded their MCL's during some high-flow events (alachlor has an MCL of 2.0 µg/L and atrazine has an MCL of 3.0 µg/L). The basinwide median concentration of triazine was 1.20 µg/L, but the maximum concentration measured during the project was 393.0 µg/L at Leslie Page karst window. The peak was detected in water flowing into the swallow hole of the karst window during a modest storm on April 16, 1998. The Leslie Page karst window sub-basin drains an epikarstic aquifer in an area containing only row crops. Between 1994 and 1997 most fields around Leslie Page karst window were planted with wheat and soybeans. Herbicides were only detected at low concentrations during those years. Corn was planted in the spring of 1998 in the fields surrounding Leslie Page karst window. The April 16 storm samples were also analyzed for carbofuran, alachlor, and metolachlor, but these pesticides were detected only at low concentrations.

Mass-flux results

Annual mass flux was the principal measure used to evaluate changes in water quality at Pleasant Grove Spring (Table 2). The mean annual mass flux for GC atrazine equivalent of ELISA triazine, GC metolachlor equivalent of ELISA metolachlor, nitrate-nitrogen, and total suspended solids during the project was approximately 65 kg (143 lb), 4.6 kg (10 lb), 127 metric tons (140 tons), and 2,992 metric tons (3,298 tons), respectively. Mass flux was not calculated for carbofuran, alachlor, orthophosphate, or bacteria.

Table 2: Summary of annual mass flux for GC atrazine equivalent ELISA triazine, GC metolachlor equivalent ELISA metolachlor, nitrate, and total suspended solids and related parameters for Pleasant Grove Spring.

Water - year	Precip. (cm)	Discharge at Pleasant Grove Spring (m³)	Row crops (ha)	Atrazine equivalent (kg)	Metolachlor equivalent (kg)	Nitrate-nitrogen (metric tons)	Total suspended solids (metric tons)
1991–92	100.3	NA	2,895	NA	NA	NA	NA
1992–93	87.8	10,227,000	2,899	46.01	5.97	51.13	2,124
1993–94	132.6	31,552,000	2,914	31.13	0.89	180.62	3,802
1994–95	110.0	15,292,000	2,973	62.82	3.92	75.03	2,255
1995–96*	136.8	19,775,000	2,986	36.42	2.25	98.69	1,591
1996–97*	140.2	42,544,000	3,082	70.52	6.61	197.63	4,553
1997–98*	135.8	32,914,000	3,171	142.52	7.79	157.65	3,625

NA: Not available   * : BMP period

Pre-BMP water quality

At Pleasant Grove Spring, pre-BMP (before October 1995) nitrate-nitrogen averaged 4.65 mg/L and never exceeded the MCL. The median orthophosphate was 0.05 mg/L. The median total suspended solids at Pleasant Grove Spring was 127.0 mg/L and the maximum was 3,073 mg/L in May 1995. Median triazine concentration measured by ELISA prior to BMP implementation was 1.44 μg/L, whereas the median concentrations determined for the other three pesticides were near ELISA detection limits. Bacteria counts at Pleasant Grove Spring always exceeded the MCL for drinking water and occasionally exceeded limits for drinking-water sources (2,000 col/100 ml). Pre-BMP samples collected monthly at Pleasant Grove Spring had median bacteria counts of 418 fecal coliform col/100 ml and 540 fecal streptococci col/100 ml.

Post-BMP water quality

During the post-BMP period (October 1995 through October 1998) the maximum nitrate-nitrogen measured at Pleasant Grove Spring was 8.11 mg/L in December 1996. The average nitrate-nitrogen concentration was 4.74 mg/L, essentially unchanged from the pre-BMP period. The median orthophosphate concentration was 0.20 mg/L. The median total suspended solids was 47.8 mg/L, a noticeable reduction. The maximum triazine concentration was 62.2 μg/L in May 1996 and was the highest observed at Pleasant Grove Spring. The median triazine concentration for the post-BMP period was 1.48 μg/L, nearly unchanged from the pre-BMP period. Post-BMP, the median fecal coliform count increased slightly to 432 col/100 ml, but fecal streptococci decreased to 441 col/100 ml. Shifts in the fecal coliform to fecal streptococci ratio during recession of high-flow events suggests the dominant waste source changed from animals to humans as runoff diminishes. Post-BMP, the monthly samples with fecal coliform–fecal streptococci ratios indicating an animal-dominated source decreased by 9 percent, suggesting that the relative contribution of animal waste had been reduced.

The quality of ground water discharging at Pleasant Grove Spring before and after BMP implementation was statistically evaluated by comparing the annual mass flux, annual descriptive statistics, or population of analyses for the two periods. Furthermore, graphs were made showing various measures, and the trends were subjectively evaluated (Fig. 2). Statistical tests used were the Student's T-test on normally distributed raw or transformed data, and the Mann-Whitney W-test to test annual medians. Dunnet's T-test and the Mann-Whitney U-test were also used to compare various annual statistics for the pre- and post-BMP periods. Flux and annual statistics for nitrate-nitrogen were statistically unchanged over the course of the BMP program. Increases in atrazine-equivalent flux and triazine geometric averages were not statistically significant. Total suspended solids concentrations decreased slightly, while orthophosphate increased slightly, both statistically significant. Reductions in fecal streptococci counts were not signficant.

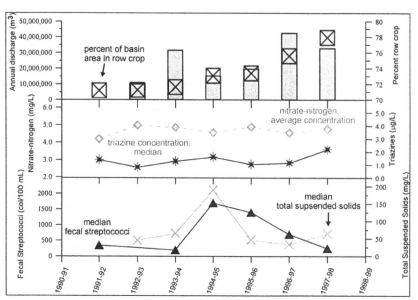

Figure 2: Trends in annual medians and means of triazine, nitrate-nitrogen, and total suspended solids concentration and bacteria counts during the Pleasant Grove Spring project.

CONCLUSIONS

Water quality, biological inventory, and land-use data all indicate the WQIP was only partially successful in protecting the quality of ground water in the Pleasant Grove Spring karst ground-water basin. Comparison of pre- and post-BMP water- quality monitoring showed that although suspended sediment and bacteria counts improved, triazine and nutrient concentrations did not change or were slightly worse. Even though the percentage of the basin in row crops steadily increased during the project, the water quality did not significantly worsen, which may indicate the BMP's were more effective than was first apparent. Although the WQIP was fully implemented, the types of BMP's funded and the rules for BMP participation resulted in less effective BMP's being chosen by farmers. Future BMP programs for the protection of ground water in karst aquifers should emphasize the installation of buffer strips around sinkholes, the exclusion of livestock from streams and karst windows, and the withdrawal of land from agricultural production.

REFERENCES

Boyer, D.G., and Pasquarell, G.C., 1994, Agricultural land use impacts on karst groundwater quality, *in* Effects of human-induced changes on hydrologic systems: American Water Resources Association, p. 791–799.

Boyer, D.G., and Pasquarell, G.C., 1996, Agricultural land use effects on nitrate concentrations in a mature karst aquifer: American Water Resources Association, Water Resources Bulletin, v. 32, no. 3, p. 565–573.

Brown, R.F., and Lambert, T.W., 1962, Availability of ground water in Allen, Barren, Edmonson, Green, Hart, Logan, Metcalfe, Monroe, Simpson, and Warren Counties, Kentucky: U.S. Geological Survey Hydrologic Investigations Atlas HA-32, scale 1:125,000.

Currens, J.C., 1994, Mass flux of agricultural nonpoint-source pollutants in a conduit-flow-dominated karst aquifer, Logan County, Kentucky, *in* Beck, B.F., ed., Karst geohazards, proceedings of the Fifth Multidisciplinary Conference on Sinkholes and the Engineering and Environmental Impacts of Karst, Gatlinburg, Tenn., 2–5 April 1995: Rotterdam, A.A. Balkema, p. 179–187.

Currens, J.C., 1999, Mass flux of agricultural nonpoint-source pollutants in a conduit-flow-dominated karst aquifer, Logan County, Kentucky: Kentucky Geological Survey, ser. 12, Report of Investigations 1, 151 p.

Dye, J.W., Barton, A.J., and Froedge, R.D., 1975, Soil survey of Logan County, Kentucky: U.S. Department of Agriculture, Soil Conservation Service, in cooperation with the Kentucky Agricultural Experiment Station, National Cooperative Soil Survey, 78 p.

Felton, G.K., 1991, Agricultural chemicals at the outlet of a shallow carbonate aquifer: American Society of Agricultural Engineers, 1991 International Summer Meeting, Albuquerque, N. Mex., Paper 91-2109, 26 p.

Gish, T.J., Isensee, A.R., Nash, R.G., and Helling, C.S., 1990, Impact of pesticides on shallow groundwater quality: Transactions of the American Society of Agricultural Engineers, v. 34, no. 4, p. 1745–1753.

Hallberg, G.R., Hoyer, B.E., Bettis, E.A., III, and Libra, R.D., 1983, Hydrogeology, water quality, and land management in the Big Spring Basin, Clayton County, Iowa: Iowa Geological Survey, Open-File Report 83-3, 191 p.

Hallberg, G.R., Libra, R.D., Bettis, E.A., III, and Hoyer, B.E., 1984, Hydrogeologic and water quality investigations in the Big Spring Basin, Clayton County, Iowa: Iowa Geological Survey, Open-File Report 84-4, 231 p.

Hippe, D.J., Witt, E.C., and Giovannitti, R.M., 1994, Hydrogeology, herbicides and nutrients in ground water and springs, and relation of water quality to land use and agricultural practices near Carlisle, Pennsylvania: U.S. Geological Survey, in cooperation with the Pennsylvania Department of Environmental Resources, Water-Resources Investigations Report 93-4172, 66 p.

Kentucky Agricultural Statistics Service, 1998, 1997–1998 Kentucky agricultural statistics: U.S. Department of Agriculture and Kentucky Agricultural Statistics Service, 144 p.

Libra, R.D., 1987, Impacts of agricultural chemicals on ground water quality in Iowa, *in* Fairchild, D.M., ed., Ground water quality and agricultural practices: Chelsea, Mich., Lewis Publishers, p. 185–215.

Libra, R.D., Hallberg, G.R., Hoyer, B.E., and Johnson, L.G., 1986, The Big Spring Basin study in Iowa, *in* Agricultural impacts on ground water, a conference: National Water Well Association, p. 253–273.

Libra, R.D., Hallberg, G.R., Ressmeyer, G.G., and Hoyer, B.E., 1984, Groundwater quality and hydrogeology of Devonian-carbonate aquifers in Floyd and Mitchell Counties, Iowa: Iowa Geological Survey, Open File Report 84-2, part 1, p. 1–106.

McMurray, S.E., 1999, Post-BMP biological survey of Pleasant Grove Spring, Logan County, Kentucky: Kentucky Division of Water, Water Quality Branch, Nonpoint Source Section Technical Bulletin, 2, 3 p.

Miller, R.C., 1968, Geologic map of the Russellville quadrangle, Logan County, Kentucky: U.S. Geological Survey Geologic Quadrangle Map GQ-714, scale 1:24,000.

Panno, S.V., Kelly, W.R., Weibel, C.P., Krapac, I.G., and Sargent, S.L., 1998, The effects of land use on water quality and agrichemical loading in the Fogelpole Cave groundwater basin, southwestern Illinois, *in* Proceedings of the Eighth Annual Conference of the Illinois Groundwater Consortiums' Research on Agricultural Chemicals in Illinois Groundwater, Status and Future Directions: Makanda, Ill., April 1–2, 1998, 13 p.

Pasquarell, G.C., and Boyer, D.G., 1995, Agricultural impacts on bacterial water quality in karst groundwater: Journal of Environmental Quality, v. 24, p. 959–969.

Pasquarell, G.C., and Boyer, D.G., 1996, Herbicides in karst groundwater in southeast West Virginia: Journal of Environmental Quality, v. 25, p. 755–765.

Ryan, M., and Meiman, J., 1996, An examination of short-term variations in water quality at a karst spring in Kentucky: Ground Water, v. 34, no. 1, p. 23–30.

Shawe, F.R., 1966, Geologic map of the Dot quadrangle, Kentucky–Tennessee: U.S. Geological Survey Geologic Quadrangle Map GQ-568, scale 1:24,000.

U.S. Environmental Protection Agency, 1983, Handbook for sampling and sample preservation of water and wastewater: U.S. Environmental Protection Agency, EPA-600/4-82/029, 402 p.

U.S. Environmental Protection Agency, 1986, Working Group on Pesticides in Ground Water: U.S. Environmental Protection Agency, Office of Ground-Water Protection, EPA 440/6-86-002, 72 p.

U.S. Geological Survey, 1982, U.S. Geological Survey national handbook of recommended methods for water-data acquisition; chapter 5, chemical and physical quality of water and sediment: U.S. Geological Survey, p. 5-1–5-194.

Van Couvering, J.A., 1962, Characteristics of large springs in Kentucky: Kentucky Geological Survey, ser. 10, Information Circular 8, 37 p.

Williams, W.M., Holden, P.W., Parsons, D.W., and Lorber, M.N., 1988, Pesticides in ground water data base, 1988 interim report: U.S. Environmental Protection Agency, Office of Pesticide Programs, 33 p.

Worthington, S.R.H., 1991, Karst hydrogeology of the Canadian Rocky Mountains: Hamilton, Ontario, McMaster University, Ph.D. dissertation, 380 p.

Geotechnical and Environmental Applications of Karst Geology and Hydrology, Beck & Herring (eds)
© *2001 Taylor & Francis, ISBN 90 5809 190 2*

Occurrence of coliform bacteria in a karst aquifer, Berkeley County, West Virginia, USA

MARK D.KOZAR & MELVIN V.MATHES   U.S. Geological Survey, 11 Dunbar Street, Charleston, WV 25301, USA, mdkozar@usgs.gov, mvmathes@usgs.gov

ABSTRACT

During the spring of 2000, the U.S. Geological Survey (USGS) in cooperation with the West Virginia Eastern Panhandle Soil Conservation District sampled 50 wells for indicator bacteria in a karst limestone and dolomite aquifer in Berkeley County, West Virginia. Wells were selected for sampling on the basis of the density of septic systems within an 80-meter radius.

Of 50 wells sampled, 62 percent contained total coliform bacteria, 32 percent contained *Escherichia coli* (*E. coli*) bacteria, and 30 percent contained fecal coliform bacteria. The high detection frequency of bacteria from wells sampled, especially fecal coliform and *E. coli*, indicate the potential for serious microbial contamination within the karst aquifer. There was no statistically significant correlation between septic density and the concentrations of bacteria detected in the wells.

The effects of agriculture were investigated to determine if they pose an increased threat of microbial contamination to the karst aquifer. Although no statistically significant trends could be determined between bacteria concentrations and land use near the wells sampled, wells with surrounding agricultural activity typically had higher concentrations of bacteria than wells surrounded by urban, rural, or commercial land use. Of wells sampled, 50 percent with nearby agricultural activity contained either fecal coliform or *E.coli* bacteria. Wells with no agricultural land use nearby had detectable concentrations of fecal coliform or *E.coli* in only 36 percent of wells sampled.

The highest concentrations of *E. coli* and fecal coliform bacteria were found in shallow wells. This trend was statistically significant at a 95-percent confidence level. Of wells less than or equal to 150 feet deep, 63 percent contained either fecal coliform or *E. coli* bacteria. For wells greater than 150 feet deep, 33 percent contained either fecal coliform or *E. coli* bacteria. Additionally, a statistically significant correlation between casing length and bacteria was also present at a 95-percent confidence level. For wells with casings 60 feet long or less, 47 percent contained either fecal coliform or *E. coli* bacteria. For wells with casings longer than 60 feet only 20 percent contained either fecal coliform or *E. coli* bacteria.

Of the 50 wells sampled, 13 were known or suspected of having microbial contamination problems, and as a result were fitted with either chlorination or ultraviolet light treatment systems. At least one of the three types of bacteria was detected in 11 of 13 (85 percent) raw water samples collected from wells equipped with treatment systems. Raw water samples from 21 of the 37 (57 percent) wells without treatment systems contained at least one of the three types of bacteria. Water samples collected from a few wells after the water was treated contained no detectable concentrations of any of the three types of bacteria.

INTRODUCTION

In many residential subdivisions of Berkeley County, West Virginia, individual wells drilled within the karst aquifer are commonly used for domestic water supply. Homes in these subdivisions also commonly have septic systems with leach fields for waste-water treatment. The high density of septic systems and individual wells located in close proximity pose a significant risk of bacterial contamination to ground water within the karst aquifer. The karst aquifer underlies more than one-third of the County (Shultz and others, 1995). In areas where overlying soils are thin, limestone aquifers with fractures or conduits allow rapid infiltration and lateral movement of ground water, which increases the risk of bacterial contamination. Newly constructed homes built on small lots could potentially have a large number of septic leach fields near drinking-water wells. The U.S. Geological Survey (USGS), in cooperation with the West Virginia Eastern Panhandle Soil Conservation District, studied the relation between microbial contamination of ground water in the karst aquifer and septic-system density as well as other factors such as well depth, casing length, and land use.

Total coliforms are a broad group of bacteria including both fecal coliform and *E. coli* (American Public Health Association and others, 1989). Although used as potential indicators of sewage or agricultural contamination, total coliform bacteria are not considered the best indicator of potential bacterial contamination of ground- or surface-water supplies. Fecal bacteria such as *E. coli* and fecal coliform grow in the intestinal tracts of both humans and animals and survive at a rate similar to that of pathogenic disease-causing organisms. Fecal coliform and *E. coli* are better indicators of potential microbial contamination than is total coliform bacteria and will therefore be emphasized within this report.

Purpose and scope

This article presents results and analysis of bacteria data collected from 50 wells (Fig. 1) sampled within the limestone and dolomite karst aquifer of Berkeley County, West Virginia, and discusses the natural and human factors that affect the microbial quality of water derived from the aquifer, which is used by many home-owners as a drinking-water source. The 50 wells sampled were analyzed for the presence of total coliform, fecal coliform, and *E. coli* bacteria. Assessment of the effects of septic density on the microbial quality of ground water from the karst aquifer was the major objective of the study.

Description of the study area

Berkeley County encompasses 325 square miles of the Eastern Panhandle of West Virginia and drains entirely into the Potomac River (Fig. 1). The western half of the County is underlain predominantly by shale and sandstone with only minor outcrops of limestone. The eastern half of the County is in the Shenandoah Valley and is characterized by gently rolling topography with altitudes ranging from 310 to 800 feet above sea level (Shultz and others, 1995). It is underlain primarily by Cambrian and Ordovician age limestone and dolomite with a significant outcrop of shale of the Martinsburg Formation. The study area for this report is primarily the limestone and dolomite dominated karst region in the eastern portion of the County, although two of the 50 samples collected were obtained from wells completed in the Martinsburg Shale Formation.

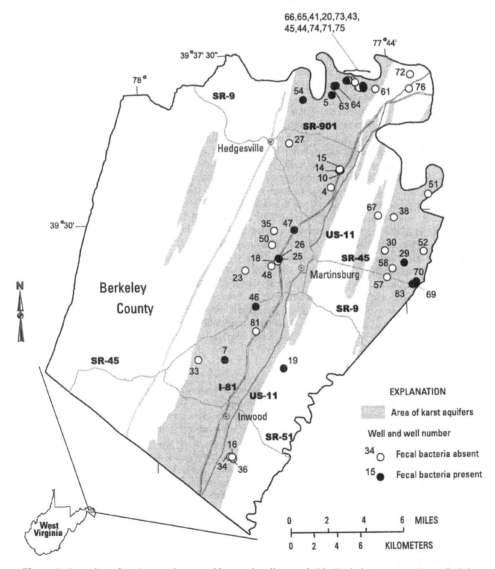

Figure 1. Location of study area, karst aquifers, and wells sampled in Berkeley county, West Virginia

Ground-water flow in the karst aquifer of the study area was believed to be predominantly diffuse, although local-scale conduit flow also occurs. Diffuse flow occurs where significant solution enlargement of fractures has not occurred and the flow is slow and laminar. Conduit flow usually occurs along faults or bedding planes, beneath losing streams, in cavernous areas, and where fractures have been enlarged by dissolution of limestone and dolomite. Conduits can range in size from less than a few inches to tens of feet in width, height, and length and are generally the main paths of flow, where present. Ground-water flow in conduits can be rapid but is much slower in diffuse-flow portions of the aquifer. In areas of conduit flow, bacteria-laden ground water can flow over distances in excess of hundreds or even thousands of feet.

Land use in Berkeley County is rapidly changing from agriculture to urban as subdivisions become more numerous. The population of Berkeley County increased 26.7 percent from 1980 to 1990 (U.S. Department of Commerce, 1991). Shultz and others (1995) report land-use statistics compiled in 1973 for Berkeley County as 46.5 percent forest, 31.4 percent pasture, 12.9 percent cropland, 4.7 percent urban or commercial, 3.6 percent orchard, 0.7 percent barren, and 0.2 percent water. Manure is applied to some of the agricultural land within the study area and is an additional source of bacteria to the karst aquifer.

STUDY METHODS

During the winter of 1999-2000, 83 candidate wells were located and identified for potential bacteria sampling. In March of 2000, the 83 candidate wells were visited by USGS personnel to obtain well-construction data, identify land use in the vicinity of candidate wells, evaluate adequacy of wells for sampling, and to select tentative wells for sampling. In June 2000, water samples were collected from 50 of the 83 wells (Fig. 1) and analyzed for indicator bacteria (Mathes, 2000).

Selection of sampling sites

Wells were selected on the basis of stringent land-use and well-construction criteria. Wells selected for sampling were typically domestic home-owner wells. Wells with adequate surface seals and grouted casings were given priority in sampling, and wells with rusty or deteriorating casings were not sampled. If a plumbing system from a well contained a chlorinator, ultraviolet light, or water softener, the plumbing system was carefully examined to assure that the water-treatment system could be easily isolated or bypassed for sampling.

Candidate wells were assigned to three categories of septic-system density – high density indicates four or more septic systems within the 5-acre circle, intermediate density indicates two or three septic systems, and low density indicates one or no septic systems. In addition to septic-system density, agricultural land use in the 5-acre circle was also determined, and included feedlots, gardens, pastures, and cropland where animal wastes may be applied as fertilizer.

Collection and analysis of samples

Ground-water samples for microbial analysis were collected according to standard USGS protocols (Britton and Greeson, 1988; Koterba and others, 1995). All wells were purged until measurements (made in a flow-through chamber) stabilized for water temperature, specific conductance, pH, reduction-oxidation potential, and dissolved oxygen. When the field measurements indicated that stable, fresh ground water was being withdrawn from the well, two sterile bottles were filled with fresh ground water for bacterial analysis. Bacteria concentrations for the water samples collected were determined by membrane filtration, and bacteria colonies were counted using methods described by Myers and Sylvester (1997).

Sample volumes of 20, 50, and 100 milliliters were filtered for analysis of indicator bacteria. Total coliform bacteria concentrations were determined by filtering sample water through a 0.45 µm (micron) membrane filter and placing the filter in a petri dish containing m-Endo culture media. Total coliform bacteria cultures were incubated for 24 hours at 35°C (Celsius). Colonies with the characteristic metallic green sheen were counted using a 10 X dissecting microscope or hand lens. Any membrane filter with total coliform growth was transferred to a separate petri dish containing NA-MUG media and incubated an additional four hours at 35°C. The cultures were examined under a 365 nanometer ultraviolet long-wave lamp. Colonies exhibiting a characteristic fluorescent halo were counted as *E. coli*. Fecal coliform bacteria concentrations were determined by filtering a separate sample through a 0.65 µm membrane filter and placing the filter in a petri dish containing m-FC culture media. Fecal coliform cultures were incubated for 24 hours at 44.5°C. Colonies with a characteristic blue color were counted using a 10 X dissecting microscope or hand lens.

Sterile dilution water was filtered through additional filters before (equipment blanks) and after (procedure blanks) environmental samples were processed. Fecal coliform bacteria were not detected in any blanks. Four procedure blanks contained fewer than seven colonies of total coliform bacteria. One equipment blank contained one colony of total coliform. These quality-assurance data did not indicate the need to adjust bacteria data collected from the 50 wells sampled.

MICROBIAL QUALITY OF GROUND WATER

The national standard for bacteria in drinking water is no detectable concentrations of total coliform bacteria in 95 percent of repeated samples (U.S. Environmental Protection Agency, 1986). Total coliform are widely present in surface and sub-surface environments. Their absence from treated drinking water suggests that treatment was effective in removing bacteria. Fecal coliform and *E. coli* are subsets of the total coliform group that grow in the intestines of humans and animals. Their survival in the environment is similar to that of pathogens. The presence of indicator bacteria in ground water is considered to be a potential indicator of microbial contamination (American Public Health Association and others, 1989). The occurrence and distribution of bacteria within the study area will be discussed in addition to the effects of septic density, land use, and well construction on the microbial quality of ground water within the karst aquifer system.

Occurrence and distribution of bacteria

Total coliform bacteria were detected in 62 percent of wells sampled, *E coli.* in 32 percent, and fecal coliform in 30 percent (Table 1). The high frequency of contamination of wells in the karst aquifer of Berkeley County suggests that the aquifer is contaminated by bacteria in many areas. Total coliform bacteria are ubiquitous, and their presence does not necessarily indicate contamination of ground water by fecal bacteria from septic systems or agricultural sources (Mathes, 2000). Therefore, total coliform bacteria concentrations will not be discussed extensively within this report. Fecal bacteria were detected in all areas of the karst aquifer but neither of the two wells sampled in shale bedrock contained fecal bacteria (Mathes, 2000).

Factors affecting the microbial quality of ground water

No statistical correlation could be determined between septic density and concentrations of fecal coliform or *E. coli* bacteria (Table 1). Approximately one-third of wells in each septic-density class contained fecal coliform or *E. coli* bacteria. The fact that bacteria concentrations in water samples did not correlate with the density of septic systems in the immediate vicinity of the wells sampled is surprising. The most likely explanation for this lack of correlation is related to ground-water flow, especially the dominance of conduit flow over diffuse flow. In the design stages of the project, it was believed that diffuse-flow processes were dominant within the karst aquifer, and conduit flow occurred only sporadically. It is plausible, however, that conduit flow within the aquifer occurs more commonly than previously thought. If conduit-flow processes are dominant or comprise a significant portion of flow within the aquifer, the basic assumption that septic density would correlate with bacteria within the aquifer would be invalid. Ground-water flow could easily occur over distances much greater than the 80-meter radius (5-acre circle) criteria used to select wells for sampling. In a previous investigation of ground-water flow in the karst aquifer of Berkeley County, ground-water flow velocity was determined to range from a minimum of about 32 feet per day in diffuse-flow portions of the aquifer to as much as 1,154 feet per day in conduit-dominated portions of the aquifer (Shultz and others, 1995). There would obviously be no correlation between septic density and bacteria concentrations within the aquifer if significant conduit flow occurs.

Table 1. Number and percentage of water samples with detections of total coliform, fecal coliform, or *E. coli* bacteria for wells sampled in a karst limestone and dolomite aquifer, Berkeley County, West Virginia.

Well classification by septic density, land use, or well construction	Number of wells sampled	-----Percent of wells with detections of bacteria-----			
		Total Coliform	Fecal Coliform	*E. coli*	Fecal coliform or *E. coli*
All wells sampled	50	62	30	32	42
Number of septic systems in 5-acres					
0-1	13	46	31	31	38
2-3	21	67	29	33	38
4-6	16	69	31	31	44
Potential agricultural land-use sources					
Yes	13	69	46	46	50
No	37	44	24	27	36
Well depth					
< 150 feet	16	69	56	44	63
150 to 300 feet	18	61	22	22	33
>300 feet	15	60	13	33	33
Casing Length					
≤ 60 feet	34	65	32	41	47
> 60 feet	10	50	10	20	20

For 13 of the 50 wells sampled additional potential sources of bacteria in the 5-acre circle were documented. These land-use related sources included gardens, pastures, and small feedlots. Although not statistically significant, wells with agricultural sources of bacteria had slightly higher detections of bacteria than did wells where agricultural sources were not present (table 1). Of wells sampled with nearby agricultural activity, 50 percent contained either fecal coliform or *E.coli* bacteria. Wells with no agricultural land use nearby had detectable concentrations in only 36 percent of wells sampled.

A statistical correlation was found between well depth and fecal coliform, total coliform, and *E. coli* bacteria at a 95-percent confidence level. Generally, deeper wells had lower concentrations of bacteria. As the majority of contaminated recharge occurs near the surface, shallower wells would normally be expected to have higher concentrations of bacteria than deeper wells. A statistically significant correlation between casing length and bacteria concentration was also present at a 95-percent confidence level. The data suggest that wells with longer casings are less likely to be contaminated with bacteria than are wells with shorter casings. Wells with longer casings can seal off near-surface zones of contaminated recharge. Longer casings however, may also reduce well yield by sealing off shallow water-producing zones. Although correlations between bacteria concentrations and well depth and casing length are evident, it must be noted that there are many exceptions to this general trend. These trends, however, emphasize the importance of proper casing installation and grouting to reduce the potential entry of contaminated surface water or shallow ground water.

For shallow wells less than or equal to 150 feet deep, there is a 63 percent chance of contamination with fecal bacteria (either fecal coliform or *E. coli*). Although wells deeper than 150 feet typically have less bacteria than shallower wells, there is still a 33 percent chance of contamination by fecal bacteria. For wells with casings 60 feet long or less, 47 percent contained fecal bacteria. For wells with casings longer than 60 feet, only 20 percent contained fecal bacteria.

Because bacteria is common in the karst aquifer, homeowners may wish to have their water tested for the presence of bacteria. A single water sample, however, may not be effective in detecting bacterial contamination. It is possible for a water sample collected from a well to test negative for the presence of bacteria and a following sample collected under a different ground-water-flow condition or during a different season of the year may test positive. The installation of chlorinators or ultraviolet light systems are an option for treating wells that are or may become contaminated with bacteria.

ACKNOWLEDGEMENTS

The authors thank Roger Boyer and Michael O'Donnell of the Potomac Headwaters Resource Conservation and Development Region, Inc. for overseeing and conducting a preliminary site analysis of 83 potential sampling sites and for obtaining permission to collect water samples, well-construction, and site-analysis data. Virginia Tabb of the U.S. Natural Resources Conservation Service is also acknowledged for her assistance in identifying potential sampling sites. Finally, appreciation is extended to the residents of Berkeley County who allowed access to their property, provided well-construction data, and allowed water samples to be collected from their wells.

REFERENCES

American Public Health Association, American Water Works Association, and Water Pollution Control Federation, 1989, Standard methods for the examination of water and wastewater (17th. Ed.): Washington, D.C., 1,391 p.

Britton L.J., and Greeson, P.E., 1988, Methods for collection and analysis of aquatic biological and microbiological samples: U.S. Geological Survey Open-File Report 88-190, p. 3-95.

Koterba, M.T., Wilde, F.D., and Lapham, W.W., 1995, Ground-water data collection protocols and procedures for the National Water-Quality Assessment Program – Collection and documentation of water-quality samples and related data: U.S. Geological Survey Open-File Report 95-399, 113 p.

Mathes, M.V., 2000, Relation of bacteria in limestone aquifers to septic systems in Berkeley County, West Virginia: U.S. Geological Survey Water Resources Investigation Report 00-4229, 12 p.

Myers, D.N., and Sylvester, M.A., 1997, Fecal indicator bacteria, in Myers, D.N., and Wilde, F.D., eds., Biological Indicators: U.S. Geological Survey Techniques of Water-Resources Investigations, Book 9, chapter A7.

Shultz, R.A., Hobba, W.A., and Kozar, M.D., 1995, Geohydrology, ground-water quality of Berkeley County, West Virginia, with emphasis on the carbonate-rock area: U.S. Geological Survey Water-Resources Investigations Report 93-4073, 88 p.

U.S. Department of Commerce, 1991, Comparison of 1980 and 1990 population data West Virginia and counties: U.S. Department of Commerce, Bureau of the census, Washington, D.C., 9 p.

U.S. Environmental Protection Agency, 1986, Quality criteria for water—1986: United States Environmental Protection Agency Office of Water, EPA 440/5-89-001 [variously paged].

Geotechnical and Environmental Applications of Karst Geology and Hydrology, Beck & Herring (eds)
© 2001 Taylor & Francis, ISBN 90 5809 190 2

Structural and karst impacts on groundwater flow and contaminant transport

EDWARD G.MILLER   Pape-Dawson Engineers, Inc., San Antonio, TX 78216, USA, emiller@pape-dawson.com

JOHN L.LUFKIN   Baer Engineering & Environmental Consulting, Inc., Austin, TX 78723, USA, baereng@aol.com

ABSTRACT

Structural and karst features controlled the groundwater flow and contaminant transport at a 900-gallon gasoline spill, which impacted 22 residential and commercial wells in a sole-source karst aquifer. Water wells for local residents and businesses in northern Bexar County, Texas were impacted when a car backed into a gasoline dispenser, resulting in a break of underground piping. Released gasoline moved vertically to the water table and within days the impacted groundwater migrated rapidly within karstification on the up-thrown side of a fault and became trapped on the down-thrown side within less karstified fractures. The structural and karst features controlled the contaminant transport and migration that resulted in the need for recovery efforts at three locations.

Six pressure transducers were installed to monitor groundwater depression in the area of two groundwater recovery and treatment systems. Data from the transducers showed that water levels in the area were subjected to pressure waves during rainfall events. Water levels rose as much as 25 to 30 feet when subjected to pressure waves and remained elevated for 9 days or less during these events. Distribution of contaminants after these events suggests that impacted groundwater was moved around within the fracture system/karst features by each event. This resulted in some contaminated groundwater being trapped in less transmissive areas on the down-thrown side of a fault.

GEOLOGIC SETTING

The geology of the site vicinity, known as Ram Store, has been characterized in regional studies of central Texas conducted by Giles, 1998; Ashworth, 1993; Hammond, 1994; and Stricklin, et al., 1971. A regional map of the area, prepared by the Bureau of Economic Geology, University of Texas, Austin, features surface exposures of dominantly Glen Rose and Edwards formations, which are cut by several faults of the Balcones Fault zone. This zone is comprised of several, northeast-southwest trending normal faults, downthrown to the southeast. One of these faults trends northeast through the site, from the vicinity of the release northeast toward Cibolo Creek at the Comal County line. Figure 1 shows the release site, the Spring Creek fault, and the impacted wells.

Field mapping traced the Spring Creek fault from about one-half mile southwest of the release site to about one-half mile northeast of the release site. The fault was also identified in geophysical logs for several wells, which confirmed a throw of about 50 feet. The fault and associated fractures trend N50E to N70E, and dip 70 to near 90 degrees to the southeast. Figure 2 presents a geologic cross-section that shows the subsurface geology, regional aquifers, location and depth of wells with geophysical logs and water levels.

This site is underlain by the Glen Rose formation, which is about 850-feet thick, and is subdivided into upper and lower members. The Glen Rose in turn is underlain by the Pearsall formation, which is subdivided into the Bexar Shale, the Cow Creek Limestone, and Hammett Shale. Collectively, these units comprise the Upper and Middle Trinity Aquifers, the sole source of groundwater for the area. The fuel release had impacted the saturated zone above and below the contact of the upper and lower Glen Rose.

The upper Glen Rose is about 500-feet thick and features bedded, gray to yellowish limestone, underlain by a more massive, gray limestone. A 5-feet thick, fossiliferous bed, know as the "Corbula bed", subdivides the two members of the Glen Rose. This Corbula bed contains abundant steinkerns of *Corbula harveyi*. Poor quality water, high in sulfates and iron, is often found associated with gypsum beds found in the basal portion of the upper Glen Rose. The lower Glen Rose, as much as 350-feet thick, is more massive and reef bearing. The lower zones of the lower Glen Rose typically produce higher yields and better quality water.

Numerous wells have penetrated the Glen Rose, in addition to the 22 water wells that were impacted by the release. Records for these wells indicate that the Glen Rose features numerous, but thin, scattered zones of solution, or karst features, including small vugs, enlarged fractures, and flowstone. As a result of the increased permeability, wells located near zones of karst development may yield several hundred gallons of water per minute (gpm), compared with rates of 20 gpm or less where solution features are not extensively developed.

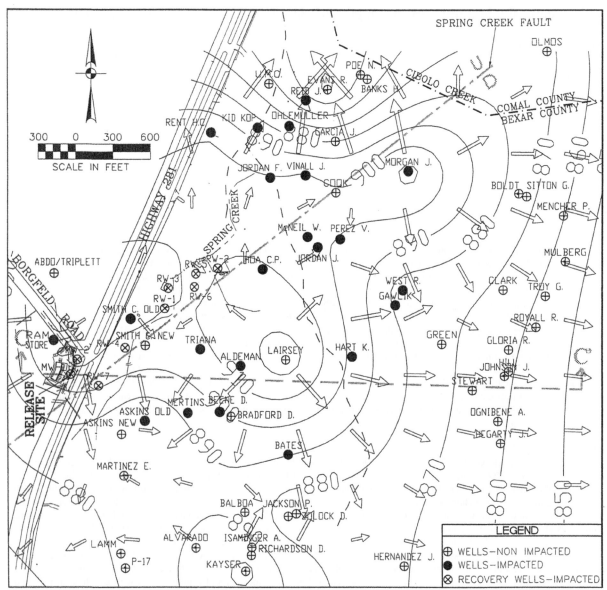

Figure 1: Site vicinity map for the release site and impacted area in northern Bexar County, Texas, showing Spring Creek fault, groundwater contours for October 2, 2000, non-impacted wells, impacted wells, and recovery wells. Water treatment systems (WTS-1 & 2) were set up in the vicinity of MW-2 and C. Smith's old well. Groundwater was also pumped from the Mertins' well to WTS-2.

HYDROGEOLOGY AND CONTAMINANT MIGRATION
Groundwater Occurrence and Wells

Groundwater in the site vicinity occurs within the upper, middle and lower Trinity aquifers. The upper Trinity aquifer consists of the upper member of the Glen Rose formation. The middle Trinity aquifer includes the lower member of the Glen Rose, Bexar Shale, and Cow Creek stratigraphic units. The lower Trinity aquifer is comprised of the Sligo and Hosston stratigraphic units as shown on Figure 2. Depending upon faulting and relative elevations of the various geologic units, the water table may occur in either the upper Glen Rose or the lower Glen Rose members. The geologic cross section for the site vicinity presented on Figure 2 shows the relative elevations for the aquifers, stratigraphic units, and water levels. The location and orientation for the geologic cross section C-C' are shown on Figure 1. Most domestic wells in the site vicinity are producing from the base of the upper Glen Rose and/ or the upper portion of the lower Glen Rose. The vast majority of wells are open across the gypsum beds found in the basal portion of the upper Glen Rose and, consequently, are high in sulfates and iron.

Subsurface geologic conditions shown on the cross section were determined from geophysical logs run on each of the wells shown. Gamma, spontaneous potential (SP), single point resistance (SPR), and caliper logs were run on most of these wells. Gamma logs were used to identify the geologic units, and SP and SPR useful in picking geologic contacts and water levels. Caliper logs were run to locate the depth of casing. The Spring Creek fault, as shown on Figures 1 and 2, was confirmed in the geophysical logs of Recovery Well No. 7 (RW-7) and Mr. C. Smith's new well.

Of the 22 wells impacted, 17 wells are known to be cased to depths above groundwater. Seven additional wells were expected to be impacted based on their location within the plume. Of the 7 wells expected to be impacted, one is reported to be cased above the water table and three are reported to be cased to elevations below the water table. This suggests that a well cased to below the water table, within the Trinity aquifer system in the site vicinity, is less subject to being impacted by an incident similar to the one that occurred at the Ram Store.

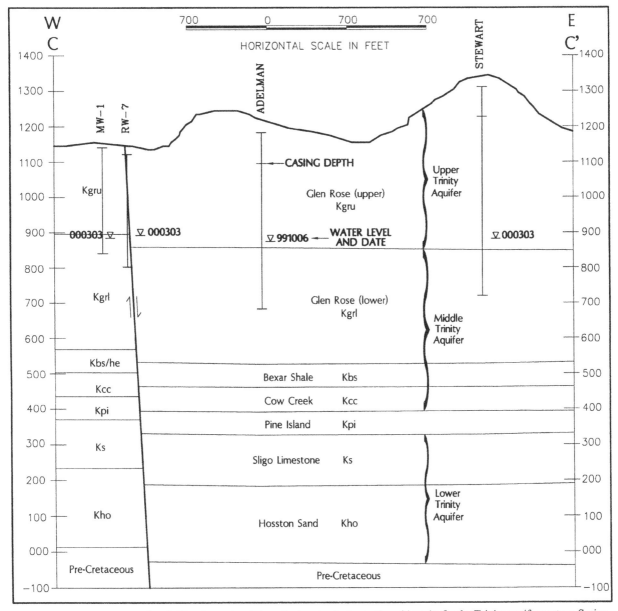

Figure 2: Geologic cross-section C-C' as located on Figure 1, showing hydrostratigraphic units for the Trinity aquifer system, Spring Creek fault with relative displacement, well information, and water level data. Dates are reported in reverse order, (yymmdd).

Groundwater Monitoring and Flow Patterns

Groundwater contour maps were prepared to predict the direction of groundwater flow and to monitor depressions in the water table caused by pumping at Water Treatment System No. 1 (WTS-1) and Water Treatment System No. 2 (WTS-2). WTS-1 included several vapor recovery wells and monitor wells MW-1 and MW-2. WTS-2 incorporated recovery wells RW-1, RW-3, RW-4 and Mr.

C. Smith's old domestic well. Groundwater contours and flow vectors for a typical monitoring event, without the recovery systems in operation, are shown on Figure 1.

Groundwater flow vectors on the groundwater contour maps represent the implied direction of groundwater flow at a point in time, as determined by water level measurements conducted in wells that may have been influenced by pumping, and natural conditions of recharge and discharge. In a limestone aquifer, such as the middle Trinity aquifer depicted on the groundwater contour maps, groundwater flow is influenced by structural and karst conditions. Therefore, groundwater flow may be in a circuitous path via fractures, faults and karst features, in a general down-gradient direction and may not follow the implied pathways represented by groundwater contours or flow vectors. Furthermore, the flow-directions shown on groundwater contour maps represent a single "snapshot" in time, and are subject to change almost on a daily basis, depending upon rates of precipitation, recharge, pumping, discharge, and the inhomogeneities of a fractured limestone aquifer with karst characteristics.

Groundwater contour maps for the site vicinity were prepared monthly during WTS operation. The contour maps showed a groundwater high in the vicinity of US Hwy 281 and Borgfeld Road and the groundwater gradient falling off steeply to the northeast. This suggests that recharge to the area is occurring in the vicinity of US Hwy 281 and Borgfeld Road, with flow primarily to the northeast parallel to Spring Creek fault. The sequence and pattern of impacted wells also support groundwater flow to the northeast. The groundwater high in the vicinity of US Hwy 281 and Borgfeld Road suggests that water is flowing into the system from the northwest. Fracture traces transverse to the Spring Creek fault can be seen on aerial photos of the area, and the alignment of Cibolo Creek conforms to these fracture traces in a northwest-southeast direction at two locations, one upstream of US Hwy 281 to the west and one downstream to the east.

If groundwater flows in the subsurface along the alignment of Cibolo Creek within the middle Trinity aquifer as suspected by the USGS, then it is possible that the site vicinity is recharged by this system to some degree year-round. Pressure transducer readings presented on Figure 3 show a steady increase in water levels from the beginning of October 1999 through the beginning of February 2000, even though there was only one minor rainfall event in mid October 1999 and the region was experiencing a drought. Other wells measured throughout the study area also showed an increasing trend for the same period, and springs along Spring Creek also began to flow later in the period.

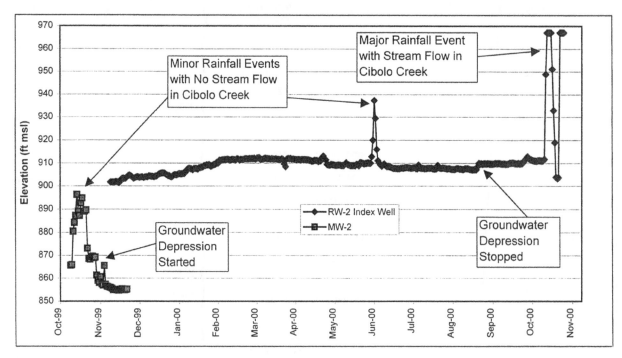

Figure 3: Daily pressure transducer readings showing water level trends and impact of minor and major rainfall events.

There are four causes suspected for the increased water levels and spring flows. They are, 1) more water is entering the system than being withdrawn from the system; 2) recharge is occurring along transverse fractures with inflow from the northwest; 3) the treated discharge water from WTS-1 and WTS-2 is recharging the system downstream along Spring Creek; and 4) septic tank/drainfield effluent is primarily going to recharge during the winter months, and less to evapotranspiration.

Release Distribution and Plume Migration

In general, the release that occurred at the Ram Store appears to have initially migrated northeast under US Hwy 281 along fractures or karst features, parallel to Spring Creek fault to the vicinity of C. Smith's old well and beyond. The release was also drawn east to Askins' well located on the downthrown side of the fault. When evaluating the sequence in which wells were initially impacted, it can be seen that wells on the up-thrown side of the fault along the fractures or karst features parallel to the fault, were rapidly impacted in sequential order from southwest to northeast. This suggests that the fractures or karst features parallel to the fault are very transmissive, at least to the vicinity of Vinall's well where the groundwater gradient steepens dramatically.

The release did not move as rapidly or sequentially to the northeast on the downthrown side of the fault. This suggests that fractures may be tighter and less transmissive on the downthrown side than the up-thrown side of the fault. It also suggested that groundwater movement follows more circuitous paths and that it may take longer to reach non-detect levels of BTEX contamination on the downthrown side. Three wells consistently remained impacted on the downthrown side of the fault in the vicinity of the wells that were initially impacted. This suggests that some residual of the release has remained in the vicinity of the Adelman, Mertens and Triana wells.

Further support for this is suggested by the sequence of impact on the Adelman, Hart, Gawlik and West wells, which occurred after a two (2) inch rainfall event on October 16, 1999. The impact of the rainfall event on groundwater levels was recorded by a pressure transducer located in MW-2 as shown on Figure 3. These readings show that the water levels went up approximately 25 feet between October 16 and October 17, and stayed up until October 28 when they returned to near the October 16 level. This suggests that a wave-like event occurred that pushed the release to the southeast along transverse fractures, trapping some of the release behind parallel fractures, and the release then migrating northeast along the parallel fractures.

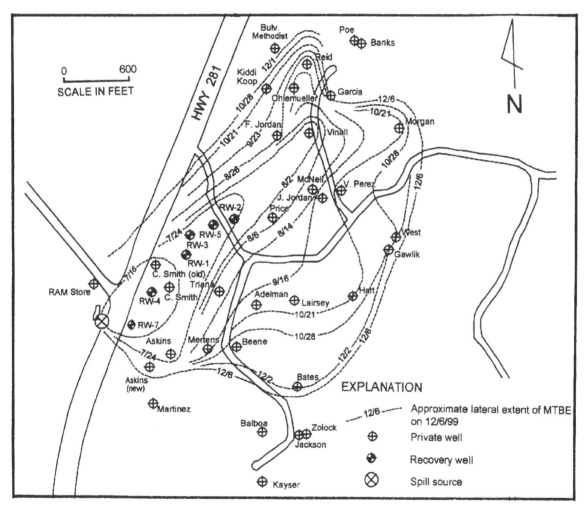

Figure 4: Chronological advance of MTBE plume, showing spill source and sequence of impacted wells.

This is also confirmed somewhat by the groundwater contour map flow vectors and sequence of impact on the Bates and Beene wells. The flow vectors and well impact sequence for October 21, November 23, and December 20, 1999 suggests that the release was pushed to the southeast and split going first to the northeast and then to a lesser degree to the southwest.

The general northeast flow pattern of groundwater is supported also by the chronological advance of the MTBE plume, shown in Figure 4. Weekly measurements of MTBE concentrations in groundwater suggest that the plume (as defined by concentrations exceeding 2 micrograms per liter MTBE) had migrated systematically down-gradient on the up-thrown side of the fault to the northeast, largely in a linear fashion, reaching the Vinall well by early August 1999. Following this period, and due to several factors discussed above, the plume continued northward, but also began spreading more easterly, affecting the wells of Hart and Morgan by October 21, 1999. By early December 1999, the plume reached its maximum size, affecting the wells of Beene, Bates, and Gawlik.

Groundwater Depressions

Groundwater depressions were created by recovery efforts at both WTS-1 and WTS-2. Groundwater depression can first be seen on the groundwater contour map for October 21, 1999, where the WTS-2 cone of depression is well developed, and by November 23,1999, in the area of WTS-1. The WTS-1 cone of depression and WTS-2 cone of depression were monitored by up to six pressure transducers so that predictions could be made with regard to how each cone of depression would react to re-injection of clean water in the vicinity of the release. This was planned for flushing of product that may have been trapped in the vadose zone. The behavior of the cone of depression at WTS-1 and WTS-2 was critical to understanding how groundwater impacted by re-injection would behave, and to be sure that any impacted groundwater would be recovered.

Re-injected water was successful in flushing some product from the vadose zone. Higher concentrations of MTBE were found in RW-4 within hours after re-injection. This further supports high transmissivity parallel to the Spring Creek fault on the up-thrown side, and slower groundwater movement on the down-thrown side of the fault.

CONCLUSIONS

Karst development along fractures on the up-thrown side and parallel to Spring Creek fault allowed a gasoline release at the RAM Store to migrate rapidly down-gradient to the northeast. Groundwater movement and contaminant migration on the down-thrown side of Spring Creek fault was much slower, suggesting less karst development on the down-thrown side. Well construction also played an important role in the number of wells impacted. Shallow casing depths on the order of 100-feet contributed to wells being impacted where the depth to water was on the order of 200 to 300 feet.

Other observations also suggest that State of Texas residential well completion standards are inadequate for the Trinity aquifer system. Existing well construction standards require minimal lengths of casing and cementing to only 10 feet. Seepage entering the Stewart well from around the outside of the casing at 100 feet, after months of drought conditions, suggests a water source other than precipitation. Water entering around the outside of the bottom of the casing was observed during a TV survey of the well. The well as shown on Figure 2 is located on top of a hill. Given the drought and no other water source on the hill, the suspected source is septic tank/drainfield effluent, which appears to be primarily going to recharge by migrating to the water table in the annular space between the casing and borehole wall.

ACKNOWLEDGEMENTS

The authors would like to acknowledge Ram Stores, Inc. for their permission to publish this paper and the Texas Natural Resource Conservation Commission for their cooperation and support in remediation of the site.

REFERENCES

Ashworth, J.B., 1983, Groundwater availability of the Lower Cretaceous Formations in the Hill Country of south-central Texas: Texas Department of Water Resources Report 273.

Barnes, V.E., 1983, Geologic Atlas of Texas, San Antonio Sheet: Austin, TX, University of Texas, Bureau of Economic Geology, scale 1:250,000.

Giles, Michael, 1988, Groundwater of the Glen Rose, M. S. Thesis, The University of Texas at San Antonio.

Hammond, Weldon W., 1984, Hydrogeology of the Lower Glen Rose Aquifer, south-central Texas, Ph. D. Dissertation, The University of Texas at Austin.

Stricklin, F.L., Jr., Smith, C.I., Lozo, F.E., 1971, Stratigraphy of the Lower Cretaceous Trinity Deposits of Central Texas, Bureau of Economic Geology, Report of Investigations No. 71.

Geotechnical and Environmental Applications of Karst Geology and Hydrology, Beck & Herring (eds)
© 2001 Taylor & Francis, ISBN 90 5809 190 2

Mapping karst ground-water basins in the Inner Bluegrass as a nonpoint-source pollution management tool

RANDALL L.PAYLOR & JAMES C.CURRENS   Kentucky Geological Survey, University of Kentucky, Lexington, KY 40506, USA, rpaylor@kgs.mm.uky.edu, currens@kgs.mm.uky.edu

ABSTRACT

The ground water in karst aquifers of Kentucky's Inner Blue Grass Region is regularly exposed to nonpoint-source pollution from agriculture and urbanization. The karst aquifers of the Inner Blue Grass are developed in Middle Ordovician carbonates that cover approximately 3,800 km² (1,500 mi²). The Kentucky Division of Water implements a program to mitigate nonpoint-source pollution, which is funded through an EPA Section 319(h) grant. As part of this program, the Kentucky Geological Survey is conducting a 3-year project to delineate karst ground-water basins in the region.

Ground-water basin boundaries are being determined by dye tracing, topographic analysis, normalized base-flow measurements, and sinkhole lineament analysis. Thirty-one dye traces have been completed, and approximately 140 more are planned. The completed dye traces support past research that indicates karst ground-water flow in the Blue Grass is primarily controlled by fractures and faults. Analysis of normalized base flow has revealed significant increases in base discharge from urbanized ground-water basins, increasing the potential for nonpoint-source pollution in those basins. The results of this project will be used to target the basins of degraded springs in which to implement best management practices, and to update publicly available 1:100,000-scale karst basin maps.

INTRODUCTION

Nonpoint-source (NPS) pollution is the primary contributor to water pollution in Kentucky, accounting for approximately two-thirds of the water-quality impairments in Kentucky streams and lakes (Ky. Division of Water, 2000). The Commonwealth of Kentucky implements the NPS Pollution Management Program as required by Section 319 of the Federal Clean Water Act, and the Kentucky Division of Water (DOW) is the lead oversight agency for the program. DOW conducts the program primarily through the activities of cooperating agencies, institutions, and organizations. Through a Memorandum of Agreement with the DOW, the Kentucky Geological Survey (KGS) is mapping karst ground-water basins in the Inner Blue Grass Region as part of the program. The primary purpose of the 3-year project is to provide a better balance to the NPS pollution-control program by allowing the watersheds of degraded springs to be targeted for best management practices (BMP's), rather than spreading limited funding over large areas. The majority of the project will be accomplished by conducting ground-water dye tracing and karst inventory field work throughout the region.

PROJECT OBJECTIVES

The NPS project goals are to attempt approximately 170 dye traces and delineate at least 20 new karst ground-water basins over the duration of the project. Ancillary data for undocumented springs will also be gathered for inclusion in the statewide spring database. KGS will produce updates to both the Lexington and Harrodsburg 30 x 60 minute karst ground-water basin maps, which are currently available to the public. Geographic information system (GIS) coverages of the updated trace routes and basin boundaries will be made available to the public on the KGS web site. KGS will also conduct two karst hydrology workshops designed to introduce local officials to the fundamentals of karst hydrology and instruct them in the use of the new maps.

KARST SETTING

The karst aquifers of the Inner Blue Grass Region occur in an area of approximately 3,800 km² (1,500 mi²) in central Kentucky (Figure 1). Parts of the Licking River, Kentucky River, and Salt River basins are located in the Inner Blue Grass. The region is largely a gently rolling upland at an average elevation of 250 m (820 ft) with generally less than 45 m (150 ft) of local relief. Most of the streams that drain the region are on the upland, but the Kentucky River has incised a gorge more than 90 m (300 ft) deep. Although the streams on the upland surface appear to provide normal surface drainage, numerous karst landforms are present, especially sinkholes. Locally, areas in excess of 8 km² (3 mi²) have no surface drainage.

Middle to Upper Ordovician rocks have been exposed in the Inner Blue Grass by erosion on the crest of the regional Cincinnati Arch. Regional dip is gentle, and beds appear nearly horizontal in outcrop. The boundaries of the region coincide with depositional or

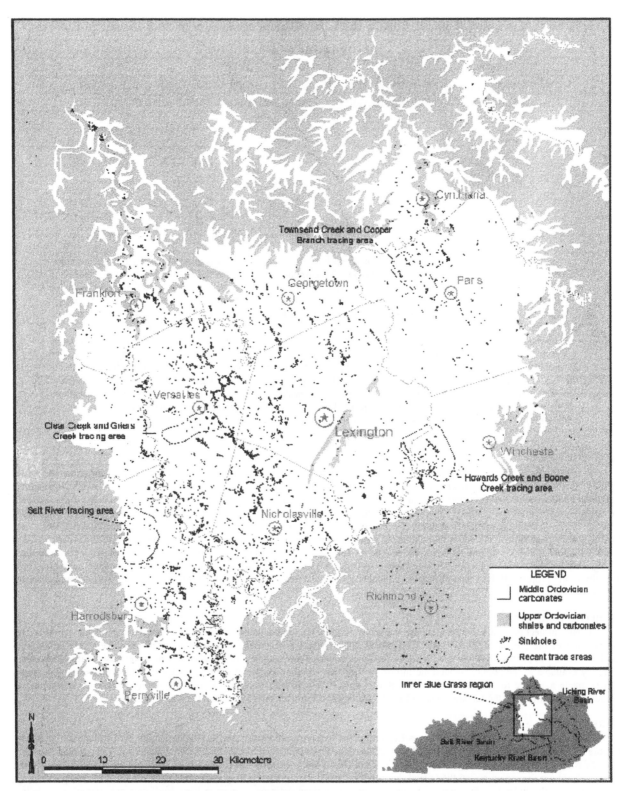

Figure 1: The Inner Blue Grass Karst region of central Kentucky showing area of Middle Ordovician carbonates and distribution of sinkholes (derived from 7.5-minute topographic quadrangle maps).

fault contact of relatively pure carbonates with the overlying, thinly interbedded Upper Ordovician limestones and shales (McDowell and others, 1988).

The region's major karst-forming units are Middle Ordovician in age. The lowest formations of the Middle Ordovician carbonates are exposed only in the gorge of the Kentucky River and its incised tributaries. Exposed over a much larger area, the Lexington Limestone is the most significant karst-forming unit of the region. Less developed karst occurs in the thinner-bedded and more argillaceous Upper Ordovician units at the outer edges of the region. Structural elements such as faults and fractures contribute largely to the pattern of karst development in the region, creating significant sinkhole alignments and ground-water flow routes.

PREVIOUS INVESTIGATIONS

A number of previous dye-tracing studies have been conducted in the Inner Blue Grass. From 1976 to 1981, Dr. John Thrailkill of the University of Kentucky conducted the most comprehensive trace study of the area. A total of 96 successful traces were accomplished and the results published through the Kentucky Water Resources Research Institute (Thrailkill and others, 1982). Area-specific trace investigations were conducted for graduate theses in Woodford County (McCann, 1978), Fayette and Scott Counties (Spangler, 1982), and Mercer and Boyle Counties (Hopper, 1985).

Later investigations were conducted for sinkhole flooding issues (Currens and Graham, 1993), environmental problems (Ewers and others, 1991; Keagy and others, 1994a,b; Felton and others, 1995), and karst ground-water fracture flow (Taylor, 1992; Elvrum, 1994). Other formerly unpublished trace data were documented from personal communications and included with all previous studies on the Lexington and Harrodsburg 30 x 60 minute karst ground-water basin maps by Currens and Ray (1996, 1998).

METHODS

The objectives of the tracing project have been pursued through a four-step process. The process includes field investigation and inventory, dye injection and monitoring, spring discharge and water quality measurements, and map and database construction. Initial areas of investigation were chosen based on previous dye-trace results, level of karst development, and priority of ground-water issues. The karst basin areas being investigated concurrently are also broadly separated to avoid risks of cross contamination between basins.

A major effort of the project has been field reconnaissance. Spring and swallet locations in the Inner Blue Grass can be unpredictable. Landowners have provided the most detailed information and saved much time that otherwise would have been spent searching large areas. Boat trips along regional base-level streams have also been effective at locating large karst springs. A number of springs have been successfully located by following sinkhole alignments to the nearest perennial surface stream. All undocumented springs that are visited are recorded on detailed DOW spring inventory forms, and all other pertinent karst features are documented and marked on 7.5-minute topographic quadrangle maps or located with handheld GPS receivers.

The dye-tracing techniques used closely follow those outlined by Quinlan and Rowe (1977) and Quinlan and Ewers (1981), with modifications by Thrailkill and others (1982). The details of these techniques are well documented and will not be repeated here. The main modification from Thrailkill is the use of unbleached cotton fabric instead of surgical cotton. The fabric has been found to be less susceptible to contamination and loss, and is just as effective at capturing dyes. Five different dyes have been used in the project, including optical brightener (tinopal CBS-X), direct yellow 96, rhodamine WT, fluorescein, and eosine. All five dyes have been used simultaneously in certain basins, with successful results. In general, optical brightener and direct yellow were used for the shorter traces to avoid the higher risk of loss by clay and organic adsorption. The activated charcoal receptors used to capture the rhodamine WT, fluorescein, and eosine dyes are analyzed on either a fluorometer or a spectrofluorophotometer, depending on the length of the trace and the amounts of dye injected.

Natural inflow to sinkholes and swallets has been relied upon for most of the traces conducted so far, although a few induced traces were conducted in sinkholes near possible ground-water divides or during dry conditions. In addition to the tracing, spring discharge measurements have been conducted at numerous springs in the Blue Grass in an attempt to relate spring base-flow discharge to basin size. If a correlation can be made, normalized base flow can be a useful tool in determining where to concentrate dye-injection efforts.

The field inventory data and dye-trace results are entered into Access databases and ArcView GIS coverages. This information is then used to construct updates of the publicly available karst ground-water basin map series, available through the KGS at 1:100,000 scale. The GIS coverages are made publicly available through the KGS Web site. To increase the detail and usefulness of the public karst maps, KGS is updating current data and GIS coverages at a scale of 1:24,000.

TRACING RESULTS

As of November 2000, 31 dye injections have been completed, and approximately 140 are still planned over the remaining period of the project. Twenty-nine of the traces were successfully retrieved, one trace had results that were too slight to call positive, and one trace was never recovered. Modifications of formulas by Aley and Fletcher (1976) were used to determine dye injection amounts, and concentrations of recovered dyes remained below the visible threshold in surface water for all dyes and all traces, but were well recovered by the charcoal and cotton receptors.

The completed traces have been concentrated in four areas throughout the region (Figure 1). Seven traces were completed in the Salt River Basin in Mercer County, all to springs located directly on the river. Seven traces were completed in the Clear Creek and Griers Creek areas of the Kentucky River Basin in Woodford County. Five traces were completed in the Townsend Creek and Cooper Branch areas of Bourbon County in the Licking River Basin. Ten traces were completed in the Howards Creek and Boone Creek areas of the Kentucky River Basin in southern Clark and Fayette Counties.

The average length of the traces for the current project is 2.3 km (1.4 mi). The longest trace was 5.3 km (3.3 mi) along a partially mapped fault in southern Clark County. Most injections were recovered within the first week, but a few took 2 or 3 weeks to appear, partially because of dry conditions. Overall, successful traces appeared at 18 separate springs, 13 of which were not fully documented before the project. Two of the traces went to distributary springs, where the dye appeared almost simultaneously at two widely separated outputs in the same basin.

Background contamination that could mask the dyes was minimal. In the more urbanized areas, especially in Woodford County near Versailles, the background of optical brightener was moderately high. Background contamination strong enough to mask fluorescein was present in one spring in Woodford County. No significant background was detected for rhodamine WT, eosine, or direct yellow at any of the springs investigated.

GROUND-WATER BASIN DELINEATION

The determination of a dye-trace vector is only the first step for delineating ground-water basin boundaries in the Inner Blue Grass. The important and well-established practice of concurrently drawing a potentiometric surface in the traced basins (Quinlan and Rowe, 1977) is unfortunately beyond the resources of this project. In the Inner Blue Grass, however, ground-water basin boundaries are coincident with topographic divides in many cases. Ground-water flow near divides is shallow, frequently because of the perching effect of argillaceous layers interbedded with the carbonates (Thrailkill and others, 1982). Ground water in these perched zones resurfaces at ephemeral springs along hill flanks, but then sinks again as flow approaches the interior of the ground-water basin into tributary conduit systems. Therefore, many Inner Blue Grass karst ground-water basin boundaries may be drawn with some confidence using topographic divides.

In a significant minority of the cases, however, major fracture systems (faults and linearly continuous joints) promote conduit development, which routes ground-water flow beneath major topographic divides (Thrailkill and others, 1982; Taylor, 1992; Elvrum, 1994). These authors considered regional shear forces the principal cause of the fracturing. Typically, a larger number of traces are needed to define a ground-water basin influenced by lineaments. Lineament analysis is therefore a second important step for both planning ground-water dye-tracing experiments and interpreting the results.

Another step that can help delineate a ground-water basin is measurement of spring base-flow discharge. Quinlan and Ray (1995) conducted work in several karst basins that showed normalized spring base-flow measurements have an almost flat trend when compared to basin size. They calculated normalized base flow for three springs in the Inner Blue Grass, but these springs did not show the flat trend that other karst basins did. Subsequently, KGS personnel have measured base flow at 14 Blue Grass springs that have well-delineated basin boundaries. The results of the measurements showed a flat trend among many of the springs, with the normalized base flow consistent at approximately 0.33 L/s/km^2 (0.03 ft^3/s/mi^2) (Figure 2). A significant number of springs were off the trend, however, with some springs having normalized base flows as high as 4.15 L/s/km^2 (0.38 ft^3/s/mi^2).

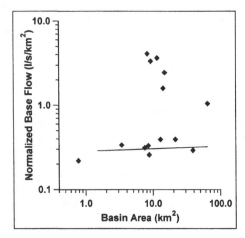 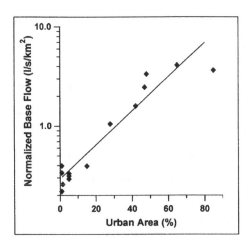

Figure 2: Inner Blue Grass normalized spring base flows as a function of basin area, showing flat trend of subset.

Figure 3: Inner Blue Grass normalized spring base flows as a function of urbanized area, showing strong positive correlation ($r^2 = 0.91$).

All of the higher normalized base-flow measurements appeared to be in more urbanized areas of the region. To test this hypothesis, the measurements were plotted against percent of urbanized area within each basin (Figure 3). The urbanized areas were derived from a Kentucky Office of GIS land-use coverage and modified slightly for recent development. The plot shows a significant relationship between urban development and increased base flow. The cause of the increased base flow is unknown, but leakage from wastewater and water-supply systems is one likely culprit, pointing to the stronger possibility of NPS pollution at these springs. Whatever the cause, the urban trend of increased normalized flow can be used to adjust base-flow measurements at other springs in the region to give a better idea of ground-water basin size.

CONCLUSIONS

Nonpoint-source pollution issues in the Inner Blue Grass can only be fully addressed by defining the karst ground-water basins of the region. To this end, the Kentucky Geological Survey has begun a 3-year program to study and delineate Inner Blue Grass basins. In addition to the scientific study, KGS is also conducting educational workshops and distributing karst basin maps to raise awareness of karst and NPS issues in the region.

KGS hopes to conduct approximately 170 dye traces over the 3-year period, and will delineate at least 20 new karst ground-water basins. Current trace results have defined karst flow to at least 18 springs, most of which are in previously uninvestigated basins. To assist in basin delineation, techniques in addition to traditional dye tracing have been used. Fracture and sinkhole lineament analysis, normalized base-flow calculations, and water-quality analysis are some of the other methods that have assisted in defining karst basins in the Inner Blue Grass. At the conclusion of the project in 2003, Kentucky will have a much more comprehensive understanding of karst flow in the Inner Blue Grass and will be able to address regional NPS issues more effectively.

REFERENCES

Aley, T., and Fletcher, M.W., 1976, The water tracer's cookbook: Missouri Speleology, v. 16, no. 6, 36 p.

Currens, J.C., and Graham, C.D.R, 1993, Flooding of the Sinking Creek karst area in Jessamine and Woodford Counties, Kentucky: Kentucky Geological Survey, ser. 11, Report of Investigations 7, 33 p.

Currens, J.C., and Ray, J.A., 1996, Mapped karst ground-water basins in the Lexington 30' x 60' quadrangle: Kentucky Geological Survey, ser. 11, Map and Chart Series 10, scale 1:100,000.

Currens, J.C., and Ray, J.A., 1998, Mapped karst ground-water basins in the Harrodsburg 30' x 60' quadrangle: Kentucky Geological Survey, ser. 11, Map and Chart Series 16, scale 1:100,000.

Currens, J.C., and Ray, J.A, 1999, Karst atlas for Kentucky, in Beck, B.F., Pettit, A.J., and Herring, J.G., eds., Hydrogeology and Engineering Geology of Sinkholes and Karst – 1999, Proceedings of the 7th Multidisciplinary Conference on Sinkholes and the Engineering and Environmental Impacts of Karst, April 10 – 14, 1999, Harrisburg-Hershey, Penn.: Rotterdam, A.A. Balkema, p. 85–90.

Elvrum, C.D., 1994, Relationship of fracture traces and sinkholes to stratigraphy and ground water in the inner Blue Grass karst region, Kentucky: Lexington, University of Kentucky, master's thesis, 139 p.

Ewers, R.O., Onda, A.J., Estes, E.K., Idstein, P.J., and Johnson, K.M., 1991, The transmission of light hydrocarbon contaminants in limestone (karst) aquifers, in Proceedings of the Third Conference on Hydrogeology, Ecology, Monitoring and Management of Ground Water in Karst Terranes: U.S. Environmental Protection Agency and Association of Ground-Water Scientists and Engineers, December 4–6, Nashville, Tenn., p. 287–306.

Felton, G.K., Sendlein, L.V.A, Dowdy, T., and Hines, D., 1995, Ground-water study of the Toyota Motor Manufacturing USA plant, Georgetown, Kentucky: University of Kentucky, Kentucky Water Resources Research Institute, Report 194, 80 p.

Hopper, W.M., Jr., 1985, Karst hydrogeology of southeastern Mercer County and northeastern Boyle County, Kentucky: Lexington, University of Kentucky, master's thesis, 122 p.

Keagy, D.M., Dinger, J.S., Fogle, A.W., and Sendlein, L.V.A., 1994a, Interim report on the effect of pesticides, nitrate, and bacteria on ground-water quality in a karst terrain – The Inner Blue Grass Region, Woodford County, Kentucky: Kentucky Geological Survey, Open-File Report OF-93-04, 46 p.

Keagy, D.M., Dinger, J.S., Hampson, S.K., and Sendlein, L.V.A., 1994b, Interim report on the effect of fractures on the quantity of ground water and the occurrence of pesticides and nutrients in the epikarst of the Inner Blue Grass Region, Bourbon County, Kentucky: Kentucky Geological Survey, Open-File Report OF-93-05, 31 p.

Kentucky Division of Water Nonpoint-Source Section, 2000, Guidelines for developing a competitive nonpoint-source project: Frankfort, Department for Environmental Protection, 96 p.

McCann, M.R., 1978, Hydrogeology of northeast Woodford County, Kentucky: Lexington, University of Kentucky, master's thesis, 101 p.

McDowell, R.C., Grabowski, G.J., and Moore, S.L., 1988, Geologic Map of Kentucky: U.S. Geological Survey, scale 1:250,000.

Mull, D.S., 1993, Use of dye tracing to determine the direction of ground-water flow in karst terrane at the Kentucky State University Research Farm near Frankfort, Kentucky: U.S. Geological Survey Water-Resources Investigations Report 93-4063, 21 p.

Quinlan, J.F., and Ewers R.O., 1981, Hydrogeology of the Mammoth Cave region, Kentucky: Geological Society of America, Field Trip Guidebook p. 457–506.

Quinlan, J.F., and Ray, J.A., 1981, Ground-water basins in the Mammoth Cave region, Kentucky, showing springs, major caves, flow routes, and potentiometric surface: Friends of the Karst, Occasional Publication 1, scale 1:138,000.

Quinlan, J.F., and Ray, J.A., 1995, Normalized base-flow discharge of ground-water basins: A useful parameter for estimating recharge areas of springs and for recognizing drainage anomalies in karst terranes, *in* Beck, B.F., ed., Karst geohazards: Proceedings, 5th Multidisciplinary Conference on Sinkholes and the Engineering and Environmental Impacts of Karst, Gatlinburg, Tenn., p. 149–164.

Quinlan, J.F., and Rowe, D.R., 1977, Hydrology and water quality in the central Kentucky karst: Phase I: University of Kentucky, Kentucky Water Resources Research Institute, Research Report 101, 93 p.

Spangler, L.E., 1982, Karst hydrogeology of northern Fayette and southern Scott Counties, Kentucky: Lexington, University of Kentucky, master's thesis, 102 p.

Taylor, C.J., 1992, Ground water occurrence and movement associated with sinkhole alignments in the inner Blue Grass karst region of central Kentucky: Lexington, University of Kentucky, master's thesis, 112 p.

Thrailkill, J., Dinger, J.S., Scanlon, B.R., and Kipp, J.A., 1985, Drainage determination for the Boone National Guard Center, Franklin County, Kentucky: Kentucky Geological Survey, University of Kentucky, contract report to Kentucky National Guard, 24 p.

Thrailkill, J, Spangler, L.E., Hopper, W.M., Jr., McCann, M.R., Troester, J.W., and Gouzie, D.R., 1982, Ground water in the Inner Bluegrass karst region, Kentucky: University of Kentucky, Kentucky Water Resources Research Institute, Research Report 136, 108 p.

Geotechnical and Environmental Applications of Karst Geology and Hydrology, Beck & Herring (eds)
© *2001 Taylor & Francis, ISBN 90 5809 190 2*

Spatial interpretation of karst drainage basins

JOSEPH A.RAY Kentucky Division of Water, Frankfort, KY 40601, USA, joe.ray@mail.state.ky.us

ABSTRACT

This paper proposes standardized methods to interpret and illustrate karst drainage basins. Dissemination of karst-drainage information is more efficient when lucid karst maps are widely used by karst investigators and educators. Kentucky's Karst Atlas map series is referenced as a model for providing important spatial information to professionals and the public about karst aquifers in Kentucky.

Karst features such as sinkholes, caves, springs, streams, marshes, and intermittent lakes are discussed. Interpretation and illustration of inferred conduit flow routes as curvilinear dendritic networks, is a preferred graphic method over straight-line, hydrologic-connection vectors. Karst-basin boundaries should be verified by tracer testing as much as reasonably possible and should be mapped with broken or dashed lines which infer their approximate nature.

Common karst features, traced groundwater flow routes, and the basin boundary, are interpreted and mapped in the Boiling Spring drainage basin of north-central Kentucky. One of the largest karst basins in Kentucky, it was investigated with 15 tracer tests to delineate an approximate watershed area of 326 km² (126 mi²). The Boiling Spring drainage system is classified as *Overflow Allogenic*, the most common type of large fluviokarst basin.

INTRODUCTION

Knowledge about karst drainage basins tends to lag behind that of most other types of groundwater systems. Conceptualization of karst flow systems is often generalized and vague because major karst drainage evolves as a complex dual system of epikarst storage and regional conduit flow. Also, differing degrees of hydrologic development may occur within contrasting geologic settings. Investigation and assessment tend to vary from site to site and with individual investigators. The cryptic nature of karst, with preferential recharge, flow, and dispersion potential, tends to thwart investigation with conventional tools and techniques such as water-well analysis and groundwater modeling. For these reasons, knowledge and instruction is often simplistic and focused on spectacular but less functional aspects of karst (Mylroie, 1984).

This paper suggests standardized methods in which to interpret karst at the scale of functional drainage basins. Three main components used to graphically construct karst drainage basins are addressed: *hydrologic features, flow routes, and boundaries*. These components are interpreted and illustrated in the Boiling Spring basin of north-central Kentucky and employed in a functional *classification of karst drainage basins*.

When color is used to illustrate regional karst maps, surface hydrologic features are traditionally blue, subsurface features are red, and groundwater divides are green. In both color and black and white maps, a feature's distinctive symbol or line character, as identified in the map legend, must clearly distinguish between various data and generate a coherent image of a functioning drainage basin. In the digital age, where mapping and illustration of hydrologic systems are evolving rapidly, we need to maintain consistent cartographic techniques which can be applied to differing methodologies and geologic settings.

The karst features and basins discussed herein generally apply to dissected carbonate rocks such as thick-bedded Mississippian and Ordovician-aged limestones and dolomites of Kentucky and adjacent states in the interior low plateau physiographic region of the USA. The main focus is a discussion of the preferred methods used by investigators to interpret and spatially portray karst drainage basins and their various hydrologic features on regional study-area maps. Part of the goal is to describe and assess methods in which karst features are shown on standard United States Geological Survey (USGS) 7.5 minute (1:24,000 scale) topographic and geologic quadrangles. Kentucky is fortunate to have complete coverage of these indispensable maps (McGrain, 1979).

The style of illustration used in the Kentucky Karst Atlas map series, published by the Kentucky Geological Survey (KGS), is cited as a preferred method. In the following discussion, a subscript "K" references current cartographic methods used in the atlas. These digital maps are printed on demand and will be periodically updated and corrected as data become available. Maps are available for the following 30 x 60 minute quadrangles: *Lexington* (Currens and Ray, 1996), *Harrodsburg* (Currens and Ray, 1998), *Campbellsville* (Ray and Currens, 1998), *Somerset* (Currens and Ray, 1998), *Beaver Dam* (Ray and Currens, 1998), and *Bowling Green* (Ray and Currens, 2000).

KARST HYDROLOGIC FEATURES

At least six types of natural hydrologic features are found in major karst regions. *Sinkholes, caves, springs, streams, marshes,* and *intermittent lakes* are variously illustrated on maps and discussed in the literature. An additional broad-scale feature is the distinctive land surface itself. Although landscapes or epikarst zones are very important physical and hydrologic components of karst terrane, they are large-scale phenomena and are not typically illustrated on a map of groundwater basins. As pointed out by Aley (1977), the majority of recharge to large karst basins in Missouri is by discrete infiltration through soil macropores and other features with little or no surface expression. Interpretation of major karst flow and boundary features is done with the recognition that these components are recharged and maintained by less mappable features of the landscape such as the epikarst.

Sinkholes

Sinkholes are probably the most commonly cited diagnostic features in soluble-rock terrane. Although karst is not dependent on the existence of sinkholes, their occurrence in these landscapes is unambiguous evidence to all trained observers that karst hydrology is present. Sinkholes may be absent in some areas containing well-developed karst drainage. Small or shallow sinkholes observed in the field are often not shown on topographic maps of the area. Because of these variations, sinkholes should not be regarded as definitive features of karst.

Most sinkholes are formed by localized dissolution of bedrock by groundwater drainage and the resulting displacement of overlying soils by subsidence, slumping, or collapse. Sinkholes are illustrated on maps by a variety of symbols, including a closed topographic contour or circle with interior tic marks, triangle, dot, circle, or circle with a central dot$_K$.

A sinkhole is often identified on a map as a tracer-injection point and is therefore the origin of a mapped groundwater flow route. For clarity, a slight separation should be maintained between an illustrated karst feature and its related subsurface flow line. Use of a triangle as an injection-point symbol is discouraged because of the potential confusion with an arrow, indicating flow direction and dye recovery point. Stream swallets are technically sinkholes, but they will be addressed under *streams.*

Caves

Groundwater conduits, whether actively transporting water or abandoned, are termed caves if they are large enough to be physically explored. Tracer dyes are often introduced into or recovered from an accessible cave stream. Monitored cave streams are valuable for documenting internal flow routes within a karst basin. On standard edition USGS topographic maps, cave entrances (and mine tunnels) are shown with a black "Y" adit symbol. This symbol should be distinguished from a blue, forked swallet symbol which indicates the point at which a stream disappears underground. On regional maps, caves are illustrated as a generalized facsimile$_K$ of a cave map or as a schematic representation such as an open circle or a circle with a bisecting line$_K$.

Springs

Large springs are classic karst features which, by their volume, require sizable conduit-network delivery systems. Abnormally large springs are essential to karst terrane, although they may be infrequent and seemingly random. Springs are generally indicated on 7.5 minute topographic maps as an open blue circle with a short meandering stream segment or squiggle. This pattern is not universal, however. Some springs are simply connected to a blue-line stream. On particular quadrangles, springs are shown on the topographic map but not on the corresponding geologic map, and visa versa. Generally, both maps correspond. Regardless, only a fraction of actual springs or swallets are typically depicted on these maps. Groundwater investigations are rarely undertaken without additional field survey for existing karst features in the study area.

Perennial or *underflow* springs are usually shown on regional karst maps as a solid blue$_K$ or black dot tangent to a stream or connected by a spring run. A large spring sometimes forms the head of a significant stream. Variable-sized dots may be used to distinguish between major and minor-volume springs. Non-perennial status can be shown with an open blue circle$_K$ for intermittent, seasonal, or overflow springs. Because traditional spring symbols are long established and universally recognized, springs should not be mapped with asterisks, X's, triangles, or other unorthodox symbols.

A karst window is a composite feature usually exhibiting both a spring and stream sink within a depression. A sinking spring is a small perched or epikarst flow that may not be related to a sinkhole or depression. Both features should be shown with dual spring and swallet symbols. A possible exception is a bluehole or cenote where bedrock conduits are obscured below water or talus. On topographic maps, karst windows are either absent or represented by a spring symbol without the down-gradient swallet symbol. Karst windows tend to be common in shallow karst aquifers but rare where water levels are deep or soils are thick. Because they may be hidden or obscured, a complete study-area inventory of these features may require the daunting task of physically searching all existing sinkholes and wooded areas.

An *estavelle* is a unique karst feature which, depending on hydrologic conditions, may function as a stream swallet or overflow spring. When located along a dry channel, it may resemble a cenote or bluehole during low flow. Its primary function is to discharge overflow water from an underlying groundwater conduit. An estavelle's function as a stream sink-point is fortuitous since it is located along an active stream channel. Therefore, it should be illustrated with an overflow spring symbol including an attendant swallet symbol directed toward the feature.

Streams

A general lack of streams is often considered a characteristic of well-developed karst. Depending on the locale, such as a sinkhole plain, this qualifier is accurate. In a majority of Kentucky karst, however, spring-fed streams are common, although not equally distributed. Karst aquifers tend to be more prolific than other bedrock aquifers in Kentucky in storing recharge and discharging sustained groundwater runoff to streams. This is reflected in a streamflow-variability index map generated for Kentucky by the USGS (Ruhl and Martin, 1991).

Spring-fed streams obviously occur in the karst basin discharge zone. Streams are also located in the recharge zone as losing and sinking (disappearing) streams, and sometimes in an intermediate portion of a basin. In the latter case, a short stream segment flowing across a sinkhole or depression, is called a karst window, as mentioned above. As originally defined, these cryptic features result from cave-roof collapse where a segment of groundwater flow is open to the surface (Malott, 1931, Von Osinski, 1935). However, mid-basin karst streams that are not the result of obvious collapse may be much more extensive than a classic karst window.

For example, Sinking Creek, a meandering perennial stream, flows for 18 km in the western mid-section of the Boiling Spring basin of Breckinridge, Meade, and Hardin counties, Kentucky (Fig. 1b). The Garfield 7.5 minute Topographic Quadrangle shows most of this watercourse as a double blue-line stream. In fact, Sinking Creek's normal sink point is not mapped on the quadrangle, and the creek is shown as a double-line stream continuing an additional 19 km to Boiling Spring (Fig. 1a). However, this lower 19 km reach, shown on four topographic quadrangles, is actually a dry overflow channel most of the year (George, 1976). In this case the local topographic maps are quite misleading about the area's highest-order karst stream.

Although USGS topographic maps are indispensable, high-quality products, several subjective factors influenced the depiction of hydrologic features. Chester Bojanowski, a USGS field cartographer for 30 years, states that solid blue lines were never intended to confer perennial-flow status to streams (oral comm., Nov., 2000). The criteria for solid blue-line status was flow for at least 8 months of the year. Variation between maps resulted from limited hydrologic observations in differing seasons and preference of individual quadrangle cartographers. Also, stream status underwent further modification in the final production stage of quadrangle sheets, by preference of individual map reviewers.

Because of these variables, some inconsistencies and hydrologic inaccuracies should be expected. Blue-line stream status, as shown on quadrangle maps, should not be referenced in regulatory guidelines or specific hydrologic assessments. When the actual perennial-flow status of a stream or spring is required, field verification in summer dry-season conditions is critical. A stream that is observed to maintain flow year-round, even during drought conditions, is perennial and should be illustrated with a solid blue$_K$ or black stream line. Streams that are observed to cease flow after storm-water recession or run dry during low flow conditions should be illustrated with a traditional dash and three dots representing intermittent flow.

Swallets, swallow holes, or ponors of sinking streams are usually shown on topographic maps with a blue fork symbol$_K$. The fork symbol is absent on some maps which may cause confusion about flow direction. Some swallet areas are much more complex than a topographic map may depict, with numerous unmapped sinks accepting recharge depending on hydrologic conditions (Ray, 1997).

Marshes

Wetland areas such as marshes, swamps, or fens are not commonly considered karst features. They do, however, occur in certain karst regions, in both basin recharge and discharge zones. Small marshes and ponds occasionally occupy poorly drained sinkholes, especially if they contain seeps or are subject to livestock wallowing and trampling. Larger marshes are typically located on shaley, less soluble carbonate rocks in the vicinity of sinking streams (Quinlan and Rowe, 1977). They also occur on flat-lying sandstone uplands which drain into karst basins. When relevant, marshes are typically shown with a bunch grass or sedge pattern and horizontal stippling.

Some marshes have expanded in recent decades with the recovery of beaver populations. In low-gradient areas, beaver dams have been observed to flood sinking stream channels and bottoms just above swallets as well as back-flood creeks and springs in discharge areas. Beaver-dam flooding can be a persistent logistical problem in certain karst study areas because of high-water interference with locating, monitoring, and gaging springs.

Intermittent Lakes

Few perennial lakes of significant size are related to karst of the interior low plateaus. However, many sinks or depressions are subject to brief inundation by flash floods (Crawford, 1981; Currens and Graham, 1993) and prolonged flooding by groundwater rises. Many intermittent ponds and lakes result primarily from groundwater rising into surface depressions as the shallow karst aquifer becomes flooded during excessively wet periods. Where the lateral transport capacity of shallow aquifers is limited by constrictions or immature conduit development, these lakes are a pervasive karst phenomenon (Aley and Thomson, 1981).

Also called transient or "phantom lakes", these features are not necessarily seasonal in frequency but may occur over periods of years or decades (White and White, 1984). Numerous broad depressions, south of Hopkinsville, Kentucky, were flooded for several months after 24 cm (9.5 in) of rainfall during the first three days of March, 1997 (Kentucky Climate Center, Western Kentucky University). Sections of KY and US highways were locally inundated for prolonged periods. Storage and gradual transfer of flood-crests in a series from one lake to another was observed in the Oak Grove area (Peter Idstein, oral comm., April, 1997). This behavior resembled the hydrology of interconnected poljes in Yugoslavia (Ford and Williams, 1989). Intermittent lakes are usually mapped with a dashed boundary$_K$ to indicate their variable and transient status.

Unwise development within these flood-prone areas has been an increasing problem in some urban areas of Kentucky. Many of these features are broad level areas that do not appear to be depressions. The lack of frequent or seasonal inundation of a land tract has often deceived developers about its long-term flood hazard.

Assessment of flood potential within some karst features is difficult and often misleading. In areas where rising groundwater inundates the depression, bored storm-water injection wells do not function as designed during large floods and may exacerbate the problem. In these areas, traditional development should be prohibited since periodic flooding cannot be economically mitigated. Targeted aerial images of floods may help delineate these intermittent lakes. Otherwise, local land owners and farmers who are often well aware of flood-prone areas should be consulted prior to development. Intermittent lake boundaries shown on Kentucky's Karst

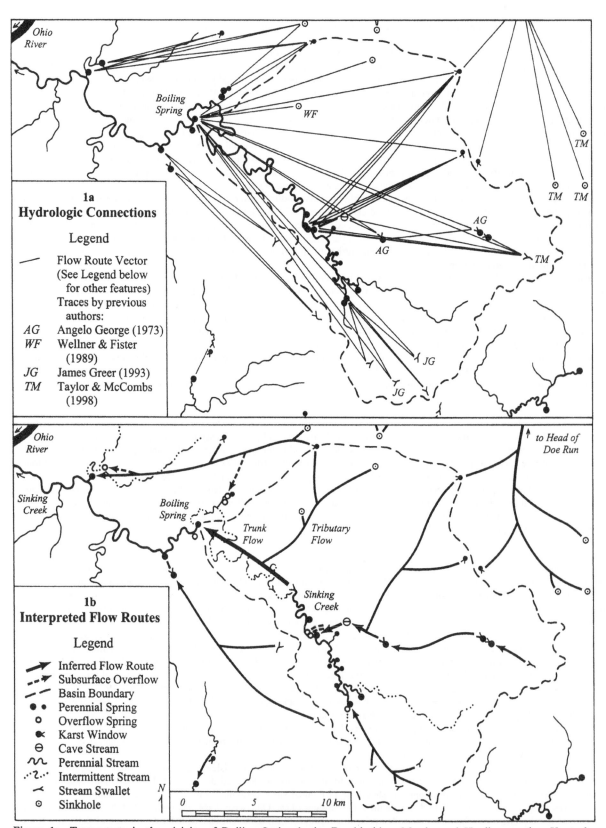

1a
Hydrologic Connections

Legend

—— Flow Route Vector
(See Legend below
for other features)
Traces by previous
authors:
AG   Angelo George (1973)
WF   Wellner & Fister
(1989)
JG   James Greer (1993)
TM   Taylor & McCombs
(1998)

1b
Interpreted Flow Routes

Legend

➤ Inferred Flow Route
➤ Subsurface Overflow
Basin Boundary
•• Perennial Spring
○ Overflow Spring
☚ Karst Window
⊖ Cave Stream
∿ Perennial Stream
∴⌇∴ Intermittent Stream
ᛕ Stream Swallet
⊙ Sinkhole

Ohio
River

Boiling
Spring

WF

TM

TM   TM

AG

AG

TM

JG

JG

JG

Ohio
River

Sinking
Creek

Boiling
Spring

Trunk
Flow

Tributary
Flow

Sinking
Creek

to Head of
Doe Run

N

0   5   10 km

Figure 1: Tracer tests in the vicinity of Boiling Spring basin, Breckinridge, Meade, and Hardin counties, Kentucky. Diagram 1a shows hydrologic-connection vectors. The same data are interpreted as inferred flow routes in 1b.

Atlas maps are generally the larger features known to occur. These approximate boundaries are generalized and should not be considered a limit to inundation. Many additional flood-prone depressions certainly exist.

KARST FLOW ROUTES

Tracer-confirmed karst flow routes are necessarily *inferred* on basin maps. The hydrologic connection between tracer injection and recovery points is reproducible and real, but in most cases the actual path is unknown. The inferred route is drawn as the most reasonable map interpretation of the proven connection. The term *hypothetical* flow route is discouraged in order to avoid the impression that the hydrologic connection is also hypothetical.

Illustrating the most reasonable inferred flow-path poses a challenge for hydrologists and, in practice, is mapped by variable methods. Many use straight-line vectors to show traced hydrologic connections, which is useful for preliminary data or initial illustrations. However, interpreted flow lines are more appropriate for complex data or finished maps.

Figure 1 demonstrates that interpreted, curvilinear$_K$ flow paths between tracer injection and recovery points is the most reasonable and accurate method available for illustrating complex hydrologic routes within karst basins. When numerous traces are involved, vectors lines may become confusing and counterproductive (Fig. 1a). Conversely, interpreted flow paths simplify the depiction of flow routes because they resemble natural systems and eliminate superfluous vector lines (Fig. 1b). Drainage networks are usually dendritic (branching) or trellised in pattern, which resembles many mapped cave systems (Palmer, 1991).

Some cartographers may be hesitant to interpret a dendritic flow route without at least some location data in addition to tracer injection and spring recovery points. However, interpreted traced flow routes are inferred and specifically *do not* imply exact location. The dendritic coalescing of tributaries into master trunk conduits is an important karst phenomena, especially in larger basins (White and Schmidt, 1966). This concept is of greater relevance to hydrologic knowledge than exact junction points.

Large springs could not exist without the development of trunk flow. In well-developed karst, drainage from a distant recharge area rarely maintains an independent flow route to a discharge point without collecting intervening tributaries into a master trunk flow. Analogous to surface drainage, the largest, deepest subsurface flow routes usually establish local base level and the hydrologic gradient of tributaries. Complex or unexpected flow paths are occasionally maintained by unusual geologic structure or hydrologic perching (Meiman and others, 2000). In most cases, however, an interpreted conduit network will be nearer the truth and more comprehendible than the same data conservatively shown as hydrologic-connection vectors.

Extensive cave data, laboriously gathered by numerous cavers over many years, have recently become available for the Boiling Spring basin (Angelo George, written comm., Nov., 2000). These data, which include mapped cave streams, are not shown in Fig. 1 but will be included in the forthcoming Tell City Karst Atlas map.

Basin flow patterns

As stated above, tracer recovery at karst windows and cave streams is very useful for interpreting tributary and trunk flow routes within interior portions of karst basins. Some basins, however, lack known intrabasin flow locations. Without this information the relative position of the interpreted trunk may vary between right, center, or left positions within the basin, depending on the location of surface features (right or left refers to a traditional down-gradient orientation, as in referencing stream banks).

Precise flow paths are not always required or vital information and will not be available for most karst drainage. The unusual case of two sub-parallel conduits that are tributary just up-gradient of the discharge spring, resembling low angle hydrologic connection vectors, is possible but much less likely than their junction at a high angle. Based on numerous cave and tracer data, the odds heavily favor a dendritic configuration that resembles ordered stream networks.

Specific cartographic methods which help clarify complex subsurface flow-route data have been used in Kentucky's Karst Atlas map series since 1998. These standard techniques include solid red flow lines used for perennial flow and graduated-width flow lines to represent increasing discharge. In Kentucky, an arbitrary rule of thumb for distinguishing trunk flow from minor tributaries is a low-flow volume in excess of about 0.1 m^3/sec (3.5 ft^3/sec). Bold arrows (if space permits) are used for showing flow direction solely at dye recovery points. Use of various tracer-injection features such as stream swallets, sinkholes, or wells, identify the origin of mapped flow paths. Red dashed lines are used for subsurface overflow routes and open blue circles for related overflow springs. Blue dashes indicate surface overflow paths in dry valleys. The precedent for most of these cartographic techniques is the regional karst map: *Groundwater Basins in the Mammoth Cave Region, Kentucky* (Quinlan and Ray, 1981; revised 1989). Other basin or regional karst maps using interpreted flow lines and other recommended techniques include Jillson, 1945, Crawford, 1985, Alexander and others, 1996, and Taylor and McCombs, 1998.

An additional technique used on the Kentucky Karst Atlas maps is a distinction between soluble and less soluble rocks. Soluble-rock terrane, where karst drainage is likely, is shown as unshaded areas whereas terrane with less soluble rocks, generally considered non-karst, is shaded brown. Unshaded karst regions absent of tracer data or basin boundaries indicate that no groundwater investigations have been conducted.

Subsurface Avulsion

In rare cases, perennial karst flow routes or springs may suddenly alter location due to conduit blockage. Subsurface avulsion in the past is suggested by abandoned spring alcoves and anecdotal information about altered spring locations. This phenomenon occurred at Turnhole Spring after a large flood in May, 1984, when collapsed sediment plugged the 17 m (56 ft) deep rise pit (Quinlan and Ewers, 1989; Ray, 1997). As a result, Turnhole-basin drainage was diverted for several years through a newly unplugged conduit feeding Sandhouse Cave and suddenly enlarged springs nearby. As of 1991, the spring system had largely reverted to its pre-1984 flow pattern. Avulsion is not a major variable in the documentation and interpretation of karst drainage basins, however, it is a natural process that could render established knowledge obsolete.

KARST BASIN BOUNDARIES

Basin boundaries are necessarily generalized because of aquifer scale and limited practical tracer verification. Most karst investigators will agree that mapping the lateral boundary of a karst drainage basin is the most tentative and difficult component of a study. Boundaries are costly to confirm with tracer tests. Even when a karst basin is adequately delineated by tracer testing, much of the boundary is inferred. Whereas a single test can identify a master trunk system, where flow is concentrated, several tests are required to verify the more extensive basin boundary. In the divide area less flow may be available for efficient dye transport, and the fact that tests are required on both sides of a divide further increases the time and effort required for adequate basin delineation.

A basin boundary, however, is arguably the most important data one can plot on a map. Accurate drainage boundaries in these extremely sensitive groundwater systems are vitally important to ecological management, water-supply planning, contaminant loading and engineering calculations, and emergency spill response. These boundary data essentially establish the spatial component of individual drainage basins.

Few karst studies are blessed with unlimited funding for the detailed delineation of basin boundaries. Karst basins of the Mammoth Cave region have been investigated in greater detail than most other areas of Kentucky. Nevertheless, most boundaries are only verified by tracer tests every few kilometers and large sections of boundaries are drawn on topographic divides with minimal verification. Intensive tracer work conducted in certain areas, such as the Mill Hole sub-basin of the Turnhole Spring watershed, largely focused on the geometry of tributary networks rather than basin boundaries. Unit base flow analysis and potentiometric-surface mapping can be helpful in estimating boundaries but they are no substitute for tracer verification (see discussion below).

In most cases, therefore, karst basin boundaries are "educated estimates". The confidence level in a boundary largely depends on interpretation of local hydrogeology and the number of successful tracer tests which establish the boundary position. Since the spatial distribution of tests is never totally satisfactory, basin boundaries are rarely definitive and must always be shown as broken or dashed_K lines. This character of a non-solid line should imply to all map readers that the boundary is not exact. New tracer data (or a spill) along the border of two basins might easily require a modified boundary. The potential for multiple groundwater flow directions should be assumed in the proximity of a karst boundary until otherwise verified.

Drainage divides

Karst groundwater divides may be fairly sharp, resembling abrupt topographic divides, or they may be rather broad and indistinct where groundwater is intermingled between neighboring basins. In the vicinity of basin boundaries, injected tracers that bifurcate and drain through multiple basins are not uncommon. Because of the potential for groundwater-level fluctuation and subsurface overflow routes, apparent boundaries may shift or temporarily disappear during high water or floods. For these complex reasons, approximated groundwater boundaries should represent low-flow conditions unless otherwise indicated.

Perched karst basins may overlie larger subjacent basins that are distinct from the perched water or hydrologically connected by downward leakage. Two or more basins appearing to occupy the same space are difficult to illustrate on typical groundwater maps. One solution is to employ a unique shading or hatching for the perched system to distinguish it from other basins (Ray and others, 1997).

Topographic divides are commonly used as karst drainage divides. Depending on the local hydrogeology, this method is risky without tracer verification. A comparison was made between tracer delineated boundaries and topographic divides within 650 km² (250 mi²) of karst terrane, in three portions of Kentucky's Mississippian Plateau. This comparison indicated that 15-20% of karst-aquifer drainage did not coincide with topographic watersheds (Ray and others, 2000). However, the majority of an individual karst basin's drainage may deviate from its topographic watershed. For example, about 75% of the Green River's 233 km² (90 mi²) Turnhole Spring watershed, including the Park City area, is incorrectly attributed to the Barren River basin when hydrologic units are based on topographic divides. Such karst phenomena was termed *misbehaved* subsurface drainage by White and Schmidt (1966).

Kentucky's Hydrologic Unit map was published in 1974 and the 11- and 14-digit hydrologic unit code (HUC) boundary layer was completed for the Green River basin by the USGS (Nelson and others, 1997). These hydrologic units were delineated based on surface drainage-divide criteria without consideration of subsurface conduit hydrology. Because of misbehaved hydrology in 15-20% of investigated karst areas, the USGS, Kentucky Division of Water, and others have recently agreed to discuss the creation of a new generation of watershed or hydrologic boundaries that utilize mapped karst-boundary information.

CLASSIFICATION OF KARST DRAINAGE BASINS

The interpretation of tracer data into a coherent drainage network can be substantially aided by a conceptual classification of karst basins (Ray and Currens, 1996). This classification centers on the dominant recharge component that controls the development and configuration of trunk flow within a basin, and is derived from assessing hundreds of basins compiled in the Kentucky Karst Atlas series. The classification system proposes an intuitive evolutionary sequence for fluvial networks encountering highly soluble rocks (Ray, 1999).

A karst flow route may initially develop when a fluvial system begins to incise relatively pure limestone, for example. Flow along secondary bedrock porosity evolves and a subsurface conduit such as a meander cutoff route (Connor, 1976) or a valley-paralleling conduit forms an incipient groundwater basin. In these initial cases most of the resurgence is derived from the nearby stream insurgence. The capacity of this initial groundwater conduit is limited to base or moderate flow while higher flows continue to erode the prevailing surface channel. Maintenance of a viable surface overflow channel is characteristic of an *Overflow Allogenic* or *Type I* karst drainage basin (Fig. 2a). Allogenic flow is defined as non-local stream drainage from either insoluble or soluble rock terrane. Perennial surface flow within soluble rocks may result from headwater position within a basin, perching units, or stream-channel incision to local base level.

With time and continued erosional evolution by the system, the outcrop of soluble rock may expand to encompass most or all of the drainage basin. During basin enlargement, allogenic recharge and resurgence points usually diverge. If the capacity of a trunk conduit evolves to the point that all ranges of allogenic flow are channeled underground, the surface stream is beheaded, thus creating a blind valley at the margin of a karst valley or plain. An *Underflow Allogenic* or *Type II* basin (Fig. 2b) results when allogenic overflow routes are no longer maintained across a karst basin. Most of the large basins in the Mammoth Cave region are the underflow allogenic type. Both Type I & II basins can be considered *fluviokarst* drainage systems (White, 1989).

This conceptual model not only reflects a reasonable evolutionary sequence but helps to explain flood response and water quality of the resurgent spring (Worthington and others, 1992). Suspended sediment and contaminants mobilized during flooding may partially bypass springs draining Type I basins. This overflow-route bypass is not available in Type II basins where springs drain the entire karst watershed. A similar classification was developed by Jones (1997) where *open* karst basins maintain through-flowing surface drainage networks, whereas *closed* basins do not.

A third type of karst watershed exhibits no significant allogenic recharge and is termed a *Local Autogenic* or *Type III* basin (Fig. 2c). These smaller basins are primarily recharged by direct recharge on the land surface and internal runoff into sinkholes (White, 1989). They mainly occur as a result of derangement of larger allogenic basins or they may be located on the margins of stream-less karst plateaus. Some basins will obviously develop a composite of the main elements described above.

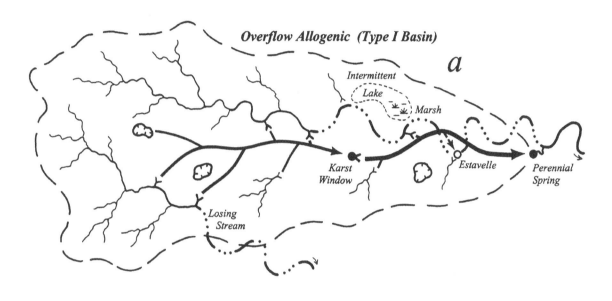

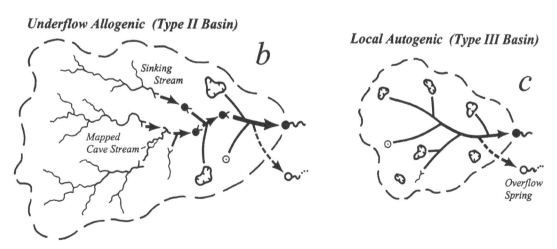

Figure 2: Major types of karst drainage basins. *Overflow Allogenic (Type I)* basins contain losing streams which maintain surface overflow channels across the groundwater basin *(a)*. These basins often develop when fluvial systems initially incise soluble rocks. *Underflow Allogenic (Type II)* basins evolve when sinking streams become blind and cease erosion of overflow channels across the basin *(b)*. *Local Autogenic (Type III)* basins lack sinking streams and are primarily recharged by infiltration through the land-surface and sinkhole drainage *(c)*. Typical karst features are identified above or in the legend of Fig. 1.

The hydrologic configuration or arrangement of most large basins in the interior low plateaus is strongly influenced by the character of allogenic recharge. When interpreting master trunk flow within a karst basin, the largest or most headward sinking stream is reasonably the beginning of the largest and most influential flow route. Smaller tributaries can be safely illustrated on a map to join that trunk at high angles.

COMPUTER GENERATED MAPS

Maps or images of karst drainage are increasingly being produced as digital data layers in a Geographic Information System (GIS). The analytical and manipulation potential of these products is undeniable, but the approach is much more expensive than previous cartographic methods and is presently available to fewer practitioners than traditional drafting equipment (although some drafting materials, which were previously used to produce well-respected maps, are becoming obsolete because of the shift to electronic media). Realizing that the 21st century demands modern digital products, we must nevertheless maintain certain cartographic standards in the electronic format. For example, when a feature is required, such as a dashed line for a basin boundary or overflow route, or a dotted line for a sub-basin boundary, the information represented by a symbol must not be altered for the relative ease of creating a solid line in the electronic program. One must not allow the "tool" to negatively influence the content of a map.

DISCUSSION

Potentiometric surface maps have been commonly produced during the investigation of karst groundwater basins in Kentucky (Quinlan and others, 1983; Schindel and others, 1994). Water-level maps are not enthusiastically recommended by the author because of numerous potential problems with their construction and interpretation (Joe Meiman, oral comm., 1994). These problems include the local use of perched water levels, variable hydrologic connection of artificial bore-holes with the major karst aquifer, poor estimation of wellhead elevation, subjective rejection of data points, and expense. The use of discharge balancing or unit base flow analysis (Quinlan and Ray, 1995) is a useful alternative method. This technique assesses perennial spring and stream-flow rather than man-made wells. With some background reference data and experience, highly useful basin estimates can be derived from minimal field measurements of base flow.

As in all scientific work, even the most experienced investigators occasionally make omissions or errors in karst fieldwork, tracer testing, or data interpretation which may impact the accurate knowledge of karst basins. In tracer testing, potential problems include failure to use adequate dye, distinguish background tracer levels from test levels, monitor all dye resurgences, monitor long enough, or correctly interpret tracer results. Other problems include contamination of samples and faulty record-keeping or illustration of results. The identification of flow in all hydrologic conditions can be difficult to nearly impossible within the timeframe of a particular study. The verification of hydrologic connections during low flow may be logistically impossible in large basins. High-flow conditions may dilute tracers designed for normal conditions. Drought conditions can delay fieldwork and affect results. Finally, dye of unknown origin may be incorrectly interpreted as a positive dye recovery. Obviously, groundwater tracing can be difficult work with many possible "pitfalls". Fortunately, most tracing work is conducted by experienced professionals who are aware of these potential problems. When possible, however, replication of previous work is a prudent idea. New details are often added by subsequent work.

In order to reduce the potential for cross-contamination of tracer dyes concurrently used by different investigators, a voluntary statewide dye-injection notification procedure is presently coordinated by the Kentucky Division of Water. This centralized clearinghouse has operated efficiently for regulators, scientists, and the public.

SUMMARY

For efficient dissemination of karst-drainage information, standardized methods are recommended to interpret and illustrate karst drainage basins. Commonly illustrated components include hydrologic features, flow routes, and drainage boundaries. Interpretation of inferred conduit flow routes, as curvilinear dendritic networks, is a preferred graphic method over straight-line, hydrologic-connection vectors. Karst-basin boundaries should be mapped with broken or dashed lines which infer their approximate nature. When data from 650 km² of karst terrane were assessed, 15-20% of mapped karst drainage basins did not coincide with topographic watersheds in well-developed karst of Kentucky's Mississippian Plateau.

Boiling Spring karst groundwater basin, in north-central Kentucky, was recently delineated with tracer testing by Kentucky Division of Water personnel and previous investigators. This large basin drains about 326 km² (126 mi²) and contains typical karst features such as sinkholes, sinking streams, karst windows, caves, perennial and overflow springs and a trunk karst stream. This basin is classified as *Overflow Allogenic*, the most common type of large fluviokarst drainage basin.

ACKNOWLEDGEMENTS

The author thanks Jack Moody, Pat Keefe, Tracy Burgess III, Bruce McKinney, Beverly Oliver, Rob Blair, Jim Currens, and Randy Paylor, who assisted with field work in the Boiling Spring study. This work was funded by a US EPA 319(h) Nonpoint Source grant and the Kentucky Division of Water. I appreciate manuscript reviews by Pete Goodmann, Jack Moody, and Phil O'dell, and Susan Tousey. Additional assistance was provided by Chester Bojanowski, Lee Colten, David Leo, Joe Meiman, Tom Mesko, and Chuck Taylor. I wish to make a special acknowledgment of the fundamental contributions provided by the late Jim Quinlan in the identification and interpretation of karst drainage basins.

REFERENCES

Alexander, E.C. Jr., Green, J.A., Alexander, S.C., and Sprong, R.C., 1996. Springsheds (Plate 9) *in* Fillmore County, Minnesota Geologic Atlas, Part B, Minnesota Department of Natural Resources, St. Paul, Minnesota (map sheet).

Aley, T., 1977. A model for relating land use and groundwater quality in southern Missouri. *in* Hydrologic Problems in Karst Regions, R.R. Dilamarter and S.C. Csallany, eds., Western Kentucky University, Bowling Green, Kentucky, p. 323-332.

Aley, T., and Thomson, K.C., 1981. Hydrogeologic mapping of unincorporated Greene County, Missouri, to identify areas where sinkhole flooding and serious groundwater contamination could result from land development. Contract report by the Ozark Underground Laboratory for Greene County, Missouri, 11 pages and 5 maps.

Connor, D.G., 1976. The lower reaches of Long Creek, Kentucky: Karst anomaly in Allen County. Masters thesis, geography, Western Kentucky University, Bowling Green, 146 p.

Crawford, N.C., 1984. Sinkhole flooding associated with urban development upon karst terrain: Bowling Green, Kentucky. *in* Sinkholes: Their Geology, Engineering, and Environmental Impact, B.F. Beck, (ed.), Balkma, Rotterdam. p. 283-292.

Crawford, N.C., 1985. Groundwater flow and geologic structure, Lost River groundwater basin, Warren County, Kentucky. Center for Cave and Karst Studies, Western Kentucky University, Bowling Green, Kentucky (map sheet).

Currens, J.C., and Graham, C.D.R., 1993. Flooding of the Sinking Creek karst area in Jessamine and Woodford counties, Kentucky. Kentucky Geological Survey, Report of Investigations 7, Series XI, 33 p.

Currens, J.C., and Ray, J.A., 1996. Mapped karst groundwater basins in the Lexington 30 x 60 Minute Quadrangle. Kentucky Geological Survey, Map and Chart Series 10, Series XI.

Currens, J.C., and Ray, J.A., 1998. Mapped karst groundwater basins in the Harrodsburg 30 x 60 Minute Quadrangle. Kentucky Geological Survey, Map and Chart Series 16, Series XI.

Currens, J.C., and Ray, J.A., 1998. Mapped karst groundwater basins in the Somerset 30 x 60 Minute Quadrangle. Kentucky Geological Survey, Map and Chart Series 18, Series XI.

Ford, D.C., and Williams, P.W., 1989. Karst Geomorphology and Hydrology. Unwin Hyman, London, 601 p.

George, A.I., Schmidt, J.B., and Weller, R.R., 1973. The Sinking Creek Cave System, north-central Kentucky (Abs.). Natl. Speleol. Soc., 30th Annual Convention, Program, p. 9-10.

George, A.I., 1976. Karst and cave distribution in north-central Kentucky. Natl. Speleol. Soc. Bull., v. 38, no. 4, p. 93-98.

Jillson, W.R., 1945. Geology of Roaring Spring. Roberts Printing Company, Frankfort, Kentucky, 44 p.

Jones, W.K., 1997. Karst Hydrology Atlas of West Virginia. Karst Waters Institute, Special Publication 4, 111 p.

Malott, C.A., Lost River at Wesley Chapel Gulf, Orange County, Indiana. Proc., Indiana Acad. Sci., 41: 285-316.

McGrain, Preston, 1979. An economic evaluation of the Kentucky Geologic Mapping Program. Kentucky Geological Survey, Miscellaneous Report, 12 p.

Meiman, J., Groves, C.G, Herstein, S., 2000. Update on dye-tracing efforts at Mammoth Cave National Park. Proceedings, 8th Mammoth Cave National Park Science Conference, p. 77-87.

Mylroie, J.E., 1984. Hydrologic classification of caves and karst. *in* Groundwater as a Geomorphic Agent, R.G. LaFleur (ed.), Allen & Unwin, Boston, p. 157-172.

Nelson Jr., H.L., Downs, A.C., Crabtree, S.D., and Hines, D.H., 1997. Development of an 11- and 14-digit hydrologic unit boundary layer for the Green River Basin using a Geographic Information System. U.S. Geological Survey Open-File Report 97-619, CD-ROM.

Palmer, A.N., 1990. Origin and morphology of limestone caves. Geol. Soc. Am. Bull., v. 103, p. 1-21.

Quinlan, J.F., Ewers, R.O., Ray, J..A., Powell, R.L., and Krothe, N.C., 1983. Groundwater hydrology and geomorphology of the Mammoth Cave Region, Kentucky, and of the Mitchell Plain, Indiana. *in* Field Trips in Midwestern Geology, R.H. Shaver, and J.A. Sunderman, (eds.), Geological Society of America and Indiana Geological Survey, Bloomington, Ind., v. 2, p. 1-85.

Quinlan, J.F., and Ewers, R.O., 1989. Subsurface drainage in the Mammoth Cave area. *in* Karst Hydrology: Concepts from the Mammoth Cave Area, W.B. White and E.L. White, (eds.), Van Nostrand Reinhold, New York, p. 65-103.

Quinlan, J.F., and Ray, J.A., 1981. Groundwater basins in the Mammoth Cave region, Kentucky: Friends of the Karst Occasional Publication, no. 1 (map sheet), (revised, 1989)

Quinlan, J.F., and Ray, J.A., 1995. Normalized base-flow discharge of groundwater basins: A useful parameter for estimating recharge area of spring and for recognizing drainage anomalies in karst terranes. *in* Karst Geohazards, B.F. Beck, (ed.), Balkema, Rotterdam, p. 149-164.

Quinlan, J.F., and Rowe, D.R., 1977. Hydrology and water quality in the Central Kentucky karst-phase 1: Kentucky Water Resource Research Institution, Research Report no. 109, 93 p.

Ray, J.A., 1997. Overflow conduit systems in Kentucky: A consequence of limited underflow capacity. *in* The Engineering Geology and Hydrogeology of Karst Terranes, B.F. Beck and J.B. Stephenson, (eds.), Balkema, Rotterdam, p. 69-76.

Ray, J.A., 1999. A model of karst drainage basin evolution, Interior Low Plateaus, USA. *in* Karst Modeling, A.N. Palmer, M.V. Palmer, and I.D. Sasowsky, (eds.) February 24-27, Charlottesville, Virginia, p. 58.

Ray, J.A., and Currens, J.C., 1996. Major types of groundwater-drainage basins in carbonate terranes of the Midwest, USA, based on recharge and key hydrogeological characteristics. Program & Abstracts, 40th Midwest Groundwater Conference, October 16-18, Columbia, Missouri, p 49.

Ray, J.A., Ewers, R.O., and Idstein, P.J., 1997. Mapping water-supply protection areas for leaky, perched karst groundwater systems. *in* The Engineering Geology and Hydrogeology of Karst Terranes, B.F. Beck and J.B. Stephenson, (eds.), Balkema, Rotterdam. p. 153-156.

Ray, J.A., and Currens, J.C., 1998. Mapped karst groundwater basins in the Campbellsville 30 x 60 Minute Quadrangle. Kentucky Geological Survey, Map and Chart Series17, Series XI.

Ray, J.A., and Currens, J.C., 1998. Mapped karst groundwater basins in the Beaver Dam 30 x 60 Minute Quadrangle. Kentucky Geological Survey, Map and Chart Series 19, Series XI.

Ray, J.A., and Currens, J.C., 2000. Mapped karst groundwater basins in the Bowling Green 30 x 60 Minute Quadrangle. Kentucky Geological Survey, Map and Chart Series 22, Series XII.

Ray, J.A., Goodmann, P.T., and Meiman, J., 2000. Hydrologically valid delineation of watershed boundaries in Kentucky's karst terrane. Proceedings, 8th Mammoth Cave Science Conference, p. 75-76.

Ruhl, K.J., and Martin, G.R., 1991. Low-flow characteristics of Kentucky streams. U.S. Geol. Survey, Water-Resources Investigations Report 91-4097, 50 p.

Schindel, G.M., Quinlan, J.F., and Ray, J.A., 1994. Determination of the recharge area for the Rio Springs groundwater basin, near Munfordville, Kentucky: An application of dye tracing and potentiometric mapping for delineation of springhead and wellhead protection in carbonate aquifers and karst terranes. Project Completion Report, Ground-Water Branch, U.S. Environmental Protection Agency, Region IV, Atlanta, Georgia, 27 p.

Taylor, C.J., and McCombs, G.K., 1998. Recharge-area delineation and hydrology, McCraken Springs, Fort Knox Military Reservation, Meade County, Kentucky. U.S. Geological Survey, Water-Resources Investigations Report 98-4196, 12 p.

Von Osinski, 1935. Karst windows. Proc., Indiana Acad. Sci., 44: 161-165.

White, W.B., 1988. Geomorphology and Hydrology of Karst Terrains. Oxford University Press, New York, 464 p.

White, E.L., and White, W.B., 1984. Flood hazards in karst terrain: Lessons from the Hurricane Agnes Storm. Internatl. Contrib. Hydrogeology 1, p. 261-264.

White, W.B., and Schmidt, V.A., 1966. Hydrology of a karst area in east-central West Virginia. Water Resources Research, v. 2, no. 3, p. 549-560.

Worthington, S.R.H., Davies, G.J., and Quinlan, J.F., 1992. Geochemistry of springs in temperate carbonate aquifers: recharge type explains most of the variation. Colloque d'Hydrogeologie en Pays Calcaire et en Millieu Fissure (5th, Neuchatel, Switzerland), Proceedings, Annales Scientifiques de l'Universite de Besancon. Geologie - Memoires Hors Serie, No. 11, vol. 2, p. 341-347.

U.S.G.S., 1976. Hydrologic Unit Map-1974: State of Kentucky, Reston,Virginia, U.S.

Geotechnical and Environmental Applications of Karst Geology and Hydrology, Beck & Herring (eds)
© 2001 Taylor & Francis, ISBN 90 5809 190 2

Geochemical methods for distinguishing surface water from groundwater in the Knox aquifer system

JAMES C.REDWINE & J.ROBERT HOWELL Southern Company Services, Inc., Birmingham, AL 35202, USA,
jcredwin@southernco.com

ABSTRACT

The Knox Group, a thick package of Cambro-Ordovician rocks, occurs over a wide geographic area in the southeastern U.S. Characteristics of the Knox Group include strong structural control on porosity and permeability; deep near-vertical solution features; great depth of water circulation; dolomite, as well as limestone, hosting the karstic features; and extreme anisotropy and heterogeneity. In this study, geochemical methods were used to distinguish ambient groundwater in the Knox aquifer from surface water, specifically, water leaking from the Logan Martin reservoir in east-central Alabama. Major cations and anions, as well as stable isotopes of hydrogen and oxygen, were used to distinguish lake water from groundwater, and to determine mixed waters. Lake water and groundwater components for mixed waters were calculated, and mapped in plan view. A relatively narrow zone of mixing occurs in the vicinity of Logan Martin dam in map view, which is consistent with the hydrogeological conceptual model of deep near-vertical solution-widened fractures (fissures), oriented east-northeast and to a lesser extent northwest, in a much less permeable dolomite matrix.

INTRODUCTION

The Knox Group of the southeastern U.S. consists of a thick sequence (up to 1280 meters or more) of limestones, dolomites, and associated cherts and sandstones (Figure 1). The Knox is an important water-supply aquifer, as well as a host for oil, natural gas, and Mississippi Valley-type ore deposits. As a substrate for diverse land uses, the Knox has received environmental pollutants, and has created numerous foundation problems for roads, buildings, dams, and other types of engineered structures.

The Knox has been extensively studied at its southeasternmost outcrop extent, in central Alabama U.S.A. The data for this study came primarily from the Logan Martin Dam site, about 56 kilometers (35 miles) east of Birmingham, which has experienced more than 30 years of foundation leakage and subsidence problems. The Logan Martin data base includes thousands of feet of core drilling, thousands of piezometer readings, grouting records, results of dye tests, photogeologic studies, geologic mapping, water chemistry studies, and other types of information. This data has been extensively analyzed, including statistical analysis, and a hydrogeological conceptual model developed for the lower Knox Group. Characteristics of the Knox aquifer include strong structural control on porosity and permeability; deep near-vertical solution features; great depth of water circulation; dolomite, as well as limestone, hosting the karstic features; and extreme anisotropy and heterogeneity (Figures 2 and 3).

At times of no generation, and after bank storage has drained, about 18 to 20 cubic meters per second (650 to 700 cubic feet per second) of seepage water discharges into the Coosa River below Logan Martin Dam. In this study, geochemical methods were used to determine the percentage of water coming from the lake (that is, leakage) compared to natural groundwater discharge from the aquifer.

MAJOR ION ANALYSIS

A comprehensive sampling program was conducted from October 27 through November 7, 1997, where water samples were taken from approximately 130 different locations. The locations included more than thirty tailrace boils, four springs along the reach of Kelly Creek, numerous site wells and piezometers, and from points at varied depths in the forebay. Samples were analyzed for bicarbonate in the field, and for all major cations and anions at Alabama Power's chemical laboratory.

An equivalents balance was performed, to make sure that no major analytical error had occurred, or that no major chemical species in the system were overlooked. In this calculation, the sum of the positive charge due to cations must equal the sum of the negative charge due to anions within an acceptable margin of error. Ninety-five percent of the samples showed agreement within 25 percent.

Major cation and anion data were plotted on Piper diagrams to show relationships among water samples or groups of water samples (Figures 4 through 6). Hill (1940) and Piper (1944) independently developed diagrams which combined the anion and cation fields, as shown in Figure 4. The anion and cation triangles occupy the lower left and lower right portions of the diagram, respectively. The anions and cations are first plotted as points in each of the lower triangles, then projected onto the diamond-shaped

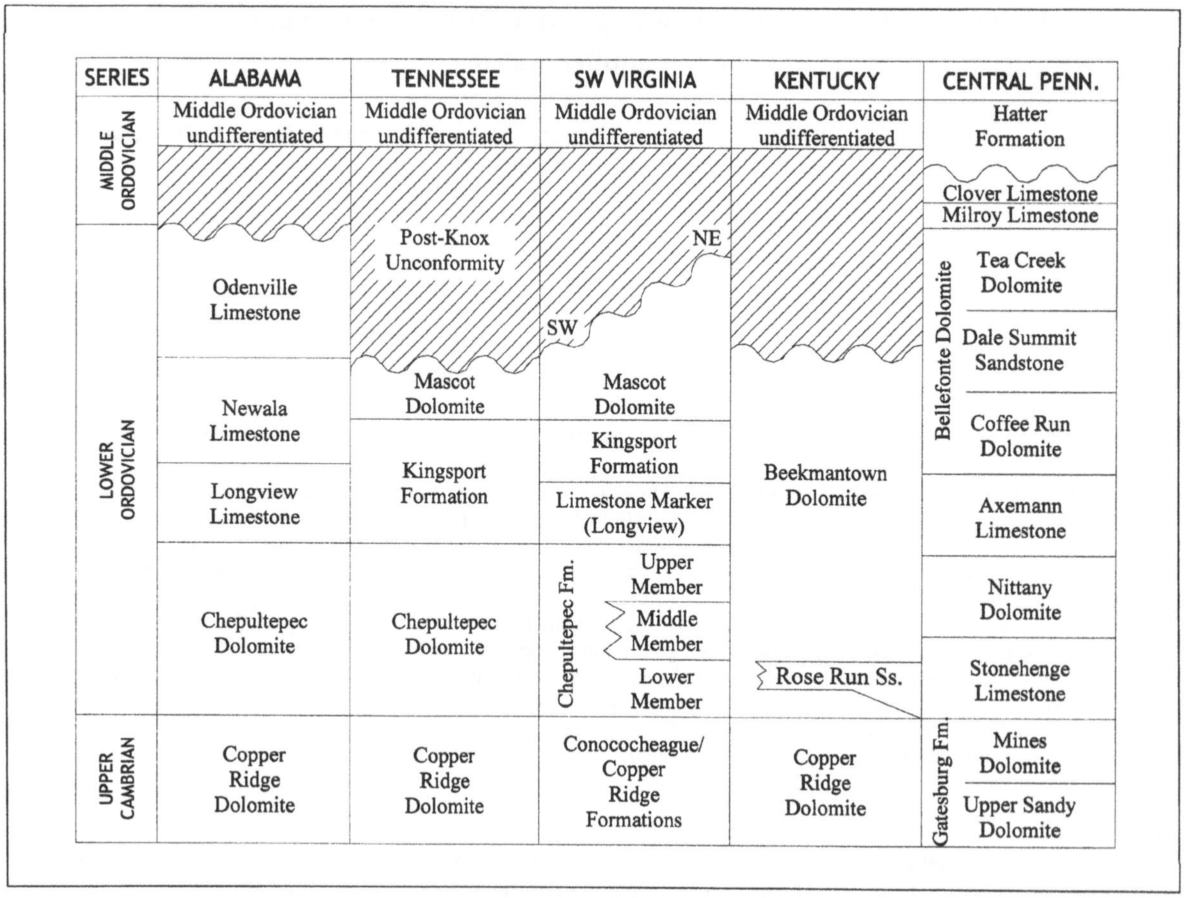

SERIES	ALABAMA	TENNESSEE	SW VIRGINIA	KENTUCKY	CENTRAL PENN.
MIDDLE ORDOVICIAN	Middle Ordovician undifferentiated	Middle Ordovician undifferentiated	Middle Ordovician undifferentiated	Middle Ordovician undifferentiated	Hatter Formation
					Clover Limestone
					Milroy Limestone
LOWER ORDOVICIAN	Odenville Limestone	Post-Knox Unconformity	NE / SW		Bellefonte Dolomite: Tea Creek Dolomite
					Dale Summit Sandstone
	Newala Limestone	Mascot Dolomite	Mascot Dolomite		Coffee Run Dolomite
	Longview Limestone	Kingsport Formation	Kingsport Formation	Beekmantown Dolomite	Axemann Limestone
			Limestone Marker (Longview)		
	Chepultepec Dolomite	Chepultepec Dolomite	Chepultepec Fm.: Upper Member		Nittany Dolomite
			Middle Member		
			Lower Member	Rose Run Ss.	Stonehenge Limestone
UPPER CAMBRIAN	Copper Ridge Dolomite	Copper Ridge Dolomite	Conococheague/ Copper Ridge Formations	Copper Ridge Dolomite	Gatesburg Fm.: Mines Dolomite
					Upper Sandy Dolomite

Figure 1: Generalized Stratigraphy of the Knox Group, Alabama to Virginia, and Age-equivalent Rocks of Central Pennsylvania (modified from Childs and others, 1984, *in* Raymond, 1993; Bova, 1982; and Parizek *in* Parizek and others, 1971)

graph, which represents sums of major cations and anions, that is, sulfate plus chloride, calcium plus magnesium, sodium plus potassium, and carbonate plus bicarbonate.

Temperature data and proximity to the reservoir were used to designate wells which contained groundwater (aquifer water) uninfluenced by reservoir water. The wells within the aquifer group showed some chemical variation, but were similar enough to plot along a best-fit line as shown in Figure 4. Lake water samples were very similar with respect to all parameters, plotting almost as a single point. However, the lake water samples were distinctly different from the aquifer water (Figure 4).

The tailrace boil data show a linear trend toward lake water, with several boil and lake water points overlapping (Figure 5). Such a trend could be created by mixing small, but varying amounts of groundwater with lake water. Figure 6 shows similar data for water from piezometers in the vicinity of the dam. Some of the piezometers, such as 301 and 156, plot on the boil - lake water trend line, though most show a slightly greater groundwater component.

Simple ratios were used to calculate the percentage lake water versus groundwater in each mixed water sample. Discharge waters (e.g., boils and seepage collected at weirs) contained primarily lake water, generally in excess of 90 percent. Percentage data were plotted in map view and contoured (Figure 7). The shaded area on Figure 7 shows where discharge waters and water from piezometers contain greater than 50 percent lake water.

STABLE ISOTOPES

Isotopes of an element have the same atomic number but different atomic weights. The small difference in mass due to the different number of neutrons in the nucleus tends to be significant in the lighter elements (e.g., oxygen and hydrogen). The resulting difference in mass, called isotopic fractionation, can result from any process that causes the ratio of isotopes in a particular phase to differ from one another. Examples of processes that cause isotopic fractionation are evaporation, radioactive decay, and chemical and biological reactions. Stable isotopes of an element are those that are not involved in any type of radioactive decay process. The stable isotopes oxygen-18 and hydrogen-2 have long been used in hydrogeologic studies to differentiate source waters, flow paths, and residence times. For example, a reservoir undergoing evaporation should become enriched with the heavier isotopes of oxygen and hydrogen, because the lighter isotopes preferentially evaporate.

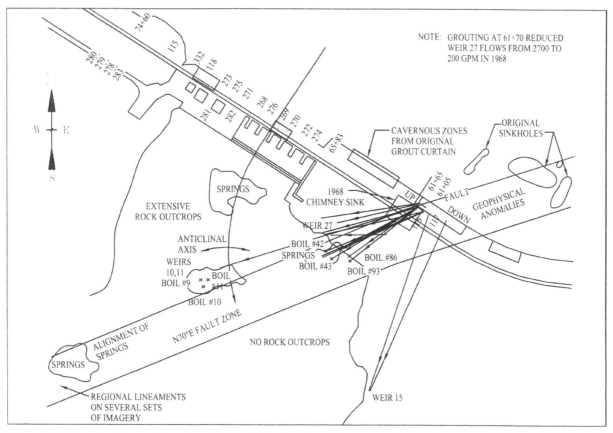

Figure 2: Evidence for Enhanced Porosity and Permeability along the N70°E Normal Fault

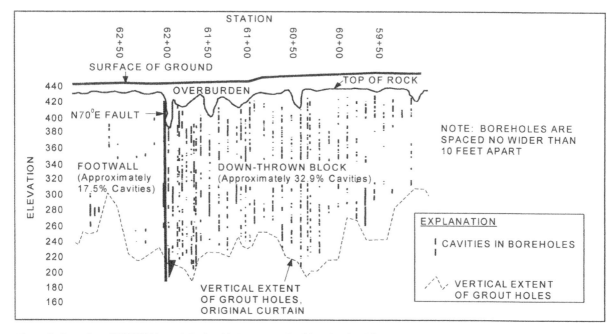

Figure 3: Location of N70°E Normal Fault with Respect to Cavities, Section View

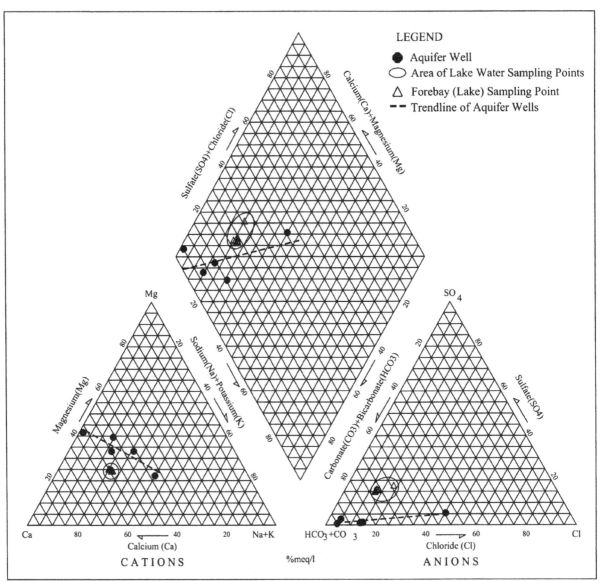

Figure 4: Comparison of Groundwater and Lake Water

For this study, water samples were taken at Logan Martin Dam to determine if the isotopic composition of the reservoir and groundwater varies enough to use the isotopic signature as a marker for determining the origin of underseepage at downstream boil locations (Weir 15). The $\delta^{18}O$ and δD isotopic compositional variations are measured with respect to the Standard Mean Ocean Water (SMOW). When compared to SMOW and plotted, meteoric water from all over the world defines a line known as the meteoric water line (Figure 8). Water with a composition falling on the meteoric water line is believed to have originated in the atmosphere, and is unaffected by other isotopic processes (Domenico and Schwartz, 1990). Numerous studies have provided insight into isotopic fractionation processes that cause specific derivation from the meteoric water line (Figure 8).

Sixteen water samples were taken for $\delta^{18}O$ and δD stable isotope analysis to distinguish between groundwater and surface water at the site. Water samples were collected during rainfall events and from various reservoir locations, groundwater wells, and Weir 15. All samples were collected in amber vials with septum lids and transported to the Environmental Isotope Laboratory at the University of Waterloo, Ontario, Canada for analysis.

Sixteen water samples were taken for $\delta^{18}O$ and δD stable isotope analysis to distinguish between groundwater and surface water at the site. Water samples were collected during rainfall events and from various reservoir locations, groundwater wells, and Weir 15. All samples were collected in amber vials with septum lids and transported to the Environmental Isotope Laboratory at the University of Waterloo, Ontario, Canada for analysis.

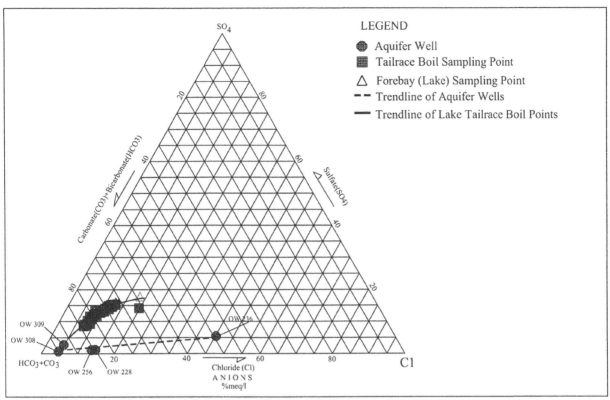

Figure 5: Comparison of Groundwater, Lake Water, and Tailrace Boils

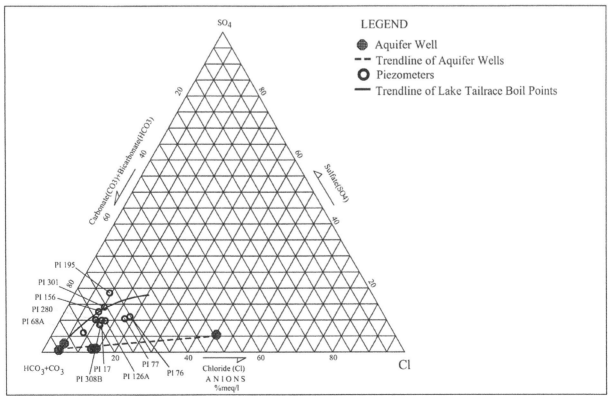

Figure 6: Comparison of Groundwater and Piezometer Water

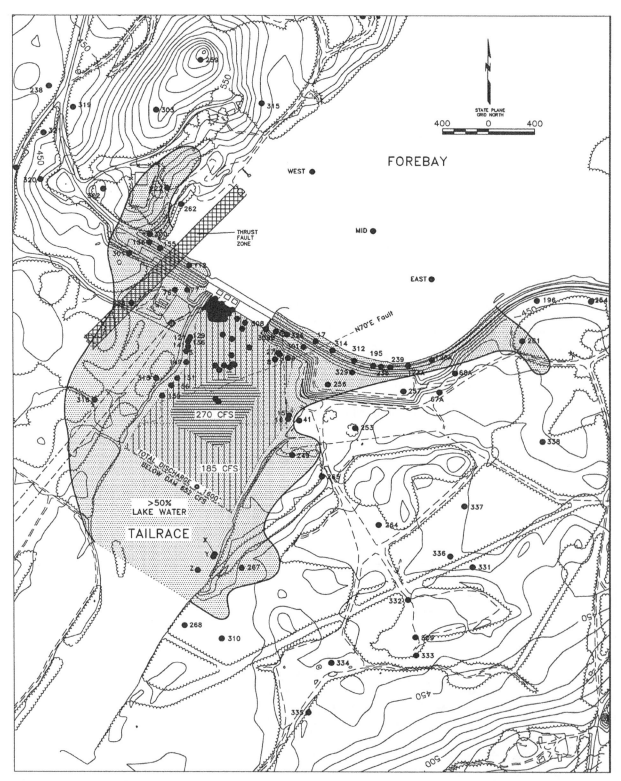

Figure 7: Areal Extent of Lake Water in Wells, Piezometers, and Boils

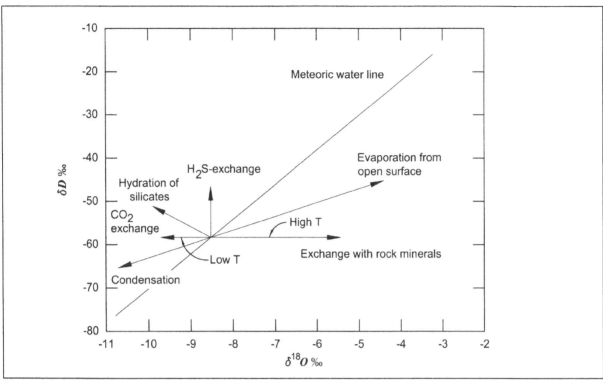

Figure 8: Derivations in Isotopic Composition Away from the Meteoric Water Line as a Consequence of Various Processes (modified from Domenico and Schwartz, 1990)

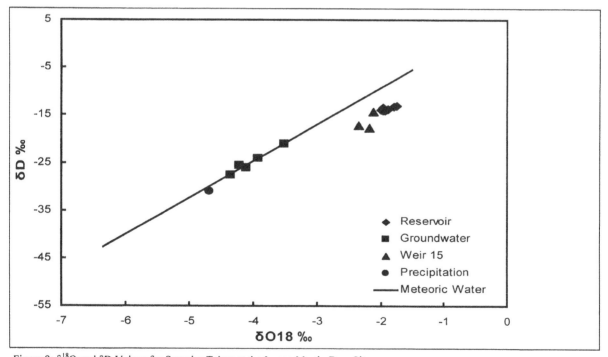

Figure 9: $\delta^{18}O$ and δD Values for Samples Taken at the Logan Martin Dam Site

Stable isotope data for the sixteen samples are plotted in Figure 9. All values in the figure are reported as delta (δ) values, indicating depletion of the isotope relative to the SMOW standard. Figure 9 indicates that reservoir and Weir 15 waters are very similar and distinctly different from groundwater $\delta^{18}O$ and δD values. The groundwater and precipitation values plot very close to the meteoric water line. The position of the reservoir and Weir 15 samples indicates that the waters have undergone isotopic fractionation due to evaporation. Weir 15 water appears to be primarily reservoir water, with a small groundwater component (Figure 9).

CONCLUSIONS

The following conclusions can be drawn from this work:

1. Comparisons of major cations and anions, as well as stable isotopes of hydrogen and oxygen, can be used to distinguish Logan Martin lake water from ambient groundwater in the underlying Knox Group aquifer.

2. Discharge waters contained primarily lake water, generally in excess of 90 percent. A narrow zone of mixing occurs, rarely extending greater than 800 feet, and usually occurring within 500 feet, of the reservoir or tailrace in plan view.

3. An isotopic contrast exists between groundwater and lake water, with the lake showing the effects of evaporation, that is, enrichment in the heavier isotopes of hydrogen and oxygen.

4. The water chemistry data are consistent with the hydrogeological conceptual model for the Knox Group aquifer in this area. The Knox Group is highly anisotropic, with near-vertical solutioned-widened fractures (fissures) capable of carrying significant quantities of water.

ACKNOWLEDGEMENTS

We would like to thank Alabama Power Company for funding this work, and Bill Garrett for leading the efforts in the field and in the laboratory. Thanks also to the other members of the Logan Martin project team for their continued support, especially Mike Akridge, Keith Bryant, Ron Robinson, Bobby Williams, and David Holland. We would like to thank Drs. Will White and Richard Parizek of the Pennsylvania State University for teaching us how to perform this work. Thanks also to Joe Abts for drawing the figures, and Dortha Bailey for preparing the manuscript for publication.

REFERENCES

Alabama Power Company, 1998, Interim report No. 13 to the board of consultants, Logan Martin Dam, Birmingham, AL.

Bova, J. A., 1982, Peritidal cyclic and incipiently drowned platform sequences: Lower Ordovician Chepultepec Formation, Virginia [M.S. thesis]: Virginia Polytechnic Institute and State University, Blacksburg, Virginia, 145 p.

Childs, O. E., Steel, G., and Salvador, A., project directors, 1984, Southern Appalachian region, Correlation of Stratigraphic Units of North America (COSUNA) project: American Association of Petroleum Geologists Correlation Chart Series, 1984.

Domenico, P.A., and Schwartz, F.W., 1990, Physical and Chemical Hydrogeology: John Wiley and Sons, New York, 824 p.

Hem, J.D., 1986, Study and interpretation of the chemical characteristics of natural water, third edition, second printing, U.S. Geological Survey Water-Supply Paper 2254, 264 p.

Hill, R.A., 1940, Geochemical patterns in Choachella Valley, California: American Geophysical Union Transactions, v. 21, p. 46-49.

Parizek, R. R., 1971, Hydrogeologic framework of folded and faulted carbonates *in* Parizek, R. R., White, W. B., and Langmuir, D., 1971, Hydrogeology and geochemistry of folded and faulted rocks of the central Appalachian type and related land use problems: Earth and Mineral Sciences Experiment Station, Mineral Conservation Series Circular 82, College of Earth and Mineral Sciences, The Pennsylvania State University, University Park PA 16802.

Piper, A.M., 1944, A graphic procedure in the geochemical interpretation of water analyses: American Geophysical Union Transactions, v. 25, p. 914-923.

Raymond, D. E., 1993, The Knox Group of Alabama: An overview: Alabama Geological Survey Bulletin 152, 160 p.

Geotechnical and Environmental Applications of Karst Geology and Hydrology, Beck & Herring (eds)
© *2001 Taylor & Francis, ISBN 90 5809 190 2*

Legal impediments to utilizing groundwater as a municipal water supply source in karst terrain in the united states

JESSE J.RICHARDSON, JR. Dept. of Urban Affairs & Planning, Virginia Tech, Blacksburg, VA 24061 USA, jessej@vt.edu

ABSTRACT

Many parts of the United States have recently encountered drought conditions. These conditions refocused the attention of many citizens, environmentalists and governmental units on sources of municipal water supply. Many jurisdictions presently utilize, or are now considering the utilization of, high volume wells as a municipal water supply source. Government officials often fail to consider legal rights to use groundwater in evaluating municipal water supply sources. In addition, withdrawal of groundwater in karst regions can result in interference with other rights of nearby landowners.

Drought conditions and increased population contribute to an increased incidence of legal disputes over the use of groundwater as a municipal water supply source in karst regions. High volume pumping of groundwater may cause land subsidence, including sinkholes, drying of springs and/or streams and loss in value of nearby land.

This paper examines the legal causes of action available to landowners adversely affected by high volume groundwater pumping. First, statutory or common law regimes grant varying rights to groundwater in each state. Second, land subsidence and sinkholes fall within the legal cause of action "loss of lateral and subjacent support". Third, nuisance entails the unreasonable interference with another's reasonable use of their property. Finally, Fifth Amendment takings claims allege a taking of private property for public purposes without just compensation.

This paper explains the possible legal attacks on high volume pumping of groundwater and examines the congruity, or lack thereof, between law and science. The author concludes that common law remedies for damages caused by high volume pumping of groundwater fail to incorporate contemporary scientific knowledge. In addition, the current remedies fail to adequately allocate risk and liability. State legislation must address these issues to balance the rights of landowners and the public.

INTRODUCTION

Recent drought conditions in many parts of the United States have caused the refocusing of attention on sources of municipal water supply. Many jurisdictions are now looking at the possibility of high volume wells as a municipal water supply source. Government officials often fail to consider legal rights to use groundwater in evaluating municipal water supply sources. "Although the law of riparian rights in surface water may be reasonably complex, it is at least relatively stable and ascertainable from the decided...cases. The law, however, concerning subterranean water is unsettled ..." (*Costello v. Frederick County Sanitation Authority*, 1999, p. 9).

State common law, which refers to rules made by judges in adjudicating individual cases, and statutory law, which refers to rules enacted by legislatures, define four regimes in relation to water rights: (1) Streams, lakes, and similar related waters; (2) Diffuse surface waters; (3) Percolating (non-stream) underground water; and, (4) Atmospheric water. (Beck, 1991a, p. 68-69). Generally, the law applies the same rules for underground streams as it applies to streams, lakes, and similar related waters on the surface. However, the law applied to other types of water is usually different, resulting in varying, and often conflicting, legal regimes depending upon the source of water. This paper focuses on the rights to use percolating groundwater.

Withdrawal of groundwater in the karst regions of the country can also result in interference with other rights of nearby landowners. "Karst" refers to a variety of carbonate landforms throughout the world. (Fischer, Fischer, Brown and Ruark, 2000, p. 16). Karst terrains are characterized by caves, springs, sinking streams, ghost lakes, bedrock pinnacles and sinkholes. (*Ibid.*). These forms are the result of the chemical decomposition of carbonate rock formations. (*Ibid.*).

This paper seeks to briefly summarize the legal pitfalls that may be encountered by municipalities that seek to utilize a percolating groundwater source for municipal water supplies in karst topography. This review is not exhaustive and should not be used as a substitute for legal advice.

THE RIGHT TO USE GROUNDWATER

Five possible systems of percolating groundwater law exist in the United States.

The "English rule" or "absolute ownership rule" grants the right to use water to the owner of overlying land who is able to withdraw the water. (Abrams, 1997, p. 3). This rule derives from the common law of England and predominated in the United States in the nineteenth and early twentieth century. Legal support for this rule appeared to derive from judicial ignorance of the science of groundwater.

Water, whether moving or motionless in the earth, is not, in the eye of the law, distinct from the earth. The laws of its existence and progress, while there, are not uniform, and cannot be known and regulated. It rises to great heights, and moves collaterally, by influences beyond our apprehension. These influences are so secret, changeable, and uncontrollable, we can not subject them to the regulations of law, nor build upon the system of rules, as has been done with streams upon the surface. Priority of enjoyment does not, in like case, abridge the natural rights of adjoining proprietors.

(Roath v. Driscoll, 1850, p. 16-17).

Texas is the only state that has adhered to the absolute ownership rule, and even Texas has made modifications. (Abrams, 1997, p. 3). Vermont recently appears to have adopted a statutory scheme that utilizes this regime. (*Ibid.*).

Most courts in the United States, beginning in the early 1900's, adopted the common law "reasonable use rule", or "American rule". This rule grants the right to use the water to the owner of overlying land who is able to withdraw the groundwater. However, the use is legally protected only if it is (1) made on the overlying tracts and, (2) a "reasonable" use. (Abrams, 1997, p. 4).

The *Restatement (Second) of Torts*, at section 858, adopts a reasonable use rule that basically mirrors the riparian rights reasonable use rule that generally applies to surface water rights in the Eastern part of the United States. (*Ibid.*). "A proprietor of land or a grantee who withdraws groundwater from the land and uses it for a beneficial purpose is not subject to liability for interference of the use of water by another, unless (a) the withdrawal of groundwater unreasonably causes harm to the proprietors of neighboring land through lowering of the water table or reducing artesian pressure, (b) the withdrawal of groundwater exceeds the proprietor's reasonable share of the annual supply or total store of groundwater, or (c) the withdrawal of groundwater has a direct and substantial effect upon a water course or lake and unreasonably causes harm to a person entitled to the use of the water." (*Restatement(Second) of Torts*, Section 858 (1)).

California has adopted the correlative rights doctrine that combines a rule of sharing for contract uses with a rule of priority for export. (Abrams, 1997, p. 4). Nebraska has a similar groundwater regime. (*Ibid.*). Correlative rights allocate rights to overlying landowners coequally and in proportion to their ownership of the overlying land. (Beck, 1991b, p. 102). For example, if Landowner A owns land overlying one-half of the aquifer, Landowner A holds the right to use one-half of the water from the aquifer. However, Landowner A's rights to the groundwater are further limited to the quantity of water necessary for use on his land.

Finally, many western states follow the doctrine of prior appropriation. (Abrams, 1997, p. 4). This doctrine is based on seniority in time based on the date of initiation of the use. (Abrams, 1997, p. 5). In other words, the senior users of the resource are fully protected to the extent of their historic withdrawals and when safe yield is reached, junior users are barred from withdrawing any water. (*Ibid.*).

Use of the American Rule for groundwater use limits the use of groundwater as a municipal water supply source.

Under the English rule for groundwater, a municipal corporation may withdraw percolating water from wells to supply water to inhabitants of the municipal corporation, and dry up wells of the neighboring landowners, without liability to them. (Beck, 1991b, p. 100). However, the municipal withdrawal of groundwater for resale was the subject area in which the first erosion of the English rule occurred. (*Ibid.*, footnote 207, citing *Forbell v. City of New York*, 164 N.Y. 522, 58 N.E. 644 (1900), and calling that case a "widely discussed and influential case"). Water use under the reasonable use rule is limited to the premises overlying the aquifer for beneficial purposes incidental to enjoyment of that land. (Beck, 1991b, p. 69). The owner of overlying land may use groundwater for reasonable uses *only if the use is made on the overlying tracts.* (Abrams, 1997, p. 4). The reasonable use rule, therefore, prohibits the use of groundwater as a municipal water supply source.

The case of *Martin v. City of Linden*, 667 So.2d 732 (Ala. 1995) involved a landowner who brought action against the city to have the city enjoined from taking an estimated 500,000 gallons of water per day from a common aquifer for use by its residents. (*Martin*, 1995, p. 732, 734). The Court held that the pumping of groundwater for use by citizens of the City was not permissible under the rule of "reasonable use". (*Martin*, 1995, p. 737).

The Alabama Supreme Court in *Martin* examined the law of other jurisdictions as to reasonable use and quoted the Supreme Court of Pennsylvania in *Rothrauff v. Sinking Spring Water Company*, 339 Pa. 129, 14 A.2d 87 (1940). "While there is some difference of opinion as to what should be regarded as reasonable use of such waters, the modern decisions are fairly harmonious in holding that a property may not concentrate such waters and convey them off his land if the springs or wells of another are impaired." (*Martin*, 1995, p. 738-739 [citations omitted]).

The Virginia Supreme Court has summarized the American rule for groundwater, stating that it "does not forbid the use of percolating water for all purposes properly connected with the use, enjoyment and development of the land itself, but it does forbid maliciously cutting it off, its unnecessary waste, *or withdrawal for sale or distribution for uses not connected with the beneficial enjoyment of ownership of the land from which it is taken.*" (*Clinchfield Coal Corporation v. Compton*, p. 451 [emphasis added]). Therefore, the offsite sale and distribution of the water, in and of itself, violates of the American rule.

Some states have adopted statutory groundwater rights.

These states take different approaches to groundwater regulation. The statutes often codify common law prior appropriation or reasonable use approaches. Several states have adopted permit systems to allocate rights to groundwater. (McEowen and Harl, 1999,

p. 13-26.4). These permit systems range from comprehensive to an area-by-area approach. In Virginia, for example, the Ground Water Management Act of 1992 (and the Groundwater Act of 1973 before it) requires a permit in designated groundwater management areas for withdrawals of more than 300,000 gallons per month or for certain other purposes. Only two groundwater management areas have been designated by the states. Common law rules apply outside of these areas.

GROUNDWATER USE AS A NUISANCE

High volume pumping of groundwater in karst terrain may also constitute an actionable nuisance. "Nuisance" may be defined as: interference with an owner's reasonable use and enjoyment of his property by means of smoke, odors, noise or vibration, obstruction of private easements and rights of support, interference with public rights, such as free passage along streams and highways, enjoyment of public parks and places of recreation, and, in addition, activities and structures prohibited as statutory nuisances.

(*Black's Law Dictionary*, p. 961). To subject a person to liability for private nuisance, his conduct must be a legal cause of the interference in someone else's private use interest and quiet enjoyment of their land. (*Restatement (Second) of Torts* § 822). An intentional interference (or invasion) with another's interest in the use of land is unreasonable if: (i) "the gravity of the harm outweighs the utility of the actor's conduct," or, (ii) "the harm caused by the conduct is serious and the financial burden of compensating for this and similar harm to others would not make the continuation of the conduct not feasible." (*Ibid.*, §826). In determining the gravity of the harm from an intentional invasion of another's interest in the use and enjoyment of land, the following factors are important: (i) the extent of the harm involved; (ii) the character of the harm involved; (iii) the social value that the law attaches to the type of use or enjoyment invaded; (iv) the suitability of the particular use or enjoyment invaded to the character of the locality; and (v) the burden on the person harmed of avoiding the harm. (*Ibid.*, § 828). The following factors weigh heavily in determining the utility of the conduct that causes the invasion of another's interest in the use and enjoyment of land: (i) the social value that the law attaches to the primary purpose of the conduct; (ii) the suitability of the conduct to the character of the locality; and (iii) the impracticability of preventing or avoiding the invasion. (*Ibid.*, § 828).

In *Henderson v. Wade Sand and Gravel Company, Inc.*, 388 So.2d (Ala. 1980), plaintiffs' houses were approximately 1/2 mile south of Wade Sand and Gravel Company, which operated a quarry. (*Henderson*, Ala. 1980, p. 900). The quarry began operating in 1957 and periodically pumped water from the bottom of its pits, emptying it into a nearby creek. (*Ibid.*). This resulted in groundwater being leeched from under plaintiffs' land, leaving large underground cavities. (*Ibid.*). Heavy rains then caused water to flow through the empty cavities at an accelerated rate, destroying the structure of the land between plaintiffs' homes and carrying away much subsoil and surface soil. (*Ibid.*). In 1977 the land on which plaintiffs' homes were situated began to sink, large sinkholes appeared, and their houses began to break up. (Ibid.).

The court in *Henderson* applied the law of nuisance in holding that when "a plaintiff's use of groundwater, whether it be for consumption or, as here for support, is interfered with by defendant's diversion of that water, incidental to some use of his own land, the rules of liability developed by the law of nuisance will apply." (*Henderson*, Ala. 1980, p. 903). The court opined that a contrary rule "could produce disastrous results today". (*Ibid.*). "Carried to its logical extension, [another rule] would allow a quarry owner to willfully sink the City of Birmingham with impunity, provided that it was done in furtherance of a legitimate enterprise and that due care was exercised in the pumping. A rule which provides no check on a land owner's ability to utilize his land to the detriment of society can not be tolerated." (*Ibid.*).

LOSS OF LATERAL AND SUBJACENT SUPPORT

High volume pumping of groundwater in karst terrain may unlawfully interfere with the rights of lateral and subjacent support of nearby landowners. The right of lateral and subjacent support is the right to have land supported by the adjoining land or the soil beneath. (*Black's Law* Dictionary, p. 795). Such withdrawal violates the damaged property owner's natural rights, *ex jure naturae*, and lack of negligence is no defense. (*Large v. Clinchfield Coal Co*, 1990).

Lateral support entails the existence of a vertical plane dividing the supported and supporting lands. (*Restatement (Second) of Torts*, Chapter 39, Scope and Introductory Note, p. 63). Support is subjacent when the supported land is above and the supporting land is beneath it. (*Ibid.*). "One who withdraws the naturally necessary lateral [or subjacent] support of land in another's possession or support that has been substituted for the naturally necessary support, is subject to liability for a subsidence of the land of the other that was naturally dependent upon the support withdrawn." (*Restatement (Second) of Torts* §§ 817, 822). One who holds the right to withdraw subterranean water, oil, minerals or other substances from under the ground of another does not acquire the right to cause, and remains liable for, the subsidence of the other's land by the withdrawal. (Ibid., § 818).

Withdrawing groundwater in karst terrain often takes water from the voids in the subsurface that characterize karst topography. The resulting air-filled void often suffers a collapse of the overlying soil, causing the appearance of a sinkhole. Alternatively, subsidence of the soil may occur without the appearance of a sinkhole. Therefore, withdrawal of large quantities of groundwater in karst terrain may result in the loss of lateral and/or subjacent support for neighboring landowners, creating liability for the withdrawing party.

In *Los Osos Valley Assoc. v. San Luis Obispo*, 30 Cal. App.4th 1670, 36 Cal. Rptr. 2d 758 (1994), the court found that the city's groundwater pumping caused subsidence that resulted in structural damage to a shopping mall's property. This removal of subjacent support constituted to taking of private property for public purposes without just compensation.

GROUNDWATER USE AS A "TAKING"

The 5th Amendment of the *United States Constitution* provides that "private property [shall not] be taken for public use without just compensation." This provision applies to State action as well by virtue of the 14th Amendment. *See, Chicago B & O Railroad v.*

Chicago, 166 U.S. 266 (1897). Note that this cause of action, unlike the others discussed, require governmental action. In other words, a landowner may not claim a takings against a private individual.

The pumping of groundwater from beneath a landowner's property may constitute a "taking". A court could find that the actual physical taking of the water itself constitutes a taking. A more likely scenario, however, would involve the pumping of groundwater causing sinkholes or depressions that render the land unusable by the owner. A court may find the interference with use so onerous as to rise to the level of a taking in this instance. In essence, this theory asserts that by causing the damage (sinkholes, etc.), the municipality is "using" the property for public purposes and so should pay compensation.

The United States Supreme Court most recently enunciated the test to determine whether a regulatory taking has occurred *Lucas v. South Carolina Coastal Council*, 505 U.S. 1003 (1992). An interpretation of the test delineated in Lucas follows:

A. Is the purpose of the regulatory action a legitimate state interest?
   1. if yes, go to B.;
   2. if no, a compensable taking has occurred.
B. Does the means used to achieve the objective substantially advance the intended state purpose?
   1. if yes, go to C.;
   2. if no, a compensable taking has occurred.
C. Does the alleged taking compel the property owner to suffer a physical invasion of his property (or the equivalent)?
   1. if yes, a compensable taking has occurred;
   2. if no, go to D.
D. "No economically viable use" test:
   1. Does the alleged taking deny the property owner of all economically beneficial or productive use of the land?
      i. if yes, go to 2.;
      ii. if no, go to E.
   2. Does the regulation simply make explicit what already inheres in the title itself, in the restrictions that the background principles of the state's law of nuisance already imposed on the landowner?
      i. if yes, go to E.;
      ii. if no, a compensable taking has occurred.
E. Apply the Penn Central balancing test, balancing:
   1. the economic impact of the regulation on the landowner;
   2. the landowner's investment backed expectations; and,
   3. the character of the government activity.
(Richardson and Feitshans, 2000, p. 131-132).

In applying the test to the context of this paper, the providing of a potable water supply to citizens is clearly a legitimate state interest. Likewise, high volume pumping of groundwater substantially advances this legitimate state interest.

Steps C. and D. of the regulatory takings test explore whether or not a categorical taking has occurred. The first type of categorical taking occurs when there is a physical invasion, however slight, onto the landowner's property. In *Los Osos Valley Assoc. v. San Luis Obispo*, 30 Cal. App.4th 1670, 36 Cal. Rptr.2d 758 (1994) (discussed above under loss of lateral and subjacent support), the court found that the pumping of groundwater by San Luis Obispo caused subsidence of property owned by Los Osos Valley Associates resulting in structural damage to a building owned by Los Osos Valley Associates. The court held that the City's actions constituted a physical taking for which compensation was required. In reference to the formulation above, the court found a taking under step C., a physical invasion. The court does not explain why it based on holding on the takings theory, when it also found a loss of lateral and subjacent support.

The next type of categorical taking, the denial of all economically viable uses, may also be implicated in a case of high volume groundwater pumping. The *Costello v. Frederick County Sanitation Authority* case in Virginia involved this theory, among others. The author represented the Lewis M. Costello and Joy H. Costello, the plaintiffs in this case. In *Costello*, defendant Frederick County Sanitation Authority withdrew over two million gallons of groundwater per day from abandoned limestone quarries adjacent to property owned by the Costellos. Most of the Costello property consisted of karst topography. The withdrawals by the Frederick County Sanitation Authority caused the expression of large numbers of depressions and sinkholes on the Costello property. The Costello's alleged that the actions of the Frederick County Sanitation Authority amounted to a taking of their property for public purposes without just compensation in part because it rendered their land valueless. The parties settled this matter, and, therefore, this legal theory was never tested in the case.

Finally, if no taking is found by application of steps A., B., C. or D. of the takings test, step E. must be applied. Step E. involves the balancing of the economic impact of the regulation on the landowner, the landowner's investment backed expectations, and the character of the government activity. This balancing entails a difficult and unclear calculus. Whether a particular situation rises to the level of a taking under this final test depends upon the facts of each particular case.

SHORTCOMINGS IN PRESENT LAW

Present law pertaining to water rights treats groundwater, surface water and other types of water separately even though where these waters are physically interrelated. Additionally, each state maintains its own legal regime of water rights.

The law on groundwater rights includes a lack of clarity and derives its principles from an early lack of judicial knowledge. The judicial principle of *stare decisis* often imposes significant barriers to courts wishing to modernize a state's law on groundwater rights since it binds the present court to often-antiquated decisions. Present legal regimes for groundwater often provide perverse incentives to landowners to pump large amounts of groundwater to maintain legal rights. Under the prior appropriation and absolute ownership regimes, landowners may lose legal rights by wisely conserving groundwater resources. Finally, the present legal rules for groundwater rights fail to properly balance the private property rights of landowners against the public interest in acquiring and maintaining a safe and reliable source of municipal water supply.

The withdrawal of large amounts of groundwater in karst terrain often leads to interference with the rights of nearby landowners. Many disputes over these groundwater uses come to the courts based on traditional legal principles like nuisance and takings. These legal principles often prove inadequate to deal with groundwater use karst terrain in modern society. Managers of municipal water supply systems often fail to consider or anticipate interferences with rights of nearby landowners.

CONCLUSION

The present legal framework for groundwater rights fails to incorporate modern scientific knowledge and fails to adequately balance private and public interests. Traditional legal remedies fall short in the quest of adequately addressing competing legal interests in this area.

Efforts should be made to coordinate regulation of groundwater, surface water and other water sources. In addition, innovative solutions to the failure of the judicial system to adequately balance risk and liability must be fashioned. History and constitutional principles indicates that the state legislative branch is the appropriate source of clarification of modernization of legal regimes addressing groundwater rights in karst terrain. State legislatures must act to coordinate, integrate and modernize regulation of groundwater withdrawal in karst regime.

REFERENCES

Abrams, Robert H., 1997, Contemporary Groundwater Issues, Course Materials from the 15th Annual Water Law Conference, American Bar Association, 24 p.

American Law Institute, 1977, *Restatement (Second) of Torts*, V. 4, St. Paul, Minn., American Law Institute Publishers, 649 p.

Beck, Robert E., Editor, 1991a, *Waters and Water Rights*, V. 1, The Michie Company (supplemented annually; last supplement, December, 1999), 768 p, supplement 276 p.

Beck, Robert E., Editor, 1991b, *Waters and Water Rights*, V. 3, The Michie Company (supplemented annually; last supplement, December, 1999), 646 p., supplement 111 p.

Black's Law Dictionary, 1979, 5th Edition, 1511 p. (A newer edition is available).

Chicago B & O Railroad v. Chicago, 166 U.S. 266 (1897).

Clinchfield Coal Corporation v. Compton, 148 Va. 437, 451 (1927).

Costello v. Frederick County Sanitation Authority, 49 Va. Cir. 41, 1999 Va. Cir. LEXIS 268 (Circ. Ct. of Frederick County) (1999).

Fischer, Joseph A., Joseph J. Fischer, Terri Brown and Vincent Ruark, 2000, "Designer Sinkholes: Managing Stormwater While Protecting Groundwater in Karst", Integrating Science into the Development and Implementation of Effective Water Resource Policy: Virginia Water Resources Research Center, Pub. No. PF-2000, pp.16-19.

Forbell v. City of New York, 164 N.Y. 522, 58 N.E. 644 (1900).

Henderson v. Wade Sand and Gravel Company, Inc., 388 So.2d (Ala.1980).

Large v. Clinchfield Coal Co., 239 Va. 144, 357 S.E.2d 783 (1990).

Los Osos Valley Assoc. v. San Luis Obispo, 30 Cal. App.4th 1670, 36 Cal. Rptr.2d 758 (1994).

Lucas v. South Carolina Coastal Council, 505 U.S. 1003 (1992).

Martin v. City of Linden, 667 So.2d 732 (Ala. 1995).

McEowen, Roger A. and Neil E. Harl, 1999, *Principles of Agricultural Law*, abridged, Eugene, Ore., Agricultural Law Press.

Richardson, Jesse J., Jr. and Theodore A. Feitshans, "Nuisance Revisited After Buchanan and Bormann", 5 *Drake Journal of Agricultural Law* 121 (Spring, 2000).

Roath v. Driscoll, 20 Conn. 533, 52 Am. Dec. 352 (1850).

Geotechnical and Environmental Applications of Karst Geology and Hydrology, Beck & Herring (eds)
© *2001 Taylor & Francis, ISBN 90 5809 190 2*

Quality of groundwater in shallow and deep karst and other associated aquifer zones in central India

YAMUNA SINGH Atomic Minerals Directorate for Exploration & Research, Begumpet, Hyderabad, India

D.P.DUBEY Government Model Autonomous Science College, Rewa, Madhya Pradesh, India

ABSTRACT

The late Proterozoic sediments of the Vindhyan Supergroup, comprising mainly sandstones, shales, and limestones constitute the aquifer system in the Rewa region of the central Indian terrain. Systematic studies for the assessment of suitability of groundwater, for various purposes, particularly from karstic area of the Rewa region (Survey of India toposheet 63H), central India, have shown that groundwaters from sandstones are remarkably less mineralised, whereas those from shales and limestones are much more mineralised. Further, waters from shallow aquifer zones are less mineralised and, with respect to this zone, waters from deep aquifer zones are 3 to 5 times more mineralised. Thus, a clear cut distinction is observed in groundwater from shallow (low- to medium-salinity) and deep (high- to very high-salinity) aquifer zones. Variation in quality of groundwater from sandstones, shales, and limestones is due to variable mineralogy of these rocks. Conspicuous variation in quality of groundwater in various aquifers, with respect to their depth, seems to be due to various factors like variable permeability, uneven karstification, downward percolation of groundwater, presence of gypsum, and anaerobic bacteria. Interestingly, veins and stringers of gypsum have been found to increase particularly in shales with depth, along with anaerobic bacteria of *Desulfovibrio desulfuricans* genus. The reduction of sulfate by anaerobic bacteria appears to have added for more mineralised nature of groundwater in deep aquifer zones. Reduction of sulfate produces carbon dioxide partial pressure, migration of which into deep aquifer zones increases acidity of groundwater present, and allows for more dissolution of carbonate and other associated minerals and, hence, high quantity of dissolved salts in groundwater from deep karst and associated shale aquifer zones, in contrast to less mineralised groundwater from shallow karst and also shale aquifer zones.

INTRODUCTION

Groundwater is valuable only when its quality is suitable for the purpose for which it is being explored. The suitability of natural water for a particular purpose depends upon the standards of acceptable quality for that use. Quality of groundwater is determined mainly by the temperature, pressure, porosity, permeability, and mineralogy of the rocks through which it flows. Presence of certain soluble salts and gases increases its dissolving capacity. Dissolved salt content of groundwater increases until a saturation level of solutes available from host rocks is attained in the solvent at prevailing pressure and temperature conditions. Also, geochemical history and climate of the area including meteorological parameters, activity of micro-organisms, topography of the area, growth of vegetation, physical, chemical, and mineralogical characteristics of the soil through which meteoric water percolates, and other miscellaneous factors are known to influence quality of the groundwater (Handa, 1975). Present study of quality of groundwater especially in shallow and deep karst aquifer zones in Bhander Group of rocks of Rewa area, central India (Fig. 1), has a special significance in view of (1) discovery of huge quantity of groundwater in deeply-buried karst aquifers, within the shale sequence, mainly under confined conditions (Singh and Dubey, 1997) and (2) growing awareness, among users, about quality and optimum exploitation of groundwater for various usages.

METHODOLOGY

To study depthwise quality of groundwater, water samples from shallow and deep aquifer zones were collected, using selected dug wells and tube wells, respectively, in the same lithologic units, in the same locality. Water samples were analysed following conventional analytical techniques. To understand role of lithology and mineralogy of the aquifers in groundwater quality, available well dumps and bore hole cores were examined megascopically and microscopically. Water from certain bore wells has been found to leave yellow encrustation on the solids due to sulfur and smell like rotten eggs due to the presence of hydrogen sulfide. Water samples from such bore wells were collected asceptically, and were used for isolation of bacteria. The anaerobic bacteria was isolated in Thioglycollate medium (Difco Manual, 1974), and same was identified on the basis of morphological, cultural, and physiological characteristics. This exercise has shown that water stratum from deep aquifer zones harbours sulfate reducing bacteria of *Desulfovibrio desulfuricans* genus.

THE AQUIFER SYSTEM

The aquifer system comprises more than one hydraulically interconnected geological units (Fig. 1), of Bhander and/or Rewa Groups of the late Proterozoic Vindhyan Supergroup, through their existing anastomosing network of fracture systems. In general, limestones possess sufficiently saturated permeable zones to provide significant quantities of water consequent to considerable karstification. On

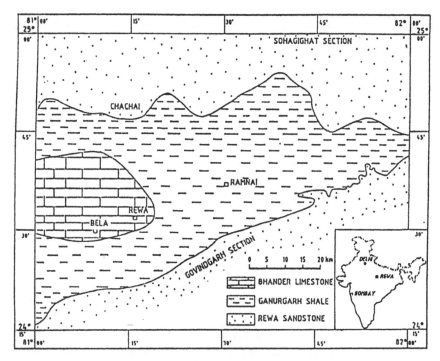

Figure 1 : Map showing geological units forming aquifer system in the Rewa area, central India

the other hand, sandstones and shales do not measure up to this requirement with relatively subdued nature of permeable zones in them. Even though all the geologic units of the aquifer system display karstification in miniature, but it is pronounced only in limestones. Average field coefficient of permeability has been found to be 10.3 m/day for a combined saturated thickness of limestone and shale aquifers of 10 to 30 m (Singh and Dubey, 1998a). Further, the approximate amount of recharge into the ground from infiltration annually per square kilometer area is reported to be 3 00,000 cubic meter (Singh and Dubey, 1999a).

LITHOLOGY AND MINERALOGY OF AQUIFER SYSTEM

Three major lithounits, which constitute aquifer system, are Rewa sandstone (Rewa Group), Ganurgarh shale, and Bhander limestone (Bhander Group) of the Vindhyan Supergroup in ascending order (Fig. 1). The sandstones are jointed, hard, compact, and quartzitic. They comprise chiefly quartz (upto 98%) with subordinate feldspar, muscovite, and clay. In transitional zone, sandstones show well-developed cross-hatched microcline (Fig. 2), and sub-rounded to sub-angular quartz grains, and relatively more clays. The Ganurgarh shales consist of laminated, micaceous, silty shales, and siltstones. They exhibit red, yellow, green, and grey color bands of varying thickness with prominent laminations. The mineralogical compositions of the shales show that their major constituent minerals are dolomite, calcite (the carbonates), and quartz (all confirmed by XRD). Evaporates (mainly gypsum) occur as veins and stringers (Fig. 3), which increase with depth. Gypsum is both fibrous (Fig. 4) and crystalline. Among the clay minerals, illite predominates along with some chlorite, mica, and iron oxides. In general, gross mineralogy of shales of different colors is the same(Singh, 1997). Within shales, profusely karstified limestone beds have been found in which huge quantity of groundwater is known to occur (Singh and Dubey, 1997). The limestones are mostly dirty-white, dirty-yellow, pink, and grey. They are massive and bedded, and are mainly of three types: argillaceous, stromatolitic, and dark grey. Mineralogically, argillaceous and dark grey limestones are made up of mainly micritic calcite, whereas the stromatolitic limestone often has alternate laminae of micritic calcite and dolomite rhombs . Limestones have well developed fibrous and dog tooth calcite cement spars (Fig. 5), besides drusy mosaic cement and syntaxial rim cement. Decreasing amounts of terrigenous material and insoluble residues have been found from stromatolitic to argillaceous to dark grey limestones (Singh and Dubey, 1998b). Occasionally, gypsum has also been found with limestones (Singh and Dubey, 1999b).

QUALITY OF GROUNDWATER IN LITHOLOGIC UNITS OF AQUIFER SYSTEM

Average concentrations of geochemical constituents of groundwater from different lithounits of the aquifer system of the study area are presented diagrammatically (Fig. 6). It is apparent from the same diagram that the quantity of dissolved solids is least in groundwater samples from sandstones, maximum in samples from shales and, closely next to shales, in samples from limestones. The solids (dissolved mineral species) are mainly carbonates, sulfates, and chlorides of calcium, magnesium, and sodium. Accordingly, the groundwater in sandstones is mostly low-salinity water, whereas waters from shales and limestones belong to the category of medium- to high- to very high-salinity waters, depending upon their depth.

QUALITY OF GROUNDWATER IN SHALLOW AND DEEP AQUIFER ZONES

Scrutiny of Fig. 7 suggests that the groundwaters from shallow (dug wells) and deep (tube wells) aquifer zones have distinct quality,

Figure 2 : Photomicrograph of sandstone from transitional zone showing microcline and sub-angular quartz grains. Nicols crossed X 66.

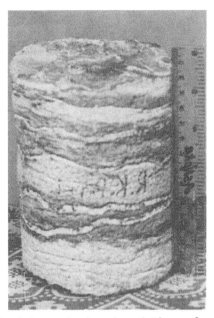

Figure 3 : Borehole core showing veins and stringers of gypsum within shales.

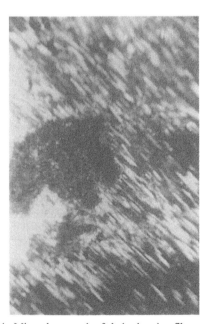

Figure 4 : Microphotograph of shale showing fibrous gypsum. Nicols crossed X 60.

Figure 5 : Microphotograph of limestone showing well developed fibrous and dog tooth calcite cement spars. The equant calcite crystals are also seen. Fibres are at right angle to the grain boundary. Nicols crossed X 99.

reflected in the form of increased total dissolved salts (cations and anions) with increasing depth, irrespective of lithology. This change in quality is, however, more pronounced in case of groundwaters from sandstone and limestone (about 5 times in each) than from shale (about 3 times) aquifer zones (Fig. 7). In other words, waters from shallow aquifer zones are less mineralised (low- to medium-salinity waters) and, with respect to this zone, waters from deep aquifer zones are upto 5 times more mineralised (thus high- to very high-salinity waters). Here it may be noted that even though groundwaters in deep zones of sandstones are upto 5 times more mineralised, with respect to waters from shallow zones of sandstones, but still they maintain low- to medium-salinity.

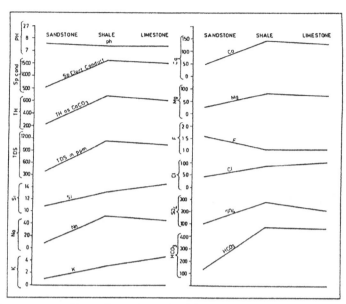

Figure 6 : Average compositions of groundwater in different lithologic units of the aquifer system

DISCUSSION

Lithology of rocks in sedimentary basins varies laterally and vertically and therefore quality of the groundwater in such basins also changes accordingly. Other variables like temperature, pressure, porosity, permeability, redox conditions, etc., also play greater role in the quality of the groundwater. Most of these factors appear to have played a distinct role in present case also. Fig. 7 suggests that the concentrations of cations (Ca, Mg, Na, K) and anions (HCO_3, SO_4, Cl), which determine water quality, are more in groundwater from shales and limestones, as compared with remarkably low concentrations of these ions in groundwaters even from both shallow and deep zones of sandstones. High concentrations of bicarbonates, sulfates, and chlorides of calcium, magnesium, and sodium are accounted for by the presence of dolomite, calcite, and veins-stringers of evaporates (gypsum) in shales; and calcite, dolomite, and some gypsum in limestones. Sandstones are almost devoid of such salt contributing minerals and, hence, low- to medium-salinity waters in them. High content of cations and anions in groundwater from shales, as compared to limestones (Fig. 7), is due to readily soluble nature of evaporates, that are abundantly present in shales (Fig. 4), as compared to less soluble nature of calcite and dolomite and also very little quantity of gypsum in limestones. Thus, it is clear that quality of groundwater in three geologic units of the studied aquifer system is controlled mainly by their respective mineralogy, besides many other factors.

Another one of the most conspicuous features is that there is a distinct increase in contents of cations and anions (TDS) in groundwater with increasing depth in all the lithologic units (Fig. 7). This change could be seen in groundwater even from sandstones where it is remarkably less mineralised (Fig. 7). Various possible reasons for contrasting quality of groundwater in shallow and deep aquifer zones are evaluated here. In the shallow zones, weathering, jointing, and karstification have generated enough secondary permeability which helps flushing of soluble compounds from rocks, as high permeability of host rocks allows for rapid circulations and proper mixing of groundwater. Such a circulation of the groundwater causes proper diffusion of salts as a result of which low dissolved salts are present in waters in contact with such rocks. In the shallow zones, pore pressure, vapour pressure, load, etc., are also low and therefore dissolved gases are low, with the result that chemical potency of groundwater at shallow zones is low which ultimately results in low dissolved salt contents in groundwater in shallow zones.

As groundwater percolates downward and laterally, it goes on dissolving calcium and magnesium carbonate until the carbon dioxide, which is absorbed from the atmosphere by the falling rainwater and from plant roots and decaying vegetation (Garg, 1978, p 253), is exhausted. Downward movement of the groundwater will go on continuously unless its movement is checked by permeability barriers. There is a gradual reduction in secondary permeability with increasing depth, especially in carbonate aquifers, due to diminishing karstification with depth (Singh, 1985). Reduced permeability impedes proper groundwater circulation, resulting in greater accumulation of soluble salts in groundwater in deep zones. Moreover, shales underlying the limestones being less permeable would also restrict the downward movement and impede proper circulation of groundwater, allowing thereby accumulation of salts in waters in carbonate karst aquifers. Increase in salinity in groundwater from deep aquifer zones in shale is due to increase in frequency in occurrence of veins and stringers of evaporates with depth. Another reason may also lie in the presence of the interstitial saline water that was never completely removed by flushing because of low permeability of the shale, as suggested by White et al (1963). Increase in chloride concentrations with increasing dissolved solids also indicates the presence of brackish saline water entrapped in pore spaces of shales at depth. Another possible reason for increase in dissolved salts with increase in depth may also lie in the downward percolation of surface water and its mixing with groundwater below water table. Indeed, it is also one of the significant reasons due to which there is a general increase in cations and anions in groundwater with increasing depth; because of this, it could be seen even in sandstone aquifers which are nearly monotonus in their lithology and mineralogy, in both shallow and deep zones.

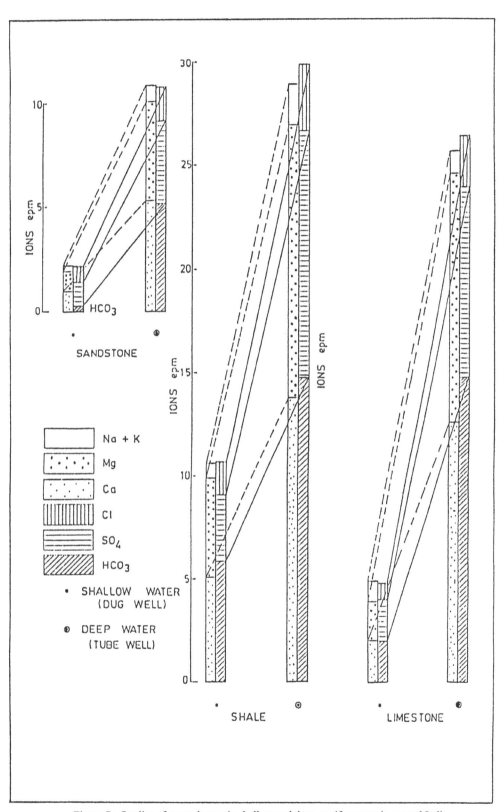

Figure 7 : Quality of groundwater in shallow and deep aquifer zones in central India.

Interestingly, certain water samples from deep zones, particularly in areas underlain by shales, showed the presence of sulfur and hydrogen sulfide, which suggest sulfate reduction (Walton, 1970, p 449) at depth. Formation of sulfur and hydrogen sulfide is known by certain bacteria through dissemination of sulfur containing amino acids, e.g., cystine and methionine, or through reduction of inorganic sulfur compounds such as thiosulfate, sulfite of sulfate (Pelczar, 1965). Absence of amino acids, and presence of anaerobic bacteria and also abundance of gypsum, especially in lower part of shales, favour reduction of calcium sulfate (gypsum), in deep aquifer zones, by anaerobic bacteria belonging to *Desulfovibrio desulfuricans* genus. Reduction of sulfate by bacteria produces carbon dioxide partial pressure, migration of which into aquifer increases acidity of water present and allows for dissolution of carbonate minerals (Henderson, 1984). Carbon dioxide in the water remained dissolved at depth due to high pressure promoting retention of higher quantity of dissolved salts, especially carbonates. Relatively higher amounts of carbonates, when compared to sulfate ions, inspite of abundance of readily soluble evaporates in shales, also support such a contention, because, unlike the carbonate minerals, the amount of sulfate dissolved in groundwater is dependent on the period of contact of the water with sulfate minerals in the aquifers (Walton, 1970, p. 472).

ACKNOWLEDGEMENTS

We are grateful to Dr. V.K. Saxena, NGRI, Hyderabad, Dr. S.K. Hansoti, and Dr. M.C. Bhurat, AMD, Hyderabad for their valuable suggestions and help during the course of this study. Dr. B.F. Beck, the conference organizer, has always been a source of constant encouragement and support to us, without which this paper would not have been published. We thank him profusely for the same.

REFERENCES

Difco Laboratories Michigan, 1974, Difco Manual of Dehydrated Culture Media and Reagents: Thioglycollate Medium, p. 199-200.

Garg, S. P., 1978, Groundwater and Tube Wells. IBM Pub. Co., New Delhi, 333 p.

Handa, B. K., 1975, Natural waters, their geochemistry, pollution, and treatment with a chapter on the use of saline water. CGWB, Min. of Agr. and Irrig. Tech. Manual No.2, 246 p.

Henderson, T., 1984, Geochemistry of groundwater in two sandstone aquifer systems in the northern Great Plains in parts of Montana, Wyoming, North Dakota and South Dakota. USGS Prof. Paper 1402-C, 84 p.

Pelczar, Jr. Michael, 1965, Laboratory Exercises in Microbiology. McGraw Hill Co., New York, p.141-144.

Singh, Y., 1985, Hydrogeology of the karstic area around Rewa town, Madhya Pradesh, India, In : Karst Water Resources. IAHS Publ. No. 161.p. 407-416.

Singh, Y., 1997, X-ray mineralogy of the Precambrian Ganurgarh shale, Bhander Group (Upper Vindhyans) around Rewa, Madhya Pradesh and its significance. Proc. of National Seminar on Precambrian Geology, Department of Geology, University of Madras, Chennai, p. 91-97.

Singh, Y., and Dubey, D. P., 1997, Deep zone karst aquifers as a boon in central India, In: Beck, B.F., and Stephenson, J. B. (eds), The Engineering Geology and Hydrogeology of Karst Terranes. A. A. Balkema, Rotterdam, p.245-249.

Singh, Y., and Dubey, D. P., 1998a, Analysis of pumping test data from Vindhyan aquifers in Rewa, central India, and its implications for groundwater supply. Jour. of the Indian Academy of Geoscience, V. 41, No.1, p.57-60.

Singh, Y., and Dubey, D. P., 1998b, Carbonate petrology of Bhander limestone of Rewa area, Madhya Pradesh, India. Gondwana Geol. Mag., V.13, No.1, p.21-38.

Singh, Y., and Dubey, D. P., 1999a, Hydrologic cycle and its effects on the flowing wells in the Rewa area, Madhya Pradesh, India, In: Inland Water Resources: India. Discovery Publishing House, New Delhi, p, 528-544.

Singh, Y., and Dubey, D.P., 1999b, Nature of cementation and its role in the diagenetic history of late Proterozoic Bhander limestone of Rewa area, central India. Jour. Indian Assoc. of Sedimentologists, V.18, No.1, p. 111-122.

Walton, W.C., 1970, Groundwater Resources Evaluation. McGraw Hill Co., New York, 664p.

White. D.E., Hem, J. D., and Waring, G. A., 1963, Chemical composition of subsurface water. USGS Prof. Paper-440F, 67p.

Geotechnical and Environmental Applications of Karst Geology and Hydrology, Beck & Herring (eds)
© 2001 Taylor & Francis, ISBN 90 5809 190 2

An evaluation of the performance of activated charcoal in detection of fluorescent compounds in the environment

CHRIS SMART & BRAD SIMPSON   Dept. of Geography, University of Western Ontario, London, ON, N6A 5C2, Canada, csmart@julian.uwo.ca

ABSTRACT

In order to systematically investigate the capability of granular activated charcoal in contaminant screening and tracer adsorption, a number of detectors were deployed for two to six days in a range of urban waters. Simultaneous grab samples showed a diverse range of fluorescence environments, ranging from relatively clean, steady groundwater discharge, to highly concentrated and variable treated municipal sewage. A wide variety of organic compounds and dyes were found in the waters, as chronic and acute contaminants. The relationship between charcoal and water depended on exposure time and loading. Short exposure times emphasized short wavelengths, longer exposure times emphasized longer wavelengths at the expense of shorter wavelengths. The magnitude of the effect depended on loading, being greater in enriched waters. In general, charcoal elutant shows a significant gain in fluorescence intensity over charcoal. However, there may be a loss at shorter wavelengths for samples with long exposure times. A similar bias was also discovered with the elution time of activated charcoal. Short wavelength fluorescence intensity peaked after a few minutes of elution; longer wavelength fluorescence increased over many days. These results show that charcoal is a reasonably effective material for adsorption of longer wavelength compounds. However, the ubiquity of fluorescent dyes in the environment, and the complex relationship between the waters and the elutant spectrum suggest that considerable care is required in the interpretation of elutant spectra.

INTRODUCTION

Granular activated charcoal (GAC) is widely used as a semi-quantitative sampling medium in fluorometric dye tracing, generally deployed in streams and springs in discrete packages know as "bugs" (e.g., Alexander and Quinlan, 1992.). The advantage of GAC bugs lies in their efficiency; they are considered to scavenge dye from water to give amplification of the ambient signal (~1 dB i.e. 10 times, e.g., Matisova and Skabakova, 1995); they are integrative, capturing transient dye breakthrough which discrete water samples might fail to capture; they are relatively immune to disturbance and vandalism; and they are inexpensive, most of the cost being associated with their distribution, collection and analysis.

GAC bugs are generally eluted in an alkaline alcohol mixture which is subjected to spectrofluorometric analysis (e.g., Alexander and Quinlan, 1997). The resulting spectra are amenable to quantitative analysis to permit objective identification of tracers with a known spectral signature (e.g., Tucker and Crawford, 1999) . However, it is generally accepted that differences in the adsorption efficiency of various dyes, uncontrollable flow in the field, and imprecision in most elution procedures make the absolute concentration of tracer less reliably ascertained. Nevertheless, relative concentration is indicated by spectral peak height, and this is commonly used as an indicator of quality assurance.

Granular activated charcoal is a highly adsorptive medium, prepared by roasting of coarse, high grade (coconut) charcoal. Fresh GAC contains a wide range of adsorption sites which vary in their strength and accessibility. Older charcoal tends to be less adsorptive, because of denaturing of the more energetic adsorption sites, and capture of organic molecules from the atmosphere. When deployed in the environment, GAC captures a broad range of organic molecules, including fluorescent dyes. A complex hierarchy of adsorption occurs based on the range of adsorptive sites, their accessibility, and the loading (composition and duration of the flow). Similarly, the desorption process targets more weakly bound, accessible species, with somewhat different results with different elutants (e.g., Cheremisnoff and Cheremisnoff, 1993).

These remarks might indicate a sophisticated grasp of the adsorption-desorption process in fluorometric dye tracing. However, these processes are of such complexity in the environment, that most practitioners have adopted a conventional "best practice" which appears to work well and is generally accepted as reliable. While accepted, such practice is not well-founded on either theoretical understanding nor controlled experiments.

PROBLEM DEFINITION

The characteristics of granular activated charcoal noted above suggest that it may be used in combination with the fluorescence spectrum as an integrative, low cost screening agent for organic contaminants. Many organic contaminants are fluorescent organic molecules which are adsorbed by GAC. While fluorescence spectroscopy is analytically indiscriminate, it provides a very inexpensive, broadly based diagnostic tool, suitable for efficient targeting of more costly discriminatory techniques such as HPLC and GCMS. The supposed ability of detectors to act as integrative sentries is particularly appealing in providing scope for capture of acute (transient) contamination episodes.

In general, we can report that fluorescence spectroscopy and charcoal are very effective in their respective screening and sentry roles. However, we were concerned with apparent inconsistency of the results obtained with GAC detectors. The organic molecules contributing to the fluorescence signature are those which contribute to fluorescence background, so our work is also relevant to tracing with fluorescent dyes in urban environments.

Once the basic spectral characteristics of site waters were assessed from water samples, three questions were addressed. First, what was the relationship between the spectra of waters and associated GAC-elutant spectra. Is the accepted "gain" of ~ 1 dB correct? The second question addressed the influence of exposure time and environment ("loading") on the resulting spectra. The third question concerned the influence of elution time on the resulting spectrum. This involved slightly more controlled elution of a standard charcoal sample.

METHODS

The city of London, Ontario, Canada occupies the confluence of the north and south forks of the Thames River which drains an area of 1800 km^2 and has a mean annual discharge of 37.6 m^3 s^{-1} and drains a Mid-west cord belt-style catchment. London is fairly typical of rejuvenated "rust-belt" cities, and the river is flanked by a number of former industrial sites, capped landfill sites and a major sewage treatment plant ("Greenway Pollution Control Plant" *sic*). Coal tar is known to be leaking into the river bed and tributary creeks are recognized as contaminated with PCBs.

Five gram aliquots of fresh granular activated charcoal were placed in nylon screen packages which were deployed just below water surface at a range of sites for between two and six days. Detectors were placed at the outlets of storm drains, former industrial sites and miscellaneous seeps along the south branch of the Thames River and downstream of its confluence with the North Branch. A number of detectors were placed in the river itself. Subsurface water samples were collected during deployment and retrieval of detectors.

Detectors were rinsed and about 2.5 g charcoal eluted using ~15 ml "Smart" solution 1-propanol, de-ionised water and 30% NH$_4$OH in the ratio 5:3:2 (Alexander and Quinlan, 1997, Smart 1972). Elution time ranged from 40 min to 5 hours within batches. In one case a series of analyses were made on a bulk preparation of elutant at intervals from a minute to several days. Waters and eluents were decanted into Spectrasil (UV-transparent) 3 ml fluorometric cuvettes which were analyzed on a PTI QM-1 spectrofluorometer. The instrument was calibrated using a Rhodamine B photon source and a NBS traceable lamp, and excitation and emission corrections were applied. Fluorometric dye standards were used to ensure continuity and replicate analyses performed to ensure adequate precision. Field (handled, but not deployed) and laboratory blanks (detectors soaked in deionised water or eluted without soaking) were run to test for contamination. Analysis of variance experiments suggested that the environmental signal was generally >>machine noise>>handling and sampling artifacts (Smart and Karunarantne 2000)

Synchronous scan spectra were run from excitation wavelengths (λ_{ex})=250-600 nm with wavelength offsets ($\Delta\lambda$)=20 and 90 nm with 2nm slit settings. Scan rates were 600nm min.$^{-1}$ at 1 nm intervals. These $\Delta\lambda$ provide identification of common fluorescent dyes (Käss 1999), petroleum compounds (Pharr, et al., 1992), Polycyclic aromatic hydrocarbons (PACs;Kumke, et al., 1995) and natural organic matter (Mobed, et al.,1996). Data were smoothed with a triple pass Hanning filter in Excel spreadsheets, and visualisation developed in Surfer and Grapher (Golden Software 2000).

WATERS

A diverse subset of a range of contaminated sites is drawn on here to characterise water and charcoal elutant signatures. Groundwater springs discharging from a sand and gravel aquifer provided a relatively clean, consistent signature comparable to deionised or tap water (Figure 1). Greenway sewage treatment plant showed much greater fluorescence and variation, with the appearance of miscellaneous peaks, presumably reflecting the composition of the feed water at the time. (Figure 1). The composition of the Thames river was relatively consistent at a number of sampling sites, and a representative sample is used (Figure 1).

The ubiquitous peak at 300 nm ($\Delta\lambda$=20) is ambiguous. It is generally regarded as indicating miscellaneous hydrocarbon fuels. Controlled extractions of various fuels into cyclohexane are reported to give exquisite differentiation at astonishingly small $\Delta\lambda$ (Pharr, et al., 1992). In our experience, the complex composition of fuels and their varied fate in the environment, Rayleigh scattering arising from the small $\Delta\lambda$, and the coincidence of the Raman peak for water make such characterization difficult to say the least. In general, volatilisation and degradation appear to result in displacement of the hydrocarbon peak to higher $\Delta\lambda$ and longer emission wavelengths. Many other peaks in this region the spectrum are transient, indicating acute (i.e. short-lived) contamination, typically in response to runoff events. Other peaks are repeatedly associated with particular sites and indicate a characteristic chronic contamination problem. Such features are not of primary interest in this paper.

The broad peak at around 400 nm ($\Delta\lambda$=20 and 90 nm) is characteristic of natural organic matter. This peak arises from a complex of humic and fulvic acids which shift in composition and expression as the complex degrades or shifts in composition. This ubiquitous peak compromises identification of many other compounds in this region of the spectrum, by masking and absorption (Roch, 1997, Mobed, et al., 1996).

Waters frequently show a peak at 512 nm (Δλ=20) which is presumed to be Uranine, presumably indicating the presence of tinted ethylene glycol antifreeze. The peak is acute in rivers (where it is indicative of road runoff, or a vehicle accident), but chronic at sites such as landfill sites, former wrecking yards and vehicle maintenance compounds. It should be no surprise that many tracer compounds commonly encountered in the environment; they are readily available, widely used, and freely disposed of. The critical question is whether their presence significantly compromises fluorescent tracing.

WATER AND ELUTANT SPECTRA

Any comparison of water and the elutant spectra, presupposes that the loading of the latter is adequately characterized by the former. This may be the case in the laboratory, or for groundwater and river sites which can be accepted as having a reasonably consistent chemical composition. More variable creeks, sewage outfalls, storm drains and intermittent flows are poorly characterised by occasional water samples. The nature of our investigation left this factor inadequately controlled.

Figure 1 shows the GAC elutant and water spectra for the three selected sites. In general, the charcoal elutant provides clear amplification of the water spectrum, and reasonable correspondence of spectral form . As anticipated, the sewage treatment plant shows the poorest match, reflecting the idiosyncratic loading of the detectors. A "Rhodamine" peak is observed in sewage discharge (λ_{em}~580 nm, Δλ=20) charcoal, but is absent from water samples, suggesting that it is more intermittent in appearance than the Uranine peak. Slight red shift of the Uranine peak reflects the elevated pH of the elutant solution (Käss 1998).

Linear gain of the GAC-elutant expressed as a spectral ratio (Figure 2) shows performance approaches and exceeds the 1dB (×10) expectation over much of the spectrum. However, the gain spectrum is far from flat, with significant gains only at higher wavelengths, and significant losses at shorter wavelengths, especially for longer exposure times. Furthermore, the magnitude of the gain appears to reflect the composition of the water, with the strongest gains occurring at peaks in concentration,

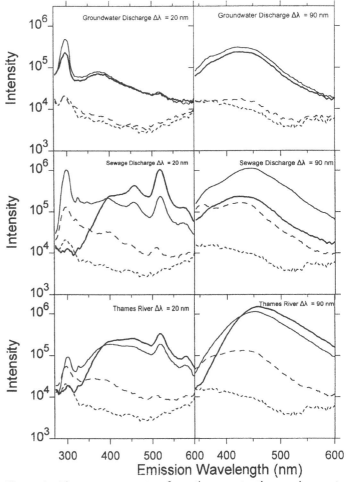

Figure 1. Fluorescence spectra from three contrasting environments. Groundwater discharge from a spring-fed pond, discharge from a sewage treatment plant, and a large surface river. Thin line represents two days exposure (four days for Thames River), Thick line represents four days exposure (six days for Thames River). Dashed line is arbitrary water sample during period of sampling, fine line is the spectrum for standard deionised water.

such as the Uranine peak. This is of concern as it suggests that dominant fluorophores appear to be preferentially adsorbed. The effect of exposure time is evaluated in the next section.

The gain from charcoal is important where fluorophores are at the detection limit. Thus, subtle peaks become more marked in elutant spectra. Note the use of a logarithmic intensity scale to show that this is not simply a scale effect. A result of the variable gain spectrum is that subtle and equivocal peaks in water are clearly defined, e.g, the appearance of a "Rhodamine"(577 nm) and 455 nm peaks in elutants of sewage and river water. However, we reiterate that the water sample does not adequately characterise the loading environment, especially at the sewage outfall.

EXPOSURE TIME

In dye tracing, control detectors are commonly placed in sites unlikely to encounter tracer, in order to provide quality assurance concerning the validity of a sample spectrum with respect to contamination and background. Replicate detectors are less commonly used, and provide an indication of sampling precision. Replicate sample spectra are expected to posses similar form, although differences in strength (fluorescence intensity) are anticipated from contrasts in environmental exposure time, elutant volume-GAC mass, and elution time. Our replicates exhibited startling differences in spectral form, arising from at least two factors: environmental exposure time, and elution time. The former is considered here.

Figure 1 demonstrates that two and four day replicates are very similar in groundwater samples, but radically different in sewage discharge. In part, this reflects the inherent instability of composition of the sewage treatment discharge. However, many other samples (e.g., Thames River samples in Figure 1) indicate a similar systematic bias, with longer exposure times reducing shorter wavelength intensity and enriching longer wavelength intensity. Unfortunately, the magnitude of this bias depends on the site; highly enriched waters (e.g., Sewage Discharge, Figure 1) showing more marked transition.

The more volatile, shorter wavelength compounds are initially scavenged very efficiently by the charcoal. Longer wavelength compounds are adsorbed less rapidly, but cumulatively. Are short wavelength compounds truly "lost" from the charcoal, or are they progressively bound more tightly? Certainly, the nature of activated charcoal provides a range of binding energies, with a range of accessibility. The tracer Rhodamine WT provides an example of a compound so strongly bound that conventional elution releases only a small fraction of the dye. However, the sensitivity to loading, indicates an exchange process (Smart 1972)

These results significantly compromise the use of activated charcoal for fluorescence screening. From the dye tracing perspective, the strongest effects appear to occur over the initial few days of exposure; typical weekly deployment may reach a quasi equilibrium. (A limited set of tests suggest that Uranine concentrations decline

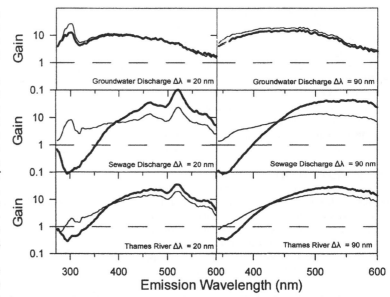

Figure 2. Gain in fluorescence intensity for activated charcoal detectors over water. The thin line represents two days exposure (four days for Thames River), and the thick line represents four days exposure (six days for Thames River). Dashed lines indicate a reference gain of unity.

at times approaching one week.) However, it is obvious that the scavenging performance of activated charcoal shifts over time. We are considering artificially preloading detectors in order to provide as flat an adsorption curve as possible. At tracer wavelengths, the exposure time effect may peak at a few days. However, it is not clear how broad wavelength-based generalization might be; the behavior of tracers may not follow this trend, and may depend on the particular loading environment. Further experiments are needed to follow up on exposure time effects. Meanwhile, a further constraint is placed on quantitative analysis of activated charcoal elutant spectra.

ELUTION TIME

A large GAC sample was obtained from the Brickpits, a known contaminated site with various hydrocarbon, lubricant and degraded products present. A single charcoal-elutant mixture was established, and aliquots were drawn at intervals of 1.7 minutes, 17 minutes, 2.7 h, 8.4 h, 27.8 h, 5 d and 13 d. Spectra were computed at $\Delta\lambda$=10,20, 50 and 90 nm on each aliquot. For each $\Delta\lambda$, the spectra were projected against log time, gridded and contoured (Figure 3). Replicate experiments were performed.

The resulting elution contour plots varied significantly with $\Delta\lambda$. At longer $\Delta\lambda$, the peak attributed to natural organic matter rises monotonically over ten days. There is slight decline at the lower limit of the spectrum after ~1 hour. Shorter $\Delta\lambda$ are more complex. Very short wavelength peaks, presumably representing hydrocarbon fuels ($\lambda_{em}\approx$290nm) rise rapidly peaking between 10 and 60 minutes, after which they decline. Less volatile hydrocarbons rise monotonically to give 380 and 390 nm peaks. Similarly, Uranine (~520 nm) increases monotonically across the 10 day period. There seems to be little loading at the wavelengths of Rhodamine (~580 nm), but the concentration seems to decline over a period of hours and then increase slightly over days.

The result displayed in Figure 3 are conditioned by the initial loading on the charcoal. We selected a reasonably rich spectrum for the evaluation. There are clearly volatile compounds still present on the charcoal. Generally, more volatile compounds appear to be very rapidly eluted, but subsequently lost from the elutant, presumably by evaporation into free air. NOM desorption is better behaved; the kinetics appear to be of similar order to that of Uranine. It is implied that approximately one day of elution achieves about 80% of 10-day concentration.

These tests were performed with the objective of assessing the potential of GAC for contaminant screening In this context, GAC is wickedly intransigent, possibly unusable. The situation appears more reasonable for apparent fluorescent dyes. However, it is clearly important to establish the elution kinetics of tracer dyes in use, and to design elution protocol accordingly, if any quasi quantitative analysis is to be attained.

CONCLUSIONS

The work we have undertaken to date has been exploratory in character, and not sufficiently well controlled to allow categorical conclusions. The complexity of the fluorophore-organic environment makes controlled experiments desirable, but difficult to design for general applicability. The type of organic loading occurring in the Thames river, however, is probably typical of that encountered in many tracer tests in North America and Europe.

In general, granular activated charcoal provides amplification of the fluorescence signature of waters. However, the gain is non-linear and conditioned by the loading environment and exposure time. Note that with GAC we found Uranine peaks in *all* the environments

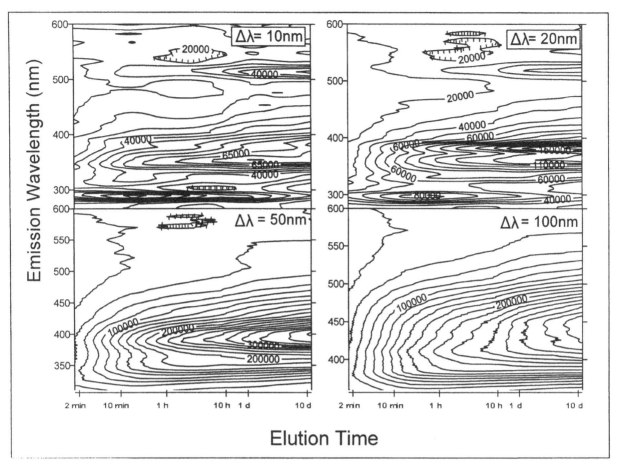

Figure 3. Evolution of synchronous scan spectra with elution time.

sampled. Only a mains water leak and controls remained clean! Rhodamine-type peaks, along with many others appeared quite frequently enough to make the issue of background definition a serious concern (See Smart and Karunaratne, this volume).

The gain of GAC detectors is sustained and improved at longer wavelengths with exposure time, while shorter wavelength compounds are lost. The reasons for this are unclear, but general volatility of many of these compounds may render them weakly adsorbed, and readily displaced as higher concentrations of larger molecules, fluorescing at longer wavelengths are accumulated. The adsorption story is also conditioned by the overall loading environment, with transitions occurring much more rapidly in enriched sites. Things seems to be more settled after a number of days. This may have lead to the general practice of deploying GAC detectors for around one week (Alexander and Quinlan, 1997.) It may be possible to stabilize the condition by preloading GAC detectors with organic matter prior to their field deployment.

It is important to match the elution period to the compounds of interest. The rate of elution of organic materials from GAC and the loss of more volatile compounds appears to follow reproducible forms. First, it is important to determine the elution rate constants for principle tracers. Second, it may be possible to strategically design elution protocols which allow preferred extraction of tracers and suppressed elution of background. Of course, if the background problem arises from tracer compounds, this will be of little use.

No investigation of particular tracer dyes was undertaken. However, some tracers, or close analogues were present in the environment. The relationship between tracer and organic adsorption and loading history are not understood. Overall, GAC detectors appear reasonably well suited to qualitative dye tracer applications, but should be used with a clear understanding of the non-linearities and artifacts inherent in their use.

REFERENCES

Alexander, C., and Quinlan, J.F, 1992. Practical tracing of groundwater with emphasis on karst terranes. A short course manual. Geological Society of America, Boulder Colorado.

Cheremisnoff, N.P., and Cheremisnoff, P.N., 1993 Carbon Adsorption for Pollution Control. PTR Prentice Hall, Inc., Toronto, 216 p

Golden Software 2000. Surfer and Grapher, Golden Colorado

Käss, W., 1998. Tracing Technique in Geohydrology Balkema, Rotterdam, 581 p.

Kumke M.U., Lohmannsroben, and Roch, Th., 1995 Fluorescence spectroscopy of polynuclear aromatic compounds in environmental monitoring. Journal of Fluorescence 5, 139-153

Matisova, E., and Skrabakova, S., 1995 Carbon solvents and their utilization for the preconcentration of organic pollutants in environmental sampling. Journal of Chromatography A 707 145-179.

Mobed, J.J., Hemmingsen, S,L., Autry, J.L., McGowan, L.B., 1996. Fluorescence characterization of IHSS humic substances: Total luminescence spectra with absorbance correction. Environmental Science and Technology 30 3061-3065.

Pharr, D.Y., McKenzie, J.K., Hickman, A,B., 1992. Fingerprinting petroleum contamination using synchronous scanning fluorescence spectroscopy. Groundwater 30 484-489.

Roch, Th., 1997. Evaluation of total luminescence data with chemometrical methods: a tool for environmental monitoring. Analytica Chemica Acta., 356 61-74.

Smart, P.L., 1972. A laboratory evalution of the use of activated carbon for the detection fo tracer dye Rhodamine WT. Unpublished MSc Thesis, University of Alberta.111p

Tucker, R.B., and Crawford, N.C., 1999. Non-linear curve fitting analysis as a tool for identifying and quantifying multiple fluorescence tracer dyes etc. 307-314 In Beck B.F., Petit, A.J., and Herring, J.G.,. (Eds.) Hydrogeology and Engineering Geology of Sinkholes and karst-1999, Balkema, Rotterdam. 477 p

Geotechnical and Environmental Applications of Karst Geology and Hydrology, Beck & Herring (eds)
© *2001 Taylor & Francis, ISBN 90 5809 190 2*

Statistical characterization of natural background fluorescence as an aid to dye tracer test design

C.C.SMART & K.C.KARUNARATNE   Dept. of Geography, University of Western Ontario, London, ON N6A 5C2, Canada, csmart@julian.uwo.ca

ABSTRACT

Tracing with fluorescent dyes is a core technique in karst hydrogeology. The critical criterion for evaluation of a dye trace is whether a tracer can be demonstrated to significantly exceed background concentrations. The critical level of significance is generally based on the judgement of the operator, rather than any objective protocol. Such an objective criterion might seem desirable, but the dependence of background on the environment, place and time of operation makes it impossible to implement. Background fluorescence arises from overlapping spectra of arbitrary fluorophores, and from contamination by the actual tracer material. The purpose of this paper was to develop a statistically valid sample of background variation at a single site over a sustained period of time.

One hundred and sixty eight hand samples were collected approximately daily from a gauging station on Medway Creek, Ontario (mean discharge 2.47 m^3s^{-1}), a surface drainage system draining an agricultural/suburban catchment on glacial till. Samples were analyzed by synchronous scanning on a PTI QM-1 spectrofluoro-photometer at $\Delta\lambda$ of 20 and 90 nm, and slit settings of 2nm.

There was considerable variability in both synchronous spectra, and poor serial correlation in time. A statistical analysis suggested that select areas of the spectrum were prone to greater increases in background: ~380nm@$\Delta\lambda$ =90 and ~300,~325 and ~510 nm @$\Delta\lambda$ =20nm. The shorter wavelengths do not interfere significantly with conventional fluorescent tracers. However, 510nm@$\Delta\lambda$ =20 corresponds exactly with the fluorescence peak of Uranine (sodium fluorescein, Acid Yellow 73), probably the most popular tracer dye. The distribution of fluorescence intensity by wavelength shows a number of regions of transient fluorescence. This can be best portrayed in terms of the skewness of the distribution.

Projection of fluorescence spectra in time allows an assessment of temporal variations in fluorescence. Based on daily samples, there is a strong, inconsistent and partial relationship to stream discharge. Fluorescence arises from release, flushing, exhaustion and dilution of natural and artificial compounds.

The results reported here can only be associated with the particular environment, and are based on a relatively small sample. However, they are probably indicative of background behavior in any moderately intensive agricultural/suburban environment. Broad organic background appears to vary considerably in composition over time, making routine extrapolative correction risky. Most tracer dyes, especially uranine can appear as acute contaminants. Background problems are most simply solved by using overwhelming quantities of dye, which may not be acceptable. More sophisticated injection encoding may be necessary.

INTRODUCTION

Background interference is a common problem in environmental analysis, in which a component of the observed values arises from sources external to the investigation. Background may constitute a consistent signal, or random noise, and may arise at any point in the sampling-analytical pathway. Proper identification, suppression and elimination of background interference is a well-known component of analytical procedures in the laboratory. Environmental background is much less studied, and may require elaborate sampling designs for its characterization and elimination.

Environmental dye tracing involves the injection and subsequent detection of fluorescent dyes for the identification and characterization of flow routes, primarily in karst terrain. Fluorescent dyes can be detected using filter and spectrofluorometers at concentrations below parts per billion. Filter fluorometers provide a measure of collective fluorescence emissions over a broad spectral window selected for the particular tracer of interest. Spectrofluorometers generate the fluorescence intensity for much narrower, traveling windows of a few nm width, and are able to provide resolved fluorescence spectra.

Fluorometric spectral signatures of common fluorescent dyes generally have well-known spectra and they are generally fairly narrow peaks. In contrast, the environmental fluorescence spectrum consists of a mix of broad overlapping peaks and local spikes depending on composition. Organic complexes such as humic acids present a broad peak, whereas organic pollutants may sharper peaks and peak clusters to the fluorescence spectrum, primarily at ultraviolet wavelengths. The form of the environmental fluorescence spectrum varies with the concentration and presence of contaminants and dissolved organic matter. In general, organic background peaks at ~300 nm emission and falls steadily up to 600 nm, depending on excitation-emission separation (8).

The background fluorescence in a dye trace therefore depends on the organic chemical hydrology of a stream, and will tend to vary in response to runoff processes and contaminant history. Dyes in the orange area of the spectrum (540-580nm) fluoresce in regions least prone to organic background interference. Green and blue dyes, however, suffer from an increasing threat of organic background interference.

Separate sampling and analytical procedures have been established for background correction, with significant differences between filter and spectrofluorometers. Sampling to characterize background can be either premonitory, or use analogue sites. Premonitoring implies the collection of samples from a site for some time prior to injection and arrival of any tracer. The average, or extrapolated trend of background is defined and subtracted from observations to obtain a "true" tracer concentration. Analogue sampling involves the collection of samples from a site very similar to a tracer-positive site at the same time as tracer is being detected at the latter. The background fluorescence at the analogue site is assumed to mimic the trace site, where true concentrations are estimated by simple subtraction. Postmonitoring for background is not recommended as samples can never be assumed free of tracer.

Analytical procedures for background correction depend on the instrumentation. Single filter fluorometers can only be corrected by subtraction of the premonitory or analogue background. Multiple filter fluorometers (or multiple analysis achieved by resetting the fluorometer) can be used to provide fluorescence in the uv-blue band, that is uniquely background. The ratio between this reading, and the green and red wavebands can be established by premonitory sampling or at analogue sites. Readings in the blue can then be used to predict background fluorescence in the green and red while dye is appearing in those wavebands. It is tempting to correct for background in situ, especially if real-time analysis of the breakthrough curve is necessary. However, it is preferable to retain raw and background data series to allow due inspection and consideration, as the pattern and style of background may change especially under variable flow conditions.

With a spectrofluorometer, the spectral form of background can be established in the UV-blue region of the spectrum and extrapolated into the blue-green-red sector and subtracted to obtain corrected dye peaks. This procedure can be visual, graphical or numerical. In particular, packages such as "Peak-Fit" (SPSS 1999) provide tools for simple characterization and extrapolation.

A number of concerns arise from these procedures. Sampling strategies require the extrapolation of premonitory values without a model for flow dependence or non-consistency. Analogue sites assume an appropriate match without clear criteria for successful matching. Multiple filter fluorometers are satisfactory if only the amplitude and not the form of the background curve changes. Spectrofluorometers permit more flexible correction, but require demonstration of the appropriateness of the extrapolation model which is seldom provided.

A distinction must be made between tractable *background* arising from a miscellany of non-tracers, and the greater difficulty posed by contamination by extraneous tracers. Stable tracer background implies a chronic source of contamination, and can be eliminated by simple premonitoring and extrapolation. Transient dye contamination may be impossible to distinguish from tracer breakthrough. None of the background correction procedures are effective in identifying the transient manifestation of extraneous tracer material. Extraneous tracer may arise from "autocontamination" or "allocontamination". *Autocontamination* arises from accidental spread of an injected tracer into samples (i.e. operator error). The risk of autocontamination is strongly linked to the sensitivity of the analytical system and is a significant risk when working at very low concentrations. *Allocontamination* arises from a tracer being present in the environment; in other words the appearance of "someone else's dye".

Uncertainties in background and correction procedures pose a threat to contemporary dye tracing normally countered by a combination of generous quantities of tracer and the assured "experience" of the practitioner. This is undesirable from an aesthetic, environmental and professional standpoint. Before improved procedures can be developed, it is necessary to better understand compositional and concentration changes of background in time and space, in the absence of an attempted dye trace. Two approaches can be envisaged: empirical monitoring of background to establish statistical properties of background, and more focused investigation of the fluorescence spectra of materials and sites likely to present background problems.

The purpose of this paper is to take the former course and to provide a statistical description of background spectra for a single site for a sustained period. The objective is to determine whether allocontamination ("someone else's dye") is occurring and whether the form of non-tracer background is consistent, varying only in magnitude with concentration.

METHODS

The approach was to sample waters routinely in the absence of any active tracer investigations. Thus "allocontamination", the accidental contamination by the operational team is reduced (but not eliminated) and is only briefly considered here. Waters were collected adjacent to a Water Survey of Canada gauging station (02GD008) located on Medway Creek, Ontario, Canada. The creek drains ~180 km^2 of agricultural-suburban land with mean annual discharge of 2.47m^{3}s^{-1}. The river is not controlled and tile drainage is extensive. Therefore, runoff is flashy with a mean annual flood of 58.9m^{3}s^{-1} and summer base flow ~ 0.06m^{3}s^{-1} (Water Survey of Canada). While there is no karst in the catchment, water quality is probably quite representative of rural-suburban areas of much of North America.

One hundred and sixty eight grab water samples were collected on a roughly daily basis, refrigerated and analyzed on a PTI QM-1 spectrofluorometer, a research grade instrument operated with Spectrasil quartz cuvettes and a xenon arc source. The instrument was calibrated using a Rhodamine B quantum source and an NBS traceable lamp. Real time corrections were applied for both emission and excitation variations. Some care is necessary in determining appropriate slit settings. Relatively narrow slits (2 nm) slits were used as they permit most detailed definition of the spectrum, and effective operation at small Δλ. Four slits are adjustable on the QM-1. Table 1 provides a qualitative outline of their impact. In most cases, paired slit widths should be used (i.e. A=D, B=C), unless anisotropic total luminescence peaks are anticipated. Dual synchronous scan spectra were produced at Δλ=20 and 90 nm from

Table 1. Implications for slit width settings in spectrofluorometer

Slit Location	Effect
A. Lamp-excitation monochromator	Sensitivity, chromatic coherence
B. Excitation monochromator-sample	Sensitivity, band width
C. Sample-emission monochromator	Sensitivity, bandwidth
D. Emission-monochromator-photomultiplier	Sensitivity, chromatic coherence

Table 2. Sampling-analytical precision tree for Medway Creek waters and PTI QM-1 fluorometer.

Source of variation (Sample size)	Mean Fluorescence Intensity	Standard Deviation	Coefficient of Variation (%)
Replicate analysis of 1 cuvette (10)	46450.	1861	4.0
Replicate decanting of one water sample (15)	47900	1898	4.0
Replicate sampling (20)	57300.	2278	4.0
Hourly serial samples (base flow) (24)	81000.	3391	4.2
Total daily samples (168)	127000	121064	95.4

excitation range 250-600 nm and emission 270-600 nm. These wavelengths were selected as providing appropriate matching to the Stoke's shift of common fluorometric tracers (Käss 1997), common hydrocarbon fuels and "humics". Selecting $\Delta\lambda=20$nm also allowed the possibility of opening up the slits for greater amplification without introducing Rayleigh scattering. Step length was one nanometer and integration time 0.1 s.

Calibration against unknown substances is not possible, but the known peaks of the tracer dyes Uranine (AY 96) and Rhodamine WT (AR96) were used to confirm spectral accuracy. The sensitivity of the instrument was demonstrated by assessing the signal to noise ratio across the Raman peak for water (PTI 1998). A precision tree for the sampling procedure was also developed to explore sources of variance in the data (Table 2). The total fluorescence emission between 300 and 390 nm was compiled and compared for the range of analytical and handling procedures. The fixed coefficient of variation in all pure replicates suggests that analytical noise dominates handling effects. The slight rise over 24 h and massive rise over the year indicate that a very strong environmental signal is present at the daily scale. This justifies focusing on the daily samples in the first instance. (A number of hourly sample suites were also collected, but will be reported elsewhere.)

RESULTS

The daily spectra were compiled by emission wavelength (λ_{em}) in two Excel spreadsheets for $\Delta\lambda=20$ (λ_{em},270-600nm) and $\Delta\lambda=90$ (λ_{em},340-600nm). Descriptive statistics by wavelength and spectral graphs were generated and inspected. Considerable variation was observed in both concentration and composition. The latter can be highlighted by taking a ratio between sample intensity and the total fluorescence intensity in a waveband, (Figure 1). While it is difficult to decipher the details in such a figure, it is apparent that variation in composition constitutes an acute problem for $\Delta\lambda=20$ nm. The peaks around 300 nm suggest the presence of volatile hydrocarbon fuels; the peak at 512 nm corresponds exactly to that of Uranine tracer dye.

The fluorescence intensity data were collected for each waveband, and ranked to provide a cumulative distribution, indicative of the distribution of intensity at each waveband. The aggregate approach does not allow direct application of conventional curve fitting software (e.g., Peakfit, SPSS 1999), but such analyses on background proved highly subjective. The percentiles of the distributions (Figure 2) can be used to visualize the consistency of the distribution of background, and to identify regions in which acute contamination can occur. In general, the percentile plots indicate a consistency of form up to the higher percentiles, implying relatively consistent behavior of background. If standard practice in tracing is to only accept concentrations an order of magnitude above background., then none of these observation lies at greater than 10 times the median fluorescence. However, the 390 and 300 nm 99th percentiles exceed 10 times the minimum values implying some risk with selected optical brightening agents.

The higher percentiles generally plot as translations of the median intensity. However, they depart from median form around 390 nm ($\Delta\lambda=90$) and 512 nm ($\Delta\lambda=20$). This indicates the occasional appearance of additional fluorescence in these regions. The latter coincides with the emission peak for Uranine, which seems to have a "probability of appearance" of ~5%. The distortion of a distribution of intensity caused by an acute contaminant can be indicated more directly by computation of the skewness of the distribution (Figure 3). The data are positively skewed at all wavelengths, but a number of significant areas of strong skewness are seen indicating areas of acute contamination which are not obvious in the simpler percentile plots. In particular, peaks arise at 390 and 525 nm for $\Delta\lambda=90$ nm, and at 290, 330, 410 and 510 nm at $\Delta\lambda=20$ nm. This suggests that many areas of the spectrum are prone to acute contamination of dye mimics. The orange area of the spectrum again appears most resilient.

If the synchronous scan spectra for a given $\Delta\lambda$ are projected in time, and the fluorescence intensity contoured in the resulting time-luminescence space, it is possible to explore the temporal patterns of daily background. Figure 4 shows the full array of daily florescence spectra from April to November 1999. Embedded in the figure are stream discharge and symbols indicating sampling times. Some of the structural features on the map are artifacts arising from irregular sampling or limitations of daily sampling. However, it is clear that there are episodes throughout the year when fluorescence peaks occur. Many of these are associated with runoff events, but in no consistent manner. Luminescence certainly does not scale with discharge. At $\Delta\lambda=20$, peaks at 300 and 300-400 nm are probably associated with fresh and degraded (volatilized) gasoline. It is disturbing that such features occur so frequently. From the dye tracing perspective, the relatively low background from 480-600 is encouraging. However, three 512 nm "uranine" peaks can be seen , indicating acute allocontamination with uranine. The most likely source of the gasoline and uranine is road runoff,

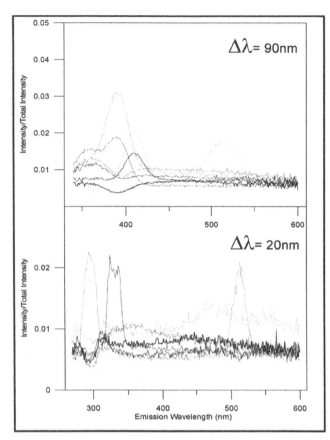

Figure 1: Fluorescence composition. Δλ=90nm and Δλ=20nm synchronous scan spectra for selected daily samples from Medway Creek. Curves corrected for concentration to show differences in composition relative to total at a given wavelength.

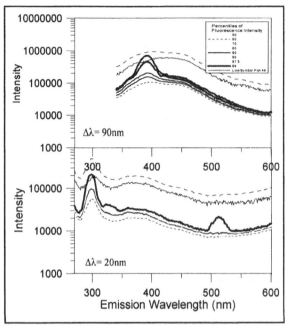

Figure 2: The percentiles of fluorescence intensity for 183 background samples analyzed at Δλ= 20 and 90 nm.

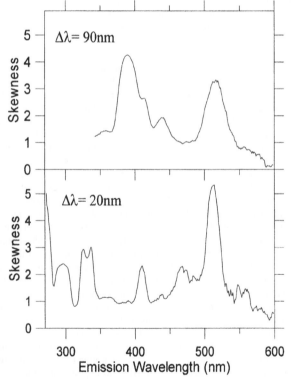

Figure 3: Skewness of fluorescence intensity for 183 background samples analyzed Δλ= 20 and 90 nm.

derived from highways in the catchment. The large peak on day 238 was preceded by a major road accident in which uranine-tagged coolant, gasoline and other fluids were released. Our work on road runoff suggests that general fluorescence builds up with time between storms, and is subsequently mobilized from the road surface, sumps and drains (e.g. Gromane et al. 2001).

Superficially, the spectral response at Δλ=90nm is similar, though characterized by a single broad peak. However, the form of the peak changes, and there is an imperfect correlation with Δλ=20nm peaks occurrence. The Δλ=90nm spectrum is ambiguous in terms of fluorophores. In general, it can be related to "humics and fulvics", the broad class of decay intermediates of natural organic matter. In natural systems the presence of a organic matter peak indicates flushing of dissolved organic matter in a catchment. Unfortunately, there are many artificial compounds also fluorescing at these wavelengths, not least optical brighteners used in common laundry detergents, and as cosmetic enhancers on vegetables. Many optical brighteners and "blue" tracer dyes fluoresce in the same region of the spectrum (Käss 1997). The convenient distinction between background, and allocontamination becomes blurred in this region. Masking of tracers and other compounds is caused by broad organic peaks, making separation impossible (Roche 1997). There are many compounds with such similar spectra, that no simple distinction can be made between tracers and background. If the tracer material has a well-defined narrow peak and is at high concentration, it may be possible to separate it from background. Overall, the instability implied by Figures 1 and 4 brings considerable uncertainty to tracer identification.

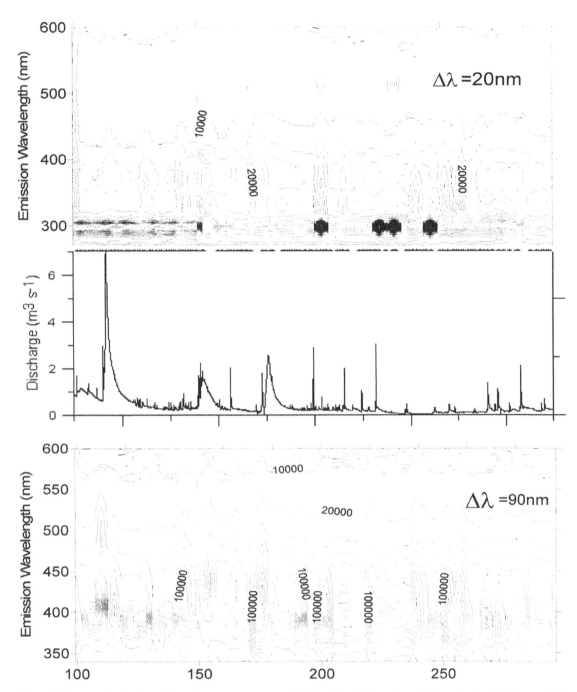

Figure 4: Temporal projection of sysnchronous scan spectra for Δλ=20 and 90 nm from Medway Creek 1999. Stream discharge shows variable correlation with fluorescence episodes. + indicates timing of samples. Irregularities in sampling are responsible for some features of the time series.

The complexity of the fluorescence-time response arises not only from the mix of artificial fluorophores, but also from the range of processes influencing their concentration in streams. Runoff events are known to constitute a complex mixture of flushing and dilution of prestorm water. The organic composition of the water depends on the degree of saturation of the source area of storm runoff, and the state of decomposition of organic material. Artificial fluorophores may be released intermittently, or flushed from major sources. In contrast, compounds mobilized by road runoff appear to be rapidly depleted by rainstorms.

CONCLUSIONS

To our knowledge this is the first systematic exploration of background fluorescence from the perspective of dye tracing. Of necessity, this work focused on a single site. This leaves open the question of the generality of our findings. However, we are satisfied that the exercise has reasonably represented the natures of background fluorescence in suburban-agricultural Midwest catchments. There appears little reason to imagine that karst aquifers in similar areas will not exhibit comparable results. However, parallel work on more industrial-urban settings (Simpson 2000), suggests that active and abandoned industrial and landfill sites may contribute strong, distinctive spectra, depending on the compounds present. Similarly, treated or untreated sewage can vary dramatically in their fluorescence properties. Knowledge of past and present land use appears to be an important aspect of quality assurance in water tracing.

While the exercise spanned the best part of one year, the sample size (183) remains relatively small for characterization of the critical high extremes of fluorescence. Acute background or contamination episodes appear to be short-lived and challenging to sample adequately. Background changes and discharge events are related, but in a complex and non-absolute manner. These feature mean that conventional premonitoring may fail to define background even should discharge be stable. Our interpretation of some allocontamination events is that they are local and idiosyncratic. Therefore parallel monitoring of non-traced analog streams may also fail.

While we have not pursued the question of the broad organic background peak, it appears to shift unpredictably which may pose difficulties in standard extrapolation correction procedures.

The present work focuses on water samples, whereas many traces are performed with activated charcoal detectors. We do not have a comparable data base for activated charcoal, but their time integrative function will make them more vulnerable to transient allocontamination. Furthermore, charcoal spectra clearly depend on deployment and elution practices (Smart and Simpson, this volume). Most worrying was the persistent presence of a uranine peak in virtually all field deployed charcoal detectors.

The results analyzed here, related works in the region and a recent editorial in Cave Science (Gunn and Lowe1999) suggest that most tracer dyes are present to some degree in the environment, albeit at relatively low concentrations. This undermines the basis for ultra-high precision tracing. The sensitivity of analytical equipment is less important than adequate background characterization. Powerful analytical software may extract subtle, even invisible peaks from a spectrum. This is to no avail if the dye is from some extraneous source.

The traditional solution to this problem is to use abundant quantities of tracer dye to produce an unequivocal positive result. Such practices are undesirable from an esthetic and ecological point of view. Tolerance of most tracers in the environment rests on the extremely low, sub-visible concentrations used. In some cases, the definition of a well-defined breakthrough curve provides some assurance. However, there is no obvious reason to suppose that extraneous dye releases (allocontamination) will not themselves produce a breakthrough curve. In environments suffering high risk of background fluorescence, it may be appropriate to develop simple tracer encoding practices. For example, two tracers may be injected in parallel, or a series of injections performed to make the labeling of the injection point less ambiguous. Such procedures work well in a general way, but the translation of injection signals to an output is not necessarily simple.

REFERENCES

Gromane, M.C., Garnaud, S., Saad, M., and Chebb, G., 2001. Contribution of different sources to the pollution of wet weather flows in combined sewers. Water Research 35: 521-533.

Karunaratne, K.C., 2000. Spectral and temporal variations in background fluorescence: implications for dye tracing. . Unpublished BSc Dissertation, Department of Geography, University of Western Ontario. 74 pp.

Käss, W., 1997. Tracer technique in Hydrogeology, Balkema, Rotterdam. 581 p

Roch, T.Ch., 1997. Evaluation of total luminescence data with chemometrical methods: a tool for environmental monitoring. Analytica Chimica Acta 357: 61-74.

Simpson, B.C., 2000. An assessment of activated carbon and spectrofluorometry in screening for environmental organic contaminants. Unpublished BSc Dissertation, Department of Geography, University of Western Ontario. 96 pp.

Geotechnical and Environmental Applications of Karst Geology and Hydrology, Beck & Herring (eds)
© 2001 Taylor & Francis, ISBN 90 5809 190 2

Effectiveness of geologic barriers for preventing karst water from flowing into coal mines in north China

WANFANG ZHOU P.E. LaMoreaux and Associates, Inc., Oak Ridge, TN 37830, USA, wzhou@pela-tenn.com

GONGYU LI Institute of Hydrogeology, Central Coal Research Institute Xi'an Branch, Shaanxi, China, igylgx@pub.xaonline.com

ABSTRACT

Coalfields in North China encompassing more than ten Provinces contain six to seven coal seams in the Permo-Carboniferous strata. The lower three seams, accounting for 37% of the total reserves, are threatened with karst water from the underlain Ordovician limestone. Hundreds of water inrush incidence have occurred in which a large amount of water suddenly flows into tunnels or working faces under high potentiometric pressure and over thirty mines have been flooded over the last twenty years. Large-scale dewatering or depressurizing of the karst aquifer was considered essential to avoid water inrushes and keep the mines safely operational. This practice has caused sinkholes, dry springs, water supply shortage, and groundwater contamination in the surrounding areas, which is environmentally not permitted. One of the alternative water control measures is to make full use of the rock layer between the coal seam and the karst aquifer as a geologic barrier.

A geologic barrier is defined as a natural rock zone between the man-made underground cavity and the water-bearing medium. Karst water inrush in mines is actually the result of interaction between the underlying confined water and the rock in the geologic barrier. Similar to the application in the nuclear industry where geologic barrier is used to contain radioactive wastes, the barrier of this application is considered as a hydraulic barrier as well with the objective to prevent or constrain water flow from the underlying aquifer into mines. Its effectiveness to constrain water flow is described by a parameter referred to as hydrofracturing pressure (P_{hf}). When the water pressure in the underlying aquifer exceeds P_{hf}, a wedging effect takes place within the fractures of the geologic barrier and as a result, water inrush occurs. *In-situ* hydrofracturing tests were used to determine P_{hf} in bauxite and silty sandstone at tunnels. The P_{hf} in the silty sandstone is larger than that in the bauxite but they both vary with depth (distance from the bottom of the tunnel). Based on the test results, a new safety criterion for water inrush was derived for mines and it has been successfully applied to mining practices with the minimum effort of dewatering in the karst aquifer. The same criterion can also be applied to tunneling and quarrying in areas with similar geologic conditions.

INTRODUCTION

In China, karst develops and distributes widely in rocks that range in age from Archaeozoic to Cenozoic, but are predominantly Paleozoic. Carbonate rocks occupy an area of about 3.25 million km^2 of the country: of this bare karst is some 1.25 million km^2 and the rest is covered or buried karst (Yu, 1994). Groundwater in the karstified carbonate rocks is a valuable natural resource for local people. Unfortunately, many mineral deposits such as coal, iron, lead and zinc, gold, aluminum, and copper are located in between, or above, or below the karst aquifers. The majority of the well-known deposits with large quantities of water (pumping water over 1 m^3/s) are karst water-impregnated deposits.

The coal seams in the coalfields of North China lie in the Permo-Carboniferous strata, as shown in Figure 1. The Taiyuan Formation of Carboniferous system has a thickness of 95 to 163 meters (m), consisting of argillaceous shale and sandstone. From the top to the bottom, the coal seams are Xia-jia, Da-xing, Xiao-qing, Shan-qing, Ye-qing and Yi-zuo and their total average thickness is 9 m. Their roofs consist of mainly thin-bedded limestone with varying thickness from 2 to 7 m. Except for the lowest layer of limestone that may have hydraulic connection with the underlain Ordovician limestone, water in the rest of the thin bedded limestone is generally static. It is relatively easy to dewater or drain. The Shanxi Formation of the Permian system has a thickness of 90 m. It includes one coal seam—Da-mei with the thickness of 6 to 7 m. Beneath the Taiyuan Formation is the Benxi Formation, which is 18 to 53 m thick and consists of arenaceous shale, bauxite shale and iron ores. The Ordovician limestone is a highly permeable confined aquifer. Its average thickness is 650 meters. Due to the potential impacts of the confined water in the Ordovician limestone on the mining activities, the three lower coal seams, accounting for 37% of the total reserve, are listed as prospective reserves.

One of the major impacts of the groundwater on the mining activities is the unpredictable occurrence of water inrush, in which a significant amount of water suddenly invades the underground working areas from the underlying aquifer under potentiometric pressure. Hundreds of water inrushes have occurred and over 30 mines have been flooded in the last 20 years. Dewatering or depressurizing of the Ordovician limestone has been essential to keep the mines safely operational. However, this practice has resulted in many engineering and environmental problems in the surrounding areas such as sinkholes, dry springs, shortage of water supply, and surface water and groundwater contamination (Zhou, 1997a).

Water inrush is the result of interaction between water and rocks in the geologic stratum between the coal seams and the Ordovician limestone. It occurs when the strength of the stratum is not strong enough to resist the water pressure. The position of the water inrush is often related to geologic structures. Adjacency, intersection and pinch of faults, anticline and synclinal axes are more susceptible to water inrush. In the studied area, over 78% of water inrushes are related to faults and the northeast fracture group controls 62% of them.

Figure 2 shows the data points of water inrushes collected at Fengfeng coalfield. The vertical coordinate represents the thickness of the rock layer between the coal seams and the Ordovician limestone, and the horizontal coordinate represents the potentiometric pressure of the karst water. The invasion points are concentrated mostly on the up right, while the safety points on the down left. Between them, there is a natural limit approaching to a straight line. Clearly, the rock layer between the coal seam and the threatening aquifer acts as a geologic barrier or a hydraulic barrier that prevents the water in the Ordovician limestone from invading and flowing into the underground cavities.

The effectiveness of the geologic barrier depends on its thickness, lithology, and integrity. Water inrushes are unlikely to occur when the geologic barrier is thick. The four upper coal seams are free from water invasion because the geologic barrier is over 100 m. Hard rocks such as limestone and sandstone, have high intensity, for example, a layer of medium-grained sandstone of 2 m can bear 7 kg/cm^2 of water pressure. Flexible rocks such as shale do not have intensity as high as hard rocks. However, they may have higher capacity of water resistance, as found in another north China's coalfield (Figure 3). The interception for shale seems to be larger than that for the hard rock, but the slope is relatively sharp, indicating water inrush could occur under higher water pressure for the same thickness of the geologic barrier. When the geologic barrier consists of inter-bedding layers of flexible and hard rocks, water invasion can hardly take place. As shown in Figure 4, the interception for such an arrangement of rocks is very small, only 3 m. Coals were mined successfully under a water pressure of 8 kg/cm^2 when the geologic barrier was 13 m thick.

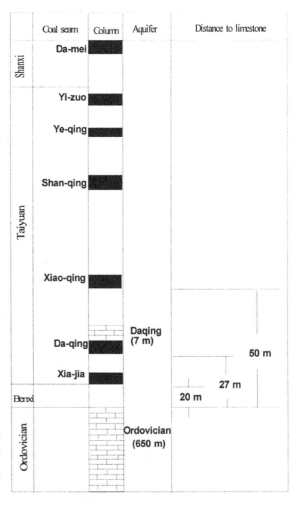

Figure 1: Geologic column in coalfields of North China

Obviously, water invasion occurs more easily under higher water pressure. When Shaqing coal seam was extracted at -90 m (above sea level) in one coalfield, water pressure in the Ordovician limestone was 22 kg/cm^2 and the protective layer was 40-45 m thick. No water inrushes took place. However, when the mining level was extended to -170 m, the water pressure increased to 30 kg/cm^2, and 6 water inrushes took place already although the protective layer remained the same.

The non-zero interception to the vertical axis implies that water inrush could take place even when the groundwater is not under pressure. We envisioned that this is the effect of mining activities. Part of the protective layer might be destroyed by mining operation. This is clearly illustrated by the different distribution of the invasion points at the working faces and in the shafts. For the same geologic barrier, water inrushes are more likely to occur at the working face than in the shaft. The space and span of the working faces has a significant influence on water-resisting capacity of the geologic layer. In addition,

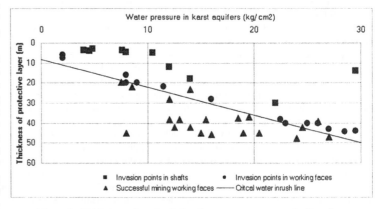

Figure 2: Thickness of geologic barrier vs water pressure in Fengfeng

water invasions in shafts might happen with a delay of one to two years after excavation of the shaft because of the long-term effect of shaft excavation on the floor.

IMPACT OF MINING ACTIVITIES ON GEOLOGIC BARRIER

With regard to the influence of mining activities on the protective layer, the data from gas-discharge in coalmines could be used as a reference to the destroyed thickness. Gas was liberated 20-80 m below the layer after mining. Due to different properties of gas and water, fractures caused by mining can conduct gas but may not be conducive to water. The thickness of the rock through which both gas and water can flow is the parameter of concern. Two water injection tests (#1 and #2) were conducted in one coalfield to investigate the destroyed thickness. In each water injection test, 5 angled boreholes were drilled into the protective layer of a mining slope (Figure 5). The vertical distance between the boreholes was 1 m and the shallowest borehole (borehole 1) was 2.5 m below the slope floor. The geologic barrier was composed of sandy shale, argillaceous limestone and coal. A small fault with 0.5 m displacement was observed in the barrier. A long-wall extraction approach was employed for the tests.

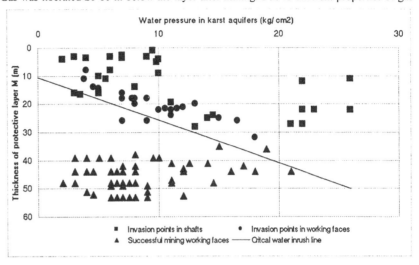

Figure 3: Thickness of Geologic Barrier vs Water Pressure in Jiaozuo

Figure 6 shows the influence of mining activities on the volume of water injected for water injection test #1. The distance to the mining face is expressed by the horizontal coordinate with zero at the face. To the left and right are the distances to the mining direction and to the extracted zone, respectively. The time (days) calculated according to the average speed of face advance during the test is shown in the figure as well. The volume of water (liter/hour) flowing through the boreholes of various depths and horizontal distance is shown on the vertical coordinate.

Borehole 5 was discharging water before the test with a water pressure of 0.68 - 0.85 kg/cm^2. The water injection pressure applied to this borehole was 1.5 kg/cm^2. The pressure applied to the other boreholes was 1 kg/cm^2. The water flow in borehole 5 increased when the central

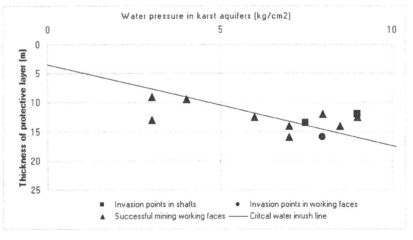

Figure 4: Thickness of Geologic Barrier vs Water Pressure in Handan

borehole was 14 m away from the working face after mining, while water flow in boreholes 2 and 3 seemed to decrease. The water flow in boreholes 2 and 3 had a slight increase at 52 m and the water flow decreased in borehole 4. Figure 7 shows the influence of mining activities on volume of water injected for test #2. Water pressure of 1 kg/cm^2 was applied to all the boreholes. In general, water flow started after water injection reached its maximum volume at the working face.

Change of the amount of water injected through the boreholes implies the effect of mining activities on the geologic barrier. The thickness of the geologic barrier destroyed by the mining could be analyzed by the water overflow from the abandoned galleries through the boreholes. Fluctuations of the water volume injected before and after mining decreased as the depth of the tested segment of the borehole increased. When the depth reached 7 to 8 m, the fluctuation became very small. In addition, the geologic barrier that was initially impermeable began to conduct water through holes at 6 -7 m below the slope after the mining. On average, the thickness of the geologic barrier destroyed by mining activity was approximately 8 m in North China.

LABORATORY EXPERIMENTS ON FAILURE OF GEOLOGIC BARRIER

Experiments were conducted in laboratory to test the failure mechanism of the geologic barrier under high water pressure in a tri-axial filter (Figure 8). The rock sample is clayey limestone with two fractures perpendicular to each other. It is 150 mm in diameter and 400 mm long. A water pressure (P_w) of 35 kg/cm^2 was applied to the sample. Stresses (σ) simulating the lateral earth stress of 200 m below the ground were applied to the cylinder of the sample. The applied stress varied to simulate three mining-related stages—compression, dilation, and recovery.

In initial phase (0—1 min), the water flow rate is greater than 300 m^3/min, where $P_w \gg \sigma$. With the increase of the stress, the rock underwent the compression process (1—23 min) where $P_w < \sigma$. The fractures were gradually forced to close and the water flow decreased to zero. At the dilation period, the earth stress decreased due to the pressure release caused by excavation. When $P_w > \sigma$ (28—29 min), the fractures opened up and water began to flow again. The maximum water flow was 120 m^3/min. During the recovery period, the roof would collapse and the stress gradually increased to its normal. When $P_w < \sigma$ (>29 min), the fractures closed again and water flow receded.

The experimental results indicate that the failure of the fractured rock sample as a hydraulic barrier depends on the relationship between the water pressure and the lateral stress. The water flow was observed when the hydraulic pressure exceeded the lateral stress. This process is very much similar to the spontaneous hydrofracturing in which the wedging effect takes place when the hydraulic pressure exceeds the hydrofracturing pressure (P_{hf}). P_{hf} is approximately the same as the minimum earth stress in value and it is a regular practice to measure the minimum earth stress by hydrofracturing tests in boreholes (Kesseru, 1997).

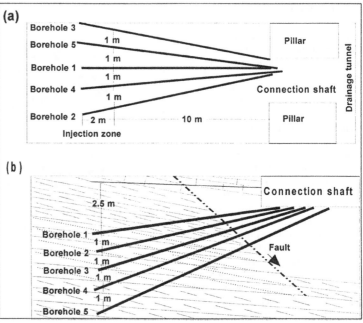

Figure 5: Water Injection Tests. (a) Plane view; (b) Profile view

Therefore, the effectiveness of the geologic barrier can be quantitatively described by measuring its hydrofracturing pressure or the minimum stress. The presence of fractures is pre-requisite to the spontaneous hydrofracturing.

INITIAL CONDUCTIVE ZONE IN GEOLOGIC BARRIER

The permeability of the geologic barrier is in general low; however, there is no distinctive contact plane between the barrier and the Ordovician limestone. The Carboniferous barrier is unconformably overlain the Ordovician limestone. There is a long period during which the limestone underwent various weathering processes. The weathering processes resulted in an irregular limestone surface. In addition, the geologic barrier has itself undergone numerous tectonic movements. Fractures exist in the barrier. When the fractures in the barrier are connected with the underlying limestone, the karst water in the limestone penetrates upward into the barrier under its potentiometric pressure. The area that has already been invaded by the karst water prior to mining is defined as the initial conductive zone. This zone cannot play an effective role in preventing karst water from flowing into the mines. However, the existence of this zone provides the essential condition for the wedging effect. At locations where this zone does not exist, the barrier would remain its integrity even the water pressure exceeds the hydrofracturing pressure. The barrier breaks only when the water pressure exceeds its shearing strength by bending of the barrier.

Continuous monitoring of the potentiometric pressure while drilling into the geologic barrier is necessary to detect the height of this zone. Table 1 lists the exploratory results at one mine.

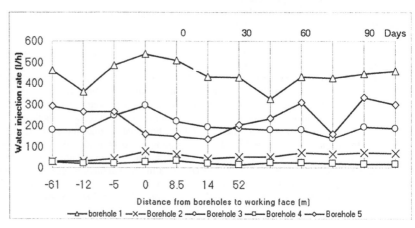

Figure 6: Water Injection Rate during Mining for Test #1

Clearly, the initial conductive zone is closely associated with the lithology and structure of the barrier. It more likely develops in harder rocks with better-developed fractures. High conductive zone may develop along fractures and around collapsed zones, which is schematically illustrated in Figure 9. Therefore, fractures and paleo-collapses are likely to become avenues for groundwater flow or contaminant transport (Zhou, 1997b)

Table 1 Measured Height of the Initial Conductive Zone

Thickness of geologic barrier (m)	Potentiometric pressure of Ordovician limestone (kg/cm2)	Height of initial conductive zone (m)	Characteristics of the geologic barrier
21.4	11.0	0.92	Fine sandstone, bauxite, clay stone, very few fractures
19.7	10.5	0.0	
23.6	12.6	0.0	Fine sandstone, coal seam, bauxite, few fractures
21.6	12.6	<7.6*	
22.4	12.0	8.5	Fine to medium-grained sandstone, fractures well developed
20.1	12.0	6.67	
17.8	12.0	7.07	
20.8		<6.3*	

*--Inferred value

IN-SITU HYDROFRACTURING TESTS

In-situ hydrofracturing tests were conducted in two geologic media—shalestone and fine sandstone, and bauxite. The test holes were drilled into the test rock from the working faces. The diameter of the test holes is 100 mm except for the last 1 m where the diameter was changed to 58 mm. Packers were used to isolate the last 1 m. Water injection rate is 0.1 m³/min, which led to the gradual built-up of the pressure. The occurrence of hydrofracturing was observed in the working face, as shown in Figure 10.

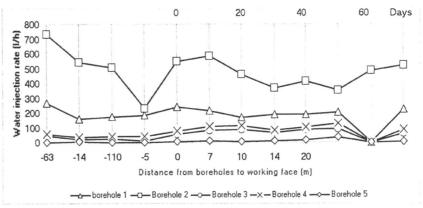

Figure 7: Water Injection Rate during Mining for Test #2

Table 2 Results of in-situ hydrofracturing tests

Geologic barrier	Vertical depth (m)	Water pressure (kg/cm²)	Duration of water injection (min)	Characteristics of Hydrofracturing	Hydrofracturing pressure (kg/cm²)
Shalestone and fine sandstone	0.5	3		Instantaneous seepage, while the water pressure decreased to 2 kg/cm².	3
	1.1	10	20	Water flow through two sets of joints.	10
	1.5	5	25	Seepage through one set of joints, while the water pressure decreased to 8 kg/cm².	15
		10	15		
		15	5		
	3	3	10	Water flow through joints	20
		5	20		
		8	15		
		10	15		
		15	25		
		20	1		
Bauxite	0.5	3	20	Seepage through one set of joints	1.5
	0.7	5	1	Seepage through two sets of joints	4
	1	5	15	Seepage through one set of joints	5

Table 2 shows the test results. In the shalestone and fine sandstone, two sets of joints develop in the rock with their densities of 14/m and 8/m, respectively. Similar joint sets develop in the bauxite. Their densities are 5/m and 3.7/m, respectively. The in-situ hydrofracturing tests indicated:

(1) The water inrush (seepage) occurred preferentially through joints or fractures.
(2) The hydrofracturing pressure of the shalestone and fine sandstone is larger than that of the bauxite.
(3) The water pressure remained at a certain level after the water inrush occurred, which is of a typical property of hard rocks.
(4) The hydrofracturing pressure decreases with depth.

Figure 11 shows the linear relationships established at shallow depths for both barriers, from which the hydrofracturing pressure at one particular point could be approximately calculated. The calculated pressure is then used to compare with the local potentiometric pressure in the Ordovician limestone to determine the risk of water inrush during mining activities. However, cautions must be taken when the equations are extrapolated to deep areas.

CONCLUSIONS

Geologic barrier is a natural rock stratum that prevents groundwater from flowing into active mines. It can also be used to contain contaminants when similar geologic conditions exist. The effectiveness of a geologic barrier can be quantitatively evaluated by measuring its hydrofracturing pressure. The hydrofracturing pressure decreases with depth due to the impact of mining activities. The initial conductive zone provides fractures for the wedging effect to take place. Water inrush occurs when the water pressure exceeds the hydrofracturing pressure in the initial conductive zone. Otherwise, minerals can be mined or shaft excavated safely without dewatering or depressurizing the underlain karst aquifer.

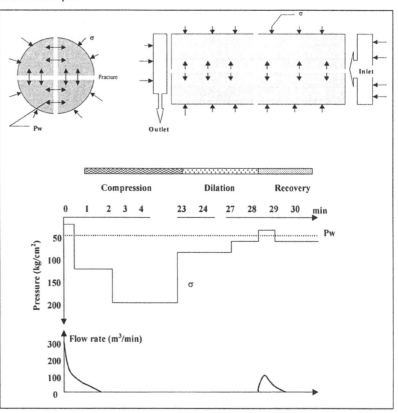

Figure 8: Experiments of Water Inrush through Geologic Barrier

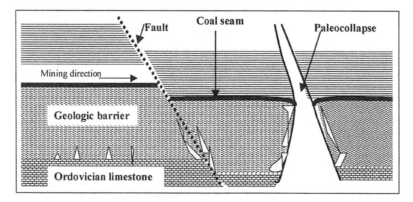

Figure 9: Schematic Illustration of Initial Conductive Zone

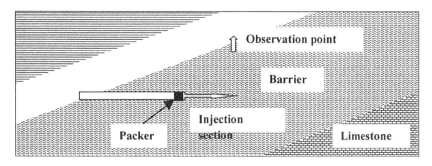

Figure 10: In-situ Hydrofracturing Tests

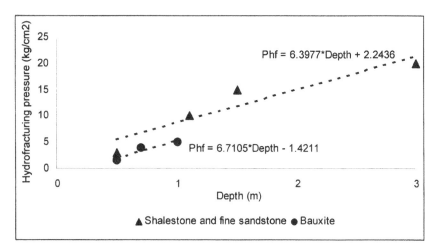

Figure 11: Hydrofracturing Pressure vs Depth

REFERENCES

Kesseru, Z., 1997, Risk and safety assessment for mining and tunneling in karst environment: J. Mine Water and Environment, 16(2): 67-82.

Yu Pei, 1994, Surface collapse in the karst mining areas in China: J. Mine Water and the Environment, 13(2): 21-25.

Zhou, W. F., 1997a, The formation of sinkholes in karst mining areas in China and some methods of prevention: Environmental Geology, 31(1/2): 50-58.

Zhou, W. F., 1997b, Paleocollapse structure as a passageway for groundwater flow and contaminant transport: Environment Geology, 32(4): 251-258.

VI Treatment of Environmental Problems in Karst

Geotechnical and Environmental Applications of Karst Geology and Hydrology, Beck & Herring (eds)
© 2001 Taylor & Francis, ISBN 90 5809 190 2

Operation and maintenance of vadose zone and groundwater remediation systems in a karst aquifer

RICHARD N.BERRY   Richard N.Berry, P.E., Inc., San Antonio, TX 78216, USA, rnbpe@aol.com

ROBERT LONG   UST, Inc., San Antonio, TX 78216, USA, rlongust@aol.com

THERESE M.BAER   Baer Engineering and Environmental Consulting, Inc., Austin, TX 78723, USA, baereng@aol.com

ABSTRACT

Three remediation methods were selected for use at a gasoline spill site near Bulverde, which is located in the Hill Country of central Texas–soil vapor extraction, groundwater recovery with treatment and surface discharge, and point-of-entry potable water treatment. Operation and maintenance issues developed immediately for all three remediation methods.

Balancing the flow rate of propane and gasoline vapors while trying to control the introduction of surface air and moisture became determining factors in the cost of operating soil vapor extraction equipment at the RAM Store No. 24 site. Two soil vapor extraction units with thermal vapor destruction capabilities were used to recover gasoline vapors from the vadose zone in this karst geology.

Scaling became the critical factor to control during the treatment of ground water pumped from the Trinity Aquifer, which is a karst aquifer. High hardness and alkalinity values characterize Trinity Aquifer water. This hard water, with its high mineral content, proved to be detrimental to the groundwater recovery and treatment systems.

Two groundwater recovery well fields and ancillary water treatment systems were used to provide hydraulic control of 900 gallons of gasoline that was released into the Trinity Aquifer. In the early stages of system operation, engineers and technicians quickly discovered that the rate of precipitation of calcium carbonate was far too great to go untreated. Deposition of precipitant was completely fouling bag filters, thus stopping operation of the treatment system every four hours. Other equipment, such as pump impellers and the air stripping trays, became fouled and required extensive cleaning two to three times a week. Clearly, long-term costs would be excessive if control of the precipitant was not achieved.

After laboratory analyses of the ground water, project engineers designed and implemented metered systems to inject various chemicals into the recovered ground water. These chemicals controlled pH and bound the calcium in solution, preventing fouling of the treatment system. After the pretreatment system was installed, bag filters were cleaned weekly, as preventive maintenance only, and treatment equipment no longer required cleaning.

Foaming was also a critical issue early in the operation of the water treatment systems. Severe foaming in the air stripper columns impeded the flow of water through the treatment system. This foam was believed to be caused by surfactants used in drilling mud during installation of the various monitoring and recovery wells. Addition of a foam control agent by metering pump solved this problem, which gradually diminished in importance.

Managing water pressure, particulate matter, and hydrogen sulfide generation at point-of-entry potable water systems proved to be a unique challenge. These issues resulted in part from the system equipment and in part from the karst water characteristics.

INTRODUCTION

Contamination of ground water is a major concern in the United States, especially in areas where inhabitants depend on an aquifer as a sole source of potable water. Hundreds of different toxic substances have been identified in aquifers, including primarily hydrocarbons, pesticides, herbicides, solvents, and heavy metals. Once an unacceptable level of a toxic substance has been detected in ground water, efforts to remediate the aquifer must begin.

The technologies used to remove chemical contaminants from ground water are numerous, with the appropriate selection based largely on the physical properties of the chemicals of concern and the hydrogeology and geochemistry of the aquifer. Frequently, multiple technologies are used concurrently to enhance contaminant removal efficiency and to reduce residence time in the water treatment system.

This paper presents the technological methods, selected equipment, and operation and maintenance practices used to remediate gasoline-contaminated ground water at the RAM Store No. 24 site in Bulverde, Texas, an area that draws water from the Trinity Aquifer.

BACKGROUND

In July 1999, while refueling his automobile at a convenience store, an elderly gentleman struck a gasoline dispenser, causing a failure in the piping system that leads from the underground storage tank to the product dispenser. Approximately 900 gallons of gasoline was released into the subsurface. The site on which the convenience store is constructed overlays the Trinity Aquifer, which is a karst aquifer located approximately three hundred feet below the surface. Over one hundred fifty residences, each of which has a potable water well completed in the middle Trinity Aquifer, surround the convenience store.

Within a matter of days, residents who live adjacent to and downgradient of the convenience store began reporting a gasoline odor in their tap water. Groundwater samples collected from these residential wells indicated unacceptable levels of dissolved-phase methyl tertiary butyl ether (MTBE) and benzene, toluene, ethylbenzene, and xylenes (BTEX). No free-phase product was encountered. Efforts to protect the residents, control migration of the plume, and remove dissolved-phase product began.

REMEDIATION METHODS, EQUIPMENT SELECTION, AND OPERATION AND MAINTENANCE

Three remediation methods were selected for use at this site–soil vapor extraction, groundwater recovery with treatment and surface discharge, and point-of-entry potable water treatment. The selected technologies included the use of two portable units for soil vapor extraction and thermal vapor destruction (SVE/TVD); two water treatment systems (WTS) with pretreatment, air stripping, granular activated carbon adsorption, and ultra-violet (UV) disinfection; and point-of-entry potable water systems (POE/PWS) with particulate filtration, granular activated carbon adsorption, and UV disinfection.

Grones, Inc., a Texas-based company, manufactured the SVE/TVD units. Project engineers hired by RAM Stores, Inc. designed the WTS units. Carbonair, a Minnesota-based company, manufactured the POE/PWS units.

Operation and maintenance issues developed immediately for all three remediation methods. The issues included difficulties associated with balancing the propane / soil vapor gas streams at the SVE/TVD units; controlling scale, foam and pH at the WTS units; and controlling water pressure, particulate matter and hydrogen sulfide (H_2S) generation at the POE/PWS units.

Soil Vapor Extraction with Thermal Vapor Destruction

Soil vapor extraction was used to remove the soil vapors trapped in the void spaces of porous rock and soil in the vadose zone beneath the convenience store. Because the extracted soil vapors contained significant concentrations of gasoline, recovery of the majority of the gasoline was accomplished using this method. These vapors were burned, with the primary air emissions being carbon dioxide (CO_2) and water, but the combustion process was not perfect.

After an intensive effort to identify and isolate zones that produced significant gasoline vapors, two SVE/TVD units were placed into operation, and the extracted vapors were thermally oxidized. Injecting propane into the oxidation unit along with the soil vapors enhanced the oxidation process. Each unit was connected to its own network of vapor recovery wells. Both units were equipped with an integral vacuum pump / blower that provided motive power for the soil vapor extraction. Propane was used as the primary fuel, and gasoline vapor in the soil vapor was used as the auxiliary fuel.

The units were adjusted to operate under stable conditions with a minimum quantity of propane. The unit control system started the vacuum pump and extracted soil vapors from the connected wells into the burner. If the temperature reached a maximum, the vacuum pump would shut down and the unit would return to propane-only operation. The operator selected which wells were connected, how much each suction line was throttled, and how much the main suction manifold was throttled in order to achieve balanced performance of the units.

The volume of gasoline removed by each SVE unit was evaluated bi-weekly by sampling the extracted vapors prior to the TVD unit. These vapors were collected in Tevlar™ bags using a vacuum pump. Once collected, the gas-filled Tevlar™ bags were submitted to a laboratory for analysis of gasoline range organics (GRO) and BTEX. Quantities of gasoline recovered for the bi-weekly period were calculated using an equation that included analytical results, gas stream flow rates, gas temperatures, and vacuum pressures. The SVE/TVD units were not decommissioned until the calculated hydrocarbon removal rate became asymptotic.

Several vapor recovery wells were connected to each of the two SVE/TVD units. Each well was equipped with valves to throttle gas flow to the respective unit. Vapor recovery wells were placed vertically around the perimeter of the site in an effort to capture fuel before it migrated off-site. Additionally, horizontal vapor recovery wells were placed under the product dispenser that had been struck. Vapor recovery was used to recover gasoline only in the immediate vicinity of the release. Field studies, using photo-ionizing detectors (PIDs) and packers to isolate vapor-producing zones, indicated elevated levels of hydrocarbons in all of the vapor recovery wells located near the source of the release, but did not indicate significant levels of hydrocarbons in wells located off site.

Operation of the SVE/TVD units required constant monitoring of the combustion chamber temperature. The optimum temperature in this chamber was determined to be approximately 900°C. As the temperature fell below this level, propane was injected into the combustion chamber to maintain the 900°C target. A temperature above about 950°C caused the SVE/TVD unit to shut down until the combustion chamber cooled.

Operating costs for the SVE/TVD unit were greatly affected by the volume of hydrocarbons in the gas stream that entered the combustion chamber. As the concentration of gasoline vapor in the soil vapor declined, more propane was needed to maintain an appropriate operating temperature. Power requirements remained fairly constant for the blowers and vacuum pumps.

In order to increase the efficiency of the SVE/TVD units, technicians at the site closely monitored the combustion chamber temperatures and adjusted the extracted gas flows, as necessary. If temperatures increased, the flow rate of propane was reduced and the flow rate from the extraction wells was increased. Clearly, high operating temperatures with greater flow rates from the vapor recovery wells, coupled with lower propane demands, indicated that mass removal of the released product was occurring. Additionally, lower propane consumption reduced operating costs.

Factors that affected the fraction of gasoline vapor in the soil vapor included ground temperature, soil moisture content, and the permeability of the soil in the area surrounding the vapor recovery wells. As the ambient temperature fell, temperature readings of the extracted gas responded by also falling rather quickly. Short circuiting, or the introduction of surface air, was suspected due to the rapid temperature responses; however, the site was completely paved with asphalt and concrete, and no surface air leaks were found.

At times, the moisture content in the vapor recovery wells would increase such that water was also extracted from these wells. Water in the vacuum lines reduced flow to the SVE/TVD units, caused higher head pressures, and reduced the combustion chamber temperatures. If water were found in any of the vacuum lines, the lines were disconnected and drained. If necessary, vapor recovery wells that produced water were throttled back until the water production subsided. Normally, groundwater levels were far below the zone of influence of any of the vapor recovery wells; however, rain events would sometimes temporarily raise water levels in the vicinity of some of the wells. Typically, the affected wells were combination groundwater / soil vapor extraction wells.

Routine maintenance of the SVE/TVD units included maintenance of air filters in front of the vacuum pumps and blowers, and monitoring of the volume of propane stored. With the high permeability of the karst geology at this site and the high vapor pressure of gasoline, removal efficiencies using SVE/TVD were excellent. Although the monitoring and operation requirements were somewhat labor intensive, SVE/TVD proved to be the most cost effective technology used at this site.

Groundwater Recovery and Treatment

Groundwater recovery and treatment was the secondary remediation method used to remove gasoline at this site. More importantly, however, the groundwater recovery and treatment systems were used to provide hydraulic control of the plume. Two groundwater recovery well fields were developed for this project—one adjacent to the point of release and one downgradient of the point of release in the affected residential area. Each recovery well field delivered ground water to its own WTS. A cone of depression was developed in the well field of each WTS, controlling migration of the plume and concentrating gasoline in the recovery well fields.

Ground water delivered to the WTSs was treated and discharged to the surface in a nearby creek. However, due to the hydrogeology and geochemistry of the ground water pumped from this karst aquifer, operation and maintenance of the WTSs became somewhat complex. Initially, the recovered ground water passed sequentially through a cascading air stripper, a bank of three bag filters, a bank of six granular activated carbon adsorption filters, and a UV disinfecting unit before it was discharged. In the early stages of system operation, engineers and technicians quickly discovered that precipitation of calcium carbonate ($CaCO_3$), pH, and foaming would require pretreatment of the ground water prior to treatment for removal of gasoline constituents.

Analysis of the ground water indicated pH values ranging from 8.0 to 8.2. Additionally, because of the calcareous composition of rock that comprises the aquifer, high hardness values [at or above 350 parts per million (ppm) as $CaCO_3$] and high CO_2 levels were characteristic of the aquifer water. Total alkalinity values ranged from 143 ppm to 300 ppm as $CaCO_3$.

As recovered ground water passed through the cascading air stripper, CO_2 was driven off, thereby raising the pH and causing precipitation of $CaCO_3$ in the stripping tray diffusers. The subsequently reduced velocity of flow through the diffusers allowed $CaCO_3$ precipitant to deposit on the interior walls of the stripping tower. $CaCO_3$ precipitant also accumulated on the impellers of the downstream pumps and bag filters. The result of this accumulation was obstructed air stripping towers and bag filters, and frozen pumps.

Design flow rates through WTS-1 and WTS-2 were 30 gpm and 60 gpm, respectively. After only four hours of operating WTS-2, the pressure drop across the three bag filters doubled, with the respective flow rates dropping significantly. The downstream pumps were freezing every two to three days due to accumulation of $CaCO_3$ precipitant on the impellers. With a projected operation period of two years, this condition was not acceptable.

Another operating parameter that affected WTS-1 and WTS-2 was foaming. During the initial operation of these systems, large quantities of foam were discharged from the vents of the two systems. The origin of the foam was believed to be surfactants used in drilling monitoring and recovery wells in the WTS well fields. Due to the high transmissivity of the formation, these surfactants apparently migrated toward the active recovery wells, where they were pumped and delivered to the WTSs. Once in the system, they formed foam as the recovered water passed through the air stripping trays. Formation of foam inside the trays held the water in suspension, thereby reducing the flow rate through the system. Additionally, the globules of foam discharged in the treated effluent were clearly visible in the creek, which raised questions from nearby residents.

Pretreatment of the ground water was implemented as soon as a process design was complete. NALCO Chemical Company, an Illinois-based company, was consulted to evaluate pretreatment options. Water samples were collected and analyzed by NALCO personnel, who recommended an anionic polymer to stabilize and disperse the $CaCO_3$. Project engineers recommended sulfuric acid (H_2SO_4) to control pH. These additives worked together to maintain suspension of $CaCO_3$ precipitant and to prevent deposition on process equipment. Once pretreatment was initiated, the bag filters no longer became obstructed and pumps ran without freezing. Bag filters were cleaned weekly as preventive maintenance only, and pressures no longer increased by the hour. In fact, upon decommissioning of the treatment systems, no appreciable quantities of scale were visible in the equipment.

The pretreatment train consisted of two high-density polyethylene tanks positioned prior to the air-stripping tower–a mixing tank and a flow equalization tank–and an in-line injection port located between the two tanks. Acid was injected into the mixing tank. A pH probe placed in this tank signaled a controller to pump metered acid into the tank whenever pH was out of range. A mixer was mounted in the mixing tank to homogenize the acid solution. The quantity of acid actually consumed in the pretreatment process was much less than what had been predicted.

Anionic polymer and anti-foaming agent were injected into the in-line port located between the two tanks. Injection of the anionic polymer and anti-foaming agent was controlled by a metering pump based on the flow rate of acid-treated water into the equalization tank. The quantity of polymer consumed in the pretreatment process was much greater than what had been predicted. The use of anti-foaming agent was discontinued after development of the recovery well field was complete.

The H_2SO_4 was stored in bulk and safely housed on site in a secured structure to discourage vandalism. The pH-controlled metering pumps and axial flow mechanical mixers were used to achieve the proper concentration and homogeneity of the acid solution. The less dangerous anionic polymer and anti-foaming agent were stored on site in their original containers. Anionic polymer and anti-foam additives were injected continuously by metering pumps whenever recovery well pumps were operating.

Pretreated water gravity-flowed from the mixing tank to the equalization tank. From the equalization tank, it was pumped to the air-stripping tower. The equalization tank was used to maintain a constant flow rate through the WTSs. Maintaining a constant flow rate at or near the maximum design flow rates of the two systems provided the hydraulic control required to mitigate plume migration.

Operational and regulatory samples were collected bi-weekly to monitor system performance and to verify compliance with the National Pollutant Discharge Elimination System (NPDES) permit. These samples were analyzed for lead, calcium, magnesium, hardness (as $CaCO_3$), alkalinity, total dissolved solids, pH, BTEX, and total petroleum hydrocarbons. NPDES permit monitoring required only a monthly average of fewer parameters for four sampling events; however, more frequent monitoring of additional parameters was necessary for operational control.

The removal efficiency for removal of volatile organic compounds (VOCs) from the recovered ground water was greater than ninety-nine percent; however, the mass quantity of VOCs removed over the life of the project was low. As previously stated, the groundwater recovery and treatment systems were used primarily to provide hydraulic control of the plume. For this purpose, the systems performed very well.

Point-of-Entry Potable Water Systems

The residential inhabitants and commercial occupants in the area surrounding the gasoline release site draw their drinking water from private and public wells located at each home or business. Ultimately, 21 wells were impacted with MTBE, and five of these wells were impacted with BTEX. To prevent gasoline constituents from reaching the service taps, POE/PWSs, manufactured by Carbonair, were positioned in line between the point-of-entry and the well, or between the well and the holding tank, whichever was applicable. Fifty POE/PWSs were installed on down-gradient and cross-gradient wells.

The systems provided primary and secondary paper filters for filtering particulate matter, and primary and secondary granular activated carbon adsorption canisters for adsorbing VOCs. A water meter was placed in line to monitor the quantity of water treated by each system. The quantity of water treated by each system was compared to the average level of contamination in each of the impacted wells to forecast the life of the granular activated carbon. System performance was monitored initially by weekly sample analyses for BTEX and MTBE. As contaminant levels declined, system performance was monitored less frequently by sample analyses.

Operation and maintenance of the POE/PWSs proved to be the most challenging task of all. The operation, maintenance and monitoring of these systems not only was very labor intensive and expensive, but also was very personally challenging to field engineers and technicians who encountered inquisitive and sometimes upset well owners on a daily basis. Particulate filters were changed weekly, or more frequently when necessary, and granular activated carbon canisters were changed every six months, or more frequently if breakthrough occurred. Water samples had to be collected each week from the wells, from between carbon canisters, and at post-treatment ports to monitor contaminant levels as well as system performance. In addition to the routine maintenance required by the systems, some residents claimed a reduction in water pressure while others claimed discoloration, a "rotten egg" odor, or concern that increased pressure head created by the systems might cause premature failure of their well pumps. The remediation team was unable to substantiate the validity of these claims (for instance, the measured pressure drop across properly operating POE/PWSs was only 2 psi to 3 psi), but each claim required evaluation.

Although problematic for the reasons stated, the remediation team believed that the POE/PWSs were necessary to provide the best available protection to the users, and that these systems performed very well. Albeit by their very number and widespread locations, these units were extremely labor intensive with respect to routine maintenance and monitoring, but frequent service calls from users increased the amount of time required for their operation. Obstruction of particulate filters, which occurred more often after significant rain events, reduced flows to the homes and were the reason for many service calls from residents at all hours of the day and night. To facilitate routine service of the systems, valves were installed that allowed the system to be bypassed during maintenance. Remediation field technicians would frequently find that residents had bypassed the systems to irrigate their lawns, and, afterwards, failed to reopen the valves. The practice by many residents of bypassing the systems to irrigate and failure to reconnect after irrigation presented a challenge to field technicians who became increasingly vigilant as the problem persisted. Maintaining the systems in the treatment mode was an ever-present challenge.

The Trinity Aquifer can also produce water that is high in iron and sulfates, depending on the geochemistry of the strata through which the well is completed and the screened interval. Ferric iron in the water caused fouling of many particulate filters. Project engineers also hypothesized that iron-oxidizing bacteria may have reduced naturally-occurring sulfates to contribute to the generation of hydrogen sulfide on the granular activated carbon in the canisters, thus compounding the "rotten egg" odor.

Claims of reduced water pressure were remedied at great expense by the installation of holding tanks at a number of residences. Discoloration was remedied by more frequent replacement of particulate filters, which were frequently fouled with iron oxides. Hydrogen sulfide odors were remedied in most cases by replacement of carbon canisters, but in more than one case by the installation of "green sand" filters, chlorinating units, or water diffusers mounted in the tops of holding tanks.

SUMMARY

Remediation of the Trinity Aquifer at this site used three technologies, each of which performed well. These technologies included soil vapor extraction, groundwater recovery with treatment and surface discharge, and point-of-entry potable water treatment. However, due to the hydrogeology and geochemistry of the aquifer formation, operation, maintenance, and monitoring of these systems was more complex than might usually be expected. Soil vapor extraction was complicated by the introduction of surface air and moisture, and rapid response to ambient temperature fluctuations. Pretreatment of ground water became necessary due to the water chemistry, and control of the plume was more difficult due to the high transmissivity of the karst aquifer. Point-of-entry systems required frequent maintenance to prevent fouling of particulate filters.

Early and quick diagnosis of the operational challenges by project engineers and field technicians, however, minimized systems maintenance. More importantly, intensive monitoring of groundwater levels, flow direction, and contaminant concentrations allowed optimal management of recovery wells, and remediation of the aquifer was achieved very quickly. Rapid response by RAM Stores, Inc. and the Texas Natural Resource Conservation Commission, proper selection of remedial technologies, and efficient operation of remediation equipment, account for the fact that the vast majority of the wells in the affected area remain uncontaminated.

ACKNOWLEDGEMENTS

The authors would like to acknowledge Mr. JD Marek and Mr. Cletus Edwards, Chairman and President, respectively, of RAM Stores, Inc., for their responsible action in the remediation of this underground release; Ms. Rosemary Wyman, Project Manager–Site Assessment and Management Remediation Division of Texas Natural Resource Conservation Commission for her assistance and cooperation in the exercise of regulatory affairs; and Mr. Scott Fisher, Vice President–Regulatory Affairs of Texas Petroleum Marketers and Convenience Store Association for his continuing efforts toward the advancement of petroleum marketing in the state of Texas.

Geotechnical and Environmental Applications of Karst Geology and Hydrology, Beck & Herring (eds)
© 2001 Taylor & Francis, ISBN 90 5809 190 2

Karst feature control on the mobility of petroleum residues in groundwater: a hydrocarbon reservoir analogue

THOMAS D.GILLESPIE Environmental Liability Management, Inc. , Holicong, PA 18928, USA,
tgillespie@elminc.com

ABSTRACT

Fuel oil released from a former storage system has accumulated in an unconfined karst aquifer in the cavernous Cambrian Allentown Dolomite Formation in the Great Valley of Western New Jersey. Free-phase petroleum product occurs in sediment-filled solution cavities, but only dissolved phase constituents are present in downgradient locations, connected via a set of joint-controlled solution channels. Residual oil from the unsaturated overburden enters bedrock and moves through three flow regimes: rapid, turbulent unsaturated flow within joints; rapid, saturated flow through solution channels, and; slow, laminar saturated flow in sediment-filled solution cavities.

The oil and water phases are emulsified by the turbulent flow in the unsaturated joints/channels, and remain emulsified in the saturated channels. In the saturated zone cavities, where flow is slow and laminar, the phases separate and the oil accumulates at the top of the zone of saturation, or at the top of the solution cavity. Once accumulated, free-phase oil can leave the cavity only if:

- A joint/channel intersects the zone of oil accumulation; *and if*
- The oil is below the zone of water saturation; *and if,*
- Flow velocity down the hydraulic gradient is sufficient to overcome the buoyancy of the oil.

Those conditions can not be satisfied under existing conditions. In this respect, the solution cavities are similar to petroleum traps in hydrocarbon reservoirs. Exploiting the conditions, the remedial action consisted of periodic passive removal of the accumulated oil. By maintaining product thickness at a minimum, residence time and dissolution were minimized and the concentrations of dissolved compounds were reduced to undetectable levels. The remedial action was terminated when petroleum recovery was consistently less than 4 litres (1 gallon) per month.

INTRODUCTION

Light-end petroleum distillates were released from an underground storage system into a residual soil above a karst bedrock aquifer system. Petroleum encountered in monitoring wells was heavily weathered and indicative of a long-term slow release over an extended period of time. Migration of the petroleum residue in the karst aquifer system, however, had not extended more than a few meters from the original release location. Dissolved compounds were detected in monitoring wells within 20 meters of the release area, but had attenuated beyond that distance. Based on the conditions, most of the light-end volatile fractions of the released petroleum had dissolved from the material and the residue was composed of the less soluble components. Petroleum residues persisted in monitoring wells at thicknesses up to a meter despite the nature of the aquifer system, which experiences rapid changes in water elevation and conduit flow conditions during periods of increased precipitation.

GEOLOGY AND HYDROGEOLOGY

The site is in the Reading Prong of the New England Province, which is composed mostly of gneissic remnants of the Proterozoic Grenville Orogeny. Bedrock below the site is the Cambrian Allentown Dolomite, which is part of a regional series of overthrust, infolded slope-rise sediments which form the Great Valley Province of Pennsylvania, New Jersey, and Maryland (Berg, et. al., 1989).

The carbonate rock was folded and faulted during the several phases of Appalachian orogenesis resulting in local metamorphism to the lower Greenschist facies, pervasive jointing and late-stage faulting. The bedrock surface is pinnacled and is covered by a blanket of clay-rich soil composed of regolith from the eroding limestone, with ground moraine from the Illinoisan ice sheet near the surface.

The rock has a well-developed karst flow system, with preferential flow directions along principal joint/fault orientations. Although conduit flow does occur, many of the larger solution openings have become partially filled with ravelled soil. Groundwater occurs in bedrock at an average depth of approximately 10 to 15 meters below grade, or five to ten meters below the soil-rock interface. Percolating water enters bedrock via solution-enhanced openings in the rock and flows down the dip of the planes until it encounters groundwater. In the saturated zone, groundwater flows through both fracture and conduit dominated flow regimes and through porous media in sediment-filled solution cavities. Because the storage capacity of the surrounding fractures/conduits is low, the sediment filled cavities provide for groundwater storage and create local groundwater highs during periods of low precipitation when water elevations drop in the adjacent open planes/conduits.

ENVIRONMENTAL FATE AND TRANSPORT OF RELEASED PETROLEUM

The released petroleum first pooled at the base of the overburden soil formation (unsaturated zone) on the bedrock surface, particularly in troughs between bedrock pinnacles which were not only low points, but are also the locations where solution-enhanced openings provide conduits for migration of the petroleum residue into the bedrock aquifer. Once mobilized with percolating water, the petroleum residues entered the bedrock aquifer and were subject to flow under three different regimes:

- rapid, turbulent flow in joint planes in the unsaturated zone, in which the oil and water became emulsified;

- rapid flow in the saturated zone along open solution enhanced planar discontinuities. Because most of the petroleum residue was mobilized during rain, when groundwater flow rates in the channels were high, the oil and water remained mostly emulsified, and;

- slow, laminar flow in saturated, sediment-filled solution cavities where the emulsified petroleum separated from the water, rose and accumulated at the top of the saturated zone, or at the top of the solution cavity.

Because the sediment filled cavities act as localized groundwater storage reservoirs, the residual petroleum which accumulated at the tops of the local saturated zones continuously leached soluble petroleum constituents into groundwater, resulting in a multisource (numerous cavities), non-gaussian plume of dissolved compounds, which spread along specific, discreet flow channels downgradient of the general release area. As groundwater reserves in the filled cavities decreased with time during dry periods, smear zones developed in the dewatered sediment, creating pulses in the concentrations of dissolved compounds in the downgradient aquifer, even after the source material in the overburden soil was depleted. The residual source petroleum did not move out of the sediment filled cavities, however, even during periods of extremely high groundwater flow rates.

Sediment-Filled Solution Cavities as Petroleum Traps

Because the petroleum residue remained less dense than water despite the loss of light end components, the residue tended to remain at the top of the local saturated zones and tended not to flow readily down the hydraulic gradient. This tendency to remain in the solution cavities was enhanced by the retentive capacity of the soil in-filling, and by the seasonal attenuation of the residue with decreasing groundwater elevations, as discussed above. Because of the difference in flow characteristics between the porous and fracture flow regimes, movement of the petroleum residues out of the solution cavities could only occur if:

- A fracture/conduit groundwater flow channel intersected the zone of petroleum accumulation during a time of the year when free-phase petroleum liquid was present; *and if*

- The petroleum residue was below the zone of water saturation; *and if,*

- Groundwater flow velocity down the hydraulic gradient created sufficient pressure to overcome the buoyancy of the free-phase petroleum residue.

The first two conditions could not occur during dry times of the year when the local water table within the cavities declines below the roof of the sediment filled solution cavities and the residual petroleum smears vertically in the sediment, thereby decreasing the volume of free-phase liquid available for transport.

The third condition could provide for movement out of the accumulation zones and into the conduit-dominated portion of the aquifer. Because no product was observed during any time of the year at any location downgradient of the accumulation zones in solution cavities at the release area, it was apparent that those conditions were not met. The conditions were defined as part of the demonstration required to obtain regulatory closure for the release.

As stated, free-phase residue could move out of the zone of accumulation if the velocity of groundwater (V_w) at any time (t) exerted sufficient pressure (P) to overcome the buoyancy (B) of the residue. Because the force exerted by flowing groundwater would need to be applied from above the residue to initiate movement, the second condition for mobility could not be satisfied during dry periods when the local water table was lower in the adjacent fractures/conduits. The instantaneous velocity of groundwater down the hydraulic gradient in planar openings at any location, ignoring aperture and wall roughness effects, can be expressed as:

$$dv_w = \rho_w gh \cos\theta \, (t) \, dt;$$
where:
θ = the slope of the water table in degrees from vertical;
ρ_w = the density of water, taken as 1.0 g cm^{-3};
g = the acceleration due to gravity
h = the change in height of the water column over distance, or the hydraulic gradient

In this case, velocity is expressed in terms of acceleration because the pressure exerted by the water is a stress (σ) expressed as:

$\sigma = F/A$;

      where F = force and A = area, and

$F = m \cdot a$;

      where m = mass, and a = acceleration.

Therefore, the velocity of water which imparts the stress can be derived as:

$dv/dt = \rho gh \cos\theta \, (t)$;

$dv = \rho gh \cos\theta \, (t) \, dt$;

$\int dv = \int \rho gh \cos\theta \, dt$;

$v = \rho gh \int \cos\theta \, (t) \, dt + C$

The buoyancy of the petroleum results from the difference density between the two immiscible liquids expressed as:

$B = \rho_w \, gh - \rho_p \, gh$

      Where ρ_p = density of petroleum

The stress, or pressure (P) exerted by the flowing groundwater on a unit mass of petroleum can be expressed as:

$P = \rho gh \cos\theta \, (m_w)/A$

      where m_w = the mass of the water

The opposing stress, buoyancy, exerted by the petroleum residue can be expressed as:

$B = (\rho_w \, gh - \rho_p \, gh)(m_p) \, /A$

Setting Area at unity leaves an unbalanced set of instantaneous forces at time (t):

$dF_v = (m_w) \, \rho gh \cos\theta \, (t) \, dt$ (the force exerted by flowing water);

and

$dF_B = m_p \, (\rho_w \, gh - \rho_p \, gh) \, (t) \, dt$ (the opposite buoyant force of petroleum).

Combining the forces into an expression of conditions, free-phase petroleum could move out of the accumulation zones only if:

$$\frac{m_w \, d(\rho gh \, \cos\theta) \, (t)}{dt} - \frac{m_p \, d(\rho_w \, gh - \rho_p \, gh) \, (t)}{dt} > 0$$

Because of the $\cos\theta$ term, the first factor in the expression describes the intuitive condition that the hydraulic pressure exerted by the flowing water increases with increasing hydraulic gradient, with the maximum (although unlikely) velocity under vertical groundwater flow conditions ($\theta = 0$, or $\cos\theta = 1$) . However, for all hydraulic gradients except vertical, $\rho_w gh \cos\theta < \rho_w gh$, so the critical gradient at which the buoyancy of the free petroleum phase could be overcome occurs where:

$\rho_w gh \cos\theta - \rho_w gh = \rho_p gh$,

assuming that the mass of both water and petroleum are unity and, to be conservative, that surface tension and other retentive forces between the various phases are negligible.

Although this would need to be derived empirically for each site, an estimate of the parameters supports a conclusion that the petroleum would not become mobile and move out of the solution cavity unless the hydraulic gradient is greater than approximately 0.6, or a groundwater slope of approximately 30^0 from horizontal. The only times when this condition might be met in typical karst terrains is between recharge events when sediment filled cavities act as local reservoirs which discharge into the adjacent conduits where groundwater storage is low and substantial groundwater elevation changes are common. However, at those times the floating free-phase petroleum would not be below the zone of saturation, and the free-phase volume is decreased because of smearing, so the petroleum would not be available for movement out of the cavity. During recharge events, the gradient between the cavities and adjacent conduits would be far less than the requisite angle for mobility.

As a result of this combination of conditions, the sediment filled solution cavities act in a manner similar to petroleum traps in hydrocarbon reservoirs, in which the interaction of buoyancy, hydraulic flow conditions and geology provide for the accumulation and long term containment of petroleum. In the current circumstance, conditions differed slightly from petroleum reservoirs in that the aquifer was not under confined conditions. However, the containing properties resulting from the natural conditions had prevented the movement of the residues from the immediate vicinity of the original release area. The corrective action for the release consisted of a passive petroleum recovery system using the existing monitoring wells. By maintaining the thickness of the petroleum residue to a minimum, the contact time with groundwater was minimized and the concentrations of the dissolved compounds in downgradient groundwater decreased. The entire corrective action was terminated when the rate of petroleum recovery declined to less than four liters per month and measurable free-phase petroleum no longer accumulated when the groundwater was left undisturbed.

REFERENCES

Berg, T.M. Barnes, J.H., Sevon, W.D., Skema, V.W., Wilshusen, J.P., and Yannacci, D.S., 1989, Physiographic Provinces Map of Pennsylvania: A Revision Geological Society of America, Abstracts, Northeasten Section Meeting.

Geotechnical and Environmental Applications of Karst Geology and Hydrology, Beck & Herring (eds)
© 2001 Taylor & Francis, ISBN 90 5809 190 2

Characterization and grouting beneath an impending landfill cap

MARK R. MIRABITO   Merck & Co., Inc., Somerset, NJ 08873, USA, mark_mirabito@merck.com

DIMITRIS DERMATAS   Stevens Institute of Technology, Hoboken, NJ 07030, USA, ddermata@stevens-tech.edu

MARK JANCIN   USFilter Groundwater Services, State College, PA 16801, USA, jancinm@usfilter.com

ABSTRACT

The variable depth to bedrock and the potential for subsidence make foundation design in karst areas extremely challenging. A subsurface characterization and compaction grouting case study is presented for the Merck landfill cap project. The 13-acre landfill in the Appalachian Great Valley was closed and capped in the summer of 1999. This moderately mature, covered karst setting is predominantly a mantled karst because most of the overburden is transported alluvium. However, the setting is locally transitional to a subsoil karst where clay layers, residual from dissolution of the underlying Paleozoic carbonate bedrock, occur along the top of bedrock and beneath the alluvium. There was concern about potential subsidence sinkhole development at the toe of the cap. No sinkholes are present in the vicinity of the landfill; however, over the years several small sinks had developed during drilling of monitoring wells along the landfill perimeter, and these were filled.

A ground penetrating radar survey prior to cap construction highlighted five potential areas of inferred subsidence along the northern edge of the landfill. Geoprobing was subsequently used to map soft zones and minimum bedrock depths in these areas, and a compaction grouting work plan was developed toward reducing the potential of subsidence sinkholes. No overburden voids were detected by ground penetrating radar or by direct penetration. A total of 154 m³ (5429 ft³) of low-slump grout was injected into 20 drillholes of various depths at target area #7, within which the depth to bedrock varied from 4.6-18.3 m (15-60 ft) bgs, locally showing very sharp and deep pinnacles. Grouting was staged upward from 1.5 m (5 ft) below the apparent top of bedrock and into the overburden until the ground-heave cutoff criterion was attained. The mean grout-take over the 20 holes was 0.7 m³/linear m (7.2 ft³/linear ft). The grouting program was terminated before other areas were tackled, because of the relatively large grout takes in the injected area, the relatively low likelihood assigned to subsidence potential, and the capacity to repair any sinkholes which may develop along the cap toe area.

Karst subsurface characterization and ground modification projects generally require a combination of surface geophysics, ground penetration-based data, an experienced karst geotechnician, and acknowledgment of an inherent degree of uncertainty in addressing small, site-specific areas. Both this case study and the geotechnical literature at large, show that current geophysical exploration practice is not adequate in fully predicting the grout-take capacity of the subsurface.

INTRODUCTION

Karst landscapes are characterized by closed surface depressions (sinkholes) and conduit-influenced subsurface drainage. In many karst settings, the bedrock surface may be highly irregular and connected to voids that have the potential to form sinkholes and cause problems with foundation design and construction. Not only limestone is predisposed to karst. Other soluble rocks include dolomite, gypsum and rock salt. The dissolution of limestone or dolomite mostly occurs on a geologic time scale and only existing voids are of concern. Solution of gypsum and rock salt occurs at a faster rate such that new cavities may form during the lifetime of a project. Although bedrock voids can be present wherever the subsurface consists of soluble bedrock, there are several carbonate-rock provinces within the United States that are particularly susceptible. These include the Appalachian Mountain region from New York to Alabama, central Florida, parts of Missouri and other midwestern states, and sporadic locations throughout the western half of the country.

Karst bedrock voids usually form by the dissolution of mineral grains from the flow of slightly acidic groundwater through pre-existing fractures. As these fractures widen they continue to be the preferential pathways for water flow, eventually growing into voids or conduits. Where a void reaches the bedrock/overburden interface, the overlying soils must bridge the gap. If the overburden soils are not strong enough to span the void, they collapse and the soils fall into the cavity. Most sinkholes in karst areas develop by downward movement of overlying unconsolidated materials (overburden or cover) into bedrock voids. Such sinkholes are called subsidence sinkholes. Sinkholes formed from the collapse of the top of bedrock (TOB), termed cave-collapse sinks, are much less common.

Subsidence sinkholes, as ground depressions, can represent the final surface "breakthrough" manifestation of a protracted, and unseen, period of movement of the overburden down into a bedrock void system. Such overburden subsidence, prior to ground breakthrough, can cause either or both of the following features in the overburden overlying (or nearby) the related bedrock voids: (1) overburden voids, which over time may grow upward and/or laterally; and (2) loose (or soft) zones in the overburden, wherein the subsidence has decreased the density of the affected soil-mantle materials. Generally, overburden voids tend to be developed in relatively clay-rich, cohesive cover, rather than in coarser-grained, non-cohesive cover.

This paper reviews recent literature on subsurface exploration and presents a case study of karst subsurface characterization and compaction grouting for the Merck landfill cap project. A primary conclusion is that current subsurface investigation practice may inadequately forecast the grout volumes necessary to stabilize an overburden, rendering difficult the cost/benefit decision of whether to use compaction grouting as a preventive measure.

LITERATURE REVIEW

Due to the great potential risks that sinkholes pose to human health, structural integrity, and the environment, an awareness of the need for stabilization measures is essential to any construction project in a karst area. Because the costs of stabilization methods are high and can escalate quickly during implementation, a well-defined understanding of the extent of subsurface problems also is vital. Researchers and field practitioners are constantly challenging the investigation techniques used to evaluate the presence of voids and to define the extent of sinkhole potential. Traditional sinkhole investigations have focused on clues given by historical evidence and numerous geologic indicators. Technological advances have brought a wide variety of geophysical techniques to the field of subsurface subsidence exploration. Currently, some combination of historical review, area reconnaissance, direct penetration, and geophysical methods is most often used.

Ogden (1984) presented a ten-step procedure, using a combination of small-scale maps and site specific data, for the forecast of sinkholes. The steps, in no particular order, include: (1) sinkhole delineation, (2) size and shape analysis, (3) lithologic controls on sinkhole development, (4) comparison of cavern occurrence to sinkhole development, (5) relationships between sinkhole development and topographic factors, (6) comparison of sinkhole development to depth to the water table, (7) search for surface signatures of newly-forming sinkholes, (8) utilization of shallow geophysical techniques (including earth resistivity and seismic reflection), (9) determine the effect of man's activities on sinkhole development, and (10) model development for collapse prediction (Ogden, 1984). The majority of these steps require evaluation of the visible surface geology.

Forth et al. (1999) used hazard mapping by dividing the area of concern into 100 m^2 grid squares, or cells, that were inventoried for surface features that might indicate sinkhole potential. The nine surface features include sinkholes, caving, gullying, slope angle, tension cracks, fissuring, rockfalls, mechanical strength and vegetation cover. Each factor was weighted according to its degree of sinkhole influence and a hazard score for each cell was generated. The cells were superimposed on a base map, giving a general understanding of sinkhole hazards that can be used to direct the next steps of a site-specific investigation or remedial action.

Upchurch and Littlefield (1988) approached the issue of sinkhole forecasting from another data analysis perspective. When performing a sinkhole risk assessment, one should consider records of modern sinkhole development, correlation of modern sinkholes with the location of ancient sinks, and the appropriate geographic scale of sinkhole occurrence. If good data are available, localized risk maps and sinkhole probability predictions can be made with reasonable confidence.

Beck (1991) also suggested using statistical analysis to determine subsidence-sinkhole risk during the site-planning phase. It is not sufficient to simply review aerial photographs of ancient sinks. One must evaluate the time dependent occurrence of sinkholes within a geologically, geomorphically, and hydrologically consistent sampling area. He argued that, "If the sinkhole hazard is significant, and if there are appropriate measures by which the hazard may be minimized or avoided, then, and only then, is it warranted to conduct a detailed, site-specific investigation" (Beck, 1991, p. 231). The decision must balance risk versus the cost of investigation and preventive measure implementation.

The current list of available techniques for the delineation of subsurface voids is long. Geophysical methods such as electrical resistivity, surface seismic, crosshole seismic, acoustic resonance, electromagnetic (radar) and microgravity are well known among geotechnical engineers. Using test sites, Cooper and Ballard (1988) evaluated the overall effectiveness of these methods and many others by taking into account practicality, feasibility and economics. While nearly all techniques are capable, their applicability is highly dependent on subsurface characteristics. The key is to choose the method or methods, " . . . which best suit site conditions and program objectives" (Cooper and Ballard, 1988, p. 36).

Walt Disney World is located within the highly karstic terrain of central Florida underlain by cavernous limestones. The methods of surface and subsurface exploration used during the construction and expansion of the theme park included review of aerial photographs, hand probes, borings, cone penetration tests (CPT), and geophysical techniques such as gravity surveys, electrical resistivity, seismic refraction and solar electromagnetic waves. Handfelt and Attwool (1988, p.52) summarized the efficacy of the various methods, concluding that, "The traditional drilled and sampled boring has proved to be the most effective investigative method."

Ballard et al. (1983) reviewed several geophysical techniques of cavity detection used for two dam construction projects in karst areas. Of the three methods used at the El Cajon (Honduras) site, electromagnetic (radar) tests showed more promise than acoustic (sonar) tests and seismic refraction methods. At the Pueblo Viejo (Guatemala) site, electromagnetic methods were used to better appreciate the size, frequency and orientation of existing voids. Ballard et al. (1983, p. 156) found that, "Under very favorable conditions (high electrical contrast) the electromagnetic method is able to detect even smaller anomalies (5-20 cm size)." They suggested a four-step investigation sequence: (1) assess general underground conditions, (2) refine the general picture, (3) perform a reconnaissance survey during construction phase, and (4) verify success of the stabilization program.

Fischer et al. (1988) also outlined four phases of construction planning and design. The first phase, pre-purchase site evaluation, might include an examination of water-bearing potential, cave and fault existence, limestone or dolomite grain size, and bedrock compressive strength. During the second, planning phase, one should consider purchasing options, design constraints, economic impact and construction alternatives. The third phase consists of a site investigation. The authors questioned the effectiveness of geophysical detection methods such as seismic refraction or reflection, gravity, conductivity, and ground penetrating radar (GPR). They argued that, " . . . there is no substitute for carefully drilled test borings, qualified full-time inspection, experienced and careful drillers, and large diameter Christensen-type double tube core barrels" (Fischer et al., 1988, p. 125). The fourth and final phase, geotechnical engineering, is an evaluation of possible foundation solutions. These include excavation and filling, pile installation and grouting.

Recently, Thomas and Roth (1999) conducted a survey of engineering firms in Pennsylvania and New Jersey for the purpose of establishing state-of-the-practice in sinkhole characterization. 12 of 60 solicited firms responded, highlighting 12 methods for evaluating subsidence potential. They are, in order of most to least commonly used: (1) borings, (2) area reconnaissance, (3) review of existing mapping, (4) review of aerial photographs, (5) resistivity survey, (6) seismic refraction survey, (7) electromagnetic survey, (8) GPR, (9) trenching, (10) microgravity survey, (11) video televiewer, and (12) tomography. The most utilized techniques are lacking in their ability to provide a complete picture of a difficult subsurface (Thomas and Roth, 1999). They recommended more sophisticated techniques of void detection should be used to avoid continued structure failure.

Kannan (1999) argued that while standard penetration tests (SPT) provide a quantitative understanding of sinkhole potential at the point of entry, geophysical tests reveal a qualitative appreciation of the broader area. He warned against over-reliance on SPT data and contended that recent " . . . case studies illustrate the use of geophysical exploration methods in adding value to the engineered design" (Kannan, 1999, p. 75). Thus, it is optimum to combine SPT and geophysical methods such as GPR, electrical resistivity tests, or seismic refraction tests. However, he concluded that, "It is not possible to (completely) quantify the potential for sinkhole development with the current level of understanding of karst terrane" (Kannan, 1999, p. 80), therein giving the technical community another call for continued improvements in subsurface investigation.

Prior to construction of a cogeneration power plant in central Florida, Chang and Basnett (1999) used a combination of GPR, SPT and Dutch cone soundings to delineate the boundary of an existing sinkhole within the proposed footprint of the facility. The Dutch CPT, " . . . consists of slowly forcing a conical point into the soil and inferring soil strength from the force required to advance the cone" (Chang and Basnett, 1999, p. 114). Because Dutch cone soundings eliminate the influence of soil friction on a resistance reading, they produce higher quality data faster and at a lower cost than SPT.

Multi-electrode earth resistivity is another recently improved exploratory method that may be useful in characterizing a site with highly variable bedrock. The method involves introducing direct current into the subsurface through a line of linked electrodes while measuring the voltage drop across any two adjacent probes. Mackey et al. (1999) presented a case study where multi-electrode resistivity was used in conjunction with soil borings and percussion rock probes. They found that the earth resistivity data correlated well with the borings and probes for depth to bedrock but were inconclusive for bedrock void location. Although relatively unproven, multi-electrode earth resistivity is a promising new tool for karst subsurface investigation.

One common theme in the conclusions of all the above research, whether explicit or implicit, is the need for a methodical, cost-efficient approach to subsurface investigation and characterization for projects in karstic regions. The steps should include a review of relevant past data, a survey of surface features, observations from direct ground penetrations, and the use of a variety of geophysical exploration techniques. As discussed later in this paper, a similar multistage approach to define the extent of soft soils and overburden voids was followed for the Merck landfill cap project. The pre-stabilization elements included historical data review, area reconnaissance, GPR and Geoprobing. Despite consistency with current practice, the Merck exploratory program did not accurately predict the grout-take capacity of the subsurface, confirming the conclusions of recent research that better techniques are needed.

CASE STUDY: MERCK LANDFILL CAP PROJECT
Preliminary Subsurface Characterization
The decision was made to close the Merck landfill and capping activities began in the spring of 1999. This 13-acre landfill in the Appalachian Great Valley is located in a moderately mature, covered karst setting that is predominantly a mantled karst because most of the overburden is transported alluvium. However, the setting is locally transitional to a subsoil karst where clay layers, residual from dissolution of the underlying carbonate bedrock, occur along the TOB and beneath the alluvium. This Paleozoic bedrock comprises folded dolostones and limestones. Along the northern flank of the landfill, the irregular TOB is typically about 6.1 m (20 ft) below ground, and the long-term water table is typically between approximately 1.2 and 3.0 m (4 and 10 ft) above the TOB. The karstic nature of the subsurface in the area and a history of sinkhole development in the vicinity of the landfill triggered, a review of subsidence potential during the final phase of cap design. Because the conceptual model of sinkhole development for the area highlighted the cover-subsidence and cover-collapse types of subsidence sinkholes, investigation focused on characterizing the overburden composition and structure. The review also included an analysis of geologic well logs, a GPR survey and collection of ground penetration data with a Geoprobe®.

A large number of groundwater monitoring wells have been installed in the vicinity of the landfill for the purpose of contaminant plume delineation. The geologic well logs indicate the overburden is primarily transported sand, gravel and cobble-sized material that has been deposited by channel migration and flooding of a nearby river. The recorded history of sinkhole formation at the plant can be attributed to well installation activities, test pumping of production wells, leaking drainage ditches or pipes, or some other man-induced hydraulic stresses. However, at the time of the subsidence investigation, there were no visible sinkholes or surface depressions around the landfill.

GPR is a technique that uses high-frequency electromagnetic pulses to produce a continuous, cross-sectional picture of the subsurface. In this setting, a GPR survey offered a reasonable likelihood of mapping the locations of overburden subsidence (prior to any ground breakthrough) for the following reasons: (1) the relatively coarse-grained, alluvium overburden should be "visible" in the GPR profiles (thick clay sequences tend to rapidly decrease the depth of the GPR electromagnetic wave penetration); and (2) the relatively shallow, expected depth to the TOB should allow visibility of much of the overburden in the GPR profiles. Under these conditions, the goal of the GPR survey was to delineate overburden layers that are either depressed (concave upward), or showing notable disruption in lateral continuity, as a result of subsidence into the underlying carbonate rock (USFilter, 1999a).

In February 1999, the northern, eastern and southern perimeters of the landfill were divided into gridlines on a 1.5 m (5 ft) spacing and surveyed. Results of the survey revealed seven areas of inferred overburden subsidence, all along the northern perimeter (Figure 1). The GPR profiles were interpreted by an experienced karst geophysicist. No overburden voids were detected by the GPR survey. Areas #1 and #6 (not pictured) were dismissed because they fell outside the limits of the final cap boundary. The remaining five areas were ranked according to risk from highest to lowest: #3, #7, #2, #4 and #5. The determination of relative risk considered the history of nearby sinkhole development, the location of known fracture traces, and evaluation of likely water table fluctuations across the TOB.

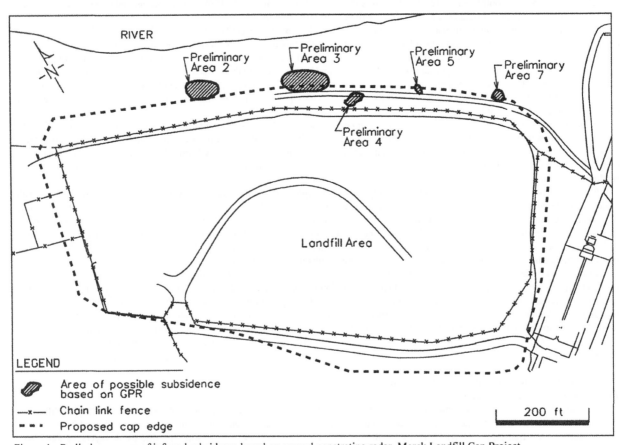

Figure 1: Preliminary areas of inferred subsidence based on ground penetrating radar, Merck Landfill Cap Project.

It was decided to supplement the interpretation of the GPR survey with a Geoprobe investigation. The Geoprobe directly pushes 5.1 cm (2 in) rods into the ground without the use of drilling to remove soil. Geoprobe offered a relatively inexpensive technique to detect any possible overburden voids, and to more clearly delineate the possible presence and boundaries of any loose soil zones, possibly reducing the extent of areas considered for stabilization measures. Also, cone penetrometers were likely to be repeatedly damaged by contact with coarser alluvium particles (locally of boulder size). A total of 137 Geoprobe penetrations were made along a grid pattern over each of the five areas in June 1999, just prior to initiation of grouting activities. As the rod was advanced, the operator made a qualitative assessment of penetration resistance, assigning integer values from 1 (extremely soft to inferred overburden void) through 5 (highest resistance to penetration), to each 0.3 m (1 ft) interval (USFilter, 1999b). Figure 2 summarizes the Geoprobe data for target area #7. Soft zones are represented by the closest 1.5 m (5 ft) depth interval.

Several Geoprobe core samples were recovered that served to confirm the predominantly alluvial nature of the overburden. Like the GPR survey, the Geoprobe investigation detected no overburden voids. Geoprobe rod penetration can be refused by larger pebbles or cobbles. (Many Geoprobe rods were bent beyond use in this program.) This means that the depths of refusal represented minimum depths to bedrock.

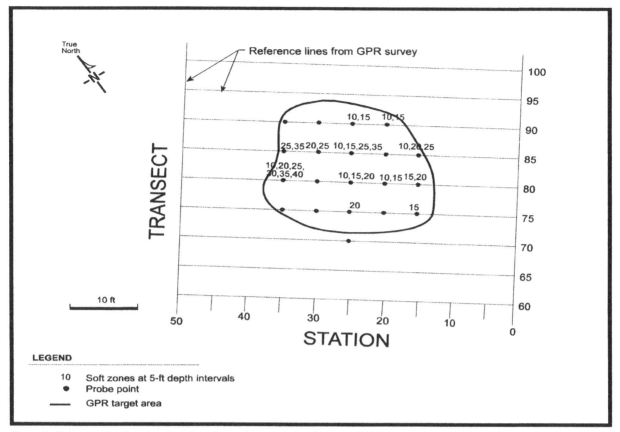

Figure 2: Softer overburden zones based on Geoprobe penetrations in target area #7, Merck Landfill Cap Project.

Compaction grouting was the preventive measure of choice for the Merck landfill cap project. Compaction grouting is an effective and widely used method of subsurface stabilization that can be used to densify loose overburden soils and to eliminate voids by filling. An industry accepted, formal definition of compaction grouting was articulated in 1980:

> Compaction grout - grout injection with less than one inch (25 mm) slump. Normally a soil-cement with sufficient silt sizes to provide plasticity together with sufficient sand sizes to develop internal friction. The grout generally does not enter soil pores but remains in a homogeneous mass that gives controlled displacement to compact loose soils, gives controlled displacement for lifting of structures, or both. (Mosely, 1993, p. 131)

This is essentially how we still define compaction grout rheology today, with the exception that slumps greater than 25 mm are often employed. The primary goal of compaction grouting is the densification and subsequent strengthening of soft soils displaced by the introduction of a mass of flowable material. Although not an engineered foundation, the bulbous columns created by compaction grouting have the secondary effect of added structural support that helps bridge gaps in the underlying bedrock.

Compaction grouting allows tremendous flexibility in making field adjustments to incoming subsurface information. With the surface and subsurface investigation completed, a compaction grouting work plan was developed as a collaborative effort between the owner, consultant and grouting contractor. Some general guidance on grout mix proportions and the results of the GPR survey were provided with the requests for proposals. The stated objective of the program was to minimize the possibility for the development of subsidence sinkholes by sealing the upper 1.5 m (5 ft) of bedrock and stabilizing soft overburden soils.

Results and Discussion

Drilling toward grouting began at target area #7, the easternmost of the five areas of concern. The area was divided into a grid with a primary grout hole spacing of 4.6 m (15 ft) and potential secondary hole locations at the midpoints. Holes along the perimeter of the target area were bored first. The contractor utilized a vibratory drilling method. Vibratory drilling " . . . utilizes a hydraulic vibratory hammer mounted on pile driving leads or a fixed beam and suspended by crane or other support such as an excavator" (Baker and Broadrick, 1997, p. 8). The injection pipe is both vibrated into the soil and extracted with this tooling. The vibratory technique is good for large volume grout projects where the injection zone contains limited rock obstructions and there is easy surface access. A 7.6 cm (3 in) diameter casing with disposable tip was pushed into the overburden until it reached refusal (possibly the TOB), at which time a 6.4 cm (2.5 in) drill bit was advanced inside the casing to the TOB and then 1.5 m (5 ft) deeper. After just a few holes, it was clear that this drilling method was unsuitable for this predominantly alluvium overburden setting. The casing installation was encountering stoppage at depths shallower than the expected TOB. Upon drilling through the level of casing refusal for the purpose of footing the grout column in the top 1.5 m (5 ft) of bedrock, breakthrough into soft soils occurred (sometimes within the first meter). Thus it was indeterminate whether the drill had encountered sediment-filled bedrock voids, or overburden sediments beneath a "floating carbonate boulder." The former case would be associated with penetration of the TOB (desirable in the work plan), while the latter is not. In some locations, the TOB was found at depths greater than 15.2 m (50 ft) – such depths were greater than those expected from the preliminary subsurface characterization.

The decision was made to switch to an alternate casing installation method that involved pre-drilling. After installation of a 1.5 m (5 ft) long section of 10.2 cm (4 in) diameter surface casing, a 8.9 cm (3.5 in) drill bit was advanced to the TOB and then 1.5 m (5 ft) deeper. The same 7.6 cm (3 in) casing with disposable tip, used for the initial drilling method, was then easily inserted into the hole for grout injection. This revised technique worked very well and, in the end, a total of twelve primary holes and eight secondary holes were drilled within target area #7. The depth to bedrock continued to be highly erratic, ranging from 4.6-18.3 m (15-60 ft) with differences of 6.1 m (20 ft) or more from one hole to the next (USFilter, 1999b). The initial drilling method had been chosen based a 6.1 m (20 ft) depth to bedrock, a cobble-filled overburden and the desire to minimize water and soil cuttings. The Geoprobe investigation suggested the potential for deeper bedrock but, because Geoprobe advancement is often stopped by large obstructions, the true TOB surface was difficult to determine. The adjustment to a substitute drilling method is representative of the critical need for flexibility in making field changes to any grouting program.

Grouting at each of these holes presented its own set of challenges. Grout injection began one week after the start of drilling activities once an adequate number of holes had been bored to make grouting efforts efficient. Sand, cement, flyash and water are the typical components of compaction grout. A suitable grout mixture contains " . . . enough fines to allow retainage of water and plasticity of flow along with enough granular components to create internal friction . . . " (Mosely, 1993, p. 141). Ready mix grout from a local supplier was delivered as needed by trucks in the following, approximate proportions: 125 kg (275 lb) cement, 272 kg (600 lb) class "F" flyash, 1361 kg (3000 lb) well graded sand, and 189 liters (50 gal) water. Daily measurements of slump varied from 1.9-8.3 cm (0.75-3.25 in), most often falling between 2.5-3.8 cm (1.00-1.50 in) (USFilter, 1999b). The grout pipe withdrawal interval was set at 1.5 m (5 ft) (accomplished by rapid 0.3 m (1 ft) incremental lifts) and volume, pressure and ground-heave cutoff criteria were established. Except for just above the TOB, where larger volumes were injected to meet the objective to seal the overburden-bedrock interface, the volume cutoff was 1.9 m^3/linear m (20 ft^3/linear ft) of drill hole. The grout delivery, or backline, pressure cutoff was set at 3447 kPa (500 psi). At depths shallower than 9.1 m (30 ft), the pressure cutoff was decreased to 2758 kPa (400 psi). Upon attainment of the volume or pressure cutoff criteria, grouting at that interval was terminated and casing was withdrawn to the next overall 1.5 m (5 ft) lift (US Filter, 1999b). Ground heave, often associated with relatively great depths in the secondary holes in comparison to the primary holes, resulted in immediate cessation of grouting at that hole.

The grout take at the first two holes was 17.1 m^3 (605 ft^3), which comprised a very large percentage of the total estimated grout volume for the entire project, including all five areas of concern. Since much more grout than originally estimated was being pumped into the ground, field cutoff criteria were reevaluated. The backline pressure criterion was lowered, the casing withdrawal interval was increased and slump characteristics were maintained closer to 2.5 cm (1 in). Changes to the fundamental approach were also proposed. Some thought was given to sealing the top of bedrock and continuing to grout only 1.5-3.0 m (5-10 ft) above. This option ignored the goal of densifying soft overburden soils. The other option involved drilling to an arbitrary depth of 6.1 m (20 ft) regardless of the actual depth to bedrock and grouting only the top part of the overburden. While this approach might have stabilized the shallow subsurface, the "floating" mass of grout might have put intensified stress on the soft soils and bedrock pinnacles. In the end, compaction grouting continued as outlined in the original work plan and a total of 154 m^3 (5429 ft^3) of grout was injected into 20 drillholes of various depths at target area #7 (Figure 3).

The average grout-take over 20 primary and secondary holes was 0.7 m^3/linear m (7.2 ft^3/linear ft) of hole (USFilter, 1999b). Table 1 summarizes the exploratory and compaction grouting data for target area #7. There is a reasonable correlation between the GPR and Geoprobe data. For the most part, at a given transect and station, where the GPR indicated inferred subsidence the Geoprobe found soft soils over a similar depth range. The data relationship falls apart when one looks at the corresponding grout-take data. For example, at transect 85/station 20, GPR and Geoprobe data indicate only a thin, shallow softer zone. The corresponding grout-take data for the closet grout injection hole, P5, reveals that 16.5 m^3 (581 ft^3) of grout was injected over a depth range of 7.6-13.7 m (25-45 ft) bgs. On the other hand, at transect 85/station 25, GPR and Geoprobe data suggest numerous soft zones over a large range of depths. However, no grout was injected into the overburden at S4, the grout injection hole nearest to this intersection. This conflicting pattern of disagreement is repeated at a majority of the locations. The same geophysical and direct penetration data that were used to draft the compaction grouting program proved insufficient in estimating the actual grout volumes. Had the preliminary, total grout volume estimate been more accurate, the grouting program might not have been implemented.

Although direct penetrations indicated bedrock would be significantly shallower at the four other target areas the decision was made, for a number of reasons, to terminate grouting activities after completing area #7. The potential total grouting cost was projected to be over 25% of the total cost for capping the landfill. A cost/benefit analysis shows that at some point, the cost of repairing a section of the cap damaged by a sinkhole is more appealing than attempting to eliminate the possibility of the sinkhole forming. As Beck (1991, p. 236) suggested, "One must balance the cost of site investigation plus the cost of engineering modifications to avoid sinkhole damage, against the potential cost and impact of the sinkhole." Areas #2, #3 and #5 (Figure 2) were positioned such that only a small portion of the each area fell beneath the outermost limit of the proposed final landfill cap. Area #4 was identified as a candidate for geogrid, a geosynthetic weave of plastic that was already being used for potential areas of subsidence on the landfill itself.

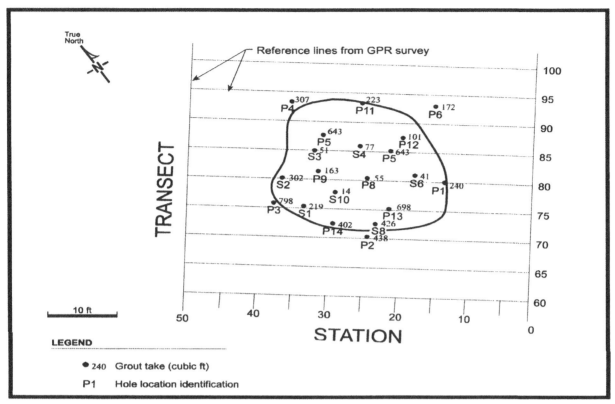

Figure 3: Grouting results for target area #7, Merck landfill Cap Project.

Table 1: Target area #7 data table, Merck Landfill Cap Project.

	GPR		Geoprobe		Grout-Take Above TOB (2)		
Transect	Station	Depth of Inferred Subsidence (ft bgs)	Station	Soft Zone Depth (ft bgs) (1)	Closest Hole	Volume (ft^3)	Injection Depth (ft bgs)
70	none	none	25	3-5	P2	392	5-50
75	15	6-12	15	4-7, 12-17	none	NA	NA
	20	6-12	20	3-5, 13-15	P13	455	30-50
	25	6-12	25	3-9, 17-21	none	NA	NA
	30	6-12	30	none	S1	219	50-55
	35	6-12	35	none	P3	798	5-25
80	15	6-10	15	3-5, 13-18, 21-24	P1	112	10-30
	20	6-10	20	3-6, 8-16	S6	0	25-30
	25	6-10	25	3-5, 8-23	P8	NA	NA
	30	6-10	30	3-8	P9	46	10-25
	35	6-10	35	3-5, 7-11, 21-35, 41-44	S2	179	30-50
85	15	none	15	3-5, 8-13, 21-27	none	NA	NA
	20	8-12	20	3-6	P5	581	25-45
	25	8-12	25	3-5, 10-16, 26-28, 34-36	S4	0	20-25
	30	8-12	30	3-5, 20-28	S3	0	25-30
	35	8-12	35	26-28, 34-38	none	NA	NA
90	15	10-15	15	NA	none	NA	NA
	20	10-15	20	3-17	none	NA	NA
	25	10-15	25	2-15	none	NA	NA
	30	10-15	30	2-8, 11-13	none	NA	NA
	35	10-15	35	3-9	none	NA	NA

(1) All depth intervals having at least three consecutively assigned penetration-resistance values of 3 or less.
(2) Grout-takes have been normalized to include volumes injected above the TOB only.

CONCLUSIONS

Karstic terrain presents a challenge to geologists trying to characterize the subsurface and to engineers attempting to construct or repair structures in such a subsurface. Variable depths to bedrock and voids can create unstable conditions in the overlying overburden sediments that can result in sinkhole formation and structural settlement. A wide variety of mechanical, hydraulic, physical/chemical and thermal modification methods exist for stabilizing an unsound subsurface. Compaction grouting is just one technique that can be used to densify loose soils in an effort to prevent future subsidence sinkholes. Consistent with current subsurface exploratory practice as published in recent literature, a historical data review, area reconnaissance, geophysics and direct penetration were used by the Merck landfill cap project. The combination of a ground penetrating radar survey and a Geoprobe investigation delineated five areas of inferred, possible subsidence. The local site geology dictated the choice of a vibratory drilling method for the purpose of grout casing installation. During implementation, it became apparent that the stepwise investigation program had not thoroughly predicted the grout volumes. When compaction grouting is proposed as a preventive, as opposed to remedial, measure, a cost/benefit analysis can be helpful in making the decision to proceed. The cost/benefit analysis relies on an accurate estimate of the total required grout volume, which is inherently difficult to predict. Better techniques of subsurface exploration are needed before compaction grouting programs in difficult karstic terrains can be executed with confidence.

ACKNOWLEDGEMENTS

The authors would like to thank Lester Urban (Merck & Co., Inc.) for review and support of the manuscript and Greg Smith (USFilter Groundwater Services) for preparation of the illustrations.

REFERENCES

Baker, A.C. and Broadrick, R.L., 1997, Compaction grouting: In: Proceedings of the 1997 ASCE Florida Annual Meeting, Ground Improvement, Reinforcement and Treatment: A Twenty Year Update and a Vision For The 21st Century, September 27, Clearwater, Florida, pp1-24.

Ballard, R.F., Cuenod, Y. and Jenni, J.P., 1983, Detection of Karst Cavities by Geophysical Methods, Bulletin of the International Association of Engineering Geology, pp153-157.

Beck, B., 1991, On calculating the risk of sinkhole collapse: In: Proceedings of the Appalachian Karst Symposium, pp231-236.

Cadden, A.W. and Traylor, R., 1998, Small project, big karst problem, solved with compaction grout: In: Proceedings of the 1998 Geo-Congress, Geotechnical Special Publication Effects of Construction on Structures, pp56-65.

Chang, K. and Basnett, C., 1999, Delineation of sinkhole boundary using dutch cone soundings, Engineering Geology, v.52, pp113-120.

Cooper, S.S. and Ballard, R.F., 1988, Geophysical exploration for cavity detection in karst terrain: In: Sitar, N. (ed.), Proceedings of a Symposium from the ASCE National Convention, Geotechnical Aspects of Karst Terrains: Exploration, Foundation Design and Performance, and Remedial Measures, May 10-11, Nashville, Tennessee, pp25-39.

Fischer, J.A., Greene, R.W., Ottoson, R.S. and Graham, T.C., 1988, Planning and design considerations in karst terrain, Environmental Geology and Water Sciences, v.12, pp123-28.

Forth, R.A., Butcher, D. and Senior, R., 1999, Hazard mapping of karst along the coast of the Algarve, Portugal, Engineering Geology, v.52, pp67-74.

Handfelt, L.D. and Attwooll, W.J., 1988, Exploration of karst conditions in central Florida: In: Sitar, N. (ed.), Proceedings of a Symposium from the ASCE National Convention, Geotechnical Aspects of Karst Terrains: Exploration, Foundation Design and Performance, and Remedial Measures, May 10-11, Nashville, Tennessee, pp40-52.

Kannan, R.C., 1999, Designing foundations around sinkholes, Engineering Geology, v.52, pp75-82.

Mackey, J.R., Roth, M.J.S. and Nyquist, J.E., 1999, Case study: site characterization methods in karst: In: Proceedings of the Third National Conference, Geo-engineering for Underground Facilities, June 13-17, pp695-705.

Mosely, M.P., 1993, Ground Improvement: CRC Press, Boca Raton, FL, 218 p.

Ogden, A.E., 1984, Methods for describing and predicting the occurrence of sinkholes: In: Proceedings of the First Multidisciplinary Conference on Sinkholes, October 15-17, Orlando, Florida, pp177-182.

Thomas, B. and Roth, M.J.S., 1999, Evaluation of site characterization methods for sinkholes in Pennsylvania and New Jersey, Engineering Geology, v.52, pp147-152.

Upchurch, S.B. and Littlefield, J.R., 1988, Evaluation of data for sinkhole-development risk models, Environmental Geology and Water Sciences, v.12, pp135-140.

USFilter Geoscience Services, 1999a, "Results of Merck landfill ground-penetrating radar survey and characterization of sinkhole potential." State College, PA.

USFilter Geoscience Services, 1999b, "Results of Merck landfill northern perimeter geoprobe investigation, slug testing, well monitoring, piezometer installation, drilling and grouting." State College, PA.

Geotechnical and Environmental Applications of Karst Geology and Hydrology, Beck & Herring (eds)
© 2001 Taylor & Francis, ISBN 90 5809 190 2

Karst void mitigation for water quality and quantity protection in Austin, Texas

SYLVIA R.POPE   City of Austin, Watershed Protection Department, Austin, TX 78767, USA,
sylvia.pope@ci.austin.tx.us

ABSTRACT

Voids are common features of the karstic limestone and dolomite formations of the Edwards Group in central Texas. They occur as interstitial cavities, solution-enlarged bedding plane cavities, solution-enlarged fractures, and cave passages. When voids are intercepted by construction activities, such as trenching or grading, the structural integrity and the hydrological function of the void may be compromised. For this reason, City of Austin regulation require that voids be reported so that they can be inspected and assessed for potential environmental impacts from the construction project. Water quality may be impacted if the void occurs within a trench for a wastewater line, stormsewer line, a petroleum products pipeline, or an onsite wastewater treatment system. Water quantity may be impacted if the intercepted feature is a conduit that conveys water to the local water table or to a nearby spring.

If a void is encountered during construction, mitigation plans to reduce potential impacts must be submitted to the City of Austin before construction activities can resume. Plans must include reports of the hydrological characteristics of the void and revised construction plans that incorporate protective measures designed to minimize negative impacts. This paper outlines common mitigation measures associated with these plans and presents case studies where these measures have been implemented.

Four examples of recent void mitigation techniques implemented on various projects are presented: encasing wastewater lines in cement for the length of a void plus 1.5 m (5 ft) on either end; installing steel plates along trench walls and encasing stormsewer or wastewater lines in cement for the length of the voids; installing PVC pipe across the base of trenches in order to maintain flow along a karst conduit; and moving infrastructure away from subsurface voids. These measures are intended to prevent the future migration of potentially contaminated groundwater into the Edwards Aquifer and to preserve the hydrologic connections between the land surface and the water table and/or springs.

WHY IS VOID MITIGATION NECESSARY?

The purpose of a void mitigation plan is to design a means for sealing off cavities within karstic rock horizons in a manner that protects the hydrologic characteristics while ensuring the structural integrity of the cavity and the manmade structures installed adjacent to such cavities.

Water quality and water quantity protection of the Edwards Aquifer is important to central Texans because the Edwards Aquifer is the most productive in the region; supplying water to domestic and municipal wells and also to springs that provide baseflow to surface water bodies. Void mitigation can preserve karst conduits that convey groundwater to springs or cave passages that are hydrologically connected to the water table. Preservation of these recharge paths helps to preserve the quantity of water entering the aquifer. Eliminating potential hydrologic connections between voids and potential pollutant sources, such as wastewater pipes or stormsewer pipes, helps to protect water quality. Rapid growth in the Austin area has resulted in a fifty percent increase in the number of voids reported annually between 1998 to 2000 (City of Austin, Watershed Protection Department internal data, 2000).

This paper describes the void mitigation process used by the City of Austin. Four examples of void mitigation plans are presented to illustrate various void mitigation techniques.

CITY OF AUSTIN VOID NOTIFICATION REQUIREMENT

The Environmental Criteria Manual (ECM) is the City of Austin's regulatory guide for implementing water quality ordinances codified in the Land Development Code. Appendix P-1 of the ECM instructs contractors to immediately report voids discovered during construction activities to a City of Austin Environmental Inspector. This requirement applies to any void that is greater than "[0.1 square meters] one square foot in total area, or blows air from within the substrate and/or consistently receives water during any rain event." Appendix P-1 also requires inclusion of a note regarding void discovery on all construction plans for projects built over the Edwards Aquifer Recharge Zone (City of Austin, 2000).

INTER-AGENCY COORDINATION

Inter-agency coordination is necessary when a construction site is located in an area of overlapping jurisdictions or regulations of federal, state, local agencies or a state-authorized groundwater district. These include areas regulated by the Texas Natural Resource Conservation Commission (TNRCC) under the Edwards Rules (Title 30 of the Texas Administrative Code, Chapter 213), or within the boundaries of the Barton Springs/Edwards Aquifer Conservation District (BS/EACD), within an area that is potential habitat for endangered karst invertebrate species under U.S. Fish and Wildlife Service (U.S.F.W.S.) regulation, or within an area designated as the Edwards Aquifer Recharge Zone by the City of Austin Watershed Protection Department (Chapter 25-8 of the Land Development Code).

Most occurrences of voids are on private development projects. Standard procedures on private projects involves City of Austin Watershed Protection Department (WPD) staff inspection of the voids, discussions of possible mitigation measures with other agency staff (TNRCC or the BS/EACD or the U.S.F.W.S.), and coordination of recommendations to the developer's consultants. A void mitigation plan is then prepared by the owner's consultants and submitted to the City of Austin Watershed Protection Department (WPD), Environmental Review and Inspection Division and the Environmental Resources Management Division for review and approval. The TNRCC requires that a licensed Professional Engineer (P.E.) prepare void mitigation plans and affix an engineer's seal to the plans. Although this is not stipulated in the City of Austin's Land Development Code or in the ECM, it is suggested practice.

VOID INSPECTION, DESCRIPTION, AND DOCUMENTATION

Inspection of the void should occur soon after notification is made in case its physical characteristics change significantly following exposure to light, air currents, heat and stormwater runoff. Important features related to the hydrological function of the void may be altered or obscured if left open for several days. Information such as speleothem activity, the presence of pools of water and the type of soil covering the void floor yield insight as to the amount of water entering and exiting the void and the possibility of nearby surface openings connecting to the void. Descriptions of the void, photographs, maps or sketches of its dimensions and a site plan showing the location and footprint of the void are to be included in the void mitigation plan submitted with revised construction plans. It is helpful if the consultants conducting the assessment of the void have been involved in earlier phases of site permitting, particularly with completing overland surveys of the karst features on the property.

Typically, voids are intercepted during utility trenching operations and inspection of them requires training in confined space entry and trench and excavation safety techniques. City of Austin staff conducting void inspections receive this training. Trench sidewall safety devices are installed by the owner's contractors prior to anyone entering the trench.

TYPES OF VOIDS AND MITIGATION PLAN STRATEGIES

Void mitigation plans typically address three types of voids:
1. A Type 1 void is less than 0.6 m (2 ft) by 0.9 m (3 ft) by 0.9 m (3 ft) in volume and is hydrologically inactive;
2. A Type 2 void is greater than 0.6 m (2 ft) by 0.9 m (3 ft) by 0.9 m (3 ft) that is hydrologically active but is an isolated feature lacking evidence of obvious connections to the water table, or a spring, or to other subsurface voids ("interstitial void"); and
3. A Type 3 void is greater than 0.6 m (2 ft) by 0.9 m (3 ft) by 0.9 m (3 ft) that is hydrologically active and is probably connected to the water table or a spring.

Caves and solution-enlarged fractures (Type 2 or 3 voids) intercepted by construction activities have the potential to be connected to the Edwards Aquifer or to a spring conduit. An assessment of the void may determine that the best mitigation strategy is to move utility infrastructure or buildings and parking areas. Otherwise, suggested strategies for void mitigation include:
1. Sealing off the opening along the face of the trench;
2. Installing durable pipe at the base of the trench to allow continued conduit flow across the trench;
3. Encasing utility pipe in concrete for the entire length of the void and a distance of 1.5 m (5 ft) on each end; and
4. Placing large diameter rock , sized 7.6 to 12.7 cm (3 to 5 inches) in diameter, by hand within the void to provide structural stability if the void will be located beneath a structure, roadway or parking area.

The void type and the layout of facilities in the site construction plan will govern which mitigation plan strategies are used.

VOID MITIGATION PLANS

Following the inspection and assessment of the void, a mitigation plan is prepared. A mitigation strategy is developed from information such as the void size, the evidence of water activity within the void, the potential for connections to the Edwards Aquifer or a spring, and the design of the utility or structure.

The void mitigation plan should provide a generalized description of how the void will be sealed and how any porous, fractured zones will be sealed. The plan should be prepared by a P.E. that has expertise in structural and geotechnical engineering and have design experience in central Texas, particularly with projects constructed over the Edwards Aquifer Recharge Zone. Information from the void assessment including photographs, field notes, maps, sketches, and geotechnical boring logs should be submitted with the plan. It is the engineer's responsibility to prepare a plan prescribing void mitigation measures that maintain the integrity of the utility lines, buildings and structures on site while preserving the hydrological characteristics of the caves/voids. The plan should provide material specifications and specific installation instructions for the closure procedure. Material specifications are required to be clear and readily apparent to contractors, particularly to personnel performing the work. Also, the engineer must revise the site plan to document the location of voids and mitigation measures used. Four recent case studies and the void mitigation plans for each are discussed below.

Example I, Lodge Cave, Parmer Lane, Austin, Texas.

The Lodge Cave was discovered on September 29, 2000 during trenching for a 1.8-m (6 ft) diameter reinforced concrete pipe (RCP) stormsewer. The cave is an isolated feature ("interstitial void") that developed vertically along a fracture and horizontally along a fossiliferous, vuggy, friable horizon within the limestone. The "footprint" of the cave is approximately 21.4 m (70 ft) by 6.7 m (22 ft). The cave roof was intercepted by trenching equipment at a depth of approximately 3.1 m (10 ft) below grade. The cave interior was fairly dry, as evidenced by the abundant cave coral and popcorn formations yet stalactites were dripping on September 29, 2000. Lodge Cave is considered a Type 2 void. Two-thirds of the cave will be beneath a future roadway.

The void mitigation strategies proposed by the engineering consultant included (Carter and Burgess, 2000):

1. Backfilling 7.6 to 12.7-cm (3 to 5-inch) diameter rock at the base of the trench to support the 1.8-m (6 ft) diameter RCP and to allow fluids to migrate along the cave floor;
2. Sealing off the cave opening at the trench face with sand bags; and
3. Pouring a concrete slab over the rock base that extends above the cave ceiling and for 1.5 m (5 ft) on each end of the cave opening. Dirt from the sidewalls of the excavation was removed by hand to ensure a strong bond between the limestone and the 17,237-kPa (2,500-psi) compressive strength concrete.

The void mitigation measures, installed on November 10, 2000, will prevent the leakage of untreated stormwater from the pipe into the cave where it could leak through fractures to the water table which is at approximately 30 m (85 ft) below ground surface in this area. Sealing off the utility trench floor and walls from the cave will allow water traveling through the vadose zone to continue to migrate through the cave.

Example II, Four Points Emergency Services and Fire Station Site, River Place Boulevard, Austin, Texas.

Voids were encountered within a trench for a stormsewer pipe and within a pit excavated for septic system holding tanks. The void horizon occurred at approximately 1.2 to 2.4 m (4 to 8 ft) below ground surface (bgs). Caves were found on the adjacent tract but no surface expression of these voids were detected during the pre-construction karst survey. Disbelievers Cave, an endangered species cave that is habitat to the Tooth Cave ground beetle *Rhadine persephone,* is located 27.5 m (90 ft) from the property boundary of the Four Points Emergency Services and Fire Station Site (Four Points AFD) on an adjacent tract. Preservation of potential recharge paths within the Edwards Group limestones are important in the Bull Creek watershed where most of the recharge occurs via micro-karst features (Johns, 1994).

A total of seven voids were intercepted along the trench for a 0.5-m (18-inch) diameter RCP stormsewer. These ranged in size from 0.5 to 1.5 m (1.5 to 5 ft) in height and approximately 1.2 to 2.1 m (4 to 7 ft) in length and 0.3 to 1.4 m (1 to 4.5 ft) in width. The voids probably developed under phreatic conditions at or near the water table and are now active vadose zone features. Stalactites, stalagmites, flowstone and cave popcorn were found in all of the voids. Tree roots, dark brown organic material and spider webs found within one of the caves prompted a biological investigation but no endangered karst invertebrate species were found.

Mitigation strategies for the Type 2 voids located adjacent to the stormsewer pipe included:

1. Bolting 1.3-cm (0.5-inch) thick steel plates to the rock face surrounding the voids;
2. Sealing off the opening between the steel plate and the rock face with cement mortar, latex caulk and/or gunnite;and
3. Backfilling the space between the steel plates and the stormsewer with an air-entrained concrete ("flowable fill"as described in City of Austin Standard Specifications Item 402) for the length of each void plus 1.5 m (5 ft) on each end and to a height 0.3 m (1 ft) above the top of each steel plate.

Two voids and a vuggy horizon were encountered in the excavation for the septic system holding tanks. These voids occur within the same general horizon as those encountered along the stormsewer trench. The vuggy horizon is nearly continuous throughout the entire width of the 6.1 m by 6.1 m (20 ft by 20 ft) excavation. The voids and vugs are potential fluid migration paths to nearby Powerline Spring and Moss Gully Spring (Veni, 1998). Six geotechnical borings were cored in the vicinity of the septic system holding tanks to determine if voids were present below the floor of the excavation (HBC Engineering, Inc., 1999). Isolated weathered seams and 5-cm to 15-cm (2-inch to 6-inch) voids were intercepted at depths of 1.8 to 4.6 m (6.5 to 15 ft) bgs. This information suggested that an extensive void horizon was not present at a depth below the base of the septic system holding tanks. Project consultants and engineers determined that proceeding with void mitigation was a better alternative than attempting to relocate the tanks and/or redesigning the septic system, given site constraints and the project deadline.

The steel plate and backfill mitigation strategies were also used to seal off the voids found in the septic system holding tanks. In addition, 15.2 cm (6 inches) of "flowable fill" were poured over the entire floor of the excavation to form a "seal." Gunnite was sprayed over the walls of the excavation to seal off the vuggy horizon.

Example III, Lone Star Natural Gas Pipeline, Parmer Lane, Austin, Texas.

Trenching for a 0.3-m (12-inch) diameter natural gas pipeline along the northern right-of-way of Parmer Lane exposed fourteen voids within a friable, sandy limestone bed and within 100 meters (305 ft) of three endangered karst invertebrate species caves. The voids occurred at approximate depths of 1.1 to 1.7 m (3.5 to 5.5 ft) below ground surface and varied in length from 0.6 to 7.6 m (2 to 25 ft). The lateral extension of a typical void beyond the trench face ranged from less than 0.3 m to 6.1 m (1 to 20 ft). Several large voids intercepted by the trench appeared to be active vadose water conduits and probably drain to nearby Yett Creek. A cave cricket and a cave spider were found in two of the voids, prompting a biological investigation for the presence of endangered karst invertebrate species. The U.S.F.W.S. Austin field office coordinated the development and preparation of the void mitigation plan and included the City of Austin WPD as a cooperating agency.

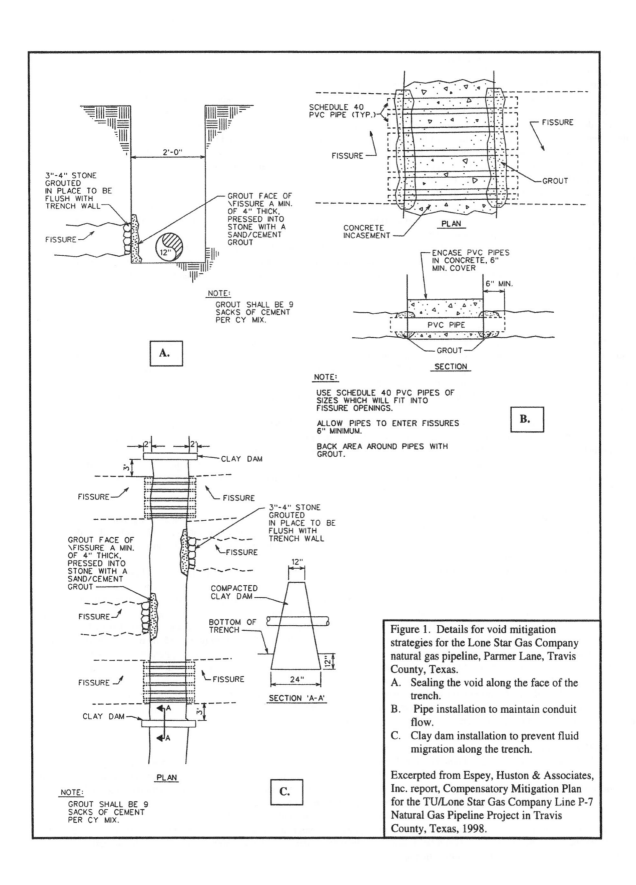

Figure 1. Details for void mitigation strategies for the Lone Star Gas Company natural gas pipeline, Parmer Lane, Travis County, Texas.
A. Sealing the void along the face of the trench.
B. Pipe installation to maintain conduit flow.
C. Clay dam installation to prevent fluid migration along the trench.

Excerpted from Espey, Huston & Associates, Inc. report, Compensatory Mitigation Plan for the TU/Lone Star Gas Company Line P-7 Natural Gas Pipeline Project in Travis County, Texas, 1998.

Development of the void mitigation plan was intended to protect the habitat for endangered karst invertebrate species but was also suitable for the protection of water quantity and water quality. Multiple void mitigation strategies were employed to protect these naturally occurring Type 3 voids from the potential impact of a future natural gas leak while preserving the hydrological characteristics of the voids (Espey, Huston and Associates, Inc., 1998). The mitigation strategies included:

1. Sealing the openings along the face of the trench with rocks and mortar;
2. Installing PVC pipe encased in concrete across the base of the trench to allow air, moisture and fauna movement between voids intercepted by the trench;
3. Constructing small clay dams within the trench to prevent fluid migration along the floor of the trench;
4. Encasing the gas pipeline with 0.4-m (16-inch) diameter polyethylene pipe for the entire length of the void horizon (732 m or 2,400 ft); and
5. Sealing the encasement pipe and venting it to the atmosphere to prevent below ground releases of a future gas pipeline leak or rupture.

Figure 1 provides illustrations of the first three strategies. Other mitigation measures that were considered but not included in this void mitigation plan were: 1. Purchasing an offsite endangered karst invertebrate species cave to compensate for a potential "take" from caves adjacent to the pipeline, 2. Moving the pipeline to the southern right-of-way of Parmer Lane, and 3. Boring the pipeline at a depth below the known cave and void-forming strata.

Example IV, Millennium Cave, La Cresada Drive, Austin, Texas.

Millennium Cave was discovered on January 31, 2000 during excavation of a stormwater treatment sedimentation/filtration basin in the Village at Western Oaks subdivision. The trackhoe operator noticed a 0.4-m (15-inch) diameter opening in the rock and stopped to inspect the opening. He immediately moved the trackhoe back from the opening after seeing that it dropped approximately 4.3 m (14 ft) and widened out to the sides 4.6 m (15 ft) or more. A thorough inspection by a cave specialist revealed that the main chamber of a cave had been popped open. The chamber is approximately 16.2 m (53 ft) wide by 13.4 m (44 ft) long by 4.3 m (14 ft) deep. The cave contains many stalactites, stalagmites and flowstone formations. No surface expression of Millennium Cave was present prior to its discovery but two caves are located within 153 m (500 ft) in a designated karst preserve.

Void mitigation for Millennium Cave was fairly complex due to the size of the cave and the proximity of the sedimentation/filtration basin. The City of Austin WPD were concerned that construction of the sedimentation/filtration basin adjacent to the cave and over undetected voids would lead to future failure of the structure and that the WPD would have to repair the structure. The owner was concerned that a proposed wet pond, to be constructed within 762 m (2,500 ft) of the sedimentation/filtration basin, would intercept a cave during excavation. The owner proceeded with exploratory geotechnical cores in the vicinity of the wet pond.

Preliminary discussions between the owner and the WPD to move the sedimentation/filtration basin were unsuccessful. Next, discussions focused on resizing the sedimentation/filtration basin, installing a structural wall adjacent to Millennium Cave, conducting a void collapse analyses, and drilling geotechnical cores in the vicinity of the sedimentation/filtration basin. The void collapse analyses relied upon the calculations presented in Ford and Williams (1989) and White (1988) and the thickness of the limestone beds measured in Millennium Cave (Trinity Engineering, 2000). Several of the cores intercepted a friable/void horizon between depths of 3.7 to 6.1 m (12 to 20 ft), including a 1.2-m (4-foot) void detected in one of the borings.

The WPD coordinated the development of the void mitigation plan with the assistance of the TNRCC. Specific measures included in the void mitigation plan are:

1. Installing a secured entry structure over the artificial skylight entrance to Millennium Cave. The structure is to preserve the existing light, humidity, temperature and air flow conditions within the cave;
2. Incorporating Millennium Cave within the boundaries of the adjacent karst preserve;
3. Installing a structural wall on the western end of the sedimentation/filtration with a vertical separation distance of 7.0 m (23 ft) from the bottom of the wall to an underlying cave passage;
4. Encasing or slip lining the 0.5-m diameter RCP segments that are located over the "footprint" of Millennium Cave;
5. Applying sealant to all joints of the splitter box structure installed west of Millennium Cave; and
6. Installing a 0.3-m (1-foot) thick clay liner (0.2 m (6 inches) is required by ordinance) below the sedimentation/filtration basin.

VOID MITIGATION IMPLEMENTATION

During installation of void barriers or other mitigation measures, photographs should be taken to document the installation and completion of these measures. Oversight during installation and construction of void mitigation measures can help ensure that the plan specifications are followed. The City of Austin retains copies of void mitigation plans for future reference on the project site and for evaluation of proposed mitigation plans for voids discovered on nearby sites. This information should also be retained by the site owner, regulatory agencies overseeing the mitigation, and should be submitted to utility companies responsible for the future maintenance of affected utility lines.

CONCLUSIONS

Construction activities occasionally intercept voids within the karstic limestone units of the Edwards Aquifer Recharge Zone in Austin, Texas. When voids are exposed, mitigation strategies are devised to protect the hydrological function of the void, to stabilize the structural integrity of the void, and to protect the structural integrity of adjacent utilities or structures. The goal is to protect these features in order to maintain the water quantity and water quality of the Edwards Aquifer in a way that is compatible with site development.

ACKNOWLEDGEMENTS

I'd like to thank my colleagues in the Watershed Protection Department for their support and encouragement, particularly David Johns, Ed Peacock and Nancy McClintock. I'd also like to thank the owners/operators of the four sites used as examples for their cooperation. They include the Hanover Company, Inc. (Lodge Cave), the City of Austin Fire Department (Four Points Emergency Services and Fire Station Site), Lone Star Natural Gas Pipeline (Parmer Lane), and Lumbermen's Investment Corporation (Millennium Cave).

REFERENCES

Carter and Burgess, 2000, Letter to the Texas Natural Resource Conservation Commission describing a void mitigation plan for Lodge Cave. Also submitted to the City of Austin, Watershed Protection Department.

City of Austin, 1995, Appendix P-1, Erosion Control Notes, of the Environmental Criteria Manual: City of Austin, Austin, Texas.

Espey, Huston & Associates, Inc., 1998, Compensatory Mitigation Plan for the TU Electric/Lone Star Gas Company Line P-7 Natural Gas Pipeline Project In Travis County, Texas: Prepared for TU Services, Dallas, Texas on September 2, 1998.

Ford, D.C. and P.W. Williams, 1989, Karst Geomorphology and Hydrology: Chapman and Hall, London, England. pp. 309 to 314.

HBC Engineering, Inc., 1999, Void Mitigation Plan, Four Points Fire Station, Austin, Texas: Prepared for the City of Austin, Department of Public Works and Transportation. Austin, Texas.

Johns, David A., 1994, Groundwater Quality in the Bull Creek Basin, Austin, Texas: in Edwards Aquifer – Water Quality and Land Development in the Austin Area, Texas" in the Field Trip Guidebook of the 44th Annual Convention of the Gulf Coast Association of Geological Societies, Austin, Texas, pp. 18 to 36.

Trinity Engineering Testing Corporation, 2000, Water Quality Pond No. 4 Report: Prepared for Lumbermen's Investment Corporation.

Veni, George, 1998, Draft Report Survey and Preliminary Hydrogeologic Evaluation for Karst Features at Proposed River Place Boulevard Fire-EMS Station, Travis County, Texas: George Veni and Associates, Inc. for Hicks and Company, Inc.

Warton, Mike, 2000, Report of Findings, Investigation and Documentation of Newly Encountered Karst Feature Found within Site Detention Pond Construction Located Adjacent to Site Karst Park Preserve Area of the "Village of Western Oaks" Residential Subdivision Development Property, Travis County, Austin, Texas: Mike Warton and Associates for Lumbermen's Investment Corporation.

White, W.B., 1988, Geomorphology and Hydrology of Karst Terrains: Oxford University Press, New York. pp. 229 to 235.

Geotechnical and Environmental Applications of Karst Geology and Hydrology, Beck & Herring (eds)
© 2001 Taylor & Francis, ISBN 90 5809 190 2

Enhanced recovery of petroleum hydrocarbon soil vapors and contaminant smear in karst

DONALD J.SCHAEZLER   ETC Information Services, Cibolo, TX 78108, USA, envmgtdoc@aol.com

THERESE M.BAER & PAUL L.SCHUMANN   Baer Engineering & Environmental Consulting, Inc.,
Austin, TX 78723, USA, baereng@aol.com

RICHARD N.BERRY   R.N.Berry, P.E., San Antonio, TX 78216, USA, rnbpe@aol.com

ABSTRACT

Structural and karst features, combined with slant borings, were used to enhance recovery of gasoline vapors and contaminant smear in the Glen Rose formation. The Glen Rose is a Lower Cretaceous formation located in the hill country of central Texas. The Trinity aquifer is a sole-source aquifer within the Glen Rose and underlying formations. When a motorist backed over a dispenser, there was a subsurface release of 900-gallons of gasoline. Emergency response measures included the installation of a soil vapor extraction unit coupled with a thermal destruction unit (SVE/TD).

A geological site investigation disclosed the presence of a significant fault located very near the source of the release. Installation of additional borings, drilled on a horizontal slant under the source area of the release, successfully increased recovery of petroleum hydrocarbon vapors in the vadose zone. Zone isolation tests using packers to isolate borehole intervals were used to identify productive zones and further enhance vapor recovery, causing a second SVE/TD unit to be installed.

After soil vapors were exhausted, a similar approach was used to enhance remediation of contaminant smear and volumetrically displace perched liquids. Near-surface injections of fresh water were conducted in increasing volumes and in increasing rates to wash, or flush, the smear zone. The success of repeated fresh water injections in remediating contaminant smear was demonstrated by analytical data in down gradient recovery wells.

INTRODUCTION

The purpose of this paper is to present the results from the application of two enhanced recovery methods for gasoline hydrocarbons released into a Karst aquifer in Central Texas. The first of these was extraction of hydrocarbon vapors from the subsurface by applying a vacuum. This technology was particularly effective early in the recovery process. The second method was flushing of the subsurface beneath the release point with clean water, with capture of the water in the groundwater treatment systems.

BACKGROUND

On July 11, 1999 approximately 900 gallons of regular unleaded gasoline were released through broken fuel piping into the subsurface at the RAM Stores, Inc., retail service station in north Bexar County, Texas. Details of this incident are provided in other papers. Fuel hydrocarbons, including benzene, toluene, ethylbenzene, and xylenes (BTEX), and methyl tertiary butyl ether (MTBE) quickly infiltrated into Karst features under the station and partitioned between free, adsorbed, and dissolved phases.

In the dissolved phase, the contamination spread eastward beneath U.S. Highway 281 and followed a generally east and northeast course. The dissolved phase was tracked by appearance of hydrocarbons in monitoring, recovery, and residential water wells. By late August, eight residential wells had been impacted. By early December MTBE had extended eastward about 3300 feet from the source. Remediation of the dissolved phase was initiated in early August. Two groundwater recovery and treatment systems were installed as part of this effort:

- WTS-1, located adjacent to the spill site
- WTS-2, located across U.S. Highway 281 near the most impacted residential wells

The locations, depths, and quantities of free and adsorbed phases could not easily be determined, but vapor phase testing in monitoring wells indicated significant materials near the release. Remediation of the release was initiated in early August. Two soil vapor extraction/thermal destruction (SVE/TD) units were installed as part of this effort:

- SVE-1, installed August 21, 1999
- SVE-2, installed October 6, 1999

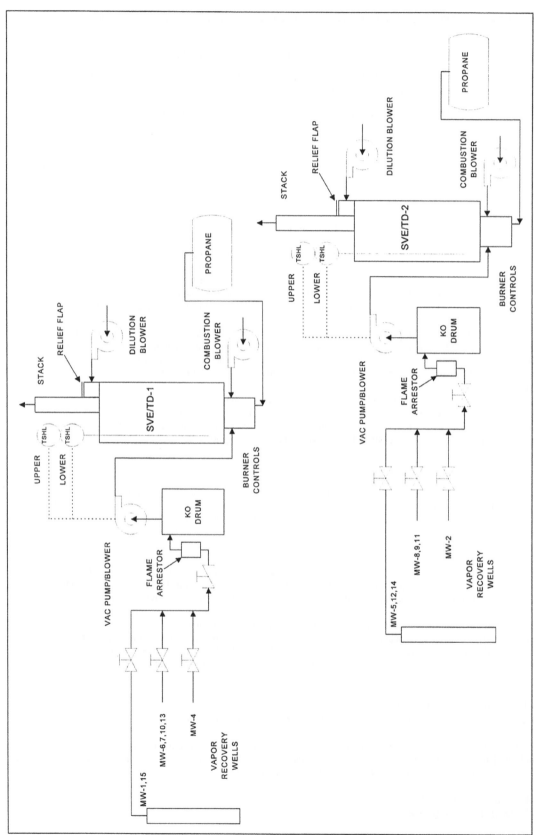

Figure 1. Simplified piping and instrumentation diagram of soil vapor extraction/thermal destruction units.

Both units were installed adjacent to the release site. They were included in a fenced enclosure with WTS-1. In early September, a new corrective action strategy was initiated, which included the following steps:

- Begin an aggressive investigation of the release site to locate product that may be located beneath the site in order that it may be recovered as soon a possible. This would include the installation of slant borings under the dispenser islands using air rotary slant drilling.
- Perform vapor testing on slant borings and other vapor wells.
- Install a second SVE/TD unit and optimize the operation of the two units.
- Obtain a standard exemption for air emissions from the two SVE/TD units.

In accordance with this strategy, three slant borings were installed. The vapor wells were tested and then connected together via a manifold to provide a flexible system for delivering soil vapor to two SVE/TD units.

VAPOR EXTRACTION WELLS

All of the initial vapor extraction wells were conventional vertical wells. These wells were completed at depths ranging from 23 feet to 300 feet below grade, and were typically screened the length of the well. Several wells were used for groundwater recovery as well as vapor recovery. These vertical wells provided good initial recovery of fuel vapors, but concentrations quickly fell.

In order to enhance vapor recovery, three slant wells were installed in October 1999. The borings were drilled using air rotary techniques. Samples were collected every five feet and scanned with a flame ionization detector for an indication of gasoline hydrocarbons. Boring angles and lengths were calculated to provide a well terminus 50 feet below the surface. All three wells were drilled 50 degrees from horizontal; lengths varied from 62 to 70 feet.

Wells were installed in the borings by placing two-inch diameter schedule 40 PVC pipe into the open holes. Screens were placed from the bottom to within 10 feet of grade. A sand pack was not placed around the screen due to the difficulty of placement at the angle involved.

Sixteen actual or potential vapor recovery wells were tested for soil gas composition from September 15 to October 12. Several were tested at multiple intervals, using packers to isolate selected screened intervals (which had been isolated during installation by use of bentonite plugs in the sand pack). The tests were performed by using SVE-1 or a portable vacuum blower to extract vapors. An instrument capable of reading the % lower explosion limit (LEL), oxygen, and carbon dioxide concentrations was used to determine compositions of the soil gas.

The tests indicated that fuel vapors were present only in the well field nearest the spill site. Wells nearest the release point (MW-1 and the three slant wells, MW-13, 14 & 15) showed the highest levels of gasoline vapor concentration, varying from 101% to 428% LEL. These wells also had relatively high carbon-dioxide concentrations (1.0 to 2.3%) and relatively low oxygen concentrations (17.5 to 18.6%). Seven wells had an average % LEL of at least 50 %. Six of these were located beneath or immediately around the release point. The seventh was approximately 150 feet northeast of that point.

MW-1 was screened from 100 to 300 feet below grade (the water level was about 250 feet below grade), whereas the slant wells were screened to an approximate depth of 50 feet below grade. The high LEL readings in all four wells may indicate significant vertical migration of the released gasoline, but much less lateral migration. The relatively high LEL readings 150 feet to the northeast and low LEL readings 50 to 100 feet in other directions indicates limited migration to the northeast. Unlike the dissolved phase, the free and adsorbed phases apparently did not migrate far from the release location.

VAPOR TREATMENT TECHNOLOGY

Each SVE/TD was manufactured by Grones, Inc. These units are similar to industrial boilers, with a vertical oxidation chamber, but without the heat exchangers for heating water. They use propane as the primary fuel, and they have an integral vacuum pump/blower that provides motive power for the soil vapor extraction. The extracted soil vapor provides the air source and secondary fuel. The propane feed rate is set manually to provide a minimum stable operating temperature. The units' automatic controls start the vacuum pump/blower, which extracts soil vapor from the wells manifolded to the unit. The units' operate in this mode subject to temperature controls. If the temperature reaches a maximum safe level, the vacuum pump will shut down and the units' return to propane-only operation.

All soil vapor entered the units through a knockout tank, where condensate and solids were removed. Condensate accumulated in the tank was bled into the burner by a small pump, so there was no liquid effluent.

The field operator used fixed flow and temperature meters and field monitoring equipment to select which wells are connected, throttling for each well, and throttling of each manifold. Samples were collected from the influent to the units weekly or semi-weekly to determine performance.

A simplified piping and instrumentation diagram for the two SVE/TD units is shown in Figure 1. Important features shown in this diagram are the flame arrestors and the relief flap on each stack. Note also the dilution blowers, which provided better dispersion of stack gas.

TREATMENT EFFECTIVENESS

Figures 2 and 3 present the concentrations of BTEX, MTBE, and gasoline range organics (GRO) in the influent to each SVE/TD. Using these data and the flow rates measured in the field or by fixed monitors, the recovery of gasoline could be calculated. The results of these calculations are shown in Figure 4. The diminishing returns of soil vapor extraction can be seen in these figures. Of the 900 gallons of gasoline released, approximately 270 gallons were recovered by the SVE/TD units.

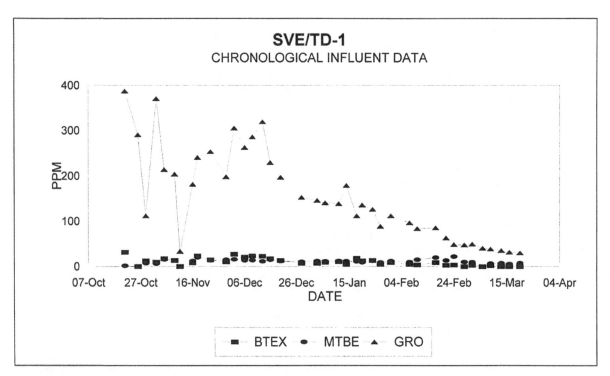

Figure 2. Influent gas composition for SVE/TD-1.

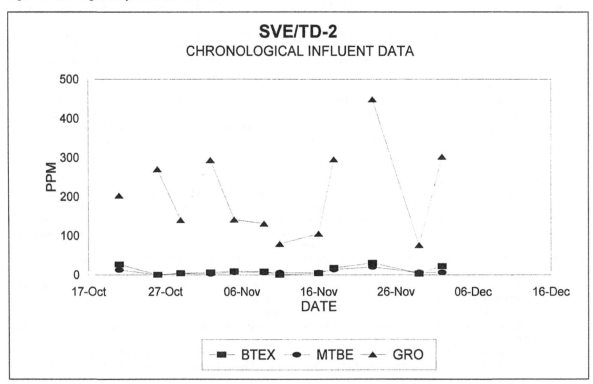

Figure 3. Influent gas composition for SVE/TD-2.

FLUSHING WITH CLEAN WATER

Dissolved-phase gasoline concentrations had diminished to very low levels, and vapor extraction was also losing its effect by February. However, rain events had had caused some increases in dissolved concentrations. This indicated there might be some

316

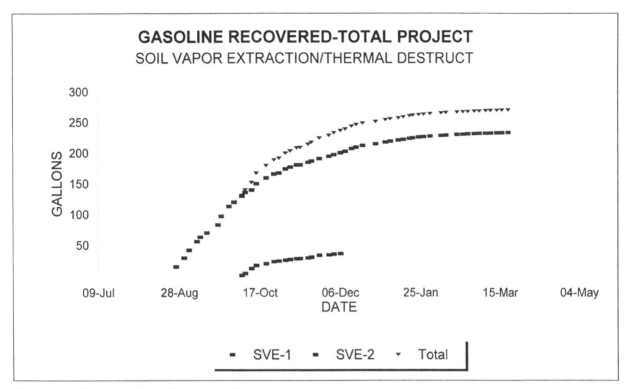

Figure 4. Gasoline recovery by soil vapor extraction.

gasoline residuals trapped above the normal water table. Therefore, we decided to inject clean water to determine if flushing of gasoline would occur. Injection of clean water at the site of the original release was accomplished February 16, February 23, August 2, and August 20, 2000. In each case, 1,000 to 10,000 gallons of tap water was allowed to percolate into the subsurface at the exact point of the original fuel spill. A monitoring plan was established for each injection. During each injection and for several days after, vapor was monitored in the two nearby recovery wells, the nearest tankhold vapor observation well, and in two recovery wells across the highway. Water samples were collected from various recovery wells in the well field adjacent to the release/injection point and the well field near WTS-2.

The first injection, using only 1,000 gallons, had very little impact, and it was decided to increase the amount of injected rinse water. The second injection, using 3,000 gallons of water, apparently mobilized some previously confined gasoline. Two recovery wells near WTS-2 had significant temporary increases in both total BTEX (up to 29 ppb) and MTBE (up to 21 ppb) concentrations eight hours after injection began. The third injection was postponed because of fluctuating cones of depression in the two well fields. On August 2, 10,000 gallons of water was injected over a 24-hour period. Moderate increases in MTBE were noted in three recovery wells near WTS-2. These experienced increases of up to 31 ppb BTEX over a 24 to 96-hour period. BTEX remained below detection limits. A final injection occurred on August 20, when 10,000 gallons of water was again injected over a 24-hour period. MTBE increases were slight, being less than 9 ppb over a 48 to 96-hour period following injection. Again BTEX remained below detection limit. At this point it was decided that essentially all recoverable gasoline had been removed in the subsurface below the spill site. On August 31, 2000, both SVE1 and SVE2 were removed from service.

CONCLUSIONS
1. Properly designed and operated soil vapor extraction and thermal destruction systems can be employed to achieve high initial recovery of released gasoline from the subsurface in a Karst system.
2. Flushing of residual free and adsorbed phase fuel with water can be effective in recovery of final residuals of gasoline form the subsurface in a Karst system.

ACKNOWLEDGEMENTS
The authors would like to acknowledge Mr. JD Marek and Mr. Cletus Edwards, Chairman and President, respectively, of RAM Stores, Inc., for their responsible action in the remediation of this underground release; Ms. Rosemary Wyman, Project Manager–Site Assessment and Management Remediation Division of Texas Natural Resource Conservation Commission for her assistance and cooperation in the exercise of regulatory affairs; and Mr. Scott Fisher, Vice President–Regulatory Affairs of Texas Petroleum Marketers and Convenience Store Association for his continuing efforts toward the advancement of petroleum marketing in the state of Texas.

Geotechnical and Environmental Applications of Karst Geology and Hydrology, Beck & Herring (eds)
© 2001 Taylor & Francis, ISBN 90 5809 190 2

Design and construction of water remediation systems used to manage karst water contamination

DONALD J.SCHAEZLER   ETC Information Services, Cibolo, TX 78108, USA, envmgtdoc@aol.com

THERESE M.BAER   Baer Engineering & Environmental Consulting, Inc., Austin, TX 78723, USA, baereng@aol.com

RICHARD N.BERRY   R.N.Berry, P.E., San Antonio, TX 78216, USA, rnbpe@aol.com

ABSTRACT

Air stripping and activated carbon adsorption were the key unit processes used for treatment of water pumped from the Trinity karst aquifer in the hill country of central Texas. Recovery and treatment of groundwater were part of an overall system used to remediate a 900-gallon gasoline spill in the Trinity, a sole source karst aquifer in the hill country of central Texas.

Two recovery fields and ancillary water treatment systems were used to provide hydraulic control during the remediation of a 900-gallon gasoline spill in northern Bexar County, Texas. The recovered water was contaminated with varying amounts of gasoline components, including BTEX and MTBE. Two separate treatments systems were designed and installed, each treating water from one of the two recovery fields. The effluent from the first system was routed under a major interstate highway and into a common holding tank. Combined effluent was pumped from this tank to surface discharge several hundred feet away.

Design and installation challenges included physical control systems for the two separate but interdependent systems, pipelines extending under the highway and through very difficult terrain, the lack of three-phase power, and the need for control of pH, scaling, and foaming.

INTRODUCTION

The purpose of this paper is to present the key problems and solutions for design and installation of groundwater remediation systems for treatment of gasoline-contaminated water in a karst aquifer in Central Texas.

BACKGROUND

On July 11, 1999 approximately 900 gallons of regular unleaded gasoline were released through broken fuel piping into the subsurface at the RAM Stores, Inc., retail service station in north Bexar County, Texas. Details of this incident are provided in other papers. Fuel hydrocarbons, including benzene, toluene, ethylbenzene, toluene, and MTBE quickly infiltrated into karst features under the station and partitioned between free, adsorbed, and dissolved phases. In the water table, the contamination spread eastward beneath U.S. Highway 281 and followed a generally east and northeast course. By late August, eight residential wells had been impacted. By early December MTBE had extended eastward about 3300 feet from the source.

Remediation of the release was initiated in early August. Two groundwater recovery and treatment systems were installed as part of this effort:

- WTS-1, located adjacent to the spill site
- WTS-2, located across U.S. Highway 281 near the most impacted residential wells

During early phases of operation, treated groundwater was reinjected via several monitoring wells. This operation was not successful, for several technical and administrative reasons, and relatively little groundwater was treated before the systems were shutdown later in August.

In early September, a new corrective action strategy was initiated, which included the following steps:

- Discontinue injection of treated water into wells that are located within the impacted zone, to minimize the spread of the groundwater plume.
- Provide discharge piping for surface discharge of treated groundwater downstream of the impacted zone, in an area where recharge of the discharged water would be maximized.
- Debottleneck the treatment systems.
- Complete installation of WTS-1 so that treated water can be pumped to WTS-2 for combined surface discharge.
- Obtain a state permit for discharge to a nearby stormwater impoundment.
- Obtain a NPDES permit for more permanent discharge to a nearby creek.
- Obtain a standard exemption for air emissions from the treatments systems.

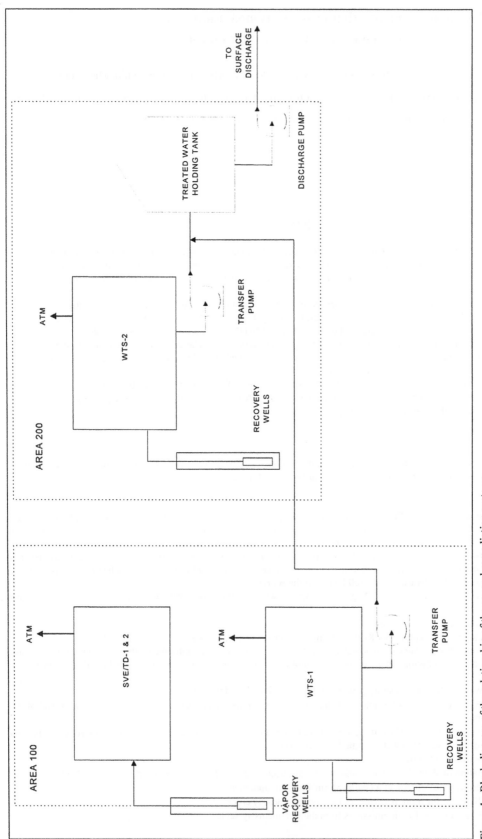

Figure 1: Block diagram of the relationship of the several remediation systems.

In accordance with this strategy, WTS-2 was upgraded and restarted on September 23. WTS-1 was upgraded and restarted on November 4. Upgrades included installation of two significant pipelines:

- 1500 feet under U.S. Highway 281 from WTS-1 to WTS-2
- 3500 feet across rugged terrain from WTS-2 to the creek.

TREATMENT TECHNOLOGIES

Summary

Two separate recovery and treatment systems were installed, and the two systems were then connected to a single discharge system. Each treatment system had the following unit processes:

- Equalization (WTS-1 also had provisions for skimming of free phase gasoline)
- Air Stripping, using perforated trays
- Filtration, using bag filters
- Activated Carbon Adsorption
- UV Disinfection

The main elements of the treatment systems were obtained from Carbonair. Tanks, pumps, piping, and controls were added to achieve the required operational logistics.

In addition chemical feed systems and related equipment were later added to overcome operational problems. The additional unit processes included:

- pH Control
- Anti-scalant addition
- Anti-foam addition

Description

In WTS-1, recovered water from MW-1 and/or MW-2 was pumped to an elevated oil/water separator, which served as a recovered water equalization tank and which could separate gasoline from water. Later, this tank was equipped with a mechanical mixer to assist in chemical control by mixing of sulfuric acid, anti-scalant, and anti-foam additives. Water overflowed by gravity through a multiple-tray air stripper into the air stripper sump. The transfer pump pumped the water through filters, activated carbon adsorbers, and UV columns into a treated water storage tank. Finally, the discharge pump transferred treated water from WTS-1 to the treated water storage tank in WTS-2, from which the water was pumped for final discharge.

In WTS-2, recovered water from RW-1, RW-3, RW-4, RW-Smith, and (later) RW-Mertens was pumped to an elevated neutralization tank, which was equipped with a mechanical mixer and designed to mix sulfuric acid according to pH control. Water overflowed into an equalization tank, with anti-scale and anti-foam chemicals being added at the overflow point. Water was then pumped to a multiple-tray air stripper and collected in the air stripper sump. A second transfer pump pumped the water through the filters, activated carbon adsorbers, and UV columns into a treated water storage tank.

Several design challenges were overcome in WTS-2. The 10 hp blower for the air stripper had a three-phase motor, and a special electrical device had to be installed to convert single-phase power into a simulated three-phase supply. The use of residential wells as recovery wells meant that complicated controls had to be installed to allow use of the well water for domestic purposes and for groundwater recovery. Level switches, solenoid valves and pump savers were used to achieve control of pumping. One steep gravity line experienced mechanical problems from air lock and water hammer. This was solved by mechanical constraint and adjustment of levels switches.

The common discharge system provided:

- Effluent Equalization
- Pumped Discharge (to surface water) on level control in the treated water holding tank

Discharge was accomplished by use of 4-inch O.D. polyethylene tubing. The tubing was unrolled and pulled through very rugged terrain with a backhoe. Joints were thermally welded in the field. The discharge was designed to pump uphill to a stormwater pond or slightly downhill to a creek some 3500 feet away. An automatic solenoid valve was used to prevent siphoning.

The arrangement of the two treatment systems, and the vapor extraction/destruction system, is illustrated in Figure 1. In this figure, Area 100 is adjacent to the release site, and Area 200 is about 1000 feet east, across U.S. Highway 281, near the most impacted residential wells. A more detailed illustration of the processes and instrumentation involved is given in Figure 2 for WTS-2. A summary of the treatment characteristics of the two treatment systems is provided in Table 1.

pH control was achieved with sulfuric acid, contained in bulk tankage and housed in a portable structure to promote safety and discourage vandalism. Metering pumps were operated by pH controllers, and mechanical mixers with axial flow impellers were used to achieve thorough mixing of the acid into the water. Anti-scalant and anti-foam additives were added continuously by metering pumps whenever recovery pumps were operating.

TREATMENT EFFECTIVENESS

The flow rates achieved in the two treatment systems are shown in Table 2. Note that WTS-1 and WTS-2 achieved maximum average monthly flow rates of approximately 21 gpm and 36 gpm, respectively. The maximum monthly total discharge rate was 74 gpm. Not all recovered groundwater was treated during some periods, although it was discharged with the effluent from both treatment systems.

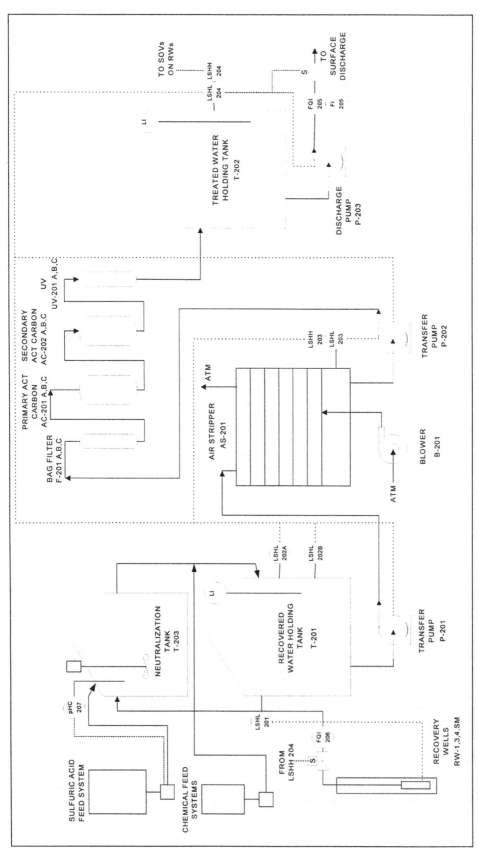

Figure 2: Simplified piping and instrumentation diagram for WTS-2.

Table 1. Summary of Treatment System Characteristics

Characteristic	WTS-1	WTS-2
Carbonair Model	STAT 30	STAT 80
Nominal Capacity, gpm	30	80
Maximum Hydraulic Capacity, gpm	25	45
Neutralization Tank, gal	150	1500
Neutralization Mixer, hp	1/12	1/2
Equalization Tank, gal	Included	3000
Stripper	6 sieve trays, 2½ hp blower, 150 cfm	6 sieve trays, 10 hp blower, 350 cfm
Filters	2	3
Activated Carbon Adsorbers	2 trains, 2 adsorbers each	3 trains, 2 adsorbers each
Discharge Holding Tank, gal	1000	3000
Discharge Pump, gpm	25	95

Table 2. Treatment System Flow Quantities

Month	Flow Quantities, Gallons			
	WTS-1	WTS-2	Untreated	Total Discharge
Pre October 99	Approx 0	203,270	0	Approx 203,270
October 99	0	545,250	0	545,250
November 99	317,281	852,060	0	1,169,341
December 99	689,067	1,021,080	0	1,710,147
January 00	942,970	1,609,210	0	2,552,180
February 00	886,157	1,517,820	61,858	2,465,835
March 00	672,948	1,432,470	361,602	2,467,020
April 00	0	817,340	2,154,684	2,972,024
May 00	0	1,422,466	2,092,105	3,514,571
June 00	0	2,144,943	1,181,190	3,326,133
July 00	0	1,402,500	1,101,686	2,504,186
August 00	0	1,278,030	1,891,764	3,169,794
TOTALS	3,508,423	14,246,439	8,844,889	26,599,751

Influent quality to the treatment systems began at high levels. During October, BTEX and MTBE entering WTS-2 were in the ranges 300–900 ppb and 150–500 ppb, respectively. Concentrations dropped into the 100–200 ppb range in November and to the 0–30 ppb range in December, for both parameters, and to even lower levels later. Concentrations entering WTS-1 were in the ranges 0–200 ppb and 100–300 ppb for BTEX and MTBE, respectively, in November and dropped rapidly later.

Discharge quality is summarized in Table 3. Note that no BTEX was ever detected from the effluent of the treatment systems. The only BTEX in discharges from the combined system was due to recovery wells that were bypassed untreated. MTBE was detected at low levels in the discharge from WTS-1 on two occasions, and it was detected in the activated carbon adsorber discharge from WTS-2 on several occasions, but it was not detected in the final effluent discharged on any of those occasions.

Table 3. Treatment System Discharge Quality

Month	WTS-1		WTS-2		Total Discharge			
	BTEX ppb	MTBE ppb	BTEX Ppb	MTBE ppb	BTEX ppb	MTBE ppb	Lead ppb	pH
October 99	NA	NA	ND	ND	ND	ND	NM	NM
November 99	ND	1.3	ND	ND	ND	ND	NM	NM
December 99	ND	0.55	ND	ND	ND	1.2	20	8.33 - 8.52
January 00	ND	ND	ND	ND	ND	ND	<3	8.52 - 8.68
February 00	ND	ND	ND	ND	ND	ND	<3	7.44 - 7.62
March 00	ND	ND	ND	ND	ND	ND	2.2	7.79 - 8.53
April 00	NA	NA	ND	ND	ND	ND	1.1	7.61 - 7.79
May 00	NA	NA	ND	ND	8.09	0.66	0.92	7.61 - 7.89
June 00	NA	NA	ND	ND	ND	ND	3.75	7.63 - 8.01
July 00	NA	NA	ND	ND	ND	ND	3.25	7.86 - 8.06
August 00	NA	NA	ND	ND	ND	ND	<3	8.00 - 8.12

ND denotes non-detect, NA denotes not applicable, and NM denotes not measured.

The WTS-1 air stripper only had BTEX to treat in the first week of operation, and it removed 100% of the BTEX at 188 ppb. The stripper removed 91– 97% of the MTBE at levels of 165–243 ppb, and, with one exception, 100% of the MTBE at levels up to 57 ppb.

The WTS-2 air stripper was 100% effective for BTEX up to 927 ppb. It was 98% effective on MTBE in the 400--535 ppb range and 100% effective at lower concentrations.

The carbon adsorbers removed 50–93% of the MTBE remaining after air stripping, but the adsorbed MTBE seemed to bleed from the adsorbers at low concentrations (2–4 ppb) over time.

CONCLUSIONS

1. Properly designed and operated treatment systems can be employed to achieve very high removals of BTEX and MTBE from groundwater.

2. Control of pH and addition of anti-scalant and anti-foam additives are important to the successful operation of such systems.

Geotechnical and Environmental Applications of Karst Geology and Hydrology, Beck & Herring (eds)
© 2001 Taylor & Francis, ISBN 90 5809 190 2

Landfill gas transport in karst

JEFF SMITH   Draper Aden Associates, Blacksburg, VA 24060 USA, jsmith@daa.com

ABSTRACT

Adjacent property owner concerns have typically excluded offsite investigations into the extent of landfill gas impacts. Direct-push gas monitoring techniques enable investigators to actively map and track landfill gas distribution and migration offsite via simple, low impact methods that respect adjacent property owner concerns. Characterizing the extent of offsite impact in karst allows the identification of previously unseen primary gas transport mechanisms and flow paths existing within voids in the bedrock. Several case studies at closed landfill sites situated in karst areas of the Valley and Ridge geologic province of Tennessee and Virginia are presented to illustrate the potential for offsite impacts resulting from the transport of landfill gas via the void network. At each site, offsite gas impacts have been remediated by accessing the bedrock voids and redirecting the flow to the surface with passive vents installed at the property line. Initially, the offsite gas monitoring completed at each case study site enabled investigators to track the gas migration to the bedrock voids. Ongoing offsite monitoring demonstrates the effectiveness of the bedrock void vents.

INTRODUCTION

As stated in "Geologic Evaluation of Sanitary Landfill Sites in Tennessee," produced by the Tennessee Division of Geology in 1972, karst terrain, characterized by sinkholes, caves, and similar solutional openings, offer poor sites for landfills and should be avoided. Existing solid waste management regulations typically restrict the development of landfills in karst areas.

Unfortunately, scores of old, closed landfills are located in karst and the post-closure monitoring and remediation many of these disposal sites remains incomplete. The remediation of landfill gas generated by waste decomposition can be particularly challenging in these areas. As explained in the Tennessee Landfill Gas Monitoring and Mitigation guidance document (March 1999), "Once landfill gas has exited the cell limits, there are numerous natural and man-made mechanisms for promoting further gas migration including, but not limited to: the presence of... (caves) in the vicinity of the cell." Landfill gas preferentially migrates via the path of least resistance. As the following case studies illustrate, this path of least resistance in karst areas is often subsurface voids. Once in the voids, the geometry and orientation of the void network govern the extent of gas migration.

Although landfill gas monitoring is generally required at landfill property boundaries, the following case studies illustrate that the potential exists for considerable gas migration offsite to go undetected in karst areas. These examples document cases where landfill gas had migrated several hundred feet offsite. At one site, landfill gas had migrated under the foundation of a home. It is apparent that gas can migrate considerable distances via voids and fractures in karst areas. Nonetheless, as the following case studies demonstrate, simple and successful gas migration characterization and gas remediation techniques are available for landfill sites situated in karst

WHAT'S THE STINK?

Adjacent property owner concerns have typically excluded offsite investigations into the extent of landfill gas impacts. The impact of offsite investigations is reduced through the use of hydraulically powered percussion/probing machines, commonly referred to as "direct-push" technology. Direct-push gas monitoring techniques allow active mapping of the distribution of landfill gas offsite, via simple, low impact methods that respect property owner concerns. Characterizing the extent of offsite impact allows the identification of primary gas transport mechanisms and flow paths existing within voids in the karst bedrock.

Recent remedial actions demonstrate that the voids can be short-circuited and landfill gas successfully vented to the surface. At each of the landfill case study sites, passive gas vents, accessing the bedrock voids, were installed at the property line. The bedrock vents remediate offsite gas impacts by venting the primary gas transport mechanism (i.e., bedrock voids) and redirecting the flow path to the surface at the property line. Ongoing monitoring at each of the landfill case study sites continues to demonstrate that passive, bedrock void gas vents can be extremely successful at remediating offsite gas impacts.

CASE STUDY SITES

The case study sites are located in the Valley and Ridge geologic province in Tennessee and Virginia. As the case studies illustrate, characterizing the extent of offsite impact has enabled the identification of primary gas transport mechanisms and flow paths existing within voids in the bedrock. By tapping into the void network along the landfill property boundaries, landfill gas has been vented from the voids directly to the surface, short-circuiting current offsite gas migration and drastically reducing and/or eliminating existing offsite impacts. Graphic illustrations of before and after offsite gas distributions demonstrating the effectiveness of the gas vents are provided below.

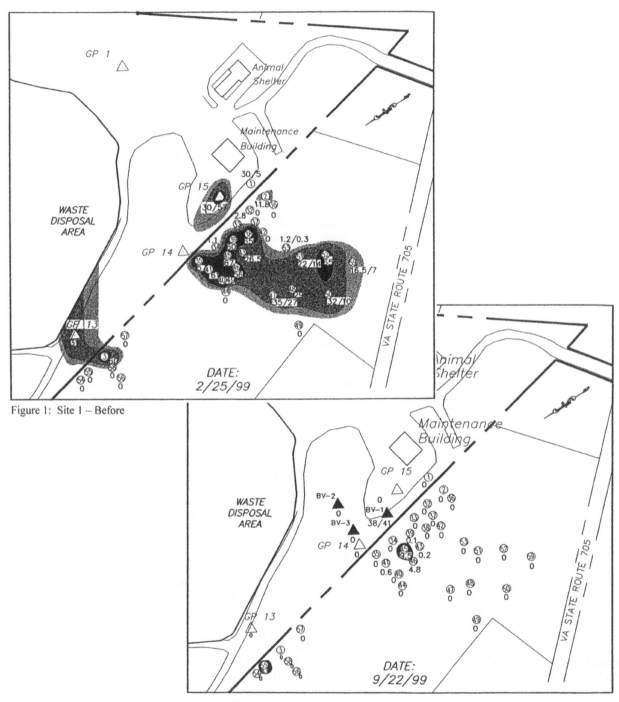

Figure 1: Site 1 – Before

Figure 2: Site 1 – After

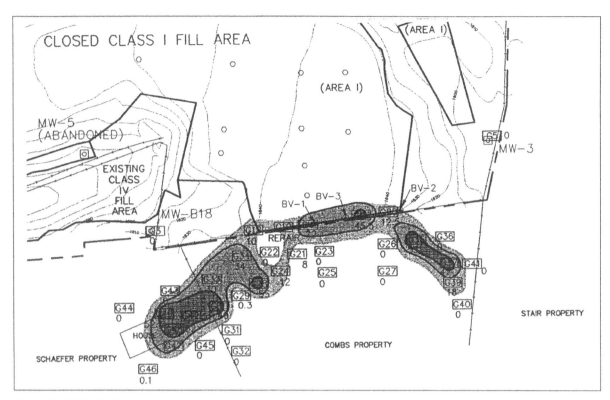

Figure 3: Site 2 – Before

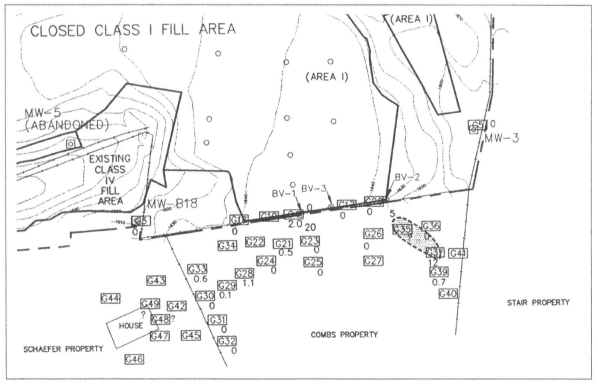

Figure 4: Site 2 – After

BASIC ELEMENTS AND PHYSICS OF LANDFILL GAS

Primary components of landfill gas are methane (40-60%), carbon dioxide (40-60%) and nitrogen (2-5%). Primary components of ordinary air are oxygen (21%) and nitrogen (78%).

Landfill gas and ordinary air are mechanical (not chemical) mixtures. Mixture density is related to the proportion of components present. Since methane is less dense (.55x), and carbon dioxide is more dense (1.5x) than ordinary air, methane-rich landfill gas mixtures, where $CH_4/CO_2 > 1$, will be less dense than air.

INFLUENCE ON LANDFILL GAS TRANSPORT IN KARST

Gas produced within a capped and inadequately vented disposal unit develops positive pressure inducing gas migration. Once gas has seeped into the void subsurface, methane's lighter density relative to air promotes transport to high points in the void network. A porous connection with the soil overburden at these high points can result in surface impacts observed some distance from the source.

Phase 1 of Investigation

Offsite probe networks indicated that concentrations of methane in the soil overburden (upper 30 feet) *increased* away from the source. If the soil medium was the primary transport pathway, methane would be expected to decrease away from the source.

Phase 2 of Investigation

At the first site, three vents were drilled into bedrock voids at 'non-impacted' locations at the landfill property boundary (1 to 5 foot diameter voids, ~ 60 to 80 feet deep). The first vent installed has consistently vented methane at greater than 50% by volume. The second two vents periodically vent methane from 5% to 55% by volume.

At Site 2, three vents were also drilled into bedrock voids along the landfill property boundary The first two vents encountered 4 to 5 ft diameter voids ~ 70 ft deep. The third vent encountered 10 ft of voids ~ 100 ft. deep. The first two vents installed have periodically vented methane at concentrations varying from 5% to 55% by volume. The third vent has never vented methane.

Soon after the bedrock vents were installed, offsite gas concentrations dropped and the extent of impact diminished. In September 1999, after tropical storms Dennis and Floyd, even concentrations at the property boundary dropped dramatically.

Phase 3 of Investigation

The monitoring networks, installed in the soil overburden, provide the data that demonstrate offsite gas distributions and the effectiveness of the void gas vents. Monitoring has shown that accessing the voids and venting gas to the surface can short-circuit the migration pathway and remediate offsite impacts. Note that the immediacy of remediation following void vent installation suggests that offsite gas existing in the karst soil overburden is not remediated by the venting of gas directly from the soil to the ground surface, but rather by the influx of fresh air in from the surface. The venting mechanisms involved are explored further below.

WHAT'S THE WEATHER LIKE DOWN THERE?

Atmospheric barometric pressure increases result in volume decreases in pockets of landfill gas in karst voids (Boyle and Mariotte's Law, i.e., air and all gases are compressible). Conversely, barometric pressure decreases result in volume increases. As experienced with the preceding case studies, the passing of tropical storms Dennis and Floyd in September 1999 coincided with a drastic off-gassing and subsequent reduction of gas levels observed in the near subsurface. These significant barometric pressure drops resulted in corresponding increases in the volume of the gas in the voids. Vents installed at these sites during the summer of 1999, promoted mobilization and enabled the release of this increased gas volume.

WHAT REMAINS AFTER THE STORM?

Although the success of the few bedrock void vents installed to date has been remarkable, not all the landfill gas residing offsite in the subsurface has vented off. As also observed prior to bedrock void vent installation, methane levels remaining in the soil overburden appear to vary with the passing of each pressure system. Gas remaining in the soil overburden and voids may reside in zones isolated from existing vents. Although additional targeted passive vents may be required for these isolated zones, recent work involving active extraction from the bedrock void vents appears promising.

SYSTEM INSTALLATION COSTS

Site 1
1994 - 15 nested wells at property boundary - $15,000.
1998 - 31 probes offsite - $3800.
1999 - 28 probes offsite - $2600.
1999 - 4 bedrock void vents - $8800.

Site 2
1998 - 16 probes on property - $1200.
1999 - 50 probes offsite in 4 phases - $8000.
1999 - 4 bedrock void vents - $7800.
Site 2 Total = $17,000

CONCLUSIONS

If a landfill's gas monitoring does not account for transport via voids in karst, property boundary monitoring may not be detecting considerable offsite impact. By missing the primary gas transport pathway, landfill gas will not be vented to the surface and offsite impacts will not be recognized and will remain. Characterizing the extent of offsite impact enables the identification of primary gas transport mechanisms and flow paths existing within bedrock voids. Short-circuiting the bedrock void transport pathway will allow landfill gas to be vented to the surface and promote the mitigation of offsite impacts.

REFERENCES

Atkinson, A. A., 1982, Mine Ventilation, The Colliery Engineering Co.

TDEC Solid Waste Technical Section, March 3, 1999, Landfill Gas Monitoring and Mitigation Guidance Document

TN Div. of Geology, 1972, Geologic Evaluation of Sanitary Landfill Sites in TN, Environmental Geology Series No. 1

Geotechnical and Environmental Applications of Karst Geology and Hydrology, Beck & Herring (eds)
© 2001 Taylor & Francis, ISBN 90 5809 190 2

In situ chemical oxidation process and karst aquifers

PAUL R.STONE   United States Army Corps of Engineers, Baltimore District, Baltimore, MD 21201, USA,
paul.r.stone@nab02.usace.army.mil

BRYAN HOKE   BRAC Environmental Coordinator, Letterkenny Army Depot, Chambersburg, PA 17201, USA,
bhoke@emh1.lead.army.mil

PAUL LANDRY & KENNETH J.COWAN   Roy F. Weston, Inc., West Chester, PA 19380, USA,
landryp@mail.rfweston.com, cowank@mail.rfweston.com

ABSTRACT

A pilot study to evaluate an innovative in situ treatment technology was implemented at Letterkenny Army Depot, Chambersburg, Pennsylvania, for a limestone bedrock aquifer contaminated with chlorinated solvents. This technology destroys VOCs in groundwater, utilizing a network of injection points to deliver a solution of hydrogen peroxide (50%), catalysts, and surfactants. The chemical reaction generated by this solution creates a hydroxyl radical, which is extremely effective in oxidizing complex organic compounds. The organic hydrocarbon contamination is rapidly oxidized by the hydroxyl radical to carbon dioxide, oxygen, and water.

This pilot study was the first to evaluate this technology in a high-flow karstic limestone aquifer. Extensive geophysical logging, packer testing, and discrete sampling were conducted to evaluate the fracturing and hydraulic properties of the bedrock for the injection design. Extensive on-site and downgradient monitoring was conducted during the pilot study to evaluate the effectiveness of this technology.

The limestone aquifer had maximum total VOC concentrations of 114 mg/L. Chlorinated compounds composed over 94% of the VOCs in the aquifer. The most prevalent VOC detected was 1,2-dichloroethene at 64% of the total chlorinated VOCs detected. Treatment was targeted at depths of 30 to 80 feet below ground surface within the source area. Over 12,000 gallons of 50% hydrogen peroxide were injected into the limestone aquifer over a 3.5-day period. The post-injection sampling results indicate that the estimated mass of total chlorinated VOCs destroyed was approximately 2,000 pounds.

The pilot study results indicate that this innovative in situ remediation technology can effectively lower chlorinated solvent concentrations in source areas located within a limestone bedrock aquifer.

INTRODUCTION

Pilot Study results indicate that chlorinated solvent source areas in karst bedrock aquifers can be effectively and efficiently destroyed by an in situ chemical oxidation technology that uses hydrogen peroxide.

An in situ chemical oxidation process that uses hydrogen peroxide was evaluated for the first time in a karst bedrock aquifer during a pilot study at Letterkenny Army Depot (LEAD) in Chambersburg, PA. The goal was to test whether this process would reduce the mass of nonaqueous phase liquid (NAPL) volatile organic compounds (VOCs) beneath a former waste disposal lagoon (the K-1 Area) at the site. The pilot was done to support a Focused Feasibility Study (FFS) of this area. The FFS is evaluating the technical feasibility and cost-effectiveness of numerous remedial alternatives. The Army Corps of Engineers Baltimore District funded the pilot study, which was performed by Roy F. Weston, Inc. and Geo-Cleanse International, Inc.

BACKGROUND

Historically, the K-1 Area was an unlined disposal lagoon that received various industrial solvents, wastes, and sludges. The impacted soil beneath the former lagoon was excavated, thermally treated, and used to backfill the excavation. The backfilled soil area was then capped. While the soil remediation removed a source of contaminants to the groundwater, contaminants remained bound within the bedrock fractures, joints, and solution features under the area. Immediately downgradient of the K-1 Area, a groundwater plume had been delineated. It contained total VOC concentrations ranging from 7 to 115,300 micrograms per liter (μg/L). Chlorinated VOCs composed over 95% of the total VOCs, consisting principally of 1,2-dichloroethene, trichloroethene, vinyl chloride, and tetrachloroethene.

THE PILOT STUDY

The in situ chemical oxidation process involved the injection of a 50% solution of hydrogen peroxide and a catalyst solution of pH amended water and soluble iron into the impacted media via a network of controlled, pressurized injection points. The chemical reaction generated by this solution, originally described by Fenton (1898), creates a hydroxyl radical that very effectively oxidizes complex organic compounds. During the reaction sequence, the organic hydrocarbon contamination is rapidly oxidized to carbon dioxide, oxygen, and water. Chlorinated VOCs are broken down in the same reaction sequence, with the chlorine ions degrading from the halogenated compounds. Residual hydrogen peroxide rapidly decomposes due to its unstable characteristics, becoming water and oxygen in the subsurface environment. The soluble iron amendments added to the catalyst solution precipitate into the groundwater during conversion to ferric iron.

Before implementing the injection, a series of six closely spaced open borehole injector wells were installed in the limestone bedrock in the vicinity of the upgradient edge of the K-1 Area. The injector wells were sampled and evaluated through the use of borehole geophysical logging and packer testing methods. This evaluation was critical to characterizing the location, orientation, and degree of hydraulic connection among the fractures, joints, and solution features of the wells, and was used to finalize the design and injection approach of the pilot study.

The injections were performed around the clock to maintain elevated hydrogen peroxide levels in the high permeability solution features common to four of the six injector wells. The injection lasted approximately 3.5 days. At this point, a total of 12,700 gallons of 50% hydrogen peroxide solution and 36,000 gallons of catalyst solution had been injected into the bedrock aquifer beneath the K-1 Area.

The mass of VOCs destroyed was estimated by use of groundwater chloride concentrations measured during the injection process. Assuming the various chlorinated compounds had dechlorinated equally during the pilot study, the mass of organics destroyed after the injection of 12,519 pounds of 50% hydrogen peroxide amounted to 1,942 pounds at the end of field monitoring round 1, a ratio of 7 to 1 mass of hydrogen peroxide injected to mass of organics destroyed. This ratio was not maintained throughout the pilot study, however, because excess hydrogen peroxide was either consumed by the hydroxyl radicals or lost to nontarget areas.

Pre- and post-injection groundwater samples were also analyzed for select inorganics to determine if the limestone bedrock was degraded by the chemical oxidation reaction and to quantify the concentration of iron remaining after injection. Based on the evaluation of the inorganic parameters, it did not appear that the injection process caused a significant increase in the dissolution of the limestone. This result also was seen in bench scale tests conducted prior to the pilot study. In addition, due to the tendency for hydrogen peroxide to break down rapidly in the environment, no hydrogen peroxide was detected in any of the off-site wells or springs monitored during the pilot study.

RESULTS

Based on field monitoring and four post-injection groundwater sampling rounds, the results of this pilot study indicate that this in situ chemical oxidation technology is capable of destroying source material and maintaining localized VOC reductions in a karst hydrogeological setting. Thus, a full-scale design and application of this chemical oxidation process should incorporate the following conclusions and recommendations:

- VOCs are mobilized. A full-scale application would require system modifications to limit mobilization of VOCs, including installation of either downgradient recovery wells or perimeter injectors, where catalyst and/or hydrogen peroxide could be injected (as needed) to form an "oxidative barrier" to treat VOCs mobilized from the source area.
- Add shallow injectors to target the overburden/bedrock interface zone. Because the overburden/bedrock interface beneath the K-1 Area was sealed off by the injector wells casings and this shallow zone was not under saturated conditions during the injection program, it was left untreated. An increase in post-injection VOC concentrations in one injector well under high groundwater conditions suggests that this overburden/bedrock interface zone may contain additional residual source material and should be targeted along with the deeper zones of concern.
- Based on the efficient destruction rations maintained during the pilot study, injection of chemical oxidation fluids into source area wells containing high permeability solution zones with shallow hydraulic gradients seems technically feasible. The pilot study clarified earlier uncertainties about the contact and residence time among chemical oxidants and contaminants, as well as the effects of dilution on the reaction.

The results from this pilot study and other full-scale applications of the technology in other hydrogeologic settings indicate that concentrations of hydrogen peroxide greater than 100 milligrams per liter (mg/L) may reduce the efficiency of chemical oxidation: as the concentration of hydrogen peroxide exceeded 100 mg/L, the hydroxyl radicals began attacking the excess hydrogen peroxide. In addition, the excess hydrogen peroxide was lost to nontarget areas downgradient, upgradient, and along bedrock strike. If the injection rate had been reduced, or the injection had been moved to alternate injectors once hydrogen peroxide concentrations reached 100 mg/L along the perimeter of the target area, it appears the competition among hydroxyl radicals and the migration to nontarget areas could have been reduced.

Because the volume of the remaining NAPL is unknown, a full-scale chemical oxidation operation would require several phases of injection with sampling between each phase to monitor the progress of the remediation.

Based on the above recommended modified injection approach, it is estimated that destruction ratios of between 7 to 1 and 10 to 1 hydrogen peroxide to contaminant mass could be maintained throughout an injection program. Thus, if destruction ratios of between 7 to 1 and 10 to 1 were maintained, the 126,000 pounds (12,700 gallons) of hydrogen peroxide injected during the pilot study would have theoretically resulted in the destruction of between 12,600 and 18,000 pounds of VOCs.

ACKNOWLEDGEMENTS

The authors would like to thank the U.S. Army Corps of Engineers – Baltimore, MD District and the Environmental Staff at Letterkenny Army Depot – Chambersburg, PA, for their financial and technical support throughout the project.

REFERENCE

WESTON (Roy F. Weston, Inc.), 2000, Summary Report for the In Situ Chemical Oxidation Remediation Pilot Study of the Bedrock Aquifer at the Southeastern (SE) Disposal Area (DA) – Letterkenny Army Depot, Chambersburg, PA.

Geotechnical and Environmental Applications of Karst Geology and Hydrology, Beck & Herring (eds)
© 2001 Taylor & Francis, ISBN 90 5809 190 2

Test methods for developing a conceptual model for a PCB-contaminated carbonate aquifer

STEPHEN R.H.WORTHINGTON   Worthington Groundwater, Dundas, ON, L9H 3K9, Canada, worth@interlynx.net

DEREK C.FORD   School of Geography & Geology, McMaster University, Hamilton, ON, L8S 4K1, Canada,
dford@mcmaster.ca

ABSTRACT

It is widely recognized that karstification can substantially influence flow and transport characteristics in carbonate aquifers. Surface features such as sinkholes are widely used to diagnose the presence of a karst aquifer, but specific borehole tests for karst have not been well defined. Such tests are especially important in glaciated areas where karst features have been eroded or buried by till. One such area is Smithville, Ontario, where more than 60 boreholes at a PCB-contaminated dolostone site provided an opportunity for a wide range of downhole tests and monitoring to be carried out. It was found that there were a number of useful tests for indicating karstification. These included 1) order of magnitude differences between pump, slug and packer test results, 2) the presence of water table troughs, 3) rapid water level response following recharge events, 4) rapid changes in water quality following recharge events 5) water undersaturated with respect to calcite following recharge events and 6) a wide range in fracture apertures along major bedding planes. Most parameters vary over a similar range at Smithville as they do at Mammoth Cave, Kentucky, and we conclude that Smithville behaves as a typical karst aquifer and that most PCB transport has been along one conduit.

INTRODUCTION

The assessment of flow and transport in carbonate aquifers is especially challenging because of the changes in porosity and permeability caused by solution. Where karst features such as dolines, sinking streams and springs are present then many hydrogeologists will be aware of the likelihood of rapid subsurface transport in conduits. However, there is often little or no such surface evidence in areas which were glaciated in the Pleistocene. This is because most surficial karst features were either eroded by glaciers or buried by till and there has been insufficient time since deglaciation for mature karst features to become reestablished. Conduits are probably widespread in glaciated carbonate aquifers, and they may be open or plugged by glacial sediments (Ford, 1983). Consequently, in glaciated carbonate aquifers it is necessary to carefully consider the available subsurface evidence for indications of the presence of hydrologically active conduits. Worthington and Ford (1997a) and Worthington (1999) suggest a number of tests methods for assessing the flow regime in carbonates. The following account demonstrates how these methods have been used at one site to develop a conceptual model of the aquifer.

SITE DESCRIPTION

The site at Smithville, Ontario, is located between Lake Ontario and Lake Erie. It is midway between Buffalo (New York) and Hamilton (Ontario). Approximately six meters of lacustrine clays, silts and sands overlie the 38 m thick dolostone aquifer, which comprises the Silurian Lockport Formation. The dolostones dip gently to the south and are on the eastern margin of the Michigan Basin. The site operated as a polychlorinated biphenyl (PCB) waste transfer facility from 1978 to 1985. During this period approximately 434,000 liters of liquid wastes were received at the site, including 266,000 liters of PCB wastes. In 1985 local environmentalists discovered that there were PCB oils in a shallow storm water lagoon at the site. The site was subsequently shut down and the surface wastes and waste-contaminated overburden were incinerated on site. A pump-and-treat system has been operating at the site since 1989.

A large number of boreholes have been drilled at the site (Figure 1). The source area in which product has found occupies an area approximately 200 m wide. Borehole sampling has shown that the dissolved-phase plumes for PCBs, trichlorobenzene (TCB) and trichloroethylene (TCE) extend approximately 50 meters, 350 meters and more than 500 meters, respectively, downgradient from the source area (Figure 1 and Golder Associates, 1995).

The overburden thins to the south of the site and is absent along part of Twenty Mile Creek. There are sinkholes along this section of the creek and tracer testing and discharge gaging have shown that all the flow of Twenty Mile Creek travels through a conduit with a cross-section of 2 m². The conduit carries all the flow of Twenty Mile Creek when the flow is less than 200 L/s but there is flow along the surface creek at higher discharges. A tributary creek flowing from the north eroded more than 2 m into the overburden to the bedrock and flowed into Smithville Cave. The cave has been mapped for 247 m to where it is flooded; the cave meanders downdip along a bedding plane and averages 2 m in width and 0.7 m in height (Figure 1 and Worthington and Ford, 1997b). Thus significant karst features are found to the south of the site. At the site itself the most significant karst features observed are in two pits excavated in the

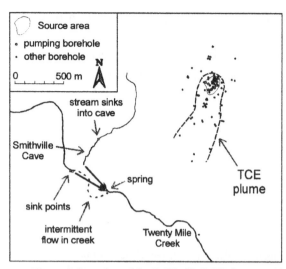

Figure 1. Location of the Smithville PCB site

overburden down to the bedrock. In both pits joints were found which had been solutionally enlarged to widths of 15 cm.

CONCEPTUAL MODELS FOR THE AQUIFER

Two possible conceptual models were considered for the aquifer by site investigators. It was realized from early testing that there were open bedding planes along which flow and solute transport would preferentially flow. One possibility was that these openings were primarily the result of stress changes. This could have been a particularly important process during the melting of the considerable thickness of ice which covered the site up to 14,000 years ago. If this were the case then the maximum aperture encountered along any bedding plane would be fairly consistent, a decrease in bedding plane aperture with depth would be expected, and there would be a low variability in head or in water quality parameters along any bedding plane. Furthermore, with fracture apertures predicted to be less than 1 mm there would be very limited movement of DNAPL and slow response in head or water quality parameters following recharge events.

The second possibility considered was that solution of the dolostone resulted in a high-permeability channel network such as is commonly found in carbonate aquifers. In this case there would be a wide variation in the maximum aperture encountered along any bedding plane, there could be solutionally-enlarged bedding plane at any depth in the aquifer, and there would be a high variability in head and in water quality parameters along any bedding plane. Furthermore, with some channels predicted to have apertures greater than 1 cm there could be substantial offsite movement of DNAPL and a rapid response in head and water quality parameters following recharge events.

BOREHOLE TEST AND SAMPLING METHODS FOR ASSESSING KARSTIFICATION

The following borehole methods were found to be the most useful in testing the two conceptual models for flow in the aquifer.

Differences between pump, slug and packer test results

Permeability is independent of the test scale in an ideal porous medium. Thus the permeability calculated from core and pumping tests should be similar. This will not be true in many fractured media, and is especially untrue where there is extensive solution (Kiraly, 1975, Schulze-Makuch, 1999). At the Smithville site there have been large-scale pumping tests (Golder Associates, 1990), slug tests (Worthington and Ford, 1999) and packer tests with 2 m and closer spacings (Novakowski et al., 1999). Results are shown in Figure 2a. There is a clear scaling effect with order of magnitude differences between the geometric means of the three test methods. Figure 2b shows a comparison between Smithville and three other well-studied carbonate aquifers using similar test methods (slug or packer testing). The three aquifers have been extensively studied and there is a consensus that karstification has been important in these aquifers (Worthington et al., 2000a; Milanovic, 1981; Smart et al., 1992). The similarity between Smithville and the other carbonate aquifers suggests that conduit flow may well be significant at Smithville.

Water table gradients

There are a considerable number of boreholes open in the top several meters of the saturated zone. The majority of these were installed between 1986 and 1989. The pump-and-treat system has been in place since mid-1989. Consequently the best water level data in the source area at the site is during the period before pumping disrupted natural aquifer heads. Figure 3 shows the water levels in the upper part of the aquifer measured on July 4th and 5th 1988. At that time there was a trough in the water table which was more than 1 m deep. Flow lines converge on the trough and then flow was along its axis, as shown by the arrow in Figure 3. Consequently the trough must be associated with a linear high-permeability feature. The trough is readily apparent in every set of monthly water level measurements made in 1988-1989. Its depth varied from just under 0.5 m to more than 1 m. The depth is defined by the difference in elevation between the base of the trough and the groundwater divide to the east. The location of the trough is essentially identical at all periods.

Karst aquifers are defined by the presence of convergent flow in networks of conduits or channels, and major conduits are associated with low hydraulic heads. This has been well demonstrated using water level measurements, cave maps and tracer data at Mammoth Cave, Kentucky (Quinlan and Ray, 1981), and the importance of channel flow there was calculated by Worthington et al. (2000a). The aquifer at Smithville is unconfined and meteoric water readily recharges the aquifer. Dreybrodt (1996) has shown that conduits develop under these conditions. Therefore a channel network is expected at the site, and the water table trough in Figure 3 demonstrates this. However, there are other hypotheses for its cause. It could represent erroneous data, but this seems most unlikely since it was defined in the same location for twelve consecutive months. Another possibility is that there is a major vertical fracture down which there is flow to the lower part of the aquifer. There are few deep monitors in the area of the trough but these do not show a consistent pattern of lower heads at depth. Furthermore, the orientation of the trough (45°) is not aligned with the major joint sets in the area (18°, 82°, 132°, and 152°). Thus none of other these explanations are satisfactory, and the best explanation is that the trough is associated with a conduit.

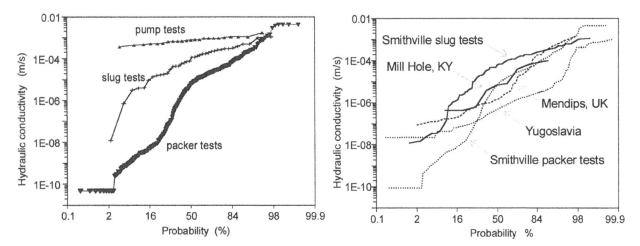

Figure 2. Hydraulic conductivity results from Smithville and other karst aquifers: a) pump, slug and packer test results from Smithville showing scaling effects, b) comparison of Smithville with three other karst aquifers showing that results are similar

It is possible to estimate the approximate size of the conduit. The size of capture zone from which flow converges on the tip of the arrow shown in Figure 3 is 90,000 m² if the area south of Spring Creek Road alone is considered, although the capture zone could be considerably larger. The discharge along the channel can be derived from

$$Q = (R-S) A \qquad (1)$$

where Q is the conduit discharge, R is the total runoff, S is surface runoff, and A is the area of the capture zone. Stream discharge data show that the runoff in this area is about 320 mm/year. Root casts, animal burrows and large dessication cracks in the clay and silt overburden promote infiltration. Surface runoff from the site may thus be in the range 50 - 250 mm. If we assume that there is a single conduit and that all recharge converges on this conduit, then the mean discharge through the conduit is 0.19 - 0.76 L/s for the range in variables given above. The size of the conduit can be estimated from tracer tests along karst conduits, since there is a large data set available which shows that most conduits have very similar mean velocities (Worthington et al., 2000a). Conduit cross-section area X can be calculated from

$$X = Q / v \qquad (2)$$

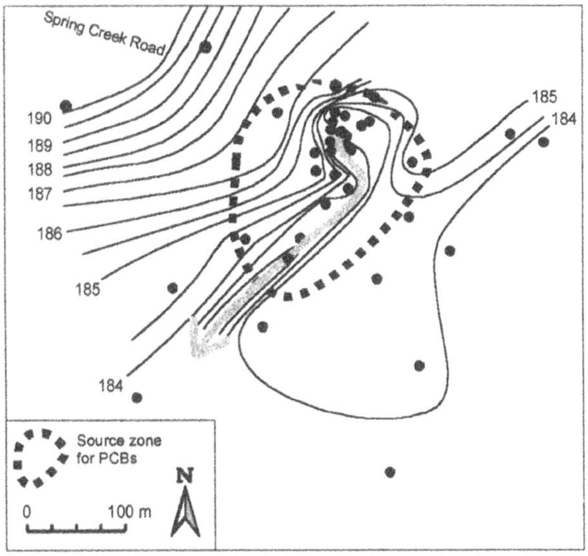

where v is mean conduit velocity, which is 0.022 m/s from 2877 tracer tests worldwide (Worthington et al., 2000a). For the discharge range of 0.19 - 0.76 L/s the cross-section is 0.0094 - 0.038 m². If we assume a circular cross-section for the conduit then its diameter is in the range 11 cm to 22 cm.

Water level and water quality response following recharge events

The interconnected network of channels in a karst aquifer typically results in a rapid water level response to recharge events. The Nyquist sampling theorem states that the sampling frequency should be at least twice the frequency of the highest-frequency cyclic phenomenon to be detected (Quinlan et al., 1993). We therefore chose to undertake continuous monitoring of water level and water quality. We found that about half the total water-level variation in the bedrock is due to rapid recharge-related infiltration and half to seasonal changes. Figure 4 shows what the effect of differing water level sampling intervals would be on an interpretation of water level change. In this borehole a transducer and datalogger were used to record water level every ten minutes. Four major rises in water level occurred during the two month period shown. Most of the water level change information is retained in daily sampling though some important transient differential heads between monitors is lost. Any less frequent sampling would be subject to serious aliasing. The

Figure 3 The water table at the site on July 4-5 1988 showing a trough to which flow converged

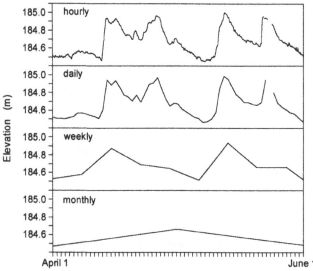

Figure 4. Heads at 5D19, a monitor with average variability, showing the effect of the sampling interval on the understanding of water level variability

high-frequency signal of four major events would be misinterpreted as only two events with weekly sampling and as only one event with monthly sampling.

Water quality also responds quickly in boreholes that are well connected to the channel network. During 1987 there was a period of daily sampling for organic compounds at some monitors. Figure 5a shows trichlorobenzene (TCB) concentrations in a monitor 14 m below the surface over a period of three weeks. Concentrations steadily rose during the recession from one runoff event and then abruptly dropped from 52 ppb to 2.2 ppb over three days and then further decreased to 0.17 ppb over the following three days. This precipitous drop occurred shortly after a major rain event which resulted in the discharge in Twenty Mile Creek rising to 18 m^3s^{-1}. Our interpretation is that rapidly infiltrating recharge water at the site diluted contaminant concentrations in conduits. We also recorded similar recharge-related drops in continuous monitoring of water level and electrical conductivity (EC). Figure 5b shows results from a monitor at a depth of 21 m to 26m below the surface, where rises in water level are synchronous with drops in EC. Such responses are typical for karst aquifers.

A number of repeat tritium measurements were made by Novakowski et al. (1999). In many cases there was little change between sampling in July 1997 and August 1998, but at a depth of 36 m in monitor 60 ^{3}H decreased from 29 TU to 0.8 TU, and at a depth of 49.5 m in monitor 53 ^{3}H increased from <6 TU to 31 TU. These large changes are consistent with the TCB and EC data described above and show how rapid water quality changes occur in monitors which are well connected to conduits.

Saturation of the water with respect to calcite and dolomite

Major ions were measured in a number of monitors at the site, and saturation indices for calcite and for dolomite have been calculated for February 1997 samples (Worthington and Ford, 1997b) and for November 1997 samples in the same monitors (Zanini et al., 2000). Most samples were supersaturated with respect to both calcite and dolomite in February but were undersaturated in November (Figure 6). The magnitude of the deviations from equilibrium are surprisingly large and exceed what is usually found in karst aquifers. On the other hand the changes from supersaturation to undersaturation are common in karst aquifers, with the undersaturated samples representing recently recharged water which has rapidly infiltrated via conduits. These data show the conduit network at the site is undergoing active enlargement.

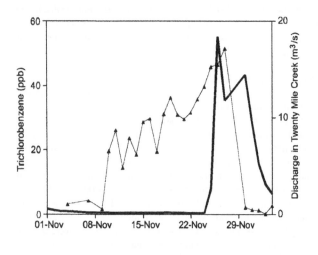

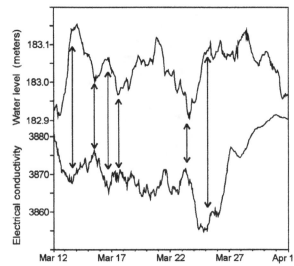

Figure 5. Examples of rapidly varying water quality at Smithville following recharge events: a) trichlorobenzene concentrations (triangles) over five weeks in 1997, showing a 300-fold dilution after a recharge event, b) inverse correlation between water level and EC at monitor 54B at a depth of 21m to 26 m below the surface

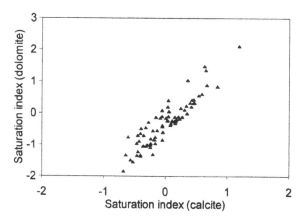

Figure 6. Saturation indices for February and November 1997, showing that substantial dissolution occurs in the bedrock.

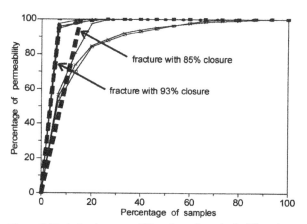

Figure 7. Variation in permeability along six major bedding planes from packer tests showing channeling

Range in aperture along major bedding planes

Packer test measurements in 15 boreholes were carefully correlated with lithology (Novakowski et al., 1999). Figure 7 shows the range in permeability at the site along the six major bedding planes dividing named members of the Lockport Formation. For three contacts about 55% of the permeability is contributed by a single borehole, while the remaining three contacts have about 95% of the permeability at a single borehole. These results show the bedding planes approximate to fractures with 85% - 93% closure, and that about 10% of each major bedding plane accounts for almost all the permeability on these planes (Figure 7). Thus there is preferential flow and the fractures do not approximate parallel-plate openings. It is likely that a larger number of boreholes would have encountered even larger channels, and most of the permeability on any bedding plane is probably focussed on <10% of the plane.

Tracer testing

Tracer testing to springs is an important method in karst aquifers of assessing groundwater flow directions and velocities along conduits. However, where tracer testing is carried out between boreholes it is unlikely than either the injection borehole or the recovery borehole will be located on a conduit. Tracer tests over longer distances between boreholes are more likely to be affected by the presence of conduits than local-scale tests. We carried out 16 local-scale tests at the site, and measured peak velocities of 15 - 120 m/day. The results indicate that open fractures in the aquifer are widespread. However the scale of the tests we were able to carry out was too small to be useful to test for channeling.

DISCUSSION

We have shown in earlier papers (Worthington and Ford, 1997c; Worthington al., 2000b) that the carbonate aquifers at Smithville and in the area of Mammoth Cave, Kentucky, are very similar in terms of matrix, fracture, and conduit porosity and permeability. They are also similar in terms of bit drops encountered during drilling. We have shown above that there is a rapid response in water level and water quality to recharge events and that groundwater is substantially undersaturated at times with respect to both calcite and dolomite. All these results point towards the presence of a conduit network at the Smithville sites. However, a very different interpretation of the hydrogeology has been proposed by Zanini et al., (2000). They attribute high-permeability open fractures solely to stress changes and do not discuss the effects of solution, the highly undersaturated groundwater, the significant temporal changes in organic compounds, tritium, major ion chemistry and electrical conductivity, or the rapid recharge-related changes in water level. Thus their analysis omits some very important data.

However, the most important finding for the characterization and remediation of PCB contamination in the aquifer is the existence of the water table trough. This trough is almost certainly associated with a conduit, and we have estimated that its diameter is at least 11 cm if it is circular. The head data suggests that the capture zone for this trough under natural gradient conditions includes essentially all of the PCB source area at the site. This one conduit is most likely to carry most off-site contaminant migration and thus is a key factor in site characterization and remedial design.

ACKNOWLEDGMENTS

This work was supported by the Ontario Ministry of the Environment through the Smithville Phase IV Bedrock Remediation Program, by the U.S. Environmental Protection Agency and by the National Science and Engineering Research Council through a research grant to Derek Ford. We thank Pat Lapcevic and Kent Novakowski for sharing results and John Voralek and Charlie Talbot for assistance in placing monitoring equipment in boreholes with packers.

REFERENCES

Dreybrodt, W., 1996, Principles of early development of karst conduits under natural and man-made conditions revealed by mathematical analysis of numerical models. Water Resources Research, 32, 2923-2935.

Ford, D.C., 1983, Effects of glaciations upon karst aquifers in Canada, Journal of Hydrology, 61, 149-158.

Golder Associates, 1990, Cleanup of abandoned PCB storage facility, Smithville, Ontario. Results of geological and hydrogeological investigations and contaminant plume delineation study, 1988 and 1989, Golder report 891-1559.

Golder Associates, 1995. Hydrogeological data compilation and assessment, CWML Site, Smithville, Ontario. Project 94-106; 41 p.

Kiraly, L., 1975. Rapport sur l'état actuel des connaissances dans le domaine des charactères physiques des roches karstiques. In: hydrogeology of karstic terrains (Eds: A. Burger and L. Dubertret), 53-67. Internat. Union Geol. Sci., Series B, 3.

Milanović, P.T., 1981. Karst hydrogeology. Water Resource Publications, Littleton, Colorado, 434p.

Novakowski, N., P. Lapcevic, G. Bickerton, J. Voralek, L. Zanini and C. Talbot, 1999, The development of a conceptual model for contaminant transport in the dolostone underlying Smithville, Ontario. National Water Research Institute, Burlington, 98p.

Quinlan, J.F. & Ray, J.A., 1981. Groundwater basins in the Mammoth Cave Region, Kentucky. Occasional Publication #1, Friends of the karst, Mammoth Cave.

Quinlan, J.F., G.J. Davies and S.R.H. Worthington, 1993, Discussion of "Review of ground-water quality monitoring network design, by H.A. Loaiciga, R.J. Charbeneau, L.G. Everett, G.E. Fogg, B.F. Hobbs and S. Rouhani". Journal of Hydraulic Engineering, 119, 1436-1442.

Schulze-Makuch, D., D.A. Carlson, D.S. Cherkauer and P. Malik, 1999, Scale dependency of hydraulic conductivity in heterogeneous media. Ground Water, 37, 904-919.

Smart, P. L., A.J. Edwards and S. L Hobbs, 1992, Heterogeneity in carbonate aquifers; effects of scale, fissuration, lithology and karstification. Proceedings of the third conference on hydrogeology, ecology, monitoring and management of ground water in karst terranes (Nashville, Tennessee), Water Well Journal Publishing Company, Dublin, Ohio, 373-387.

Worthington, S.R.H., and D.C. Ford, 1997a, Borehole tests for megascale channeling in carbonate aquifers. Proceedings of the 6th conference on limestone hydrology and fissured media. Centre of Hydrogeology, University of Neuchatel, 191-195.

Worthington, S.R.H., and D.C. Ford, 1997b, Analysis and modelling of the potential and evidence for a channel network in the fractured carbonate bedrock at Smithville. Report prepared for Smithville Phase IV Bedrock Remediation Program, December 1997, 67p.

Worthington, S.R.H., and D.C. Ford, 1997c, Strategy for evaluating channeling in the carbonate bedrock at Smithville, Ontario. Proceedings, Air & Waste Management Association Annual Conference, Toronto, June 1997.

Worthington, S.R.H., and D.C. Ford, 1999, Chemical hydrogeology of the carbonate bedrock at Smithville. Report prepared for Smithville Phase IV Bedrock Remediation Program, May 1999, 169p.

Worthington, S.R.H., 1999, A comprehensive strategy for understanding flow in carbonate aquifers. In: Karst Modeling, (Eds. A.N. Palmer, M.V. Palmer and I.D. Sasowsky), Karst Waters Institute Special Publication #5, 30-37

Worthington, S.R.H., G.J. Davies, and D.C. Ford, 2000a, Matrix, fracture and channel components of storage and flow in a Paleozoic limestone aquifer. In: Groundwater flow and contaminant transport in carbonate aquifers, Eds. C. Wicks and I. Sasowsky, Balkema, Rotterdam, 113-128.

Worthington, S.R.H, D.C. Ford and P.A. Beddows, 2000b, Porosity and permeability enhancement in unconfined carbonate aquifers as a result of solution, in Speleogenesis: Evolution of karst aquifers, Eds. A. Klimchouk, D.C. Ford, A.N. Palmer and W. Dreybrodt, National Speleological Society, Huntsville, p. 463-472.

Zanini, L., K.S. Novakowski, P. Lapcevic, G.S. Bickerton, J. Voralek and C. Talbot, 2000, Groundwater flow in a fractured carbonate aquifer inferred from combined hydrogeological and geochemical measurements. Ground Water, 38, 350-360.

VII Using Geophysics in Karst Investigations

Geotechnical and Environmental Applications of Karst Geology and Hydrology, Beck & Herring (eds)
© 2001 Taylor & Francis, ISBN 90 5809 190 2

A geoelectric investigation of the freshwater aquifer near the well fields of north Andros Island, Bahamas

ANGELA L.ADAMS P.E.LaMoreaux & Assoc., Inc., Oak Ridge, TN 37830, USA, aadams@pela-tenn.com

ABSTRACT

The freshwater lens on Andros Island, Bahamas is an important source of potable water for its citizens and the more densely populated New Providence Island. Because of the possibility that fresh water in the aquifer is currently being depleted, it is necessary to observe the rate of salt water incursion in order to prevent salt water contamination of the freshwater aquifer.

Fresh ground water lenses in the Bahamas may be mapped using electrical resistivity surveys utilizing the large contrast in resistivity between the freshwater and saltwater saturated carbonate rocks that make up the island. On northern Andros Island there have been numerous studies in the past eight years completed in an effort to understand the thickness and continuity of the freshwater lens from Conch Sound through Red Bays and onto the western tidal flat.

Results to date indicate that there is a single continuous lens from Red Bays on the west to Conch Sound on the east. Salt water is upwelling in the well fields due to pumping, as well as in the Charlie's Blue Hole area due to karst solutioning. Because of over pumping the trench wells, the thickness of the freshwater lens has thinned under the well fields.

INTRODUCTION

The Bahamas are a group of carbonate islands in the Atlantic Ocean lying east of south Florida (Figure 1). For the past eight years, students and faculty at Wright State University have been investigating the geology, hydrology, geochemistry and geophysical nature of the freshwater lens on North Andros Island. These studies have generated a clearer picture of lens development and continuity across the island. This paper presents geophysical data gathered and gives a profile of the North Andros aquifer. Several electrical resistivity surveys in the northern portion of Andros Island were completed between 1992 and 1999 and have shown that this geophysical method works well for this study area.

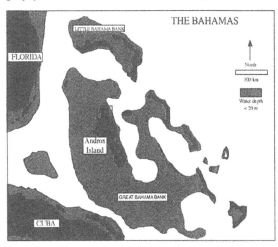

Figure 1. Map of the Bahamas

In the Bahamas, the majority of the fresh water resources occur as fresh ground water lenses that float on top of the denser salt water (Cant and Weech, 1986). Andros Island, the largest island on the Bahamian Platform, receives approximately 120 cm of rainfall per year allowing the island to develop a thick freshwater lens (Cant and Weech, 1986; Bukowski, 1999). Andros Island is sparsely populated, and the aquifer supplies the needs of the inhabitants and is used as a freshwater resource for the more populated New Providence Island and its capital city, Nassau. Pumping of Andros water for shipment to New Providence Island by the Bahamian Water and Sewerage Corporation began in 1977 and continues today. The production of large quantities of water for export raises concern for the long-term sustainability of the resource.

Data from pumping centers maintained by the Bahamian Water and Sewerage Corporation indicate that annual yields are exceeding design capacity (Lloyd, 1991). By comparing previous data prepared by the Bahamian Water and Sewerage Corporation in 1970 with the data that were collected during this project it should be possible to record the changes in the thickness of the freshwater lens caused by the pumping of the trench wells.

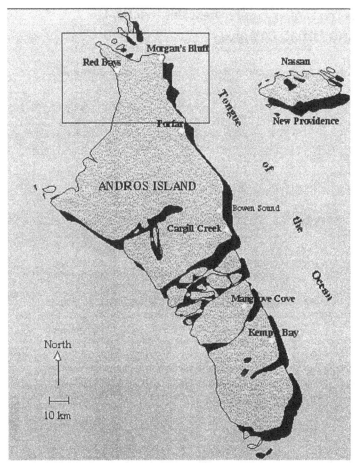

Figure 2. Map of Andros Island with the study area outlined.

ANDROS ISLAND GEOLOGY

Andros Island lies 40 km west of the island of New Providence, where Nassau, the capital city, is located (Figure 2). Andros trends north-south and has dozens of little settlements with only about 10,000 people on an island that is approximately 60 km wide and 160 km long. Andros Island is on the windward (eastern) edge of the north-central region of the Great Bahama Bank and is bounded by the Tongue of the Ocean on the east and on the west by the Florida Straits (Shinn et al., 1969; Shinn et al., 1989). It is separated from the Little Bahama Bank by Northwest Providence Channel to the north (Bathurst, 1975). Most of the island is less than 5 meters above sea level (Sealey, 1985). Along arcuate Pleistocene eolian dune ridges, that are predominantly north-south oriented parallel to the shoreline, elevations can reach up to 30 m (Bathurst, 1975).

The lithology on Andros Island consists solely of carbonate rocks. The Lucayan Formation (Late Pliocene-Pleistocene limestone) is the main aquifer in the Bahamas (Beach and Ginsburg, 1980; Cant and Weech, 1986). Two predominant facies within the Lucayan are lagoonal and eolian dune deposits which form low arcuate dune ridges (Shinn and Lidz, 1988). On North Andros Island the Lucayan has an average thickness of 43 meters and is a dull yellow to buff non-skeletal limestone. A paleosol covers the surface of North Andros below which the rocks are dominated by Pleistocene oolitic, peloidal and bioclastic grainstones and packstones (Carney and Boardman, 1991; Boardman et al., 1993). Using Joulters Cays as a modern analog, Carney and Boardman (1991) interpreted the Lucayan limestone of Andros as a Pleistocene ooid shoal complex with associated reefs, lagoons, eolian dune deposits, stabilized sand flats, and island deposits.

Discontinuity surfaces interpreted as horizons of subaerial exposure are common throughout the formation. In the subsurface of North Andros, paleosols are commonly found at depth of 2 to 4 m, 7 m, and 12 m (Boardman and Carney, 1997). The bedding is horizontal and lateral changes are generally gradual except in areas of karst development.

Andros Island hydrology

Andros Island is in a semi-humid climatic zone receiving an average of 120 cm of rainfall annually with precipitation occurring seasonally (Cant and Weech, 1986; Tarbox, 1987; Vacher and Wallis, 1992). The fresh ground water occurs as an extensive lens, which is recharged by rainfall and is bounded basally and laterally by saltwater. The freshwater lens is present because discharge of fresh water to the sea does not exceed recharge from precipitation (Bukowski, 1999). Theoretically, the thickest part of the lens should be in the center of the island (Little et. al., 1973; Lloyd, 1991). The Ghyben-Herzberg theory states that the thickness of the freshwater lens is directly proportional to the elevation of the water table above sea level (Fetter, 1994). On Andros the depth of the freshwater below sea level is approximately 10 times the head above sea level, rather than the 40 times of the typical Ghyben-Herzberg theoretical lens (Bukowski, 1999). The depth to the water table in the study area is 1 to 2 m below the ground surface, this is approximately 1 m above present sea level (Cowles, 1993; Wolfe, 1994; Jacob, 1997). The flow of water is from the thickest portion of the lens, which is toward the center of the island, outward toward the shore (Bukowski, 1999).

There are three main well fields, each consisting of a large number of trench wells, producing millions of gallons of fresh water every day for shipment to Nassau (Cowles, 1993; Bukowski, 1999). The Water and Sewerage Corporation produced a preliminary isopach map of the fresh water lens on North Andros based on boreholes and limited resistivity data gathered in 1973. An updated isopach map of the fresh water lens was also generated incorporating more resistivity data gathered during 1988 and 1989 (Figure 3). The 1973 map shows two distinctive thick areas approximately in the same areas as the three well fields. By 1989 the lens had apparently thinned and consisted of one centrally located area.

The well fields on North Andros are 305 m long inter-connected trenches approximately 1 m wide and 2 m deep. The trenches consist of four radial arms connected to a central cruciform that is connected to other cruciforms by culverts (Figure 4). The water flows to a central pump by gravity through these trenches; the pump then removes the water to a reservoir and holding tanks at Morgan's Bluff.

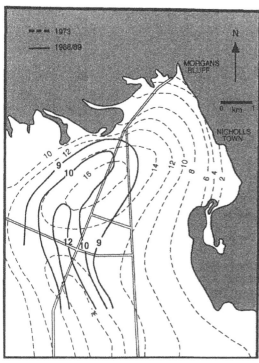

Figure 3. Isopach map of the freshwater lens on northern Andros Island in 1973 and 1988/89.

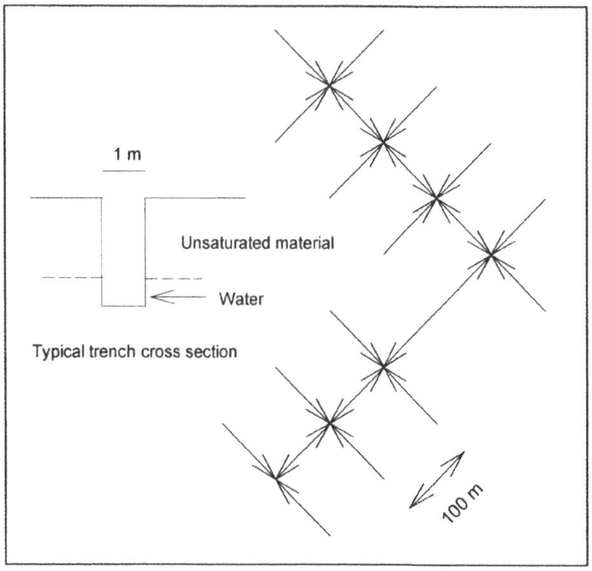

Figure 4. Schematic representation of a trench well field with an insert showing the trench cross section.

DATA ACQUISITION AND ANALYSIS

Resistivity methods involve passing an electrical current between points on the earth's surface and, at two other points, measuring the electrical potential difference produced by that current flow through the earth. From this, an apparent resistivity value can be calculated (Burger, 1992). As the distance between the current electrodes is increased, the depth that the current penetrates into the subsurface is increased. A sequence of measurements with expanding electrode spacing around a common center point is called a vertical electric sounding (VES). Resistivity surveys are effective for determining the thickness of the freshwater lens because the resistivity of the earth materials is very dependent on the pore fluid. Saltwater saturated rock has a significantly lower resistivity than freshwater saturated rock. VES methods were used to determine the depth to the interface between fresh and salt water.

Cowles (1993) completed 13 resistivity soundings along a 2.1 km profile along Charlie's Blue Hole Road. This east-west road intersects Queen's Highway and is located on the eastern side of North Andros Island. The soundings were completed using a Schlumberger electrode configuration and a Johnson IC-69 resistivity meter. His modeling with the computer program RESIX[PLUS] produced 3 and 4 layer models. As a three layer model the subsurface consists of an unsaturated zone, freshwater lens, and saltwater. The four layer model takes into account a brackish mixing zone between the freshwater and saltwater zones. His 3 layer models indicated the freshwater lens varied in thickness from 9.2 m to 14 m. His 4 layer models indicated the freshwater lens varied in thickness from 7.5 m to 11.2 m. The mixing zone varied from 8.6 m to 13.8 m thick.

Extending Cowles' 1993 research Wolfe (1994) conducted three series of resistivity soundings along perpendicular transects to Charlie's Blue Hole Road using the Schlumberger electrode configuration and a Terrameter SAS-300C resistivity meter. Wolfe used two modeling programs, RESIX[PLUS] and ATO. His 3 layer models produced a freshwater lens thickness ranging from 2 m to 19 m. His 4 layer models produced a freshwater lens thickness ranging from 1 m to 13.4 m, with a mixing zone thickness of between 1 m and 11 m.

Hodl (1997) conducted a resistivity survey that included 8 stations along Red Bays Road in the northwestern portion of Andros Island. She utilized the Schlumberger electrode configuration and collected the data with a Sting R1 resistivity meter. She processed the data using the RESIX[PLUS] computer software, and her best-fit models indicated a fresh water lens that ranged from 9 m to 30 m thick. The mixing zone ranged from 4 m to 9.7 m thick.

Jacob (1997) performed three azimuthal resistivity surveys, as well as obtaining VES data from four locations on North Andros. The three azimuthal survey sites were located on Charlie's Blue Hole Road, Main Lumber Road, and on the east side of Queen's Highway south of Red Bays Road junction. She collected the data with a Sting R1 resistivity meter employing the Schlumberger electrode configuration for the VES surveys and the Wenner electrode array (Burger, 1992) for the azimuthal resistivity surveys. Azimuthal resistivity surveys investigate the horizontal anisotropy in resistivity in order to determine the fracture orientation in the subsurface, or other preferred ground water flow paths. The azimuthal surveys were performed by rotating the electrode array about a fixed point at increments of 15°. The Paradox of Anisotropy (Keller and Frischneckt, 1966) tells that the orientation of the major axis of the apparent resistivity ellipse is the fracture orientation. Her investigation showed that a fracture system oriented approximately

N60E dominates the subsurface in the Charlie's Blue Hole area. The results for soundings made at the anisotropy test sites were included in our final interpretation.

Reinker-Wilt (1998) took 14 resistivity soundings located approximately 2 km north of Red Bays Road and included some of Hodl's (1997) stations. She utilized the Schlumberger electrode configuration and collected data with a Johnson IC-69 resistivity meter. The data were processed using the RESIX[PLUS] computer software, and generated a 3 layer model of a fresh water lens that was between 3.4 m to 18.8 m thick. The 4 layer models indicated a fresh water lens thickness that ranges from 3.4 m to 17 m. The mixing zone thickness ranged from 1 m to 7 m.

In 1999 thirty-one resistivity soundings were completed using the Schlumberger electrode configuration and a Sting R1 resistivity meter. Seventeen resistivity soundings were completed within the Bahamian Water and Sewerage Corporation's well fields. Eleven resistivity soundings provided additional information regarding a potentially anomalous area north of Red Bays road that was encountered by Reinker-Wilt in her 1998 investigation. Three resistivity soundings aided in defining the depth and extent of the freshwater lens into the western tidal flats. The data were processed using the RESIX[PLUS] computer software, and generated a 3 layer model of a fresh water lens that was between 10 m to 19 m thick in the well fields, and between 3 m to 21 m for the rest of the study area (Red Bays and the western tidal flat. Figure 5 is a summary map of the resistivity surveys completed from 1992 to 1999.

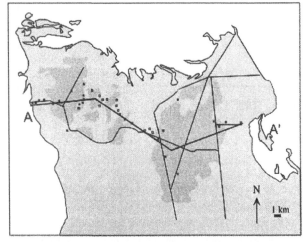

Figure 5. Map of the study area with VES sites, roads, and profile A-A'.

One of the critical areas for determining the depth to the saltwater interface is in the well fields. Access in the well fields is limited to strips along the trenches. Since the trenches cut 2 meters into the rock, the assumption of horizontal layering that is used in resistivity interpretation is violated near the trenches. Two approaches were used to determine the effect on calculated depths of a nearby trench. We used a 2.5D finite-element modeling program (Zhou, 1998; Wolfe, 1999) to calculate the effect of the trench. We also performed field measurements collecting sounding curves parallel to a trench at distances ranging from 0.5 to 2 m.

RESULTS

A single continuous freshwater lens extends across Andros Island from Red Bays in the west to Conch Sound in the east. A compilation of the resistivity data from 1992 to 1999 was used to generate the cross-island cross section shown Figure 6. The modeled depth from the surface to the top of the freshwater lens has a mean of 2.5 m. The depth to the top of the saltwater lens ranges from 6.6 m to 24 m with a mean of 14.5 m. The thickness of the freshwater lens ranges from 3.8 m to 21 m with a mean of 12.1 m. Resistivity in the fresh water lens ranged from 9.59 Ωm to 1526 Ωm with two high spikes being 2269.6 Ωm and 2678 Ωm. The unsaturated layer has significant variability in resistivity. It is commonly expected that the resistivity should be higher than the freshwater saturated zone, but some places it was lower and other places nearly equal. Initially this led to some confusion on layer assignments. We think the variability is due to organic matter and water retention on the irregular surface.

The cross section shows three areas of thinning; within the Bahamian Water and Sewerage Corporation's well fields; at Red Bays, near a well field that is utilized by the town, and at Charlie's Blue Hole, a karst solution feature. The cross section shows that the lens is continuous across the island, even in areas of low topography. It also shows that the freshwater lens thins towards the west coast.

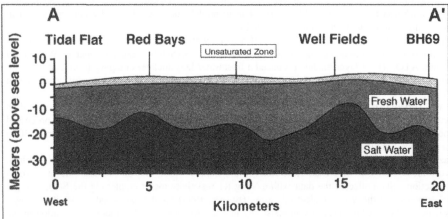

Figure 6. Cross section (A-A') of aquifer model determined from resistivity.

The finite difference modeling and data collected in the area of the well fields show that the effect of the Andros trench wells decreases rapidly with distance from the trench (Figure 7). A sounding parallel to the trench well is practical when offsets of the electrode line are greater than one meter (Figure 8). A shallow layer of water in the trench wells can be ignored.

CONCLUSIONS

North Andros Island has a well-developed freshwater lens, which is exploited by a series of trench wells. Because of the

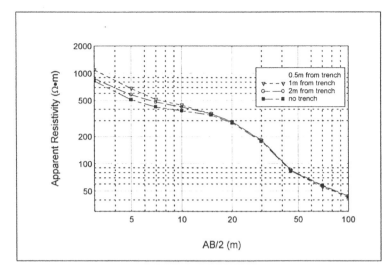

Figure 7. Schlumberger sounding curves from finite-element modeling for a 15 m deep saltwater interface with a range of electrode offsets from the trench.

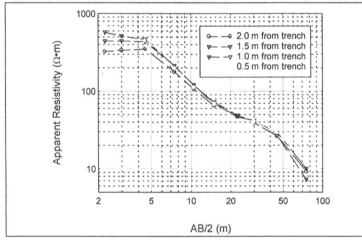

Figure 8. Measured Schlumberger sounding curves for a range of electrode offsets from the trench.

possibility that fresh water in the aquifer is currently being depleted it is necessary to monitor the rate of salt water incursion in order to prevent salt water contamination of the freshwater aquifer.

Three-layer modeling indicates a continuous freshwater lens that varies in thickness from 3.8 m to 21 m within the study area. In the well fields the freshwater lens varies from 6 to 12 m indicating a lens that is slightly thinner than the Bahamian Water and Sewerage Corporations isopach maps indicate. The three-layer model incorporates the mixing zone into the lens thickness and therefore may be a slight overestimate of the freshwater lens thickness, on the order of 1 to 3 meters. The average trend of the thickness of the aquifer decreases across Andros by 1.25 m from east to west.

The freshwater lens is thinned in the well fields due to pumping, as well as in the Charlie's Blue Hole area due to karst solutioning. In 1973, the freshwater lens was between 15 m and 10 meters. In 1988/89, the freshwater lens was between 12 m and 9 meters. Ten years later in 1999 the freshwater lens was between 12 m and 6 m.

A sounding parallel to the trench well is practical when offsets of the electrode line are greater than one meter. Trench wells of the depth seen on Andros do not violate the utility of the horizontal layer assumption of the data as distance from the trench approaches 2 m.

ACKNOWLEDGEMENTS

I thank the Bahamian Water and Sewerage Corporation for access and data, the Miami University Bahamas Workshop for financial and logistical support, the Forfar Field Station for logistical support, Cindy Carney and Paul Wolfe for endless rewrites and facilities to work at Wright State University, and the students who helped with field work.

REFERENCES

Bathurst, R.G.C., 1975, Carbonate sediments and their diagenesis: New York, Elsevier, Developments in Sedimentary Petrology, 12, 659 p.

Beach, D.K., and Ginsburg, R.N., 1980, Facies successions of Pliocene-Pleistocene carbonates, northwestern Great Bahamas Bank: American Association of Petroleum Geologists Bulletin, v. 64, p.1634-1642.

Boardman, M.R., McCartney, R.F., and Eaton, M.R., 1995, Bahamian paleosols: Origin, relation to paleoclimate, and stratigraphic significance: in Curran, H.A., and White, B., (eds.), Terrestrial and Shallow Marine Geology of the Bahamas and Bermuda: Geological Society of America, Special Paper 300, p. 33-50.

Boardman, M. R. and Carney, C. K., 1997, Influence of sea level on the origin and diagenesis of the shallow aquifer of Andros Island, Bahamas: in Carew, J. L., (ed.), Proceedings of the Eighth Symposium on the Geology of the Bahamas, Bahamian Field Station, San Salvador, Bahamas, p. 13-32.

Bukowski, J.M., Carney, C. Ritzi, R. W., Jr., Boardman, M. R., 1999, Modeling the fresh-salt water interface in the Pleistocene Aquifer on Andros Island, Bahamas: in Curran, H. A., and Mylroie, J. E., (eds.), Proceedings of the Ninth Symposium on the Geology of the Bahamas, Bahamian Field Station, San Salvador, Bahamas, p. 1-13.

Cant, R. V., and Weech, P.S., 1986, A review of the factors affecting the development of the Ghyben-Herzberg lenses in the Bahamas: Journal of Hydrology, v. 84, p. 333-343.

Carney, C., and Boardman, M.R., 1991, Petrologic comparison of oolitic sediment from Joulters Cays and Andros Island, Bahamas: in Bain, R. J. (ed.), Proceedings of the Fifth Symposium on the Geology of the Bahamas: Bahamian Field Station, San Salvador, Bahamas, p. 37-54.

Cowles, R. E., 1993, Delineation of the fresh water lens on North Andros Island, Bahamas using resistivity methods: Unpubl. M.S. Thesis, Wright State University, Dayton, Ohio, 215 p.

Fetter, C.W., 1994, Applied Hydrogeology: Macmillan College, New York, 434 p.

Hodl, S. B., 1997, Relating paleo-depositional environments to fresh water lens development across central North Andros Island, Bahamas: Unpubl. M.S. Thesis, Wright State University, Dayton, Ohio,   111 p.

Jacob, L.J., 1997, An azimuthal resistivity investigation, North Andros Island, Bahamas: Unpubl. M.S. Thesis, Wright State University, Dayton, Ohio, 113 p.

Keller, G. V. and Frischknecht, F. C., 1966, *Electrical Methods in Geophysical Prospecting,* Pergamon, New York, 519 p.

Little, B. G., Buckley, D.K., Jefferiss, A., Stark J. and Young, R. N., 1973, Land resources of the commonwealth of the Bahamas, volume 4a Andros Island: Land Resources Division, Tolworth Tower, Surrey, England, KT67DY, p. 40-110.

Lloyd, J.W., 1991, A study of saline groundwater responses to abstraction from trenches in the Bahamas: United Nations Development Programme, Department of Technical Co-operation for Development, Project BHA/86/004/A/01/01, 15 p.

Reinker-Wilt, A. S., 1998, Paleo-environments and continuity of the fresh water lens on central and western North Andros Island, Bahamas: Unpubl. M.S. Thesis, Wright State University, Dayton, Ohio, 164 p.
Sealey, N.E., 1985, *Bahamian Landscapes: An introduction to the geography of the Bahamas*: Collins Caribbean, London, United Kingdom, 96 p.

Shinn, E.A., Lloyd, R.M., and Ginsburg, R.N., 1969, Anatomy of a modern carbonate tidal-flat, Andros Island, Bahamas: Journal of Sedimentary Petrology, v. 39, p. 1202-1228.

Shinn, E.A., Steinen, R. P., Lidz, B.H., and Swart, P.K., 1989, Whitings, A sedimentologic dilemma: Journal of Sedimentary Petrology, v. 59, p. 147-161.

Tarbox, K. L., 1987, Occurrence and development of water resources in the Bahamas Islands, in Curran, H. A.. (ed.), Proceedings of the Third Symposium on the Geology of the Bahamas: Bahamian Field Station, San Salvador, p. 139-144.

Vacher, H. L., and Wallis, T. N., 1992, Comparative hydrogeology of freshwater lenses of Bermuda and Great Exuma Islands, Bahamas: Ground Water, v. 30, p. 21-37.

Wolfe, B.L., 1994, A geoelectric interpretation of groundwater near Charlie's Blue Hole, North Andros Island, Bahamas: Unpubl. M.S. Thesis, Wright State University, Dayton, Ohio, 141 p.

Wolfe, P. J., 1999, Resistivity interpretation near a trench well, Expand. Abstracts, 69th Society of Exploration Geophysicists Int'l Annual Meeting (Houston), p. 571-574.

Zhou, Bing, 1998, Crosshole resistivity and acoustic velocity imaging: 2.5D Helmholtz equation modeling and inversion: Unpubl. Ph.D. dissertation, University of Adelaide, Adelaide, Australia, 248 p.

Zohdy, A. A. R. and Bisdorf, R. J., 1989, Programs for the Automatic Processing and Interpretation of Schlumberger Sounding Curves in Quickbasic 4.0: United States Geological Survey, 64 p.

Geotechnical and Environmental Applications of Karst Geology and Hydrology, Beck & Herring (eds)
© 2001 Taylor & Francis, ISBN 90 5809 190 2

Application of thermography to groundwater monitoring at Arnold Air Force Base, Tennessee

C.WARREN CAMPBELL Engineering Div., City of Huntsville, Huntsville, AL 35801, USA, wcampbell@ci.huntsville.al.us

MIKE SINGER CH2M Hill, AAFB, Manchester, TN 37389, USA, msinger@ch2m.com

ABSTRACT

Arnold Air Force Base (AAFB) sits atop a groundwater high in the Highland Rim area of Tennessee. Consequently, groundwater drains radially away from the site. To assist in the AAFB groundwater monitoring program, aerial thermography was collected for 470 km² (180 sq mi) of AAFB and surrounding area. This thermography was color enhanced and analyzed to identify potential springs. Currently, more than 600 potential springs have been located with the thermography and more than 100 springs were field located. The discharge of these springs ranged from a few ml/s to almost 600 l/s (20 cfs). As of July 2000, the known groundwater discharges around AAFB have been doubled with this program. Recharge estimates of 30 cm/yr (12 in/yr) were developed using hydrograph separation of base flow from surface stream gage data. Also, monthly recharge estimates indicated that the average spring flow for May or December approximately equals the mean annual flow.

INTRODUCTION

Arnold Air Force Base (AAFB) near Manchester, Tennessee sits on a potentiometric high so that groundwater drains radially away from the site. AAFB is in the eastern part of the Highland Rim physiographic region of Tennessee, with the edge of the low inland plateau lying to the north and west (Figure 1). The bedrock consists mostly of impure limestones, with the Chattanooga shale present in outcrops on the slopes below the plateau top. Many springs discharge at, or just below, the upper contact with the Chattanooga. Because of the radial groundwater drainage, groundwater monitoring must include surrounding areas. In karst aquifers, springs make much better groundwater monitoring sites than monitoring wells (Quinlan, et al., 1992 and Quinlan and Ewers, 1986). Since thermography has been shown to be an effective spring location technology (Brown, 1972; Boggle and Loy, 1995; Campbell, et al., 1996), it was collected for a 470 km² area around AAFB. The thermography data provided digital thermal maps with a temperature sensitivity of 0.1°C (noise equivalent ΔT) and a ground resolution of one meter. Groundwater temperatures in the Manchester area range from 11 to 16°C. During the thermography flight, surface water temperatures ranged from 0 to 5°C, so that a good temperature contrast between ground water and surface water was obtained. Several springs with discharges as small as a few ml/sec were visible on the thermal imagery. Though the ground resolution was 1 m, many springs smaller than this were visible because they had enough temperature contrast to change the effective pixel (picture element) temperature. In Figure 2, power lines much cooler than the surface water and much smaller than 1 m in diameter are clearly visible on the thermography.

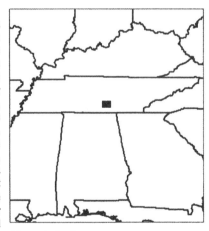

Figure 1: Study area

The sensitivity of the thermography permitted location of six hundred probable springs on the thermal images. More than one hundred springs were located in the field. These springs ranged from seeps with almost no perceptible flow, to the largest spring at AAFB with a flow of 570 liter/sec (20 cfs).

Most significant springs in the area were probably visible on the thermography. Exceptions included deeply submerged springs in local lakes and smaller springs obscured by larger, upstream springs. The following section describes some of the springs located in this study.

Figure 2: Power lines visible on thermography

SPRINGS LOCATION AROUND ARNOLD AIR FORCE BASE (AAFB)

The Negro Bottom distributary was the largest spring system located at AAFB with a total flow of 570 liter/sec (20 cfs). Figure 3 shows the thermography of the system. The discharge points included typical rise pools and submerged discharges in the surface stream (Figure 4). From recharge estimates described later, this system drains an area of approximately 60 km^2 (20 mi^2).

Several submerged springs were located. These are difficult to find in the field without thermography. Figure 5 shows one of these. This spring discharges from a solution hole in the bedrock of the stream bottom with a flow strong enough to create the surface boil shown in the figure. Though not apparent in the figure, the discharge has lower turbidity than the surface stream. This spring is clearly visible in the thermography shown at the right.

At this site, springs often came in groups. Figure 6 shows an example of a group of springs. Cascade Spring is marked on U.S.G.S. topographic

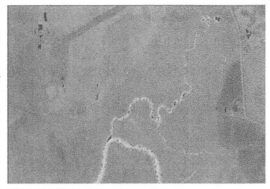

Figure 3: Negro Bottom thermography

maps, but the other springs indicated in the figure are not. The figure also shows the characteristic pattern of discharge along a line at the top of the Chattanooga shale. This occurred so frequently, that these springs could be used to assist in geologic mapping of this important unit.

Some very small springs were clearly visible on the thermography. Figure 7 shows an example of one of these. The figure shows three known springs and two unknown ones. The circle indicates a seep of a few ml/sec that can clearly be seen on the thermography.

The thermography could be used to find almost every spring at this site excepting deeply submerged dis- charges (3 m water or more) and smaller discharges obscured by large upstream springs.

Figure 4: Negro Bottom rise pool discharge (left) and sand boil (right) in creek bed

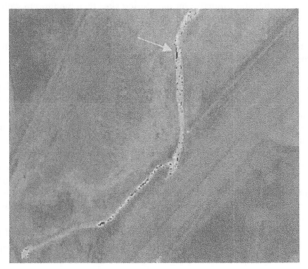

Figure 5: Tree Stump Spring surface boil (beneath the left hand) and thermography (right). The spring indicated by the arrow in the upper center of the thermal image.

354

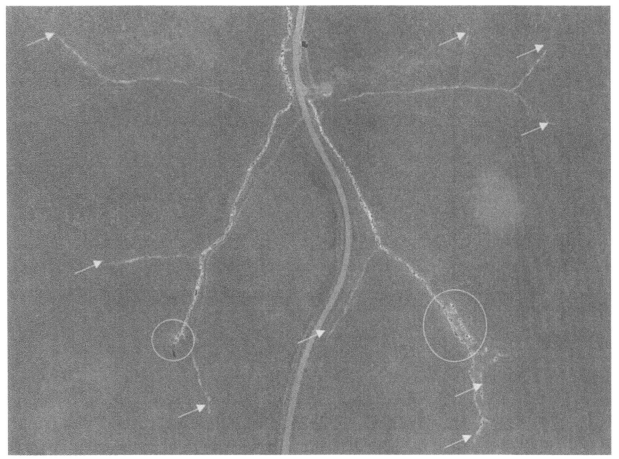

Figure 6: Cascade Spring (left circle) is surrounded by several other springs including a linear discharge along the top of the Chattanooga shale (right circle).

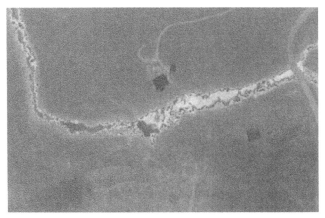

Figure 7: Large and small springs at Rutledge Falls. A seep with a flow of a few ml/sec is circled.

IMPACT OF THERMOGRAPHY ON DYE TRACING

In areas with either hardwood forests or with little vegetative cover, aerial thermography appears to be an unsurpassed spring location technology. In the right setting, almost every significant spring can be found. Submerged discharges in broad streams or shallow lakes are easily seen on thermography. Without thermography, they are difficult or impossible to find. Brown and Wigley (1969) observed that there are 5 basic karst dye trace topologies as shown in Figure 8.

Since thermography permits the location of practically every spring, Type 3 topology becomes Type 1, Type 4 becomes Type 2, and Type 5 converts to either Type1 and/or Type 2. This also simplifies QTRACER type analyses (Field, 1999) of quantitative dye traces. The QTRACER approach can be used to estimate karst conduit volumes and dispersion characteristics, but it assumes a Type 1 topology. It can be used with some stipulations for Type 2 topologies. For Types 3 – 5, its application can be very complex and difficult to interpret. However, with thermography Types 3 – 5 effectively become Types 1 or 2, and the interpretation is simplified. Consequently, the availability of thermography can provide a much better understanding of the plumbing of karst aquifers.

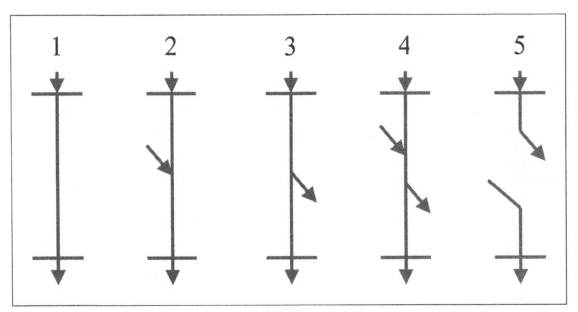

Figure 8: Possible dye trace topologies (Brown and Wigley, 1969)

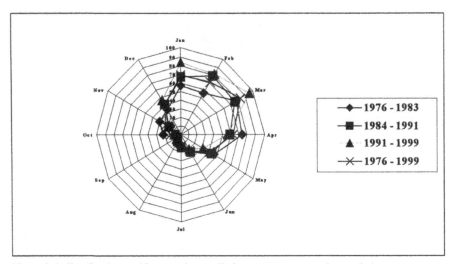

Figure 9: Indian Creek monthly groundwater discharge averages over 8-yr periods

RECHARGE ESTIMATES FOR ARNOLD AIR FORCE BASE

Recharge estimates for AAFB were obtained by hydrograph separation of surface stream data. Unfortunately, stream gage records for this site were relatively short so that recharge estimates based on these methods have significant errors. For example, Indian Creek in north Alabama has 24 years of available record. Figure 8 shows the result if monthly groundwater discharges are averaged over 3 eight-year periods. The three 8-yr periods are all plotted with averages for the full 24 years. The variability is obvious for these short records. The data available for the AAFB area are similarly short and resulting estimates of recharge are subject to errors.

The recharge estimates for AAFB are based on hydrograph separations for Boiling Fork near Winchester, TN and Bradley Creek near Prairie Plains, TN. Eight years of data were available for each of these streams. The estimated recharge for these streams was 300 mm/yr (12 in/yr). Radar plots of monthly average groundwater discharge (see Figure 10) show that the mean annual discharge for springs near AAFB occur in May and December. These are the months when the mean annual recharge crosses the monthly recharges. These "magic months" are similar for those in Madison County, Alabama (Campbell and Keith, these proceedings). Quinlan and Ray (1995) recommended measuring normalized base flow or base flow per unit area, and using this to estimate recharge areas of springs. They believed using base flow data would reduce the required flow data by 75 percent. By measuring flow during the magic months of December and May (for this sites), the amount of data required is reduced even more. The magic months vary from site to site. For example, the magic months for Big Spring in Missouri are January and June-July. The months when ground water discharge equals the mean annual discharge varies widely, but can be determined from surface stream flow data.

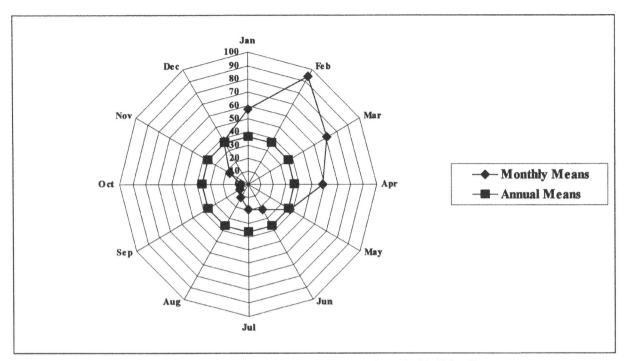

Figure 10: Monthly and annual mean groundwater discharge from Bradley Creek near Prairie Plains, Tennessee

SUMMARY

This study showed that, in the right setting, thermography is a useful tool for spring location. This setting is in temperate, hardwood forested areas or in areas with little tree cover. At Arnold Air Force Base near Manchester, Tennessee the thermography was used to field locate more than 100 springs including the Negro Bottom distributary with a combined discharge of 570 liter/sec (20 cfs). More than six hundred potential springs were indicated on the thermal images. The thermography was also helpful in locating outcrops of the Chattanooga shale where characteristic linear discharges were indicated. Small seeps with discharges of a few ml/sec were also clearly visible on the thermal images.

Springs often occurred in groups and many submerged springs were found. The submerged springs were often difficult to locate in the field even with good thermal locations. The only springs believed to have been missed were those submerged in more than 3 m of water, and smaller springs obscured by large upstream springs.

Recharge for the AAFB area was estimated using surface stream hydrograph separation. The best estimate of recharge was 300 mm/yr (12 in/yr). Monthly averages of base flow in surface streams indicated that, on average, in May and December, springs discharge at their mean annual flow. These "magic months" vary from site to site.

ACKNOWLEDGEMENTS

Thanks are due to Arnold Air Force Base and Pam King who provided support and encouragement for this task. We also wish to thank Keith Dobson of ACS, and Linda Balckwelder and Greg Schaefer of CH2MHILL for encouragement and field support during this work. We thank Steve Bong and Tom McLaughlin of JAYA Corporation in Huntsville, Alabama for support of the first author during the work reported here. Connor Haugh of the USGS provided us with much information on spring and surface stream flows in the area. We also thank Dave Massey of JAYA Corporation who accompanied the first author during almost all of the field work.

REFERENCES

Brown, Michael C. (1972). "Karst Hydrogeology and Infrared Imagery: An Example," *Geological Society of America Bulletin*, Volume 83, pp 3151 – 3154.

Bogle, F.R., Loy, K. (1995). "The Application of Thermal Infrared Photography in the Identification of Submerged Springs in Chickamaugua Reservoir, Hamilton County, Tennessee," *Karst Geohazards, Engineering and Environmental Problems in Karst Terrain, Proceedings of the Fifth Multidisciplinary Conference on Sinkholes and the Engineering and Environmental Impacts of Karst*, Gatlinburg, Tennessee, pp. 415-424.

Brown, M.C., and Wigley, T. M. L. (1969). "Simultaneous Tracing and Gaging to Determine Water Budgets in Inaccessible Karst Aquifers," *Proceedings of the 5th International Congress on Speleology*, Band 5, Hydrologie des Karstes, HY 3/1 – 3/5.

Campbell, C. Warren, El Latif, Mohamed Abd, Foster, Joseph W. (1996). "Application of Thermography to Karst Hydrology," *Journal of Cave and Karst Studies*, Volume 58, No. 3, pp. 163-167.

Field, Malcolm S. (1999). "The QTRACER Program for Tracer-Breakthrough Curve Analysis for Karst and Fractured-Rock Aquifers," U.S. Environmental Protection Agency, EPA/600/R-98/156a, 137p.

Quinlan, James F., Davies, Gareth J., and Worthington, Stephen R. H. (1992). "Rationale for the Design of Cost-Effective Groundwater Monitoring Systems in Limestone and Dolomite Terranes: Cost Effective as Conceived is Not Cost-Effective as Built if the System Design and Sampling Frequency Inadequately Consider Site Hydrogeology," Eighth Waste Testing and Quality Assurance Symposium, Washington D.C., pp. 552-570.

Quinlan, James F., and Ewers, Ralph O. (1986). "Reliable Monitoring in Karst Terranes: It Can be Done, but Not by an EPA-Approved Method," *Ground Water Monitoring Review*, Volume 6, Number 1, pp. 4-6.

Quinlin, James F., Ray, Joseph A. (1995). "Normalized base-flow of groundwater basins: A useful parameter for estimating recharge areas of springs and for recognizing drainage anomalies in karst terranes," *Karst Geohazards*, Balkema Press, Rotterdam, pp. 149 – 176.

Geotechnical and Environmental Applications of Karst Geology and Hydrology, Beck & Herring (eds)
© *2001 Taylor & Francis, ISBN 90 5809 190 2*

An application of cone penetration tests and combined array 2D electrical resistivity tomography to delineate cover-collapse sinkhole prone areas

OLIVIER KAUFMANN & YVES QUINIF   CERAK, GEFA, Faculté Polytechnique de Mons, Mons, B7000, Belgium, olivier.kaufmann@fpms.ac.be, yves.quinif@fpms.ac.be

ABSTRACT

In urbanized covered karst terranes where sinkhole collapses occur, there is a need to develop affordable and reliable investigation methods to delineate areas of potential sinkhole collapse. This paper reports the results of a geophysical and geotechnical survey conducted over such terranes.

In the 1.5 ha area surveyed, the limestone bedrock is overlain by 3 to 10 meters (10 ft to 30 ft) of cover mainly consisting of clayey sands. Forty-six cone penetration tests (CPT) and fifteen borings were conducted in this area to check unconsolidated cover thickness and depth to bedrock.

The geophysical part of the survey consisted of 2D electrical resistivity tomography (2D-ERT). Electrical resistivities were collected along eight lines using 2.5 m (~8 ft) electrode spacing and tests with several electrode arrays were conducted. Simulations and field tests led to the use of dipole-dipole and Wenner-Schlumberger arrays in a combined array inversion procedure to obtain a better image of the subsurface.

A 3D model of limestone bedrock was built using 2D-ERT results. This model was then compared to CPTs and borings results as well as locations and alignments of known cover-collapse sinkholes in and around the survey area. This comparison showed the validity of the proposed model and its usefulness to infer potential sinkhole collapse areas.

INTRODUCTION

In covered karst terranes, cover collapse or cover subsidence sinkholes usually occur over cutters affecting the limestone bedrock. Thus, characterization of subsurface conditions is essential to delineate sinkhole prone zones on a local scale. Using borings to assess the geometry of the top of rock is generally expensive because of the amount of borings required to delineate the pinnacle and cutters over the area. Moreover, insufficient coverage may be misleading and result in misinterpretations. In order to overcome these drawbacks, geophysical methods have long been tested in such contexts. Recently, electric resistivity tomography has proven to be a valuable tool to identify and delineate covered karst features particularly where a conductive (clayey) cover prevents the use of GPR.

STUDY SITE

The 1.5 ha study site lies in the Tournaisis area (southern Belgium). It is located in Gaurain-Ramecroix, to the south of the Warchin brook where many cover-collapses have occurred since 1984. Borings, a digging site and observations in former collapses near or within the study area have shown that the siliceous limestones of Warchin member are overlain by 3 to 10 m of sandy silt, clayey sand and silty residuum. The limestone is dipping approximately 5° to the North and is affected by cutters filled with residuum. Locally, sinkholes tend to line up on directions that are N105°E and N20°E.

Figure 1 shows the extent of the study area and the locations of former collapses. Two houses, one of which has been ruined by a cover-collapse sinkhole, lie in the study area. The figure also presents the layout of the lines along which electric resistivity measurements were taken.

ELECTRIC RESISTIVITY TOMOGRAPHIES

2D Electric resistivity tomography is an improvement over previous resistivity profiling techniques. It consists of an inversion of an apparent resistivity pseudosection in order to infer the distribution of the underground resistivities. Commonly, in covered karst terranes the allochtonous and residual sediments covering the limestone are usually significantly less resistive than the limestone rock. The tomographs should reveal the hidden geometry of the bedrock and make it possible to identify pinnacles and cutters. It should be stressed however that the solution to the inversion of such a pseudosection is not unique and that, at least with 2D-ERT, the model is 2D and cannot therefore fully represent complex 3D structures. All tomographs presented in this paper have been computed with RES2DINV software (Loke, 1996).

To produce a good tomograph, several hundred resistivity measures are commonly required. This is why the resistivities are usually measured with an automated acquisition resistivity meter linked with a multi-electrode system.

Choice of arrays

Most of the classical electrode arrays are in-line quadripoles consisting of two current electrodes and two potential electrodes. They include dipole-dipole, Schlumberger and Wenner (alpha, beta and gamma) arrays. Each array has its advantages and disavantages in terms of depth of investigation, sensitivity to horizontal or vertical structures, signal strength and horizontal coverage along a line. Thus, the choice of an array may be of great importance.

In urbanized areas, it is common that obstacles make it impossible to extend the lines outside the area of interest. The poor horizontal coverage of the wenner arrays is then a major drawback because only resistivities associated to very shallow zones can be measured near the ends of the lines. This means that only the center of the

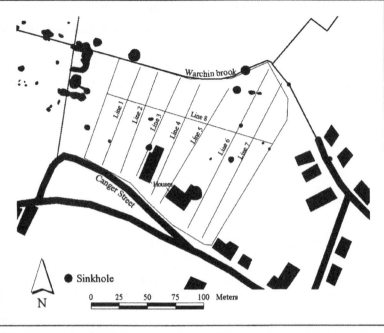

Figure 1: Location of sinkholes around and within the study area (shaded area) and layout of the lines of the electrical resistivity survey.

area of interest can be investigated in depth with such arrays. Dipole-dipole and Wenner-Schlumberger arrays (Figure 2) have a better horizontal coverage than wenner arrays.

The dipole-dipole array is often preferred because of its good sensitivity to vertical structures, which makes it appropriate to detect cutter-like features. However, its signal strength is pretty weak. This can result in poor quality readings as the separation factor increases, particularly in noisy environments. Such poor quality readings may lead to poor quality tomographs especially in depth where the most interesting features usually lie.

On the other hand, the Wenner-Schlumberger array shows a fair sensitivity to horizontal and vertical features. It also has a slightly better depth of investigation and a stronger signal than the dipole-dipole array. However its weaker sensitivity to vertical features makes it less appropriate to detect narrow cutters or pinnacles than the dipole-dipole array.

In the application presented in this paper, rather than choosing a single array type, apparent resistivities were measured with both dipole-dipole and Wenner-Schlumberger arrays. Then, the inversion procedure was performed on all these apparent resistivities. This makes it possible

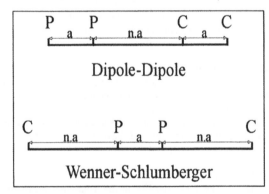

Figure 2: Dipole-dipole and wenner-schlumberger arrays; current electrodes are noted C and potential electrodes are noted P; a is the minimal electrode spacing and n is the separation factor.

to take advantage of the high sensitivity of the dipole-dipole array to vertical structures and the stronger signal of the Wenner-Schlumberger array. Moreover, these two arrays have very different sensitivity patterns (Loke, 1999). Therefore, more constraints are added to the model by using the apparent resistivities measured with these arrays in the same inversion procedure.

Layout of the lines

The direction of the main cutters being N105°E, seven lines were set perpendicular to this direction in order to intersect such cutters. As the maximum depth of investigation of the arrays used in this survey is about 10m (~33ft), the lines have been spaced by approximately twice this distance. One more line has been laid perpendicular to the others in the northern part of the area.

CPTs and Borings

In order to check the results of the 2D-ERTs, 46 cone penetration tests and 15 borings were conducted in the study area. During CPT tests, shocks were carefully recorded. Depth of refusal was also carefully monitored as it may indicate depth to bedrock or very stiff residuum. Remolded samples of the cover were gathered from auger drillings. After hard rock was reached, the drilling went on with air track for at least one meter in order to make sure that massive sound rock had been encountered.

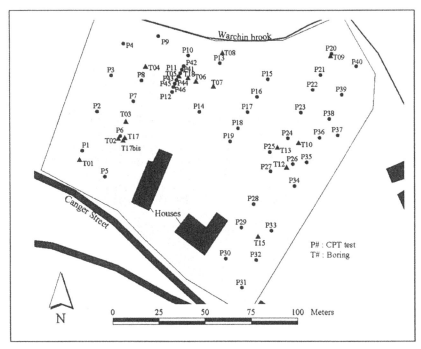

The locations of the CPT tests and borings are shown on Figure 3. Most of them are situated on the geophysical lines. CPTs were evenly distributed along the lines. In the northern part of line #3, penetration tests were closed up to delineate a wide cutter-like feature. No borings or CPTs were done in the gardens surrounding the houses.

Figure 3: Enlargement of the study area showing the locations of CPT tests and boreholes.

The map of Figure 4 shows the result of an interpolation of the depth of refusal of the cone penetration tests below the ground surface. This map shows that although the topography of the area is flat refusal of CPT take place at a depth of around 3m in most of the northern half of the study area while it is recorded below 6m in the southern half.

Comparisons between borings and CPTs have shown that refusal of CPTs correspond either to top of rock or to very stiff residuum. It could also be observed that the upper part of large cutters is filled with a few meters of soft allochtonous green clay covering stiff residuum with cherts. Thus even if CPT performed straight above a cutter refused on the residuum, its refusal occured a few meters below the surrounding ones.

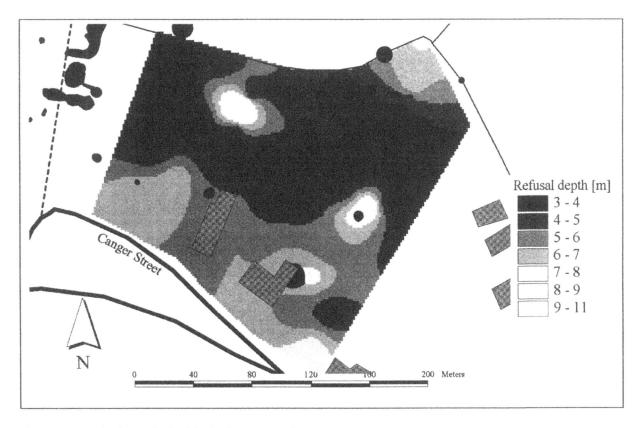

Figure 4: Interpolated map of refusal depth of cone penetration tests.

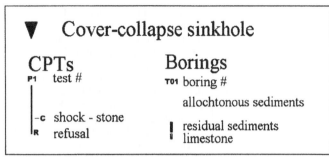

Figure 5: Symbols used on tomographs.

Tomographs and resistivity maps

Figure 5 shows the symbols used on the tomograph of Figure 6 to summarize the information gathered during CPTs and borings. Locations of former cover-collapse sinkholes are also shown on this tomograph.

Figure 6 is an example tomograph corresponding to line #6. CPT and borings suggest that the top of rock should correspond on the picture to resistivities between 100 and 250 ohm.m. Like the CPTs, the tomograph shows a greater cover thickness in the southern part than in the northern one. However, the tomograph also shows in some detail a very irregular top of rock. A pinnacle is clearly visible between P30 and P29, then follows a large cutter which might be associated with the sinkhole that ruined the nearby house. A narrower cutter encoutered by boring T12 is also visible. A nearby sinkhole could have formed on the northen edge of this cutter. After boring T13, the bedrock is nearer to the ground surface.

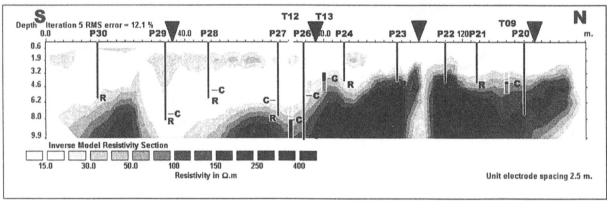

Figure 6: 2D electric resistivity tomography combining dipole-dipole and Wenner-Schlumberger arrays along line 6.

A cutter is clearly visible between CPT tests P22 and P23 straight below a sinkhole. At the northern end of the line, the resistivities drop slightly but no clear cutter can be identified on the tomograph below the 0.25m deep collapse sinkhole located near test P20. This might be due to side effects in the lower part of the tomograph.

In order to summarize all tomographs, the modeled resistivities were interpolated to produce a 3D model of the area. This model is presented as a series of resistivity maps corresponding to different depths. Figure 7 shows such a map at a depth of 6.2 meters while Figure 8 corresponds to resistivities at a depth of 9.9 meters.

First of all, a good accordance between these maps and the one of Figure 4 should be noted. It should also be recognized that the extension and orientation of features like the pinnacle and the large cutter in the SW corner of the map are much more obvious on Figure 8. Smaller features not visible on Figure 4 are also identifiable on the resistivity maps.

CONCLUSIONS

The results presented in this paper show the suitability of electric resistivity tomography to delineate buried cutters above which cover-collapse sinkholes occur in the Tournaisis area. The use of combined dipole-dipole and Wenner-Schlumberger array proved to be effective in a case where measurements with dipole-dipole arrays alone was not satisfying.

With well-oriented profiles covering the study area, a valuable 3D model could be built from the 2D tomographs. This model is presented as a series of resistivity maps, each one corresponding to a particular depth. These maps give a good understanding of the hidden underground conditions and make it possible to identify and delineate the main pinnacles and cutters affecting the bedrock. Therefore, this technique is of very useful to infer potential sinkhole collapse areas.

ACKNOWLEDGEMENT

This study is part of a research program with the support and funding of the 'MINISTERE DE LA REGION WALLONNE, DIRECTION GENERALE DES RESSOURCES NATURELLES ET DE L'ENVIRONNEMENT'.

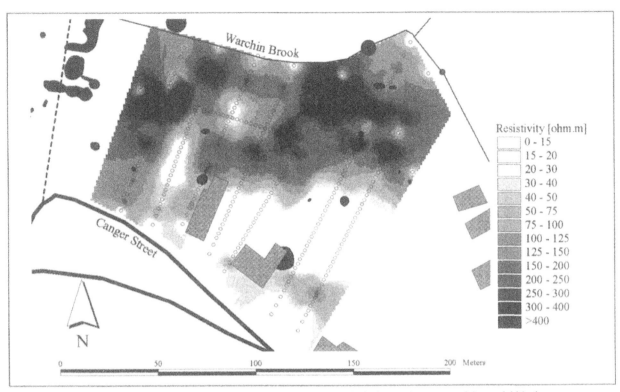

Figure 7: Map of interpolated modeled resistivities at 6.2 meters deep, small circles are the centers of the model blocks.

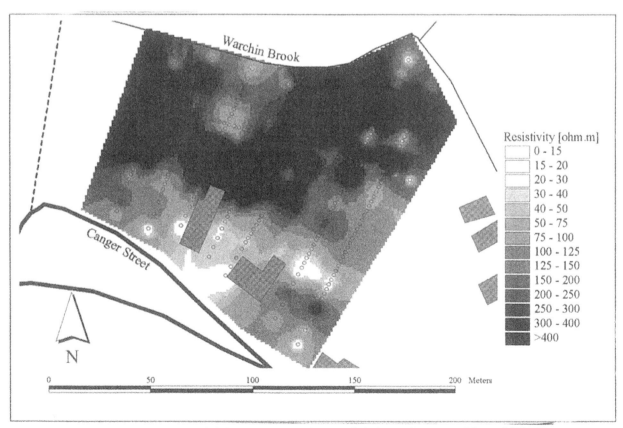

Figure 8: Map of interpolated modeled resistivities at 9.9 meters deep, small circles are the centers of the model blocks.

REFERENCES

Kaufmann O., 2000, Les effondrements karstiques du Tournaisis : genèse, évolution, localisation, prévention, Faculté Polytechnique de Mons, Belgium, in press.

Loke M.H., 1996, Manual for RES2DINV, ABEM France, France.

Loke M.H., 1999, Electrical imaging surveys for environmental and engineering studies - A practical guide to 2-D and 3-D surveys, http://www.abem.se, 63 p.

Loke M.H. and Barker R.D., 1996, Rapid least-square inversion of apparent resistivity pseudosections by a quasi-Newton method, Geophysical prospecting, v. 44, pp. 131-152.

Roth M.J.S., Mackey J.R., Mackey C. and Nyquist J.E., A case study of the reliability of multi-electrode earth resistivity testing for geotechnical investigations in karst terrains, Hydrogeology and Engineering Geology of Sinkholes and Karst, Beck, Pettit & Herring (eds) Balkema, Rotterdam, pp. 247-252.

Zhou W., Beck B.F. and Stephenson J.B., 1999, Application of electrical tomography and natural potential technology to delineate potential sinkhole collapse areas in covered karst terrane, Hydrogeology and Engineering Geology of Sinkholes and Karst, Beck, Pettit & Herring (eds) Balkema, Rotterdam, pp. 187-193.

Geotechnical and Environmental Applications of Karst Geology and Hydrology, Beck & Herring (eds)
© 2001 Taylor & Francis, ISBN 90 5809 190 2

The effect of groundwater pollution on the natural-electric potential in karst

KEVIN T.KILTY   Industrial Physicist, La Center, WA 98629, USA, kkilty@ix.netcom.com

ARTHUR L.LANGE   Karst Geophysics, Inc., Golden, CO 80401, USA, karstgeo@aol.com

ABSTRACT

The investigation of groundwater contamination in a karst environment is complicated by a substantial component of flow along fractures and solution openings. This behavior contrasts with the distributed flow pattern that characterizes the diffuse-flow environment. In many situations, boreholes for monitoring pollution in a karst terrain must intercept fractures or conduits in order to effectively sample active groundwater flow. In response to this need, the authors have employed the geophysical method of natural potential (NP) during the past fourteen years for detecting and mapping karst features and for identifying locations for successful monitoring wells. The technique has accomplished these objectives rapidly and economically in diverse carbonate environments across the U.S. and overseas.

In the course of conducting these surveys, we have noted enhanced natural-potential responses to the presence of organic pollutants in groundwater. These electrical effects have also materialized within laboratory models involving VOCs and fresh water. There are several possible reasons for expecting such responses. One mechanism involves the mixing of polluted groundwater with fresh water occurring through a diffused front, wherein the diffusion itself produces an NP signal. Another possibility is that the situation resembles the refining of petroleum products or the handling of powdered materials, where friction and flow create large, and sometimes dangerous, static electrical potentials. This is especially true when the petroleum product contains a contaminant comprising minor amounts of an aqueous phase, solid material, or gas bubbles.

Although the actual mechanism may differ from one site to the next, it is likely that enhanced potentials from the presence of VOCs in groundwater will be encountered at the ground surface. In practice, elevated NP signals have been registered from contaminants within carbonate aquifers, as well as in the associated epikarst and overlying residuum.

INTRODUCTION

The risk of groundwater contamination is greater in karst terranes than in diffuse-flow environments. In karst, subsurface drainage occurs primarily through caves and solution conduits, while runoff flows vertically toward the underground drainage system via sinkholes and open funnels, suffering little or no filtration or sorption along the way. Water can travel rapidly through the underground conduit network as a free-flowing or tube-full stream to its eventual discharge at springs, wells and rivers. Pollutants draining into sinkholes from industrial sites or landfills can move rapidly along a karst system and contaminate groundwater over a large area, and at the same time can become trapped in pools or pockets resistant to extraction. Boreholes for dye injection or groundwater monitoring and treatment must intercept the pathways of karstwater flow in order to be effective. Boreholes that penetrate the carbonate rock but miss the conduit are likely to be dry or at best encounter water residing in storage.

The geophysical method of *natural-potential* (NP)[1] affords an economical means for detecting and tracking karst systems from the surface, hence, facilitating the placement of monitoring wells and boreholes (Corwin, 1990; Lange & Quinlan, 1988). The method utilizes the *natural d.c. electric field* present everywhere in the ground and produced in part by chemical reactions between groundwater and the soil/rock environment. A major contribution to this natural field comes about as a consequence of the movement of groundwater—the streaming phenomenon, or *streaming potential.*

G. Quincke first reported streaming potentials in 1859, but a cogent theory to explain them was not available until Helmholtz proposed an electric double layer in 1879. In 1931 Onsager further developed our understanding of the streaming phenomenon through his formulation of non-equilibrium thermodynamics (Adam, 1968). Nourbehecht (1963) applied this theory to geological problems, but Sill (1983) extended the theory to geophysical interpretation by showing how to model natural-potentials in terms of primary flows. While the theoretical development and laboratory testing focused on capillary flow in granular media, Bogoslowsky and Ogilvy (1972) performed tests on flow through narrow fissures, and Binder & Cernak (1961) investigated turbulent flow in tubes as large as one inch (2.54 centimeters). Krajew (1957) extended theory to large flow pathways and provided examples from Caucasian rivers. Very recently Zhou, et al (1999) apportioned natural-potentials from karst environments into their various causative components.

---

[1] Also referred to as *self-potential (SP).* For a historical review of the method see Rust (1938).

NATURAL-POTENTIALS AND KARST

Between 1986 and the present time, Lange and his coworkers have observed hundreds of natural-potential anomalies above and within karst features (Lange & Kilty, 1991, Lange, 1999; Lange & Barner, 1995). The features involved a depth range between 250 meters and the ground surface. A theory to account for these observations involves streaming potentials (Kilty & Lange, 1991), among other mechanisms. We have used the method to site test wells for numerous environmental and engineering projects. Independently, Quarto and Schiavone (1996) also observed natural-potential anomalies over shallow caverns. Their work showed the influence of seasonal streaming potentials and thermal effects, which several investigators had suggested might be important (Ernstson & Scherer, 1986; Kilty & Lange, 1991), and demonstrated that natural-potential anomalies in karst do not necessarily arise from variations in the overlying soil. There is no doubt that karst conduits, sinkholes, and caverns produce natural-potential anomalies.

Caves frequently occur along joints, fractures, or faults in bedrock, and are associated with enhanced hydraulic permeability[2]. The consequent nature of groundwater flow makes it a source of natural-potential. A mathematical theory to describe this situation couples together the equations for groundwater flow and electric current. The linear equations describing this in homogeneous material and in matrix form are:

$$\begin{bmatrix} j \\ v \end{bmatrix} = - \begin{bmatrix} \sigma & C_{\sigma K} \\ C_{K\sigma} & K \end{bmatrix} \cdot \nabla \begin{bmatrix} \Phi \\ P \end{bmatrix} \quad , \tag{1}$$

where j = electric-current density,

σ = electrical conductivity,

$C_{\sigma k}$ and $C_{k\sigma}$ are cross-coupling (Onsager) coefficients,

∇ = gradient operator,

ϕ = electric potential (natural-potential in this instance),

v = Darcian velocity,

K = hydraulic conductivity, and

P = non-hydrostatic fluid pressure.

This equation is nothing more than Ohm's Law and Darcy's Law coupled together through a linear relationship utilizing cross-coupling coefficients. The cross-coupling coefficients might have a zero or negligible value. Only experiment and observation can prove their importance in particular situations. If other primary flows, such as heat flow, are important, we can add them in a similar manner.

The cross-coupling coefficients arise from a disturbance to the electrical structure of the interface where two material phases meet. For example, in a fractured carbonate solid saturated with an aqueous solution of nearly neutral pH, anions in the solution are adsorbed at the solid surface, leaving a diffuse region in the adjacent fluid slightly enriched in cations[3]. The electrical potential across this diffuse zone is known as the *zeta-potential*. As groundwater flows through this material, viscosity prevents fluid at the solid surface from moving with the bulk fluid flow; hence, the cations, being distributed farther from the solid interface, travel more readily with the flow. This leads to electric charge separation, which, in turn, produces an electrical field. This electrical field, when integrated between two points in the flow, is the streaming potential. In porous silicate materials the situation is similar, except that mineral grains present particular chemical bonds to the fluid which attract cations. Thus anions are typically the more mobile species in porous silicate material (Rivel, et al., 1999). Anything that affects the zeta-potential necessarily affects the streaming potential.

Generally, the coefficients in these equations are functions of position. However, to illustrate how natural-potentials arise, we assume that the material is homogeneous. Taking the divergence of both sides of Equation 1, focusing attention on the equation for electric-current density, and using the fact that $\nabla \cdot j$ is zero or nearly so, results in the following equation for the electric potential (NP):

$$\nabla^2 \phi = (-C_{\sigma K}/\sigma) \nabla^2 P \tag{2}$$

Therefore, the Laplacian of the non-hydrostatic fluid pressure (P) is a source, or primary potential, for natural electrical potentials. People often speak of the groundwater flow itself as a *primary flow*. However, it is more convenient to use a scalar field equation like Equation 2 rather than the vector equation that a primary flow would require. One may regard pressure as the source of an electromotive force (EMF) and natural-potential as the voltage drop accompanying the flow of current produced by the EMF[4]. A similar but reciprocal equation exists for fluid pressure (P) having electric potential as a primary potential, but this phenomenon—*electro-osmosis*, has an insignificant effect in normal groundwater flow.

---

[2] A concise introduction to the complex hydrology of caves and karstwater flow can be found in Chapters 6 and 7 of the text on karst by Ford and Williams (1992).

[3] Most laboratory studies suggest that carbonate rocks typically have cations bound at the surface, implying that the zeta-potential is negative. However, we have observed evidence of both positive and negative zeta-potentials in our field work, and the laboratory results which we discuss later definitely imply a positive zeta-potential.

[4] People sometimes argue that the ohmic potential drop *is* the streaming potential. However, in natural earth media ohmic potential drop can occur—and natural-potential can be measured—in places where there is no streaming at all. Therefore, the streaming potential is the cause (EMF) and natural-potential is the effect (ohmic potential drop from return current flow).

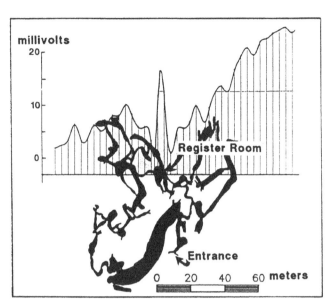

Figure 1. Natural-potential profile over Glenwood Caverns, Garfield County, Colorado. The cave passages shown in black in plan view lie within approximately 30 meters of the surface. The prominent central "sombrero" anomaly overlies the Register Room. Smaller positive anomalies of the profile correspond to other passages. The Register Room anomaly disappeared temporarily after heavy rains saturated the soil near the surface. [Map adapted from compass and tape survey by the Colorado Grotto of the National Speleological Society]

Some confusion arises over the use of piezometric head (h), rather than P, as a primary potential. For example, Nourbehecht (1963), recognizing that $\nabla^2 h$ is zero in homogeneous material and h itself is zero at the ground surface, found that $\nabla^2 \phi$ is zero throughout a homogeneous material. He concluded from this that ϕ is always zero in homogeneous material, even when there are primary flows, because electrical potential has to satisfy the condition wherein $\partial\phi/\partial n=0$ at the ground surface. People have taken this literally to mean that natural-potential anomalies are not possible except where there are material boundaries in the subsurface. This would place natural-potential sources of a limestone cavern only at the near-surface boundary between bedrock and soil; something that both Lange and Quarto & Schiavone have ruled out by studying caverns in outcropping carbonate rock.

Nourbehecht's theorem is not universal for several reasons. First, h is not necessarily zero on the same boundary at which $\partial\phi/\partial n=0$. In particular, caves often occur above the saturated zone in the vadose zone. Electric current passes from the free water surface, where h=0, into the capillary fringe without notice.

Second, $\nabla^2 h=0$ does not describe groundwater flow in many important situations. For example, $\nabla^2 h$ is not zero when there is explicit time dependence. Propagation of a diurnal or annual temperature wave into the soil, or seasonal components of soil-moisture percolation contribute to natural-potentials. For these reasons it seems better to use non-hydrostatic fluid pressure (P) instead of piezometric head (h) as the primary potential.

Finally, as Nourbehecht recognized, when material is inhomogeneous, $\nabla^2 P$ is not zero, but is proportional to $\Sigma\nabla X\cdot\nabla H$ where the X's are material properties and H's are the related potentials.[5] This applies both to gentle gradients of material properties and to abrupt boundaries.

The interpretation of natural-potential observations boils down to finding a reasonable distribution of P that explains ϕ; and then deciding what this means in terms of the site geology. Modeling could proceed with computer programs written originally for other potential fields, such as magnetics. However, some analysts find that a few current bipoles provide an adequate model, because ϕ is often explained well enough by the first few moments of a current source distribution (Corwin & Hoover, 1979; Kilty, 1984).

EXAMPLES OF NP OBSERVATIONS

Figure 1 shows a typical natural-potential profile superimposed on a cavern map. Directly above the Register Room of Glenwood Caverns, the natural-potential exhibits a characteristic anomaly consisting of a central peak flanked by side lobes of opposite sign—descriptively characterized as a "sombrero-type" anomaly. The peak in this instance has remarkably sharp flanks. This anomaly also disappeared temporarily after heavy rains saturated the soil, showing that seasonal factors can alter NP observations. An anomaly of similar shape but having a notch in its top (*fedora-type* anomaly) occurs over a major stream cave in Texas (Figure 2). Here the inflection points of the profile occur above the cave walls, making the anomaly equivalent to the case of Sill's (1983) vertical dike. Once again the sharpness of the anomaly is surprising. We commonly observe that the flanks of natural-potential anomalies above caves are too abrupt for their source

---

[5] This sum of inner products follows directly from the divergence operator acting upon the product of a scalar and a gradient. It is the equivalent of the product rule for differentiation in vector calculus.

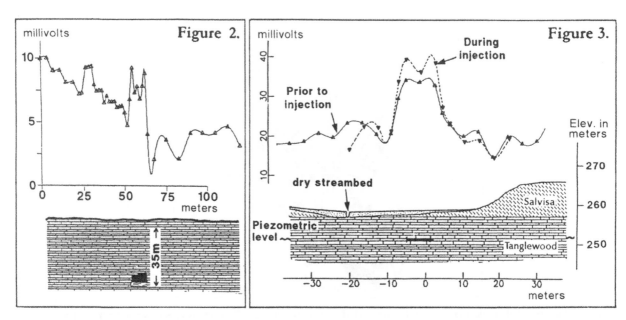

Figure 2. An M-shaped natural-potential anomaly over a stream passage of Honey Creek Water Cave, Comal County, Texas—the longest cave in the state. The stream flows at a depth of approximately 35 meters. The lesser anomaly to the left may indicate a smaller unmapped passage.

Figure 3. A flat-topped "fedora" anomaly (solid curve) over a deduced karst stream passage in the Tanglewood Limestone, Mercer County, Kentucky. The anomaly is part of a linear anomaly trend that passes through an estavelle, or alternating spring/swallet, 23 meters south of the profile shown. Injection of water into an upgradient dye-injection trench resulted in the enhancement (dashed curve) of this and parallel profiles, demonstrating the effect of lateral conduit flow.

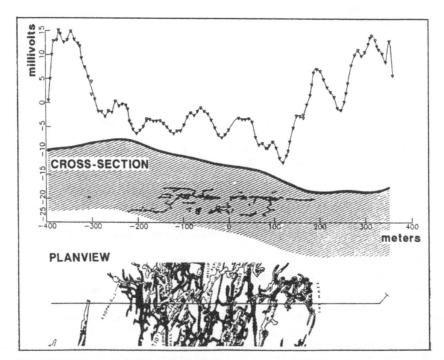

Figure 4. A comparison between a natural-potential profile above Jewel Cave, South Dakota and a cross-section and partial plan of its underlying labyrinth. Passages here are as deep as 120 meters below the ground surface. The overlapping anomalies lead to a compound anomaly generally negative in amplitude. The profile has been smoothed by a five-point moving average and a linear regional trend has been removed. [Cave map courtesy of Jewel Cave National Monument]

to occur entirely at depth near the cave ceiling. We infer from this that the source of natural-potential is distributed above the cave; almost to the ground surface in some instances.

Figure 3 shows a similar fedora-type anomaly centered over a deduced karst conduit in Harrodsburg, Kentucky. The response increased by about one-third when water was injected into the system. At Jewel Cave National Monument, South Dakota the lower levels of the labyrinthine cave are 120m below ground surface. Great depth in this instance results in the coalescence of individual negative anomalies giving rise to a compound profile at the surface (Figure 4).

Figure 5 demonstrates that even man-made tunnels produce natural-potential responses at the surface. This NP profile was observed 33 meters above a railroad bed in the Cricket Tunnel at Omaha, Arkansas, where water drips continuously from the ceiling.

We have many more examples of natural-potential anomalies associated with caverns as well as tunnels, but these few illustrate their characteristic features. Typically the natural-potential anomaly is directly over the cavern; however, on a hillside the anomaly may be shifted downslope. It can exhibit a central positive or negative anomaly with sharp edges, flanked by anomalies of opposite sign. Deep caves typically, but not always, show broad anomalies. The amplitude, and perhaps even the sign of the anomaly, may vary over different parts of the cave. Laboratory studies (Rivel, et al., 1999; Ishido & Mizutani, 1981) suggest that the sign of an anomaly depends, in a complex way, on pH, temperature, and the nature of ions present in the groundwater; and that an anomaly may, therefore, change form according to season or current soil conditions.

EXPLANATIONS OF NP ANOMALIES

Voids, fissures, or conduits in the subsurface cause large and strangely-directed pressure gradients nearby. In particular, infiltration in the material above a cave enters it via the ceiling. This does not mean that water will necessarily drip from the cave ceiling–it may simply evaporate and/or release CO_2; but the pressure gradient is undoubtedly oriented vertically. The cave ceiling and adjacent rock acquire an excess charge from ions that flow most readily with the groundwater. Farther upstream in the infiltrating roof water there is a distributed charge made up preferentially of ions bound to rock (Rivel, et al., 1999; Ishido & Mizutani, 1981). Therefore, between the ground surface and the ceiling of a cave a vertically oriented current bipole develops in which the groundwater flow pulls ions toward the cave ceiling, and the resulting electric field pushes ions back toward the ground surface in a return current. This is the principle reason for the natural-potential anomaly, but it is not the entire story.

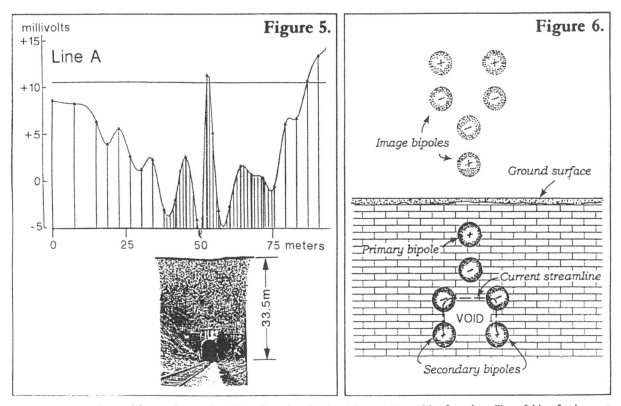

Figure 5. A natural-potential anomaly over the Cricket Tunnel at Omaha, Arkansas. Water drips from the ceiling of this—for the most part—unlined tunnel. The anomaly is similar to that in Figure 1, and has been demonstrated to be not the effect of the tracks.

Figure 6. A simple model of natural potential consisting of three current-source bipoles in the subsurface plus their three images above the ground surface. The images constrain the ground surface as a current streamline; while the three underlying bipoles maintain the dashed line, which approximates the cave boundaries, as a current streamline.

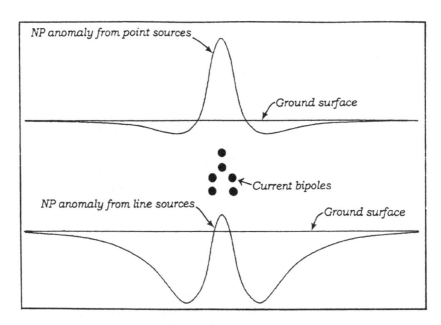

Figure 7. The natural-potential anomaly resulting from the model of Figure 6: The upper curve derives from point sources, while the lower curve is the effect of line sources.

The cavern itself is an insulator that prevents passage of the return current. Instead, the return current must flow parallel to the cave ceiling, which is to say that there must be an additional current bipole near each of the cavern's flanks. Likewise, current cannot pass through the ground surface. Mirror images of each subsurface current bipole placed above the ground surface model this condition exactly. Figure 6 shows a model consisting of three simple current bipoles in the subsurface and their images, simulating the salient features of natural-potential expressions over caves. Figure 7 shows two natural-potential profiles–one resulting from treating the current poles as points, and the other from treating the current poles as lines normal to the page. These correspond to a simple void, and to a cylindrical cavern having linear extent, respectively. Each model anomaly consists of a central peak flanked by peaks of opposite sign. While the form of the model anomaly is basically correct, we often observe flanking peaks sharper than those in this figure. This leads us (Lange & Kilty, 1991; Kilty & Lange, 1991) to propose that a portion of the primary flow is distributed above the cave almost to the ground surface. However, another reasonable explanation is that anisotropy in the hydraulic and electrical properties of karst compresses   lateral electrical fields into a narrow region directly above the cave.

NATURAL-POTENTIAL AND ORGANIC SOLVENTS

We typically observe NP anomalies above caverns that have 5 to 40mV amplitudes. Quarto and Schiavone (1996) observed somewhat larger amplitudes, 30mV to 100mV, over very shallow cavities. At sites where the groundwater contains localized concentrations of organic liquids normal NP responses tend to be enhanced and prominent, with amplitudes sometimes exceeding 100 millivolts (Kilty & Lange, 2000). This occurs often enough at hazardous waste sites (Stierman, 1984; Nash, et al., 1997) to indicate that the streaming potential is somehow amplified or changed in the presence of organic pollutants.

Figure 8 shows in profile form a 35mV natural-potential anomaly conspicuously located on the leading flank of a trichloroethylene (TCE) contaminant plume in Tennessee. Figure 9 depicts a different plume in plan view whose TCE concentration peaked over 32,000ppm, exceeding its solubility in groundwater. Therefore, the TCE can probably occur as an independent liquid phase in addition to being dissolved in the groundwater and adsorbed on rock and soil.

Besides direct observation in the field, two other observations suggest that organic chemicals should affect the streaming potential. First, large and dangerous electric charge accumulates during the pumping of refined petroleum products (Klinkenburg & van der Minne, 1958; Crowley, 1986). People refer to this as *static electricity*, but it is actually a streaming potential resulting from friction of fluid against pipe walls or the settling of particulate in tanks. Highly refined product is an electrical insulator which allows large quantities of charge to accumulate, and which also has a long decay time that is measured in hours. The magnitude of charge accumulation increases when an aqueous impurity contaminates the refined product.

In groundwater contamination, the situation is inverted; i.e., a non-conductive organic pollutant is now dispersed in a conductive aqueous solution. The conductivity of the solution leads to a short decay time, which, in turn, attenuates the charge build-up, compared to the analogous *static voltages* attainable in refined petroleum.

Second, downhole SP logs show a substantial signal increase when gas bubbles entrained in aqueous liquid pass through rock (Rivel, et al., 1999; Sprunt, et al., 1994). Thus, the presence of a third phase—gas–can enhance the natural-potential over what would normally

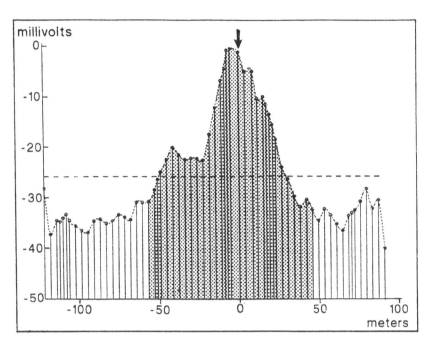

Figure 8. A natural-potential profile (5-point moving average) located fourteen meters downgradient from a cluster of monitoring wells containing TCE and other volatiles in groundwater of Tennessee. The arrow marks the location opposite the wells. The broad nature of the deduced plume (shaded) suggests that the contamination occurs within the residuum above the carbonate rock.

develop from a water/rock system alone[6]. Probably the same is true if the gaseous phase is replaced with immiscible organic liquid. In fact, volatile organic liquids themselves become a source of gas bubbles as they evaporate within the saturated rock and soil.

EXPERIMENTS AND MODELS

We became interested in investigating further the natural-potential in carbonate rock through laboratory study, and constructed an apparatus that allowed us to measure the streaming potential generated when water percolates through crushed carbonate rock . This apparatus is similar to those built to study the streaming potential in fissured or porous silicate minerals (Bogoslovsky & Ogilvy, 1972; Ishido & Mizutani, 1981). It consists of a PVC pipe column one meter in height, packed with crushed limestone, and fitted with copper/copper sulfate electrodes placed 70 cm apart. We measure potential of the upper electrode with respect to the lower one.

Figure 10 shows results from a cycle of percolation through the apparatus. It displays a large potential anomaly, up to +220mV, that appears the first time water percolates through freshly crushed limestone. The observed potential in each percolation cycle is time dependent. It increases rapidly as percolation begins and attains a peak value just as the percolating front reaches the lower electrode. It then declines rapidly and approaches an asymptotic value that pertains as long as percolation is steady. If we stop pouring liquid into the apparatus, so that percolation ceases, then the potential declines eventually to its original value. While this general pattern of behavior occurs over and over with each cycle of percolation, the particular details of each percolation depend on other circumstances.

Typically, successive cycles of percolation achieve smaller peak amplitudes. Eventually the apparatus matures and each cycle of percolation from this time on produces the same peak potential anomaly. This is typically +80mV or a little more. The transient behavior in any single cycle of percolation indicates that mobile ions travel in a sharp front on the leading edge of the fluid slug. Less mobile ions spread into a diffuse zone in the trailing fluid. Possibly air that is entrained in the flow also has an influence. The evolution from one percolation cycle to the next is more puzzling. One possibility is that fine limestone dust, left over from crushing, is being washed from the rock and is producing a Dorn effect (Klinkenburg & van der Minne, 1958). We have observed that the effluent does contain very fine limestone sediment. Subsequent to these experiments we discovered that Schwartzendruber and Gairon (1975) performed similar experiments. They flooded a column of sand from below, but otherwise obtained similar results.

The potential maintained in steady percolation depends on the water chemistry, or more directly on the zeta-potential of the rock/water interface. Distilled water with near neutral pH produces a potential of about +20mV. Increasing the water's pH by adding sodium bicarbonate does not reverse the streaming potential of percolation (Ernstson & Scherer, 1986), but simply reduces its magnitude. At a pH of about 8.0 the potential we achieved during repeated cycles of percolation was a mere +1 to +2mV. This agrees with a zero point of charge in calcite that occurs at a sodium carbonate-derived pH between 8.0 and 9.0 (Somasundaran & Agar, 1967). However, what affects the

---

[6] "...In the course of conducting the measurements as a function of permeability, it was observed that electrokinetic potential increases drastically when two phases are flowing" (Sprunt, et al., 1994).

streaming potential is not pH *per se*, but the accompanying concentration of ions in the water. More concentrated ionic solution compresses the diffused ion distribution toward the solid surface, decreasing the zeta-potential. (Bogoslovsky & Ogilvy, 1972; Kilty, 1984; Dakhnov, 1962).

When we used distilled water to flush an apparatus currently saturated with sodium bicarbonate solution, we produced a temporary increase in potential of 13 to 30mV, which decayed as the percolation continued. This large magnitude apparently resulted from pushing an interface between waters of different chemistry through the apparatus, because subsequent cycles of flushing with distilled water failed to show the same transient.

Light oils or solvents percolating through a water-saturated column produced a small effect, decreasing the potential by no more than 3mV. However, when water pushed oil through the apparatus the potential jumped 30mV to 70mV, and diminished to half this value as the effluent changed from oil to water. The potential difference across a fluid interface has three separate parts. The chi-potential at the interface arises from oriented dipoles and adsorbed ions. A zeta-potential in the aqueous phase depends on the distribution of counter ions. A zeta-potential in the organic liquid does not exist if that liquid contains no ions, but otherwise depends on the distribution of ions on this side of the interface. When this interface moves because one fluid pushes the other, the zeta-potential changes in an asymmetric way. When oil pushes water, the distribution of counter ions in the water becomes compressed. This reduces the zeta-potential and the result is a very small voltage drop as the interface passes a measuring electrode. When water pushes oil the limited mobility of some aqueous ions is accentuated, drawing out the distribution of these ions, increasing the zeta-potential, and causing a large voltage change as the interface passes an electrode. Perhaps this is also related to the extreme streaming potential associated with a gas kick in oil wells (Sprunt, et al., 1994).

A few milliliters of methylene chloride or TCE, both of which are soluble in water, added to the apparatus during percolation of distilled water causes the potential to drop immediately. Pouring enough to create a slug of pollutant that flows through the apparatus causes the potential to drop further and then remain constant as long as TCE is pushing water out the bottom of the apparatus. As the effluent switches from water to TCE, the natural-potential again declines. The entire decline in potential involved here is only 12 to 13mV, but it is sufficient at times to reverse the polarity of the potential.

Methanol, which is completely soluble in water, causes a dramatic decline of potential by 46mV as it percolates through an apparatus saturated with distilled water. The potential remains constant as long as methanol displaces water. Once the apparatus becomes saturated with methanol, the potential is only a few millivolts negative during each successive percolation. Eventually, the methanol-soaked apparatus

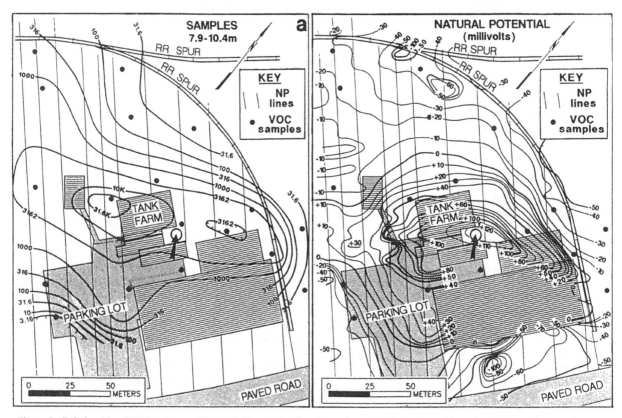

Figure 9. Relationship of VOC patterns of Geoprobe® samples (*a*) to natural-potential contours at the surface in Knox County, Tennessee (*b*) The TCE concentrations occur in groundwater within the residuum above Lenoir Limestone. The logarithmic concentrations were derived from the eighteen sample sites shown by the solid circles. The NP millivolt contours were generated from 424 stations in soil at 3.8-meter intervals, or closer, along the traverses shown. Station intervals in pavement were 7.62 meters. The most probable source of the contamination was a previously-removed underground tank at the location of the open circle (arrow).

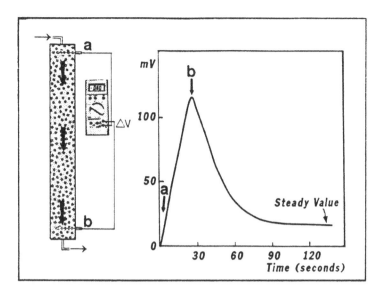

Figure 10. A typical voltage transient arising from the percolation of water through a limestone-packed column. The voltage rises suddenly (a) as water pushes air toward the outlet, and then decreases rapidly as the water slug passes the lower electrode (b). Eventually the voltage approaches a steady value of about +20 millivolts, depending on water chemistry, so long as the percolation continues.

becomes so resistive that we cannot measure a stable potential value. The correlation between solubility and enhancement of potential suggest that these soluble organic liquids augment the streaming potential through diffusion.

DISCUSSION

Laboratory studies as well as field observations of natural-potential anomalies coincident with contaminant plumes of organic liquid show that organic liquids affect the streaming potential. This suggests that the presence of organic contaminants alter the electrical structure of the water/rock interface. How does this occur? We can organize possible mechanisms into four categories, and illustrate each with a particular example.

First, immiscible liquids in contact with one another, and which share an ion species, form an electrical interface like that of a metal placed in an aqueous solution (Koryta, 1987). For example, TCE is a common pollutant. TCE often has ammonium carbonate added to sop-up the HCl released as it decomposes. Karst groundwater contains carbonate ion, and the two phases possibly share a common ion. The interface between TCE and carbonate groundwater then behaves like the interface between a metal and a solution containing a salt of that metal. Nernst's equation provides a way to calculate the interfacial potential. Alternatively, the organic liquid and aqueous solution may share no common ion and the interface between them may be quite impermeable to the passage of ions in general. The interface between groundwater and fuels or oils is an example of this. There is yet a potential difference even though it is not stable and one cannot calculate or measure it. It behaves like the zeta-potential in an aqueous solution in contact with rock, in which ions are adsorbed and polar molecules are oriented at the interface. There may be a separate zeta-potential on each side of the interface, both of which can be sheared by a groundwater flow, and both of which contribute to a streaming potential. This is essentially what happens when oil displaces brine in sediments surrounding a borehole. In contrast to what happens at rock/solution interfaces, the interface between immiscible liquids maintains a diffused-ion distribution even in highly concentrated solutions (Vanysek, 1996)

Second, organic liquids often have an affinity for the solid matrix where they become a separate adsorbed phase. This is true whether or not the organic liquid is soluble in water. Adsorption has two effects. It alters the zeta-potential typical of the interface between groundwater and rock. It also prevents water from wetting the rock. Thus, adsorbed organic liquid alters the streaming potential directly through the cross-coupling coefficient, and through its effect on the primary flow. Moreover a volatile pollutant, like methylene chloride, evaporates within the groundwater and produces bubbles of vapor phase. As Sprunt, et al. (1994) found, a gas phase traveling through rock pores causes a large potential kick.

Third, when an organic liquid is soluble in water, it travels with groundwater surrounded by a diffused front consisting of more mobile ions and species traveling ahead of less mobile ones. Our laboratory experiments have shown that the diffusion potential might be quite large. Methanol provided the best example of this behavior in our laboratory tests, but TCE behaves similarly. TCE has a solubility of one part per thousand, and is alternatively considered insoluble in water or highly soluble depending on one's point of view[7]. TCE is a dense hydrocarbon (having a specific gravity of 1.49) which settles into low places in an aquifer collecting in a pool and providing a steady source

[7] Consider the varying characterization of solubility in *Environmental Toxicology Data Sheet* by Alison Hardy at http:www.science .mcmaster.ca/Biology/4S03/AH2.htm" or in the *Merck Index*, 1983. 10th Edition. Merck and Co.. Inc.

of pollution as fresh groundwater flows past. Thus, as long as the pool of contaminant TCE lasts, it sets up a static diffusion front and produces a natural-potential anomaly.

Finally, imagine that contaminant disperses into droplets or vapor bubbles as it flows though the subsurface. As these droplets break free, they possibly acquire a surface charge, which they carry in their travels, and which makes them act like ions of very low mobility. Along its flow-line, a droplet either accepts or rejects ions to maintain equilibrium with its surroundings. This may also play some role in the gas kick on borehole SP logs and in experimental results (Sprunt, et al., 1994).

We performed several experiments with a falling droplet electrometer to measure this behavior. Droplets of aqueous liquid breaking free and flowing through TCE/methylene chloride carry with them an average charge amounting to 2.5×10^{-10} coulombs per square meter of surface. This is a small amount of charge, but within a karst environment, where flow takes place in fissures and open conduits, it may be a more important charge-separating mechanism than the normal streaming potential itself, which is confined to the rock walls and operates on a relatively small surface area.

CONCLUSIONS

The flow of groundwater within and around caves produces natural-potential anomalies, which aid in tracing the course of conduits and fissures, and facilitates the placing of exploratory boreholes. Field observations, and experimental and theoretical results, suggest that organic pollutants in groundwater have a substantial effect on surface NP anomalies as well. In this regard observations of natural potentials can provide a valuable tool for detecting the presence of pollutants and for tracing their subsequent movement.

ACKNOWLEDGMENTS

We thank E. Michael Riggins of Environmental Strategies Corp. and Michael Singer of CH2M Hill, Inc. for their assistance in projects associated with VOC anomalies. In addition, we thank those property owners who granted access to their lands, including Peter Prebble (Glenwood Caverns, Colorado), Jewel Cave National Monument, South Dakota, and John Gass (Honey Creek Water Cave, Texas).

REFERENCES

Adam, N.K., 1968. *The physics and chemistry of surfaces*. Dover, Inc., New York. 436p.

Binder, G.J. & Cermak, J.E., 1961. Streaming potential fluctuations produced by turbulence. *Physics of Fluids*,6: 1192-1193.

Bogolslovsky, V.A. & Ogilvy, A., 1972. The study of streaming potentials on fissured media models. *Geophysical Prospecting*, 20 (1): 109-117.

Corwin, R.F., 1990. The self-potential method for environmental and engineering applications. *Society of Exploration Geophysicists*: Investigations in Geophysics No. 5: Geotechnical and Environmental Geophysics: Vol. 1: 127-145.

Corwin, R.F. & Hoover, D.B., 1979. The self-potential method in geothermal exploration. *Geophysics*, 44: 226-245.

Crowley, J. M., 1986. *Fundamentals of Applied Electrostatics*. Wiley Interscience, 255p.

Dakhnov, V.N., 1962. Geophysical well logging. *Quarterly of the Colorado School of Mines*, 57 (2): 287-361.

Ernstson, K. & Scherer, H.U., 1986. Self-potential variations with time and their relation to hydrogeologic and meteorological parameters. *Geophysics*, 51: 1967-1977.

Ford, D. & Williams, P., 1992. *Karst Geomorphology and Hydrology*. Chapman & Hall, New York, 601p.

Helmholtz, H., 1879. Studien über electrische Grenzschichten. *Annalen der Physik und Chemie*, new series 7 (7): 338-383.

Ishido T. & Mizutani, H., 1981. Experimental and theoretical basis of electrokinetic phenomena in rock-water systems and its application to geophysics. *Journal of Geophysical Research*, 86: 1763-1775.

Kilty, K.T., 1984. On the origin and interpretation of self-potential anomalies. *Geophysical Prospecting*, 32: 51-62.

Kilty, K.T. & Lange, A.L., 1991. Electrochemistry of natural-potential processes in karst. *U.S. Environmental Protection Agency and Association of Groundwater Scientists and Engineers: Third Conference on Hydrology Ecology Monitoring and Management of Ground Water in Karst Terranes, Proceedings*: 163-177.

Kilty, K.T. & Lange, A.L., 2000. Groundwater Pollution, Fracture Flow, and Natural Electrical Potential (Abstract). *American Association for the Advancement of Science, Pacific Division: Proceedings*, 19 Part 1: 50-51. June 11, 2000.

Klinkenburg, A., & Van der Minne, J.L., 1958. *Electrostatics in the Petroleum Industry*. Royal Dutch/Shell Research and Development Report. Elsevier Publ. Co., New York. 191p.

Koryta, J., 1987. In: *The Interface Structure and Electrochemical Processes at the Boundary between two Immiscible Liquids*. V. E. Kazarinov. Springer-Verlag, New York. 246p.

Krajew. A.P., 1957. *Grundlagen der Geoelektrik*. VEB Verlag Technik. Berlin, DDR. 357p.

Lange, A.L., 1999. Geophysical studies at Kartchner Caverns State Park, Arizona. *Journal of Cave and Karst Studies*, 61 (2): 68-72.

Lange, A.L. & Barner, W.L., 1995. Application of the natural electric field for detecting karst conduits on Guam. *Fifth Multidisciplinary Conference on Sinkholes and the Engineering and Environmental Impacts of Karst*: 425-441.

Lange, A.L. & Kilty, K.T., 1991. Natural-potential responses of karst systems at the ground surface. *U.S. Environmental Protection Agency and Association of Groundwater Scientists and Engineers: Third Conference on Hydrology Ecology Monitoring and Management of Ground Water in Karst Terranes, Proceedings*: 179-196.

Lange, A.L. & Quinlan, J. F., 1988. Mapping caves from the surface of karst terranes by the natural-potential method. *National Water Well Association: Second Conference on Environmental Problems in Karst Terranes and their Solutions, Proceedings*: 369-390.

Lange, A.L. & Wiles, M., 1990. Mapping Jewel Cave from the surface! *Park Science*, 11 (2): 6-7.

Nash, M., Atekwana, E. & Sauck, W., (1997). Geophysical investigation of anomalous conductivity at a hydrocarbon contaminated site. *Symposium on the Application of Geophysics to Engineering and Environmental Problems, Proceedings 1997*: 675-683. Reno, NV.

Nourbehecht, B., 1963. *Irreversible Thermodynamic Effects in Inhomogeneous Media and their Application to certain Geoelectric Problems*. Massachusetts Institute of Technology, PhD Dissertation. 121p.

Quarto, R & Schiavone, D., 1996. Detection of cavities by the self-potential method. *First Break*, 14: 419-431.

Quincke, G., 1859. Über eine neue Art elektrischer Ströme. *Annalen der Physik*, ser. 2, 105: 1-48.

Rivel , A, Pezard, P.A. & Glover, P.W.J., 1999. Streaming potential in porous media 1: Theory of the zeta-potential, *Journal of Geophysical Research*, 104: 20,021-20,231.

Swartzendruber, D. & Gairon, S., 1975. Electrical potentials during water entry into an air-dry mixture of sand and kaolinite. *Soil Science*, 120: 407-411.

Sill, W.R., 1983. Self-potential modeling from primary flows. *Geophysics*, 48: 76-86.

Somasundaran, P. & Agar, G. E., 1967. The zero point charge of calcite. *Journal of Colloid and Interface Science*, 24: 433-440.

Sprunt, E.S., Mercer, T. B. & Djabbarah, N.F., 1994. Streaming potential from multiphase flow. *Geophysics*, 59: 707-711.

Stierman, D.J., 1984. Electrical methods of detecting contaminated groundwater at the Stringfellow Waste Disposal site, Riverside County, California. *Environmental & Ground Water Science*, 6: 11-20.

Vanysek, P., 1996. *Modern Techniques in Electroanalysis, Chemical Analysis*, series 5 139. Wiley & Sons, New York. 369p.

Zhou, W., Beck, B.F. & Stephenson, J. B., 1999. Investigation of groundwater flow in karst areas using component separation of natural-potential measurements. *Environmental Geology*, 37: 19-25.

Geotechnical and Environmental Applications of Karst Geology and Hydrology, Beck & Herring (eds)
© 2001 Taylor & Francis, ISBN 90 5809 190 2

Integrated high-resolution geophysical investigations as potential tools for water resource investigations in karst terrain

RICHARD J.MCGRATH & PETER STYLES   School of Earth Sciences & Geography, Keele University, Staffordshire, ST5 513G, UK, r.j.megrath@esci.keele.ac.uk

EWAN THOMAS   Microsearch Ltd., Surrey, GU21 3LF, UK

SIMON NEALE   Environment Agency (Wales), Cardiff, CF3 0EY, Wales, UK

ABSTRACT

Karstic aquifers can be particularly vulnerable to both pollution from surface activities and the impacts of abstraction. This is due to the enhanced vertical and lateral flowpaths, resulting from the dissolution of carbonate facies by rainfall. Often this process results in the development of voids that can range in size from several centimetres to several tens of metres.

The uncertainties that are presented by the by the presence of water carrying conduits in karst areas can be significantly reduced by the use of new techniques for detecting and delineating underground conduits. Of the available geophysics techniques that may allow for the identification of such features, microgravity and resistivity imaging are likely to be the most successful. Microgravity surveying has the potential to identify the presence and location of such voids, and with the integration of electrical tomographic work, can provide "targets" for the location of monitoring boreholes. Whilst these techniques are intensive and may not be cost effective on a regional scale, they do have the potential to provide high-resolution data over smaller areas which would be invaluable to any site or area specific hydrogeological assessment.

INTRODUCTION

The area studied in this paper is the Schwyll catchment area in the Carboniferous Limestone near Bridgend, South Wales, UK, (Figure 1). Schwyll is a large resurgence used as a public water supply that discharges to the River Ogmore. The conjectural catchment of the source is 25km^2 and includes both unconfined and confined areas (both drift and consolidated Jurassic sediments). There are known water filled conduits within the catchment, the locations of which have been surveyed by cave divers.

In gravity surveying, subsurface geology is investigated on the basis of variations in the Earth's gravitational field generated by differences of density between subsurface rocks. The underlying concept is one of a subsurface body or cavity, which is of differing density from its surroundings. A causative body represents a subsurface zone of anomalous mass and causes a very localised perturbation in the gravitational field, i.e., a gravity anomaly. Cavities show a minute (typically a few parts in a billion of g) reduction in the gravitational acceleration over them because of the missing mass associated with the void. In order to detect voids or cavities, very high precision instruments and meticulous field techniques are required. A Scintrex CG-3M Autograv Microgravity Meter, with a 1 microgal sensitivity was used for the investigation in this paper.

In the resistivity method, artificially generated electric currents are introduced into the ground and the resulting potential differences are measured at the surface. Deviations from the pattern of potential differences expected from homogeneous ground provide information on the form and electrical properties of subsurface inhomogeneties. Electrical tomography involves measuring a series of constant separation traverses with the electrode spacing being increased with each successive traverse. Since increasing separation leads to information being obtained from greater depth, the measured apparent resistivities may be used to construct vertical contoured sections displaying the variation of resistivity both laterally and vertically over the section. Recently, various techniques of electrical tomography have been refined and fast, efficient commercial systems are now available. New inversion algorithms produce electrical images that accurately model 2-D and 3-D subsurfaces. A Campus Geophysical Geopulse general-purpose earth Resistance Meter was used to profile the subsurface for the data included in this paper.

METHODOLOGY

Microgravity

The station spacing used in a gravity survey vary greatly depending on the size and depth of the cavities that are to be detected. At Schwyll, cave divers estimate the depth of the karstic conduits and channels to be between 2 and 10 metres. For this study a station spacing of 2.5m was chosen. Multiple traverses with closely spaced gravity stations, i.e., grids, result in useful spatial redundancy and improved data accuracy, which can be used to separate the anomaly caused by cavities, from the geological or topographical background noise (Bishop et al 1997). During the gravity survey the gravimeter was read at a base station at hourly intervals in order to determine instrument drift. At each survey station, the time was internally recorded in the gravimeter and the reading was automatically corrected for earth tide variations.

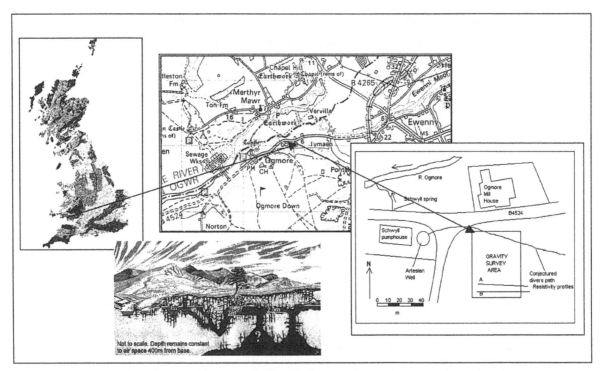

Figure 1: Map showing the location of the gravity survey and resistivity profiles, including the conjectured subsurface divers path.

Resistivity

Modern field systems use a multicore cable to which 50 or more electrodes are connected at equal spacing. With this system any electrode arrangement can be used but more often with these shallow karstic studies the Wenner array is chosen, for the reason that it provides an intermediate depth range and resolution and is moderately sensitive to near surface lateral effects.

A pseudosection is constructed by plotting each apparent resistivity value at a central point below the four electrodes at a depth equal to that of the median depth of investigation of the array. A nominal depth of investigation for the Wenner configuration equal to half of the maximum spacing between the operative electrodes is assumed here (Edwards 1977). Investigations at Schwyll targeted areas where the microgravity had revealed the existence of subterranean cavities, thus allowing a sensible choice of electrode spacing of between 1 and 2 metres, resulting in a depth of 7-14 metres with 50 electrodes.

VOID DETECTION

Microgravity

The presence of a cavity produces a reduction in the value of gravity, known as a negative gravity anomaly. Air-filled cavities provide the largest anomalies for gravity surveying, because of the complete absence of material in the target. Water-filled cavities provide anomalies 60% that of the same cavity containing air, and rubble filled cavities about 40% that of air. So an air-filled cavity has a density contrast, $\Delta\rho \approx$ -2.5gcm^{-3}, for water-filled, $\Delta\rho \approx$ -1.5gcm^{-3} and for rubble filled, $\Delta\rho \approx$ -1.0gcm^{-3}.

This simplified picture of an isolated cavity is not usually the case. The rock surrounding any cavity is often disturbed and associated fracturing may extend for two or more diameters away from the cavity (Daniels 1988). This is particularly true in karstic environments where the associated dissolution of the rock enlarges faults and fractures etc. This secondary effect is termed the 'halo' effect and normally serves to increase the effective target size.

Resistivity

Most earth materials are good insulators, i.e., they do not conduct electricity very well. The current that is artificially fed into the earth is transported via interstitial water. Rock has very few and very small pore spaces and so the resistivity of rock is very high, whereas infill rubble or clay have more pore spaces and therefore retain more water and have lower resistivities. Water filled cavities therefore have even lower resistivities.

The resolution of resistivity data decreases with depth due to the number of measured points decreasing with depth as a function of electrode spacing and electrode configuration. The normal soil/bedrock resistivity pseudosection would generally have increased resistivity with depth. At greater depth the pore spaces in rock are increasingly smaller and therefore hold less conductive water. When there is a change in rock type or a void or cavity present, there will be a contrast between the mediums' resistivities. The contrast between limestone and water is large enough to be detected.

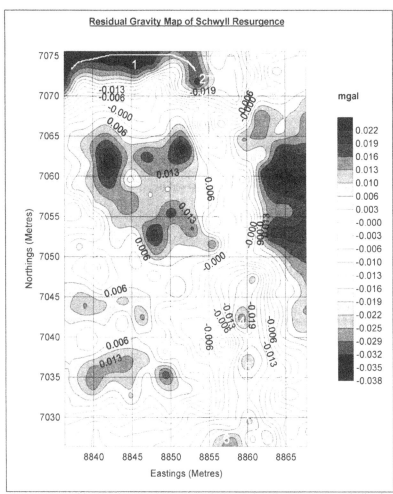

Figure 2: Residual Microgravity map of the Schwyll Resurgence, indicating anomalous areas (1-7) and probable subsurface karst channels and feeder conduits.

DATA INTERPRETATION

Microgravity

An area of land was surveyed immediately adjacent to the Schwyll Pump House (Figure 1) on a gently sloping field with stations set out on a nominal 2.5m grid. The data was reduced, with a Bouguer density of 2.2 gcm^{-3}, a polynomial trend was removed and the data presented as a Residual Bouguer anomaly map (Figure 2). The survey indicated several areas of mass deficiency, where there are low gravity values relative to the background (numbered 1 to 7). Between the numbered anomalies it is possible to infer the presence of interconnecting channels, i.e., semi-linear negative anomalies. Anomalies 2 and 4 show the largest amplitude anomalies of approximately –0.035 to -0.040 mgal, between which there is a North–South trending gravity low of approximately –0.010 to 0.020 mgal, i.e., a possible subsurface karst channel. Between anomalies 3 and 7 the same is true although the anomaly amplitude is not as large, probably due to either the interconnecting channel being deeper or thinner or both. It is conjectured that the cave system that the divers swam through is the anomaly to the North of the survey, i.e., the linear anomaly between 1 and 2. It is clear that the cavers have further exploring to do. With the knowledge from the cave dive survey of the general geometry of the subsurface cavities; position and depth modelling can be carried out using Euler Deconvolution techniques. Initial modelling using this technique provided depth estimates for these gravity anomalies of between 2 and 5 metres. Using Gauss's theorem it is also possible to calculate the mass missing beneath from the gravity anomaly measured over it. If the density contrast is known, then the volume can be easily derived. A mass estimate was derived for a quadrant spanning the anomalies 4, 5, 6 and 7 and gave a current total mass deficiency beneath this area of approximately 140 tonnes. This technique is particularly useful in the context of this study in that it can be used to estimate the total quantity of grout necessary for any remedial work required and subsequent repeated surveys can asses the efficacy of the remediation work by confirming that the anomaly has been eradicated.

For the purposes of this study, two resistivity profiles were carried out perpendicular to known negative gravity anomalies (Figure 1). The electrical image line was centered near the position of the gravity low and 50 electrodes were deployed at 1m spacing so a depth penetration of 7m could be attained. Good ground contacts were achieved and contact resistances of less than 500Ω were measured. Measurements were completely automatic and 392 readings were taken in about 90 minutes.

RES2DINV is a computer program which automatically determines 2-D resistivity models of the subsurface by inverting the data obtained from electrical imaging (Griffiths and Barker 1993). The model consists of a number of rectangular blocks tied in with the arrangement of the datum points in the pseudosection. The distribution and size of the blocks are automatically generated so that the numbers of blocks do not exceed the number of datum points.

A forward modelling subroutine is used to calculate the apparent resistivity values, and a non-linear least-squares optimisation technique is used for the inversion routine (deGroot-Hedlin and Constable 1990, Loke and Barker 1996). The purpose of this program is to determine the resistivities of the rectangular blocks that will produce an apparent resistivity pseudosection that agrees with the actual measurements. The optimisation method tries to reduce the difference between the calculated and measured apparent resistivity values by adjusting the resistivity of the model blocks. The measure of the difference is given by the root-mean-squared (RMS) error.

Figure 3 clearly indicates the presence of a low resistivity area/layer around 2-4m depth which relates very well with gravity anomaly 6. The low resistivity area tends to deepen towards the west which also corresponds well with the lower negative amplitude gravity anomaly seen there. Figure 4 shows resistivity anomalies which correspond well with gravity anomalies 5 and 7. The low resistivity areas are smaller, which explains why gravity anomalies 5 and 7 have smaller negative amplitudes than anomaly 6.

The depth from the electrical resistivity section can then be used as an independent constraint allowing inversion of the microgravity data to give thickness of the underground cave system. This requires assumptions to be made concerning the likely depth of the median plane of the cavity and density contrast between the surrounding rock mass and the cavity. The depth can be constrained at c 3 meters because of our electrical resistivity data and the density contrast has been assumed to be -1.5 gcm^{-3} (i.e., water filled cavity). The technique makes a first guess of cavity thickness based on a simple Bouguer slab formula and then iteratively adjusts the thickness at each grid node by comparing calculated gravity with observed gravity and scaling the thickness accordingly. This generally converges within a few iterations to a very good fit between the observed and calculated anomalies. The calculated anomaly from the inversion is shown in Figure 5a. Figures 5b and 5c show a contour map and three-D model of cavity thickness across the area based on this inversion. The cavity appears to be continuous across the surveyed area and to have several sub-branches. Some of these would be unlikely to be passable to cave-divers precluding physical inspection of the whole of the system. This technique which can identify and define the configuration of cave systems shows good potential for hydrogeological assessments in Karstic environments.

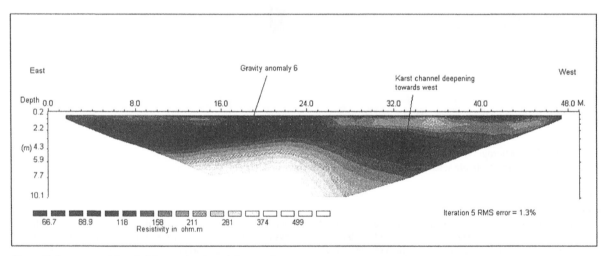

Figure 3: Inverse model resistivity section A, at 1.0m spacing.

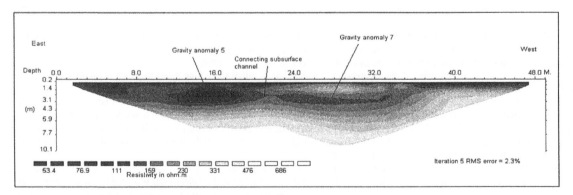

Figure 4: Inverse model resistivity section B, at 1.0m spacing.

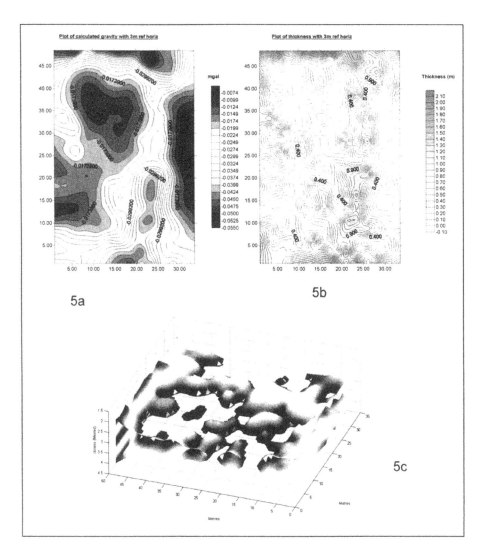

Figure 5: The figure on the top left (5a) shows the calculated microgravity field, which closely resembles the observed data of Figure 2. Top right (5b) is a contour map of the thickness of the cavity from the inversion and the lower image (5c) is a three-dimensional perspective map of cavity thickness.

CONCLUSIONS

This study demonstrates how integrated geophysical methods provide an alternative to grid (or random) drilling for the detection and mapping of subsurface karst topography. The microgravity technique is shown to be a very effective and adaptable non-destructive tool for the detection of naturally occurring cavities. Multiple traverses with closely spaced gravity stations, i.e., grids, result in useful spatial redundancy and improved data accuracy, that can be used to separate the anomalies caused by an ensemble of cavities, from the geological or topographical background noise and permit the utilisation of inversion techniques.

Microgravity surveying has been shown to be invaluable for targeting anomalous zones following initial resistivity tomography. The resistivity results can be used to further constrain the location and size of the cavities to enable robust inversions of the data for 3-D inversion to give detailed configurations of the sub-surface cave system. In the context of this study it is clear that these geophysical techniques are invaluable to site specific hydrogeological assessments in Karstic areas. By integrating the use of these geophysical methods with groundwater tracing methods the high permeability flow paths present in karst area can be identified, located and characterised with increased accuracy. In turn monitoring station locations (boreholes) can be placed with more confidence than previously.

Areas for future exploration with respect to the detection of subsurface karst flowpaths include the integration of ground penetrating radar techniques with microgravity and electrical tomography. Radar is capable of three-dimensional definition of stratal surfaces and underground cavities. In the right circumstances (sites free from clay and saline groundwater), 3-D georadar can give unparalleled images of the shallow subsurface and joint inversion methods can be developed which fully exploit the complementary capabilities of these high-resolution geophysical techniques.

ACKNOWLEDGMENTS

We would like to thank the Environment Agency (Wales) for providing the knowledge about the Schwyll site, and R.J. McGrath would like to thank Nicola Wilton for her assistance in the collection of the gravity field data for this study.

DISCLAIMER

The opinions expressed in this paper are those of the authors and may not necessarily reflect the current policy of the Environment Agency.

REFERENCES

Bishop, I., Styles, P., Emsley, S.J. and Ferguson, N.S., 1997, The detection of cavities using the microgravity technique: case histories from mining and karstic environments: Geological Society Engineering Geology Special Publication No. 12, pp. 153-166

Daniels, J., 1988, Locating caves, tunnels and mines. Geophysics: The Leading Edge of Exploration, Vol. 7; pp32-37.

DeGroot-Hedlin, C. and Constable, S., 1990, Occam's inversion to generate smooth, two-dimensional models from magnetotelluric data: GEOPHYSICS, Vol. 55; pp. 1613-1624.

Edwards, L.S., 1977, A modified pseudosection for resistivity and IP: GEOPHYSICS, Vol. 42, No.5; pp. 1020-1036.

Griffiths, D.H. and Barker, R.D., 1993, Two-dimensional resistivity imaging and modelling in areas of complex geology: Journal of Applied Geophysics, Vol. 29; pp. 211-226.

Loke, M.H. and Barker, R.D., 1996, Rapid least-squares inversion of apparent resistivity pseudosections by a quasi-Newton method: Geophysical Prospecting, Vol. 44; pp. 131-152

Geotechnical and Environmental Applications of Karst Geology and Hydrology, Beck & Herring (eds)
© *2001 Taylor & Francis, ISBN 90 5809 190 2*

Application of bistatic low frequency GPR for mapping karst features and bedrock topography

MICHAEL S.ROARK & DOUGLAS W.LAMBERT   Geotechnology, Inc., St. Louis, MO 63146, USA,
msr@geotechnology.com

ABSTRACT

The St. Louis Metropolitan Sewer District (MSD) is planning the construction of a relief sewer in the vicinity of Grand Avenue and Bates Street in south St. Louis, Missouri. The project would include construction of a cast-in-place concrete, 7-foot diameter, machine-bored tunnel generally along Bates Street that terminates at the Mississippi River. The invert of the tunnel is expected to be approximately 55 feet below surface level (bsl). Due to the proximity of karst features to the survey area, MSD was concerned that bedrock depressions and solution features may be present at the depth of the proposed sewer.

A low-frequency bistatic ground penetrating radar (GPR) survey was conducted along a 1,290-foot section of Bates Street from South Grand Boulevard to Colorado Avenue. The GPR survey was conducted in an attempt to image karst features and map the bedrock topography beneath the street. The antenna dipoles were mounted on a rolling sled and advanced along the survey line. Discrete readings were collected at a center frequency of 40 MHz and were stacked 64 times every 10 feet along the survey line. The processed and interpreted data were constrained at the beginning of the line with an existing borehole. According to the boring log, the depth to bedrock at the beginning of the survey line was 48 feet bsl. The GPR data suggested the presence of a depression near the end of the survey line. Subsequent drilling in the suspect area encountered bedrock at 68 feet bsl.

INTRODUCTION

The St. Louis Metropolitan Sewer District (MSD) is planning the construction of a relief sewer in the vicinity of Grand Avenue and Bates Street in south St. Louis, Missouri. The project would include construction of a cast-in-place concrete, 7-foot diameter, machine bored tunnel generally along Bates Street between the Union Pacific Railroad and the Mississippi River. The invert of the tunnel is expected to be approximately 55 feet bsl. Due to the proximity of karst features to the survey area, MSD is concerned that bedrock depressions and solution features may be present at the depth of the proposed sewer.

Geotechnology, Inc. was contracted to conduct a GPR survey along a section of Bates Street between South Grand Avenue and Colorado Street in South St. Louis, Missouri. The objectives of the survey were to attempt to image karst features beneath the survey line and to profile the top of bedrock. Previous borings along the proposed tunnel path encountered the top of bedrock between 30 and 73.5 feet below surface level. In order to collect data through clay-rich overburden to these depths, bistatic low-frequency GPR was used to collect the data.

A 1,290-foot survey was conducted along Bates Street between South Grand Avenue and Colorado Street. A confirmation boring was drilled following the survey. The locations of the survey line and borings are shown on Figure 1.

GEOLOGIC BACKGROUND

The stratigraphic succession of the survey area in south St. Louis consists of limestone and dolomite of the Salem Formation and St. Louis Limestone of the Meramecian Series, Mississippian System.

The Salem Formation is comprised of cross-stratified, fossiliferous limestone and dolomite (Thompson, 1995). A "cannonball" chert zone is located near the top of the formation and indicates the contact between the Salem Formation and St. Louis Limestone. The formation ranges between 100 and 160 feet thick and becomes increasingly dolomitic in the St. Louis area. Several types of fossils are found in the formation including blastoids, crinoids, echinoids, and bryozoa (Thompson, 1995).

The St. Louis Limestone is comprised of fine to medium grained, medium- to massive-bedded limestone and dolomite (Thompson, 1995). Shale partings commonly occur between beds. The thickness of the St. Louis Limestone is generally less than 50 feet in the St. Louis area.

Karst features are common in south St. Louis and include large caves, springs and sinkholes. Two well-known caves, Cherokee and Cliff Cave, are located within 2 miles of the survey area. The majority of the karst features in the south St. Louis area are situated in the St. Louis Limestone (Lambert *et al.*, 1998).

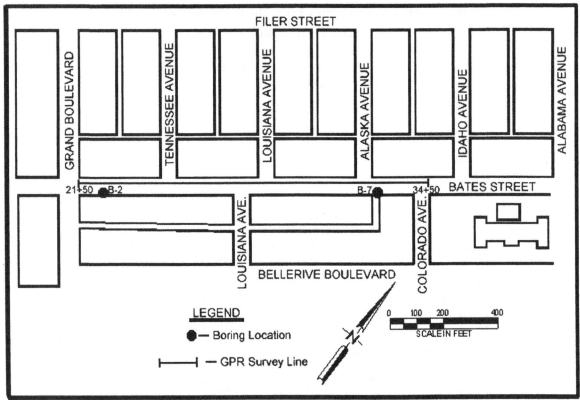

Figure 1. Map of the survey area, line and boring location.

BISTATIC GROUND PENETRATING RADAR

Bistatic GPR differs from the traditional monostatic GPR method in that the transmitting and receiving antennae are separated by a specific distance. The data can be acquired continuously at a fixed offset, at discrete intervals at a fixed offset, or in common depth point mode. The latter is most similar to the seismic reflection method in that common shot gathers may be acquired.

Radar propagation velocity is affected by the conductivity (σ) and dielectric constant (ε) of the subsurface. When the radar wavefront encounters an interface of differing electrical properties, some of the energy is reflected back to the surface. The GPR system measures the travel time from transmission to reception, and a two-dimensional profile is generated. The bistatic GPR method is illustrated on Figure 2.

The effective depth of penetration is a function of the moisture content, conductivity and clay content of the subsurface materials. Soils with high moisture and clay content rapidly attenuate the radar energy, thus greatly reducing the depth of penetration. Lower frequency bistatic GPR systems allow the collection of deeper data in clay-rich environments. Vertical resolution, however, is decreased as the radar frequency is lowered. Depending on site-specific conditions, depths of greater than 100 feet may be attained with a low-frequency GPR system.

DATA ACQUISITION AND PROCESSING

The data were collected using a Geophysical Survey Systems, Inc. (GSSI) SIR-2 System with Model 3200MLF Bistatic Antennae. The antenna dipoles are unshielded and allow the user to vary the center frequency between 16 and 80 MHz by changing the dipole length. Lower frequencies and larger antenna offsets provide deeper radar penetration. Stacking the data at each shot location provided a greater signal-to-noise ratio.

The GPR data were acquired in the discrete-point, fixed offset mode. A single radar trace was recorded every 10 feet and stacked 64 times at each station within a 1,000-nanosecond (ns) time window. The antenna dipoles were mounted on a rolling wooden sled to allow rapid data collection. A center frequency of 40 MHz was used to provide the necessary resolutions required at the target depths. A fixed offset of 8 feet was used for the entire survey.

The recorded data were downloaded to a PC and processed with GSSI's RADAN for Windows. The processing steps included amplitude normalization, vertical bandpass filtering, horizontal filtering and horizontal stretching. Due to the relatively planar surface of Bates Street, elevation corrections were not applied to the data.

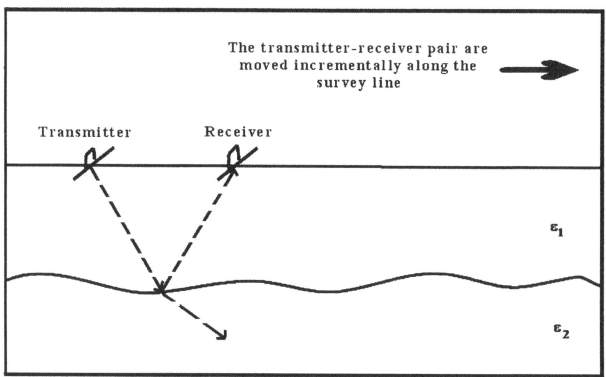

Figure 2. Bistatic GPR showing reflected and refracted raypaths.

DATA INTERPRETATION

Presented in Figures 3 and 4 are non-interpreted and interpreted versions of the GPR data, respectively. Depth control was established near Station 22+50 using Boring B-2. The top of bedrock was encountered at 48 feet bsl in Boring B-2. Unconsolidated sediments above bedrock consisted of clay, silty clay, sand and gravel. Based on a depth of 48 feet bsl for the top of bedrock, the relative average dielectric constant for the sediment package was calculated to be 42.

Several shallow reflectors are apparent on the GPR profile between 100 and 350 ns. The noisy region between Stations 21+50 and 23+00 is due to shallow utilities beneath the street. The shallow planar reflections around 190 ns between Stations 21+50 and 30+50 represent interbedded clays, silts, sands and gravels. Reflections from the water table, shallow sediment and top of bedrock are indicated on Figure 4 as "Event A," "Event B" and "Event C," respectively.

The top of bedrock reflection is characterized by a semi-continuous, high-amplitude event at 640 ns near Station 23+50 and at 730 ns near Station 32+50. Due to the highly weathered nature of the limestone in the survey area, the bedrock reflection is not continuous. Based on the GPR data, the top of bedrock falls from approximately 48 feet bsl near Station 29+00 to 62 feet bsl near Station 33+00. Confirmation drilling near Station 32+30 (Boring B-7) encountered bedrock at 68 feet bsl. Larger amplitude events below the interpreted bedrock reflection may be attributed to voids, fractures and vugs in the limestone. "Event B" becomes disrupted between Stations 30+50 and 34+50. This phenomenon is interpreted as soil piping due to karst collapse within the underlying limestone.

SUMMARY

A bistatic GPR survey was conducted along a 1,290-foot section of Bates Street in South St. Louis, Missouri for the purpose of imaging bedrock depressions and other karst features in the area of a proposed tunnel. A karst-related bedrock depression was identified on the GPR data near the end of the survey line, and was subsequently confirmed by drilling. Low frequency bistatic ground penetrating radar was effective in imaging bedrock at depths greater than 50 feet in a clay-rich environment. The flexibility of the low frequency GPR system allowed the acquisition of data greater than the target depth with the vertical resolution required for the project.

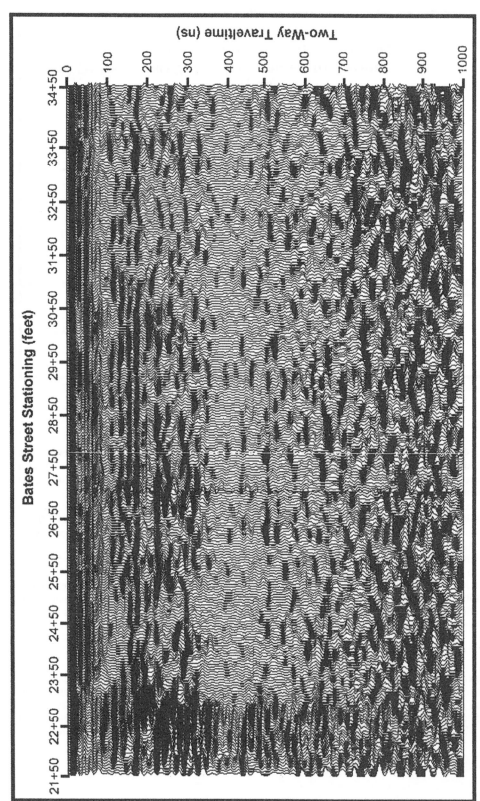

Figure 3. Uninterpreted GPR profile.

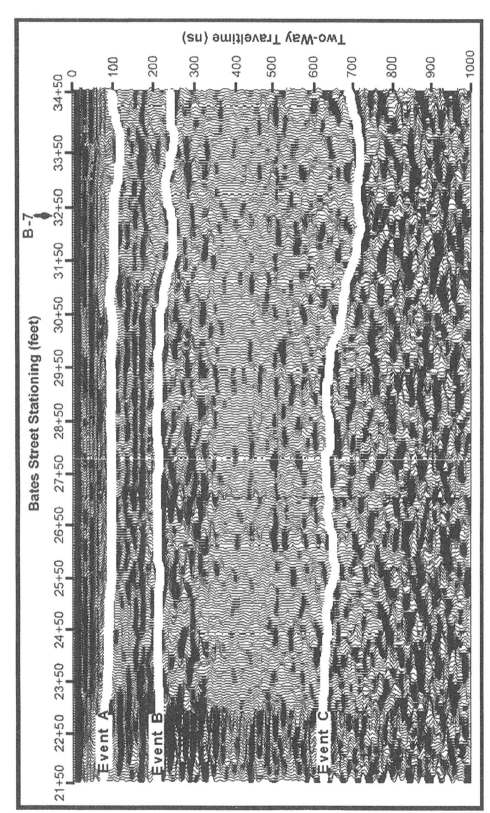

Figure 4. Interpreted GPR profile.

REFERENCES

Lambert, D., and Kreuger, D., 1998, Karst Geology in the Area of South St. Louis, Missouri: Association of Engineering Geologists-St. Louis Section, 1-16.

Thompson, T.L., 1995, The Stratigraphic Succession in Missouri: Missouri Department of Natural Resources-Division of Geology and Land Survey, 83-85.

Geotechnical and Environmental Applications of Karst Geology and Hydrology, Beck & Herring (eds)
© 2001 Taylor & Francis, ISBN 90 5809 190 2

The use of microgravity for the characterisation of karstic cavities on Grand Bahama, Bahamas

PETER STYLES   School of Earth Sciences & Geography, Keele University, Keele, Staffs ST5 5BG, UK, p.styles@esci.keele.ac.uk

EWAN THOMAS   Microsearch Ltd., Surrey, GU21 3LF, UK

ABSTRACT

The Bahamas have significant cave systems which can be extremely extensive at shallow depths. These cause major problems for engineering projects on these rock types. Cavities were encountered during investigation for construction of a container terminal at Freeport, Grand Bahama. Microgravity surveying has been successfully used to delineate a large cave system lying beneath the site. Euler Deconvolution and automated inversion of the data have allowed the detailed configuration of the cave system to be determined and this was verified by drilling and cave diver surveys. Mass deficiency calculations carried out using the microgravity map were used to determine grout takes for remediation prior to construction.

INTRODUCTION

The Bahamas are formed on a Carbonate Platform composed of a variety of carbonate sediment types including coralgal and Oolitic Limestones. The exposed rocks on the Islands have been deposited in the relatively recent geological past, (Pleistocene and Holocene). Although potential evaporation at *circa* 1250 to 1375 mm/yr exceeds annual precipitation at 1000 to 1250 mm/yr the distribution of rainfall leads to a temporary freshwater surplus during May to January. This leads to a freshwater lens overlying saline waters existing beneath the islands but additionally saline lakes and tidal linkage have a significant effect on hydrogeological pathways. The caves fall into four main types following the classification of Wilson, Mylroie and Carew (1995) of the modes of cave formation and the statistics of cave occurrence on San Salvador. These are:

1.   Banana Holes: Oval chambers, generally less than 12m diameter and 4 m deep, They form at the top-surface of a fresh-water lens where vadose waters (occurring above the water table, meet the phreatic flows within the lens. They are tabular in aspect with a depth/width ratio of less than one.
2.   Flank Margin Caves: These can be very large (100m by 17m by 5m on San Salvador) and are considered to form at the point of maximum dissolution at the edge of the freshwater lens. They generally occur on the margins of Pleistocene dune ridges.

3.   Pit Caves: These are circular to elliptical shafts with a depth to width ratio greater than one. They range from *ca* 1 metre to *ca* 7 metre in diameter and with depths up to 10 metres. They are relatively recent features formed since the last inter-glacial *ca* 125000 years ago by downward percolation of surface vadose waters and as such should have a surface expression.
4.   Blue Holes: These features for which the Bahamas (especially Andros) are famous are deep water-filled open pits, some of which occur inland but others connect with lagoons or open ocean.

This classification has been defined for San Salvador and there may be variations in cave type and mode of formation from island to island. However, it gives a useful framework in which to evaluate the likely sizes and architecture of caves and cavities to be found on Grand Bahama.

A major container Port is under construction at Freeport, Grand Bahama (Figure 1). During the construction of Phase I, cavities were encountered in the sub-surface during site investigation which showed the presence of two major zones of cavities, an area of *ca* 60m by 75m with depths between

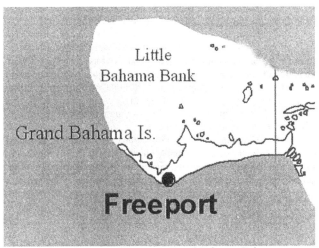

Figure 1: Location map of Freeport, Grand Bahama, Bahamas

12m to - 18m MLW and an area of 40m by 60m> with depths -3m to - 14m MLW respectively. The heights of these features were up to 5m. These features clearly have a flat-lying aspect rather than vertical and as they do not breach the surface cannot be pit-caves. A cavity uncovered within the confines of the site was an oval hole with plan dimensions of 4.5m by 3m and a depth of at least 2m. This is almost a perfect description of a *banana hole* (Wilson et al 1995). A very large N-S trending water-filled cavity of 65m by 25m by 5 m with a water-level at least 2m' above sea-level which is understood to connect with a subterranean cave system was found just off the site. The cave types which are likely to pose a hazard because they occur at depth are the uncollapsed banana holes. In order to assess the likelihood of cavities beneath Phase II of the project we undertook a major microgravity (>6000 stations) survey to detect the presence of cavities and voids.

MICROGRAVITY SURVEYING

The Microgravity technique in the context of civil engineering consists of measuring minute variations in the gravitational pull of the Earth and interpreting the presence of voids and cavities from these readings. Gravity anomalies arising from voids and cavities are superimposed upon much larger variations due to height, latitude and regional geological variations and are virtually undetectable by conventional gravity investigations. The presence of karstic features or mining related cavities in the rock mass often leads to restrictions in land utilization and can pose a variety of problems for both the current and future users of that land. Cavities constitute a hazard to development and redevelopment, and their migration to the surface may seriously damage property or services. Responsible engineering practice requires the positive identification and location of such cavities prior to final engineering design and construction. Underground features of interest consist of natural and man-made cavities, and are generically termed targets. The most common natural targets are solution cavities such as swallow holes, sinkholes and cavern systems in limestone. Man-made cavities, which include mine workings, shafts and tunnels, however, can be even more prevalent (Owen 1983).

Prior to development or re-development, the most common method of site investigation used to identify cavities has been to drill a pattern of boreholes over the target area in an attempt to firstly locate and secondly to define the spatial extent of any cavities. Indirect techniques such as geophysics give a cost effective, non-invasive method of cavity delineation with targeted drilling used as a confirmation rather than a primary search technique. Microgravity surveying has developed considerably over the last 35 years since the early work of Colley (1963) and Arzi (1975). With the development of modern high-resolution equipment, careful field acquisition techniques and sophisticated reduction and analysis, these anomalies can be detected and interpreted. Not only do the isolated anomalies reveal the location of caverns and voids, but they also provide information on their depths and shapes. The 'missing mass' associated with the void can also be calculated and by implementing a post-remediation survey the remediation can be verified. The method is becoming widely used in engineering investigations to detect natural and man-made cavities, and has the significant advantage of leaving the ground completely undisturbed. A cavity is a body of material which has a lower density that of the surrounding material. A cavity may be air filled (i.e. a void), it may be filled with water, alluvium, collapse material or a mixture of all of the above. The existence of this cavity alters the physical state of the strata, and results in a contrast between the cavity and the host stratum which can be detected using suitable geophysical methods e.g. Electrical Resistivity Mapping if the targets have adequate contrasts with their surroundings and are sufficiently large (Cripps et. al. 1988).

The Microgravity survey technique is a powerful cavity location method. The principle of the technique is to locate areas of contrasting density in the sub-surface by collecting surface measurements of the variation in the Earth's gravitational field. Because a cavity represents a mass deficiency a small reduction in the pull of the Earth's gravity is observed over the cavity. This is called a negative gravity anomaly. Although the method is simple in principle, measurement of the minute variations in the gravity field of the earth requires the use of highly sensitive instruments, strict data acquisition procedures and quality controls, careful data reduction and sophisticated digital data analysis techniques to evaluate and interpret the data. Multiple traverses with closely spaced gravity stations or the collection of data on a regular grid result in improved data accuracy that can be used to separate the cavity's effect, from the geological or topographical background effects. The techniques and processing sequences are well described in Butler (1984), Emsley et. Al. (1992), Bishop et. Al. (1997), Daniel and Styles (1997), Cuss and Styles (1999). Conventional site investigation techniques are then employed, as directed by the skillfully interpreted Microgravity results, to verify the presence of areas deficient in mass.

In order to detect a target using Microgravity there must be a difference in density (mass/volume) between the target and it's surroundings. Cavities usually present a significant density contrast with their surroundings. Air- filled cavities offer the largest anomaly condition because of the complete absence of material in the target. Water-filled cavities on the other hand offer an anomaly effect of only 60% that of the same cavity containing air, and rubble or mud-filled cavities only about 40% that of air. Although these are large density contrasts the targets are often of small volume. However, the simplified picture of the solitary cavity described above is not usually the case for natural or man-made cavities. This is because the rock surrounding the cavity, particularly in the case of natural cavities and unsupported mines is often disturbed. The associated fracturing may extend for two or more diameters away from the cavity (Daniels 1988). A similar effect is observed with caverns in karst systems due to the development of sinkholes, collapse features, caves, passages, variable porosity, dissolution, enlarged faults and fractures. Consequently, the "effective" size of the target is dependent not only on the strict volume of the cavity, but also its connectivity, the secondary effects imposed by the cavity on the surrounding rocks which arise from the genesis of that cavity in its host rock. The secondary effect (or cavity enhancement) is often termed a "halo" effect and serves to increase the effective target size, ensuring that the cavity can be indirectly sensed (Chamon and Dobereiner 1988, Bishop et. al. 1994, Patterson et.al. 1995).

Mass Deficiency Calculations

A potentially valuable piece of information about an area which is underlain by cavities, is the amount of material which has been removed or lost from this area, in other words the mass deficiency. It is particularly useful in estimating the total mass deficiencies due to the presence of subsurface voids and can be used to estimate the amount of grout required to fill the voids. This mass deficiency, which is the cause of the negative gravity anomaly, can be estimated by a mathematical theorem known as Gauss' theorem directly from the anomaly map without any prior knowledge of the exact location or nature of the targets.

This technique is particularly useful in that it can be used to estimate the total quantity of grout necessary for any remedial work and repeated surveys carried out post-grouting can assess the efficiency of the remediation work by confirming that the anomaly has been removed.

CAVITY CHARACTERISATION USING MICROGRAVITY SURVEYING

The data were collected using four Scintrex CG3-M gravimeters on a 5 metre grid. The heights were determined using a gridded network of stations established with Total Stations, and the heights of individual points were established using a rotating laser level to an accuracy of 5 mm. Several gravity base stations were used and the meters were returned to these at hourly intervals to determine earth tides and drift correction which were applied in addition to the internal Scintrex corrections. We find this essential to obtain adequate accuracy for this type of survey. The data were corrected for base station offset, earth tides and free-air, and the Bouguer correction was applied to give a raw Bouguer Microgravity map which is shown in Figure 2.

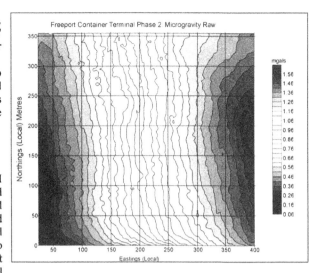

Figure 2: Bouguer Microgravity Map

This map shows a variation from *ca* 0.06 mgals in the southwest to *ca* 1.6 mgals in the northeast (Local Grid). Superimposed upon this regional trend are higher frequency variations and it is these, which contain the information concerning the presence of cavities and voids. Analysis of the data shows that a modified third-order polynomial is the most appropriate. An estimate of the terrain effect of the cut-face has also been made and removed from this map. The anomaly which remains after this process is known as the Residual Microgravity Anomaly and is shown in Figure 3.

The map now reveals a complex set of residual anomalies with amplitudes varying from *ca* +0.05 mgals to -0.06 mgals. The areas shown in blue are known as negative anomalies and are indicative of regions of mass depletion which are associated with cavities or, in some circumstances, distributed void space such as might be found in a particularly vuggy formation. However in order to be detectable a significant amount of mass depletion must take place which may compromise the engineering properties of the strata. The most prominent feature is the zone of negative anomaly which extends across most of the width of the wharf between c.100N and 150N which is linked to a second zone of more limited extent which lies between 200N and 300N. Within this region localized features are present which can be correlated with the position of known cave entrances from cave-dive surveys. Within these regions of general negative anomaly there some very significant localized anomalies, particularly at *ca*

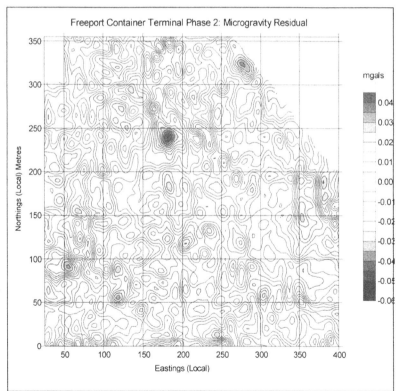

Figure 3: Residual Microgravity Map

(60,90) and at (180,245). At (60,90) a deep low (c-0.04mgals) suggests the presence of a significant cavity which has a width of some 20 metres or more and appears to link south-eastwards and southwards into a more sinuous set of anomalies which appear to

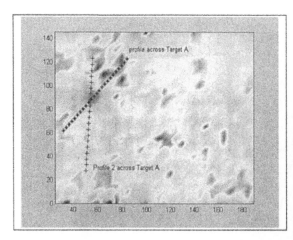

Figure 4: Location of Euler Deconvolution

connect to the cut face at 55E metres and also at 100E metres. The sinuous trend continues to the north via anomalies of *ca* 0.025 mgals amplitude and link to the very deep anomaly at (180,245) of 0.050 mgals amplitude.

AUTOMATED DEPTH ANALYSIS

The position of the bodies which cause the microgravity anomalies can be determined by a process called Euler Deconvolution. For gravity and magnetic fields (and their gradients) it is possible to determine the position of the causative body based on an analysis of the gravity field and the gradients of that field and some constraint on the geometry of the body. We have written specialist software for applying this type of analysis to microgravity data. The profiles shown in Figure 4 have been selected for automated depth analysis as they cross the significant low at 60,90. Profile 1 crosses the anomaly from southwest to northeast and the resulting depth section is shown on Figure 5a. The circles are the calculated position of the top surface of the body made as a window of calculation is moved across the profile. The main cluster of solutions occurs at between 10 and 12 metres beneath the ground surface over a width of some 15 metres. This is the short axis of this feature which clearly extends some 30 metres or more in an east-west direction. An additional Euler Deconvolution analysis has been made (Profile 2) shown in Figure 5b which crosses anomaly 60,90 in a north-south direction. This confirms the depth of the body causing associated anomaly to be at *ca* 10 metres with a shallower body near by.

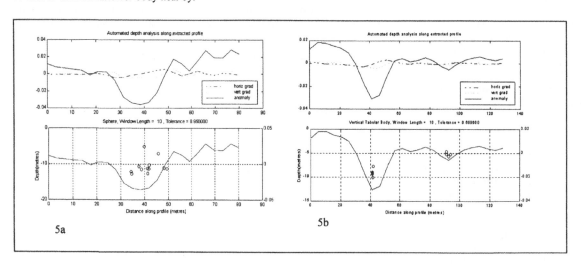

Figure 5: Euler Deconvolution of profiles1 (5a) and profile 2 (5b)

A series of probe holes were carried out, using microgravity results to target potential cavities. The probe drilling showed cavities at several levels. In all cases except one there were significant voids of up to 7 metres beneath the negative anomalies. One of the boreholes was drilled to target a gravity high (i.e. mass surplus) and that intersected hard or well cemented rock .

GRAVITY DATA INVERSION

The Euler Deconvolution together with confirmatory depth from the probe drilling can then be used as an independent constraint allowing inversion of the microgravity map to give thickness of the underground cave system. This requires assumptions to be made concerning the likely depth of the median plane of the cavity and density contrast between the surrounding rock mass and the cavity. The depth can be constrained at *ca* 10 to 12 meters below surface because of our Euler Deconvolution solutions and the drilling results and the density contrast has been assumed to be -1.5 g/cc (i.e. water filled cavity). The technique makes a first guess of cavity thickness based on a simple Bouguer slab formula and then iteratively adjusts the thickness at each grid node by comparing calculated gravity with observed gravity and scaling the thickness accordingly. This generally converges within a few iterations to a very good fit between the observed and calculated anomalies. The calculated cavity thickness across the area based on this inversion is shown in Figure 6. The cavity appears to be continuous across the surveyed area and to have several sub-branches. The cavity at (0,90) has an entrance accessible from the harbour and this has been investigated by cave divers who found a good correlation between the gravity derived map and their physical underwater mapping. Table 1 shows the relationship which was established by the divers for the area

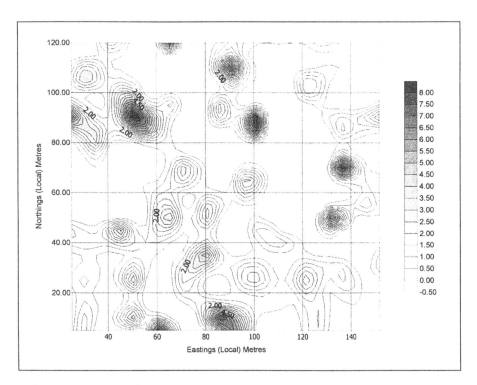

Figure 6: Calculated cavity thickness (metres) from inversion of the microgravity residual map

which they could obtain access to. Using Gauss' theorem the mass deficit for the area covered by Figure 6 was estimated to be almost 1600 tonnes.

Table 1: Relationship between microgravity amplitude and cavity height for Freeport, Grand Bahamas, microgravity survey.

Anomaly (mgals)	Cave height (Feet)
0.020 and above	Solid Ground
0.005 to 0.015	1 to 3
0 to 0.005	3 to 4
-0.010 to 0	5
-0.030 to -0.010	6
-0.045 to -0.030	7 to 8
-0.065 to -0.045	8 to 10

CONCLUSIONS

Microgravity surveying at Freeport, Grand Bahama, has enabled the detailed delineation and characterization of an extensive network of submarine caves beneath a proposed container terminal. The area shows significant negative microgravity anomalies, which are associated with large amounts of missing mass beneath the major anomalies identified from this survey. This missing mass may be caused by discrete cavities or distributed void space, but the shape and size of the anomalies suggest that cavities are present beneath many of these features. Drilling and the cave–dive survey have confirmed this. It has been calculated that some 32% of the site would therefore be expected to be at risk from significant voiding. The major anomalies appear to link form a network of linked sinuous features, which connect the major anomalies and have hydraulic connections to the sea to the south and north of the island.

The cave system crosses the western part of the area with a major feature at (60,90) which is most probably a very significant cavity, that may fall into the classification of a Flank Margin Cave. This feature alone has an estimated volume of about 1300 to 1400 m3. This appears to connect to the sea to the south, and it connects north to an even larger system at *ca* 180, 250. The continuity of the anomalies suggests the presence of a linked drainage system through the area which may extend further inland and to the east. Microgravity surveying proved a cost-effective and invaluable technique for investigating and characterizing a large area (600 metres by 600 metres) without enormous numbers of probe holes.

REFERENCES

Arzi, A. A. 1975. Microgravity for engineering applications. Geophysical Prospecting, 23, 408-25.

Bishop, I., P. Styles, S. J. Emsley and N. S. Ferguson. (1997), The Detection of Cavities Using the Microgravity Technique: Case Histories From Mining and Karstic Environments in Modern Geophysics in Engineering Geology, McCann, Fenning, & Reeves (Eds)., Geol. Soc.of London Special Publication No 12, 155-168.

Butler, D. K. 1984. Microgravimetric and gravity gradient techniques for the detection of sub-surface cavities. Geophysics, 49, 1084-96.

Chamon N. and Dobereiner L. 1988. An example of the uses of geophysical methods for the investigation of a cavern in sandstones. Bulletin of the International Association of Engineering Geology, No. 38, Paris, 37-43.

Colley, G. C. 1963. The detection of caves by gravity measurements. Geophysical Prospecting, 11, 1-10.

Cripps, J. C., McCann, D. M., Culshaw, M. G. and Bell, F. G. 1988. The use of geophysical methods as an aid to the detection of abandoned shallow mine workings. Institute of Mining Engineers, MINESCAPE 88, Harrogate.

Cuss , R. A., and Styles P., (1999),The Application of Microgravity in Industrial Archaeology: An Example From the Williamson Tunnels, Edge Hill, Liverpool, Geol. Soc. Spec Pub., 'Geoarchaeology, Exploration, Environment, Resources', Pollard (Ed)., 165, 35-40.

Daniels, J. 1988. Locating caves, tunnels and mines. Geophysics: The Leading Edge of Exploration, 7, No. 3, 32-7.

Daniel, A. J., and Styles, P., (1997). Topographic Accessibility and the Tectonic interpretation of gravity data. Geophysical Prospecting., Vol 45, 6., 1013-1026.

Emsley, S. J., Summers, J. W. and Styles, P., (1992). The detection of sub-surface mining related cavities using the micro-gravity technique. Proc. Conf. Construction over Mined Areas, Pretoria, S. Africa., May 11-12 1992., 10p.

Owen, T. E. 1983. Detection and mapping of tunnels and caves. In: Fitch, A. A., (Ed.), Development in Geophysical Exploration methods, Volume 5, 161-258. Wiley.

Patterson, D. A, J.C. Davey, A.H. Cooper, & J.K. Ferris, (1995). The investigation of dissolution subsidence incorporating microgravity geophysics at Ripon, Yorkshire., Quart. Jour. Eng. Geol. 28, 83-94.

Wilson, W . L., J. E. Mylroie & J.L. Carew, (1995), Caves as a geologic hazard: A quantitative hazard assessment from San Salvador Island, Bahamas, in "Karst Geohazards", Beck (Ed). Balkema, Rotterdam, 487-495.

VIII Field Trip Guide

Geotechnical and Environmental Applications of Karst Geology and Hydrology, Beck & Herring (eds)
© 2001 Taylor & Francis, ISBN 90 5809 190 2

Field Trip Guide, Part 1: Environmental problems associated with urban development upon karst, Bowling Green, Kentucky

NICHOLAS C.CRAWFORD   Center for Cave & Karst Studies, Applied Research & Technology Program of Distinction, Dept. of Geography & Geology, Western Kentucky University, Bowling Green, KY, USA, nicholas.crawford@wku.edu

INTRODUCTION

Bowling Green, Kentucky, with a population of approximately 50,000, is located on the classic Pennyroyal Sinkhole Plain of South Central Kentucky, a highly karstified portion of the Mississippian Plateaus physiographic province (Figure 1). It is located along the northwest flank of the Cincinnati Arch as it dips northwest toward the Illinois Basin. The Mississippian rocks that underlie The city are dipping northwest at approximately 30 feet per mile toward the Dripping Springs (Chester) Escarpment.

The city is underlain by the upper St. Louis Limestone, the Ste. Genevieve Limestone, and where several hills protrude above the sinkhole plain, the Girkin Limestone. These hills are outliers of the sandstone-capped Chester Upland. The Chester Upland is capped by the Big Clifty Sandstone and located about two or three miles to the northwest. The city is drained almost entirely by cave streams and is a focal point for the convergence of groundwater flowing beneath the sinkhole plain of Warren County, Kentucky. Almost all drainage from the sinkhole plain for about 15.5 miles northeast (Quinlan and Rowe, 1977) and 13 miles south (Crawford, 1985) issues from Graham Springs and the Lost River Rise to flow into the Barren River at Bowling Green (Figure 2). Because of its location the city has serious problems associated with urban development upon karst. This field trip will deal with the impacts karst has upon urban development and the impacts that urban development has upon karst. Four karst hydrogeologic problems will be featured: a) groundwater contamination, b) sinkhole collapses and c) sinkhole flooding and d) high radon levels in homes and buildings. The field trip will emphasize the considerable effort being made to solve karst problems in the Bowling Green area by the Center for Cave and Karst Studies at Western Kentucky University as well as various federal, state, and local agencies.

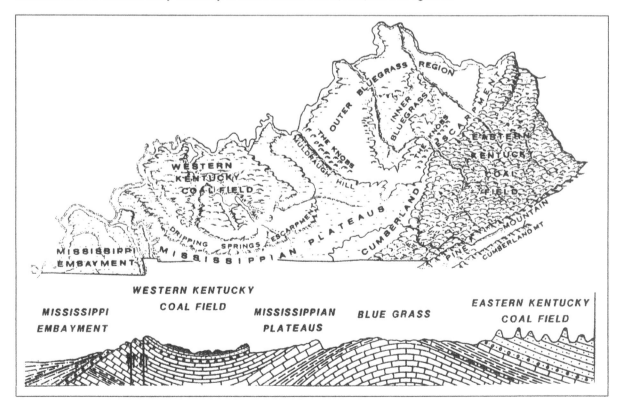

Figure 1: Physiographic map of Kentucky (Lobeck, 1932).

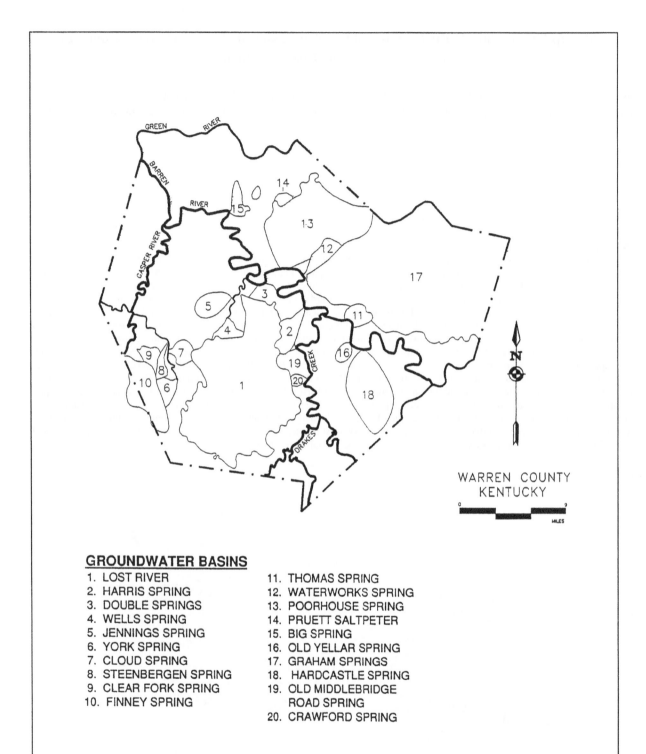

GROUNDWATER BASINS

1. LOST RIVER
2. HARRIS SPRING
3. DOUBLE SPRINGS
4. WELLS SPRING
5. JENNINGS SPRING
6. YORK SPRING
7. CLOUD SPRING
8. STEENBERGEN SPRING
9. CLEAR FORK SPRING
10. FINNEY SPRING

11. THOMAS SPRING
12. WATERWORKS SPRING
13. POORHOUSE SPRING
14. PRUETT SALTPETER
15. BIG SPRING
16. OLD YELLAR SPRING
17. GRAHAM SPRINGS
18. HARDCASTLE SPRING
19. OLD MIDDLEBRIDGE
    ROAD SPRING
20. CRAWFORD SPRING

Figure 2: Karst groundwater basins of Warren County. Sources: North of Barren River (Quinlan and Ray, 1981), South of Barren River (Crawford, 1985).

FIELD TRIP STOP . LOST RIVER CAVE AND VALLEY

Lost River Groundwater Basin

The Lost River Groundwater Basin includes most of Warren County south of Bowling Green, (Figure 2). Dye traces have revealed that the Lost River begins in uplands about 13 miles south of the city where several small streams sink and flow into the underlying St. Louis Limestone. These subsurface streams unite to become the Lost River which flows north under the city (Figure 3).

The Lost River flows across the bottom of several karst windows, the largest being the Lost River Cave Valley located at the southern edge of Bowling Green where the cave stream rises at the Lost River Blue Hole Spring, flows about 400 feet on the surface and sinks at the Historic Entrance to Lost River Cave. From the Historic Entrance the Lost River travels through large cave passages under the southwest portion of the city to the Lost River Rise (Figure 4). The Lost River drainage system has formed in the Mississippian Ste. Genevieve and St. Louis Limestone in the vicinity of two chert confining layers (Figure 5). The stream begins near the town of Woodburn south of Bowling Green as surface streams invade the subsurface upon breaching the Lost River Chert Bed (Figure 6). It then flows north perched primarily on the Corydon Member of the St. Louis Limestone and in places the Lost River Chert Bed (Crawford, 1988). The Lost River Chert Bed is named for the famous Lost River of southern Indiana. It appears that this prominent 10 to 20-foot zone of bedded, light gray, fossiliferous chert extends from southern Indiana to the Bowling Green area and probably as far south as the Mississippian sinkhole plain of the Highland Rim of Tennessee. It is somewhat ironic that the Lost River Chert (of southern Indiana) plays such an important stratigraphic role in the development of the Lost River Cave system of southern Kentucky. At every location between the headwaters and the rise where the Lost River is visible, it is flowing upon either the Lost River Chert or the Corydon Chert. It is interesting that structure and stratigraphy have influenced groundwater flow and thus cavern development to a much greater extent in the Bowling Green area than in the Mammoth Cave area only 25 miles to the northeast.

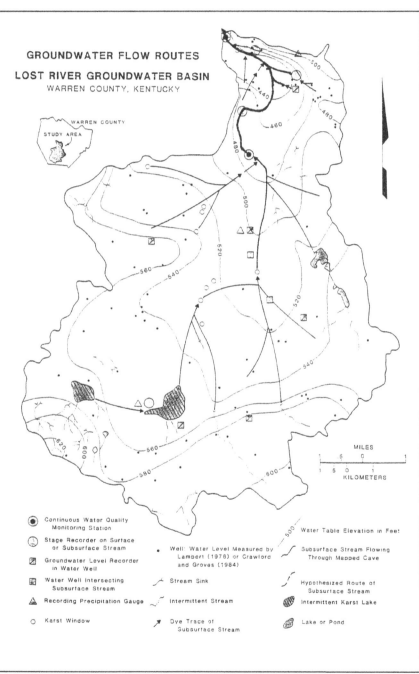

Figure 3. Groundwater flow routes, Lost River Groundwater Basin, Warren County, Kentucky (Crawford, 1985).

399

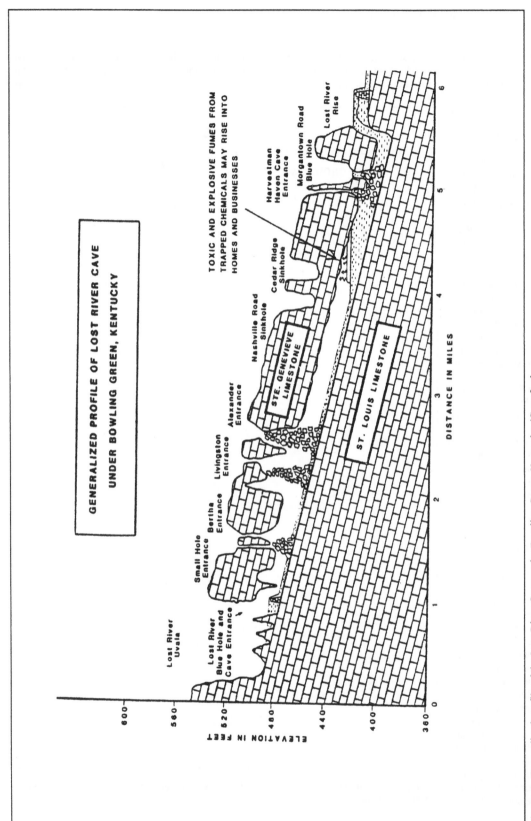

Figure 4: Generalized profile of the Lost River Cave under Bowling Green, Kentucky (Crawford, 1989a).

400

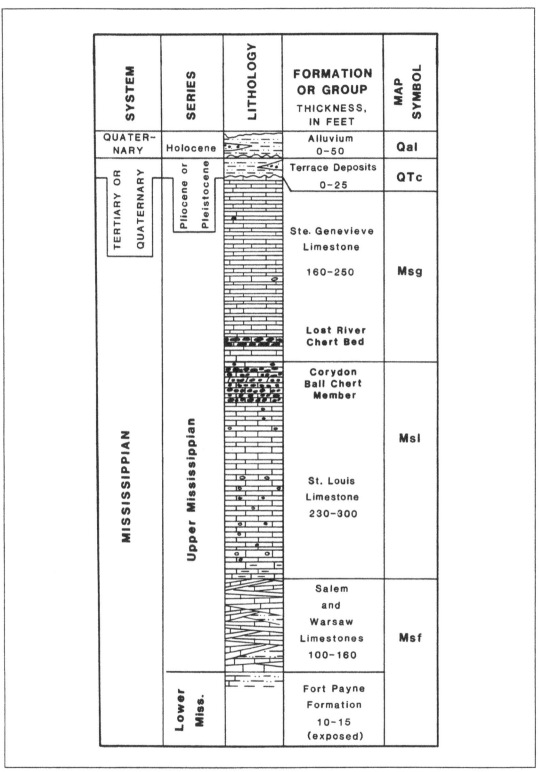

SYSTEM	SERIES	LITHOLOGY	FORMATION OR GROUP THICKNESS, IN FEET	MAP SYMBOL
QUATERNARY	Holocene		Alluvium 0-50	Qal
TERTIARY OR QUATERNARY	Pliocene or Pleistocene		Terrace Deposits 0-25	QTc
MISSISSIPPIAN	Upper Mississippian		Ste. Genevieve Limestone 160-250	Msg
			Lost River Chert Bed	
			Corydon Ball Chert Member	Msl
			St. Louis Limestone 230-300	
			Salem and Warsaw Limestones 100-160	Msf
	Lower Miss.		Fort Payne Formation 10-15 (exposed)	

Figure 5: Stratigraphic column for the Bowling Green area. The Lost River flows primarily upon the Corydon Member of the St. Louis Limestone, but in places it flows upon the Lost River Chert Bed. Source: Modified from McGrain and Sutton (1973).

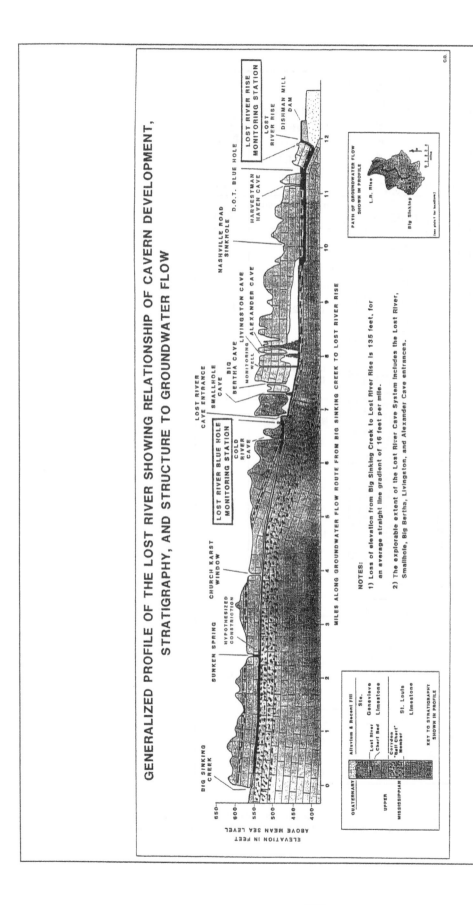

Figure 6: Generalized profile of the Lost River showing relationship of cavern development, stratigraphy, and structure to groundwater flow (Crawford, 1989a).

GROUNDWATER CONTAMINATION BY DOMESTIC SEWAGE DISPOSAL SYSTEMS

Lost River Cave is an excellent site to see the impact of domestic sewage disposal systems on the karst aquifer. At this stop the tour will float the Lost River under Bowling Green to observe Effluent Falls. This once beautiful flowstone formation has been created by the water from Jessie James Spring as it discharges from the roof of the cave and precipitates calcium carbonate (travertine) as it cascades down the breakdown wall of the cave. The formation once resembled the beautiful Frozen Niagara Formation in Mammoth Cave. Unfortunately, instead of being orange and white, it is black and brown due to untreated human waste and often one can smell the odor a hundred feet away.

Bowling Green and Warren County have a long history of groundwater contamination by improper or inadequate sewage disposal. In the 1920's the city did not have a sewage system and the accepted method of sewage disposal was to flush it directly into the caves under Bowling Green. When a new home was built, it was standard practice to employ a "sink finder" to locate a crevice in the limestone bedrock for the direct injection of sewage. The "sink finder" used a "witch stick" (a forked peach tree branch) to find a crevice under the thin soil which covers the limestone bedrock. A hole was then dug to reveal the crevice an a garden hose was used to test the fissure and prove it free from obstructions. It was then approved by the city inspector and the house was connected to a "Sewer System More Than a Million Years Old" as explained in Figure 7 (Mace, 1921).

Fortunately, most of the homes in Bowling Green are now connected to the Bowling Green Municipal Utilities sewer system. Unfortunately, there remain some homes that are still using the caves for a sewer and throughout Warren County there are homes where septic tank effluent is not being sufficiently treated by the soil before reaching the karst aquifer.

Water which has had contact with human and animal waste is one of the most serious pollutants in underground streams in the Bowling Green area. High fecal coliform and

SEWER SYSTEM MORE THAN A MILLION YEARS OLD

BY CHARLES E. MACE

THE only city in the United States boasting a sewer system in which all the "pipes" were laid by Mother Nature is Bowling Green, Ky. Although the prosperous little municipality has a population of 15,000 there is not a foot of man-made sewer pipe in any of the streets or alleys.

The explanation is that the city is built over a formation of oölitic white limestone which is a maze of connected crevices extending to a considerable depth below the surface; much the same formation as that of the famous Mammoth Cave just 30 miles distant. This limestone is said to be composed of the fossilized eggs of prehistoric marine animals. The "logs" of oil wells drilled in the western Kentucky fields, of which Bowling Green is the center, show that limestone of one kind or another is encountered as deep as drilling has ever yet been carried.

When a new residence is being built in the Bowling Green region, a sink finder" is employed, who merely goes out in the back yard and digs about in the surface soil, which is seldom more than 3 ft. deep, until he locates a fissure. A

Once a Fissure in the Limestone is Found, It is Tested with a Hose to Prove It Free from Obstructions, Then Approved by the Inspector, and the House Has Sewer Connection

garden hose is then placed in the crevice, and the water is allowed to run until it is free from obstructions. It is then approved by the city inspector, and the house has perfect sewer connection. No city has a more sanitary sys-

"Uncle" Henry Jameson, Aged Negro Specialist, with the Divining Rod. or "Witch Stick": He Uses This Peach-Tree Fork to Locate a Crevice in the Underground Formation

tem. Chemists say the sewage would be purified in a very short distance by passing through the limestone. Seepage never comes to the surface, the explanation of geologists being that it flows through these natural passageways in the stone until it finally finds an outlet in the river bed.

An interesting character is found in "Uncle" Henry Jameson, an aged negro who has specialized in locating fissures and digging "sinks" for the past 25 years. When asked just how many he had dug, he laughed and said "Lawdy, Boss, I reckon I couldn't count that many." Uncle Henry uses the divining rod, or "witch stick." as he calls it, in locating the fissures, and declares he would never dig without first employing his forked peach-tree branch. The frequency with which his attempts are successful is amazing. Although Henry is 74 years of age,

Outcropping of Oölitic Limestone in Bowling Green's City Park. Which Shows Very Clearly the Cavernous Structure of This Formation Which Underlies the Entire City, Hidden for the Most Part by a Light Covering of. Surface Soil

Figure 7: Page 687. Popular Mechanics Magazine, Vol. 35, No. 5, May 1921.

indicate that groundwater has had recent contact with human and animal waste. Runoff from heavy rain washes past livestock waste from farmland south of the city into subsurface streams at swallets. Waste from dogs and cats is a problem in urban areas. Heav rains also flush septic tank effluent from the soil down into underlying conduits in the limestones. Water samples with fecal coliforr counts of over 40,000 colonies per 100 millimeters have been taken at the Lost River Blue Hole Spring and the Lost River Rise a Lampkin Park during relatively high discharges following heavy rains. Dye traces of septic tanks by Crawford (1979) have reveale surprisingly rapid flow from septic tanks into underground streams. Rhodamine WT dye was detected at the Lost River Blue Hol Spring only ten hours after it was flushed down the toilet of a house 0.8 miles away.

Most people are not even aware that there are two kinds of septic tank system failures. There is the kind that most people ar familiar with, where the soil cannot absorb the septic tank effluent fast enough and it rises to the ground surface and "stinks-up" th whole neighborhood, and there is the other kind of failure when the septic tank effluent sinks to the water table too fast and is nc adequately treated by the soil. Both types of septic tank system failures are common in Bowling Green and Warren County. Of th two, the latter is by far the most serious yet it can go unrecognized indefinitely.

Septic tank treatment systems consist primarily of a sewage pre-treatment unit which is a watertight structure designed an constructed to receive raw sewage, separate solids from liquids, digest organic matter through anaerobic bacterial action, and allov clarified effluent to discharge to a subsurface soil absorption system.

After sewage goes through a pretreatment unit it flows into a subsurface soil absorption system (Figure 8) for further treatment b microbial plant and animal life within the soil and by filtration, chemical decomposition and bonding within the soil.

In Warren County, the subsurface soil absorption system is almost always a system of buried perforated pipes which release th effluent into the soil. The most important component of this type of onsite sewage disposal system is the soil. The soil must treat th septic tank effluent before it percolates down to the water table and becomes a potential source of potable water at a water well o

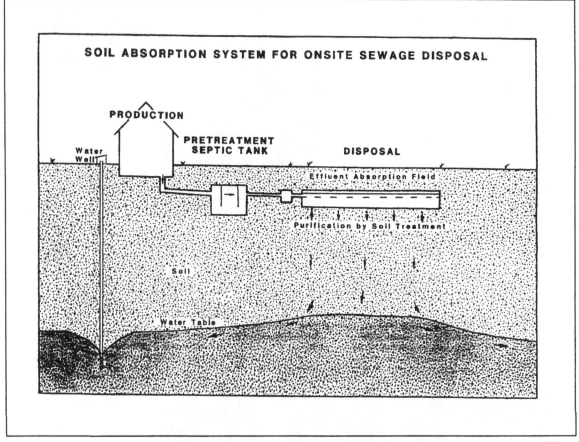

Figure 8: Septic tanks provide pretreatment by anaerobic bacteria. The soil absorption system treats the septic tank effluent with aerobic bacteria, chemical decomposition, soil bonding and slow filtration, which provides a long "die-off" time for pathogenic bacteria and viruses. The type and thickness of soil must be sufficient to treat the septic tank effluent before it reaches the water table and becomes a potential source of potable water at a well or spring (Crawford and Sneed, 1989).

spring. Possibly the most important aspect of the septic tank effluent soil absorption system is the long "die-off" time for pathogenic bacteria and viruses. Most pathogens cannot live more than a few days or weeks outside of the human body, particularly in the cold and hostile soil environment. In karst aquifers where groundwater can sometimes travel at velocities greater than one-half mile per hour, treatment of the septic tank effluent by the soil before it reaches the limestone bedrock, is critical.

In areas within thin soils (Figure 9) or areas with soils which do not adequately treat the septic tank effluent, septic tanks with soil absorption systems should not be used. They were never intended for use in these areas. Therefore, alternative types of sewage treatment should be used in these locations. In addition to thin soils, another case of septic tank system failure in the karst areas of Warren County are macropores in the soils. These openings, which are often interconnected, can range in size from barely visible to several inches in diameter. Macropores can be caused by tree roots, which after they die and decay become conduits for infiltrating water, allowing for the rapid vertical movement of water through the soil to the limestone bedrock. Even earthworms can burrow down to six feet in some soils creating small conduits through the soil.

Macropores exist in virtually all soils but they usually do not provide rapid flow routes for the downward movement of soil water. The reason being that they do not have outlets at their bottoms. If they end in the soil, even if they are several inches in diameter, rapid downward movement of soil water usually cannot occur because the macropore does not have an opening at the bottom. Even if a macropore extends to bedrock there will still not be a void at the bottom which will permit the rapid vertical flow of soil water-- except in karst areas. If a macropore intersects an opening in the limestone or even one of the small, solutionally-enlarged channels which often exist in the limestone along the soil bedrock interface, there will be an outlet for the soil water and it will flow directly, without filtration through the soil, into the karst aquifer (Figures 10 and 11). In areas of thin soils macropores can provide a direct route for surface runoff water and septic tank effluent to flow into the karst aquifer. with ditches containing the perforated delivery pipes for soil absorption fields being several hundred feet long for even small septic tank systems, the chances of intersecting one or more macropores, and thus a direct route to the karst aquifer below, are probably high.

In addition to septic tank effluent from the suburbs, human waste enters the Lost River from many homes in the older areas of town which were never connected to city sewers. The city did not require that they connect when the sewer pipes were laid years ago, and the city does not have complete records of which homes are on sewer and which are not. All homeowners are sent a sewer bill based on water usage each month regardless of whether they are connected or not. Many homeowners have not had septic tank problems because the effluent sinks into the limestone rather than appearing at the surface when the system fails. Consequently, they assume that their septic tank is working correctly and that there is not need to pay a plumber to connect their home to the sewer. Houses change owners frequently, and it is likely that some homeowners who pay their sewer bill each month believe that they are connected to the sewer, when in fact, they are injecting their waste directly into the karst aquifer.

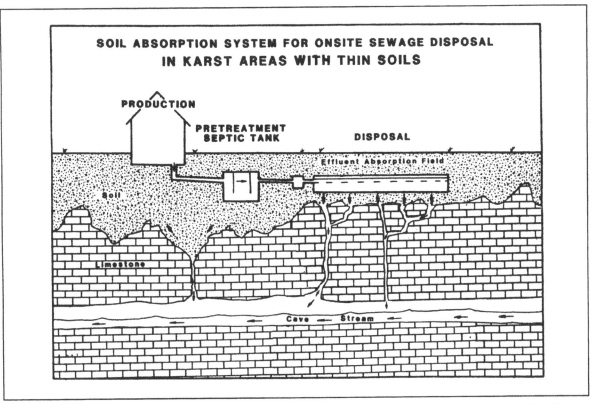

Figure 9. If the thickness of soil is insufficient to treat the septic tank effluent before it reaches bedrock, it may travel rapidly through conduits in the limestone to distant water wells and springs (Crawford and Sneed, 1989).

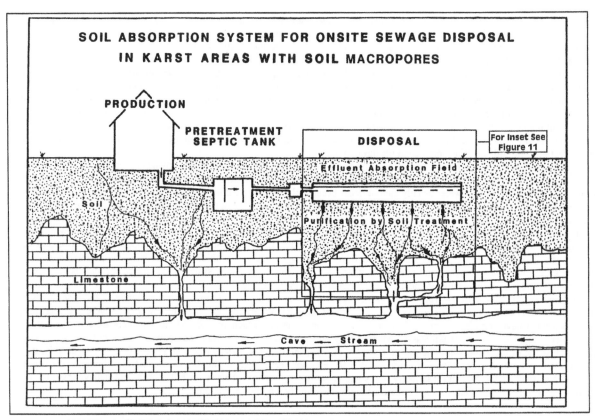

Figure 10: Septic tank effluent absorption field located in soil with macropores. Macropores permit the effluent to rapidly flow, without soil filtration, into the karst aquifer (Crawford and Sneed, 1989).

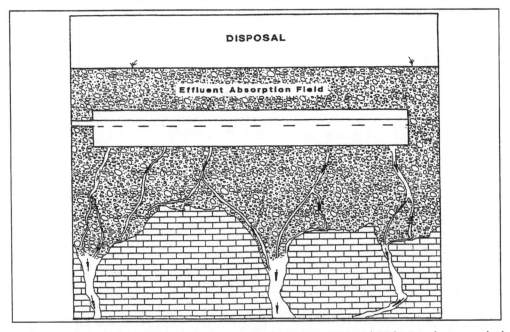

Figure 11: Macropores created by desiccation cracks, decayed tree roots and earth worm burrows exist in most soils but only in karst areas are there openings at the bottom. If a macropore intersects a void along the soil-bedrock contact, it becomes an open conduit which permits septic tank effluent to flow directly into the karst aquifer without any treatment by the soil (Crawford and Sneed, 1989).

FIELD TRIP STOP 2: I-65 GREENWOOD INTERCHANGE: SINKHOLE COLLAPSES

At this stop we will observe several sinkhole collapses and discuss a sinkhole collapse that occurred under a nearby building. Virtually all sinkhole collapses in karst areas are regolith collapses (Figures 12 and 13). Although bedrock cave roofs do occasionally collapse, these collapses are so rare that they do not constitute a serious threat. However, regolith collapses occur frequently in many karst areas.

Regolith collapses occur due to the formation and the eventual collapse of regolith arches (domes). Urban development in karst areas will often result in an increase in the collapse of regolith arches. Regolith arches form by the downward movement of unconsolidated sediments into voids in the bedrock. In areas where the water table is usually above the regolith-bedrock contact, collapses often occur when the water table drops during droughts or high-volume pumping (Figure 12). Physically, the collapses in these cases are caused by loss of buoyant support for the regolith arches which span openings in the limestone. Collapses are also caused by spalling of saturated regolith down the opening, enlarging the arch, and eventually causing collapse at the land surface.

Regolith collapses also may occur in situations where the water table is usually below the regolith-bedrock contact (Figure13). Construction and land use changes that concentrate surface runoff in sinkholes, retention basins, ditches, and ponds may locally increase the downward movement of water resulting in the piping of saturated regolith into openings in the limestone (Figure 14). Most of the sinkhole collapses investigated in the Warren County, Kentucky area by the Center for Cave and Karst Studies at Western Kentucky University (Sinkhole Collapse Inventory Vols. 1-4) are of this type (Crawford, Webster, and Veni, 1990). An estimated 75 percent are man-induced collapses of existing regolith arches (Figure 15). Changes in the surface drainage associated with farming and particularly urban development are believed to be the primary causes of most collapses.

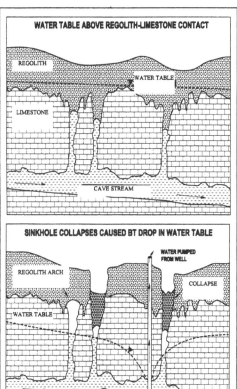

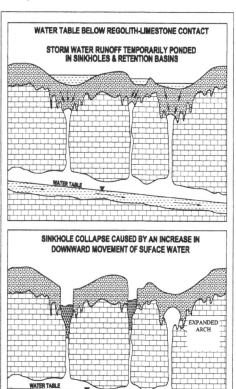

FIGURE 12, LEFT. Sinkhole collapses in areas where the water table is above the regolith-limestone contact are usually caused by a drop in the water table. Regolith arches spanning openings in the bedrock collapse because of the loss of buoyant support and because of downward-moving surface water.

FIGURE 13, RIGHT. Sinkhole collapses in areas where the water table is below the regolith-bedrock contact are usually caused by an increase in the downward movement of surface water. Land use changes and construction activities that concentrate surface water in drains, sinkholes, and impoundments may locally increase downward movement of surface water and induce the collapse of regolith arches.

Microgravity will often reveal the existence of a void in the regolith above bedrock and therefore identify a potential sinkhole collapse. Figure 16 is an example of a microgravity traverse over a regolith void that had collapsed to the surface so that a small hole was actually visible.

Microgravity traverses can be made along foundations previous to building construction to identify low-gravity anomalies. However, depth to bedrock borings into the anomalies are usually needed to establish if they are regolith voids, bedrock caves, or cutters (areas of deep regolith between pinnacles).

Crawford has used microgravity at several sites to investigate subsurface conditions in the vicinity of sinkhole collapses. It provides useful information concerning: a) depth to bedrock, b) extent and shape of the collapse area below the surface, c) location of the crevice or crevices, into which the regolith is collapsing, and d) locations of additional regolith voids in the vicinity of the initial collapse. The following case study discusses the application of microgravity techniques in greater detail.

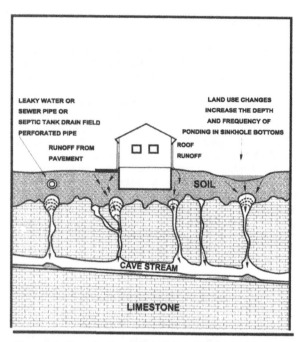

Figure 14: Growth of regolith arches toward the surface induced by modification of natural runoff and infiltration conditions.

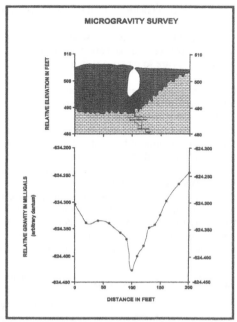

Figure 15: Low-gravity anomaly along a traverse over a regolith void that had collapsed all the way to the surface so that a small hole was actually visible.

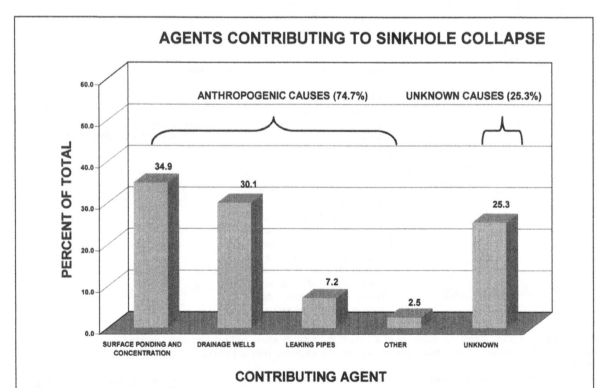

Figure 16: An investigation of 80 sinkhole collapses in Bowling Green, Kentucky shows that surface ponding and concentration is the leading cause of sinkhole collapse (Crawford, Webster, and Veni, 1989).

CASE STUDY: SINKHOLE COLLAPSE UNDER A TWO-STORY BUILDING IN BOWLING GREEN, KENTUCKY

Visible structural damage to a two-story building in Bowling Green, Kentucky consisted of extensive cracks through the front brick wall of the building, extending to the second level. Visible subsidence and ground cracks existed under the outside walls of the structure, and a 6 inch depression existed in the asphalt parking lot at the rear of the structure. A more extensive investigation from the crawl space under the building revealed cracks in footers and walls supporting the structure, deep cracks in the ground under the structure, and collapses under the two central footers and the rear footer. Extent of surface expression of the collapse was approximately 60 feet by 70 feet (Figure 17). Based on preliminary evidence, the site was at risk for a catastrophic collapse. In order to delimit the scale of collapse and determine remedial strategy, a microgravity survey accompanied by exploratory soil borings were conducted.

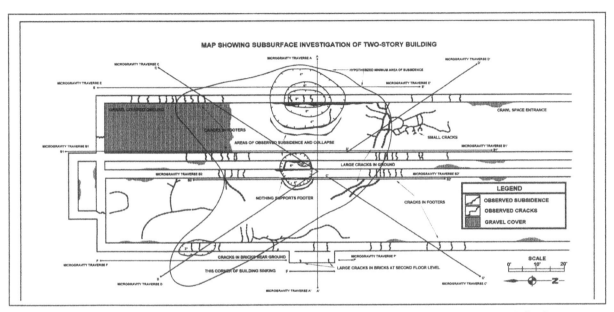

Figure 17: Plan view of site illustrating foundation supports of two-story building, location of physical features indicating collapse and/or subsidence, and location of microgravity survey traverses.

A total of 7 microgravity traverses were surveyed under and outside of the building using a LaCoste and Romberg Model D microgravity meter (Figure 17). A 5-foot station interval was used in the crawl space under the building, and a 5 to 15-foot interval was used on traverses extending outside of the building. Depth to bedrock soil borings were used to establish depth to bedrock at the front and rear of the building and to calibrate microgravity measurements. Well logs indicate that depth to bedrock varied from 34 feet to greater than 58 feet, the greatest depth achievable by the drill rig (Figure 18).

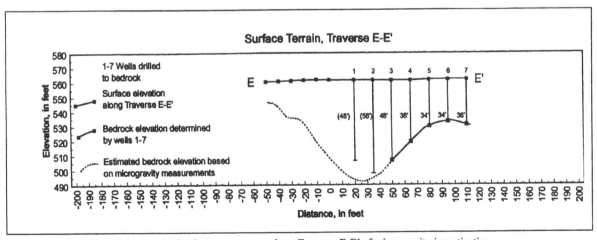

Figure 18: Elevation and depth to bedrock measurements along Traverse E-E' of microgravity investigation.

409

Traverse E-E' extended north to south for 160 feet on the east side of the building. Measured gravity (Figure 19) compared with depth to bedrock at wells 1-7 was used to estimate depth to bedrock along the entire traverse.

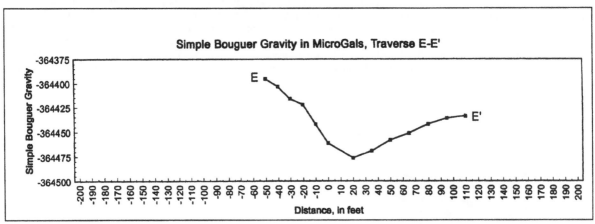

Figure 19: Microgravity along Traverse E-E'.

Traverse A-A' was surveyed under and outside of the building, running east to west for 260 feet perpendicular to the building. A profile of measured gravity was used to estimate depth to bedrock and model anomalies (Figure 20).

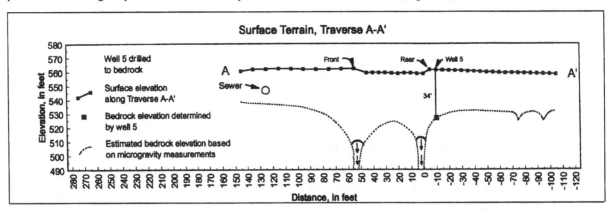

Figure 20: Elevation and depth to bedrock measurements along Traverse A-A' of microgravity investigation.

The microgravity investigation indicated a very large low-gravity anomaly under the north wing of the two-story building (Figure 21). Borings into the low-gravity anomalies indicated a depth to bedrock in excess of 58 feet. The collapse was mitigated by drilling wells into the low-gravity anomalies and injecting grout under pressure to fill the voids and compact the unconsolidated material under the building. Figure 22 is a conceptual geologic profile modeled after exploratory borings and microgravity data for the site.

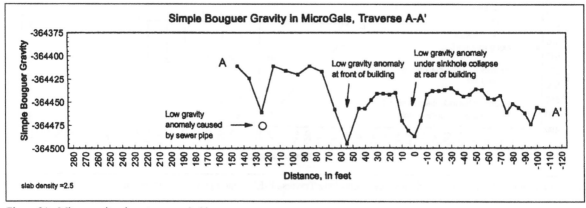

Figure 21: Microgravity along traverse A-A'.

410

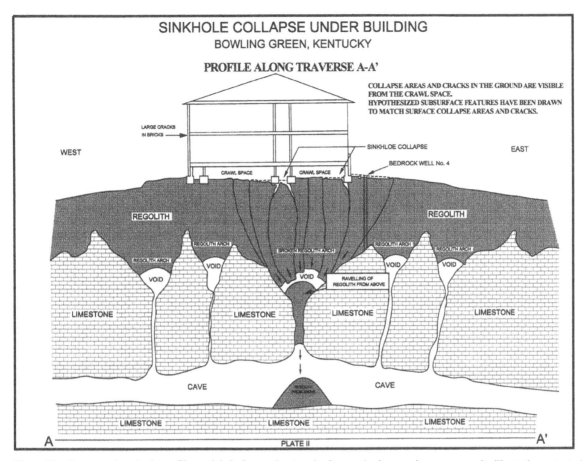

SINKHOLE COLLAPSE UNDER BUILDING
BOWLING GREEN, KENTUCKY

PROFILE ALONG TRAVERSE A-A'

COLLAPSE AREAS AND CRACKS IN THE GROUND ARE VISIBLE
FROM THE CRAWL SPACE.
HYPOTHESIZED SUBSURFACE FEATURES HAVE BEEN DRAWN
TO MATCH SURFACE COLLAPSE AREAS AND CRACKS.

WEST

LARGE CRACKS
IN BRICKS

SINKHLOE COLLAPSE

EAST

CRAWL SPACE   CRAWL SPACE

BEDROCK WELL No. 4

REGOLITH   REGOLITH

REGOLITH ARCH   REGOLITH ARCH   REGOLITH ARCH

REGOLITH ARCH   BROKEN REGOLITH ARCH

VOID   VOID   VOID   VOID   VOID

VOID   RAVELLING OF
REGOLITH FROM ABOVE

LIMESTONE   LIMESTONE   LIMESTONE   LIMESTONE

CAVE   CAVE

REGOLITH
FROM ABOVE

LIMESTONE   LIMESTONE   LIMESTONE

A   PLATE II   A'

Figure 22: Conceptual geologic profile modeled after exploratory borings and microgravity survey results illustrating extent of sinkhole collapse under two-story building in Bowling Green, Kentucky.

FIELD TRIP STOP 3. FAIRVIEW PLAZA SHOPPING CENTER: GASOLINE VAPORS RISING FROM KARST AQUIFER RESULTED IN THE EVACUATION OF FIVE HOMES

Introduction

Ground water contamination problems are particularly serious when they involve toxic or explosive chemicals. Not only are the chemicals a threat to water supplies and aquatic life, but upon vaporizing they may become concentrated in the cave atmosphere and rise into homes on the surface. Intermittent problems with toxic and explosive fumes have occurred in Bowling Green for at least thirty years with homes and buildings in various parts of the city affected.

Leaking underground storage tanks are believed to be the major source of fumes in the caves. The Bowling Green area has an estimated 500 buried tanks, most of them containing gasoline and belonging to auto service stations. Gasoline floating on a subsurface stream can travel several kilometers (miles) from the site of a leak or spill in a few hours, rapidly filling cave passages with explosive fumes. The fumes may then rise up solutionally-enlarged joints, faults, bedding plane partings, and caves into basements and crawl spaces. They may also rise into homes built over sinkholes, up water wells and storm water drainage wells, and even up abandoned wells and natural openings used for waste disposal.

Riverwood

In 1969 and again in 1981 residents of several homes in the Riverwood area were evacuated because explosive levels of fumes were detected in their basements and crawl spaces. In 1981 a trench was dug in an attempt to find the cave stream and obtain a sample of the gasoline. Since the backhoe excavated many large blocks of Lost River Chert (a light gray flint resembling limestone) mixed with the soil, it is fortunate that the caves and gasoline were not located! A survey of storm water drainage wells in the vicinity revealed high readings on a portable combustion meter for two drainage wells in the nearby Fairview Plaza Shopping Center parking lot. If someone had discarded a cigarette near one of the wells, it is conceivable that the explosion could have traveled through the cave system destroying houses some distance away. An underground explosion in the sewer system of Louisville, Kentucky in 1981 traveled along the sewer system for eleven blocks with estimated damages exceeding forty-three million dollars. The potential for such a large explosion in the caves under Bowling Green is probably remote but certainly possible (Crawford 1982).

FIELD TRIP STOP 4. BATSEL AVENUE SINKHOLE FLOODING PROBLEM

Introduction

Periodic flooding of karst depressions is a serious problem for urban areas located upon sinkhole plains. The problem is particularly serious in the city of Bowling Green, Kentucky, where homes, streets, apartments, and businesses are affected. The city is located entirely upon a sinkhole plain, with underground streams flowing through solutionally-enlarged caves in the shallow carbonate aquifer. All precipitation not lost to evapotranspiration travels by way of these streams to springs and into the nearby surface-flowing Barren River. The landscape resembles large funnels (sinkholes) which direct storm water runoff into the underlying caves. Storm water ponds at the bottom of some sinkholes and then sinks slowly through the soil into caves streams below. However, most of the larger sinkholes and many of the smaller ones have experienced sinkhole collapses which have created drains permitting storm water to flow directly into the caves. Periodically, these drains become clogged only to be opened again by collapses during later floods. This sequence repeated over thousands of years is the process by which most sinkholes have formed. Even before Bowing Green was built, storm water runoff sank directly through numerous sinkhole drains into caves below. The caves acted as storm drains for this landscape then, and they continue to serve that function today.

The flooding of sinkholes in karst regions is a part of the natural hydrologic system. Flooding occurs during periods of intense rainfall, usually of short duration: 1) when the quantity of storm water runoff flowing into sinkholes exceeds their outlet capacities, and they cannot drain into underlying caves fast enough to prevent ponding, 2) when the capacity of the cave system to transmit storm water is exceeded, and the water must be stored temporarily in sinkholes since it cannot be stored on flood plains like surface streams, and 3) when high water table results from a backwater effect on ground water flow caused by surface or subsurface streams at flood stage. Unfortunately, in the Bowling Green area houses have been built in these natural storage areas (sinkholes). The problem has been greatly aggravated by increased runoff resulting from urban development and by sinkhole filling by developers and landowners (Crawford 1981 and 1984).

The worst flooding problems in Bowling Green occur in large, shallow sinkholes with large catchment areas (Figure 23). Often individuals who build or purchase homes in such areas fail to recognize them as sinkholes and never consider the chance of flooding, especially since the nearest surface stream may be miles away. Unfortunately, many people believe that a sinkhole must be a steep-walled depression, a "hole" in the ground. In Figure 23, the steep-walled depression near Batsel Avenue is easily recognizable as a sinkhole, but most of the people who built homes on Covington Street did not realize that they were building in the upper portion of that same sinkhole. People normally do not build in the bottoms of deep, easily recognizable sinkholes, and some towns built upon sinkhole plains have relatively minor sinkhole flooding problems for this reason. Unfortunately, Bowling Green has mostly large, shallow karst depressions, and consequently flooding is a major problem.

The sinkhole flood plain

The U.S. Department of Housing and Urban Development defines the 100-year flood elevation along streams as the flood plain for flood insurance purposes. For Bowling Green, the Department has accepted the sinkhole flood plain as the three-hour, 100-year flood elevation assuming no drainage from the sinkhole (Booker 1978). This definition of the sinkhole flood plain, first suggested by Daugherty (1976) has been a part of the Bowling Green-Warren County Storm Water Management Program for establishing flood easements since 1976.

Sinkhole flooding may not be a problem when a home or business is built, but continued urbanization of the catchment results in greater areas of impervious surface and consequently an increase in storm water runoff. As land use in sinkhole catchments changes from agricultural to suburban or from suburban to commercial, the depth, area, and frequency of sinkhole flooding increases. Thus, a home built in an agricultural catchment may find itself within the sinkhole flood plain if the land use changes to suburban. In order to prevent this from occurring, developers in Warren County are required to retain on site any increased runoff during a 100-year rainfall resulting from land use changes associated with construction.

Kemmerly (1981) agrees that the definition of the sinkhole flood plain should be the 100-year flood contour assuming no outflow, but he recommends that it reflect the anticipated runoff volumes with maximum urbanization (i.e., impervious surface area = >50%). For Springfield-Greene County, Missouri, Aley and Thomson (1981) recommend a 24-hour, 100-year flood elevation assuming no drainage and 100% runoff of the rain falling within the area topographically tributary to the sinkhole. Mills, Starnes and Burden (1982) also suggest a 24-hour rainfall for Cookeville, Tennessee. They maintain that for nonkarst areas hourly rainfall intensities are the most important, but for sinkholes the time interval should be somewhat longer because they drain much more slowly than do stream channels. The definition of a sinkhole flood plain may therefore vary from one location to another.

Towns like Springfield, Missouri and Cookeville, Tennessee have large areas for growth which do not have karst topography. Considering the problems of sinkhole flooding, groundwater contamination and sinkhole collapse common to sinkhole plains, the cities should not only establish maximum levels for their sinkhole flood plains but also take other measures, such as, restricting lot sizes to a minimum of three acres (Aley and Thomson 1981) in order to encourage development in areas not having karst topography. Unfortunately, Bowling Green is located entirely upon a sinkhole plain and does not have this option.

Excavation of sinkhole drains

The first step in correcting a sinkhole-flooding problem is usually to unclog the sinkhole drain by excavating with a backhoe. Although this often reduces future flooding levels, the excavation occasionally blocks the drain further and flood levels increase. If a crevice in the bedrock can be found, a concrete box with a grate is often constructed to prevent the drain from becoming plugged with soil and debris.

Storm water drainage wells

The most effective wells intersect solutionally-enlarged bedding plane partings or joints, and occasionally they hit microcaves or even large cave passages. Other wells, often located only a few meters (feet) away from wells of high capacity, may hardly drain at all. Drainage wells help prevent storm water from ponding in streets and yards during normal rains, and some are effective in preventing or greatly reducing flooding of sinkholes with relatively small catchments even during flood–producing rains. In sinkholes with large catchments, wells do not appear to have the capacity to significantly lower the level of flooding.

Batsel Avenue Sinkhole

The Batsel Avenue Sinkhole is a good example of the kinds of sinkholes that have the most serious flooding problems (Figure 23). It is very shallow with a large catchment area. Actually, it is about one-half mile long and only four-feet deep. It floods completely during major floods and then overflows into adjacent sinkholes and causes them to flood. Notice that Figure 23 has a 2-foot contour interval. No one would build a house in the deep portion of the sinkhole along Batsel Avenue, but the topography along Covington Street is almost perfectly flat. One would have to survey the area to reveal that it is, in fact, a sinkhole. The people who build homes along Covington Street probably never considered the possibility of flooding, particularly since the nearest stream of any kind, the Barren River, is over a mile away. After 50 years of periodic flooding, the city has recently obtained a FEMA Grant to purchase some of these homes, and they are presently being torn down and removed. Homeowners in twenty sinkholes in Bowling Green are eligible for Federally-subsidized flood insurance.

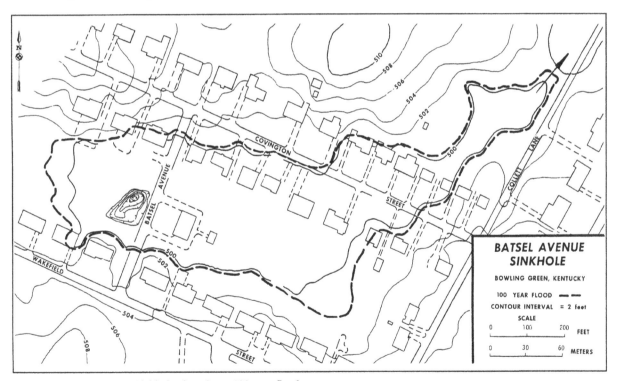

Figure 23: Batsel Avenue Sinkhole, three-hour, 100-year flood contour.

FIELD TRIP STOP 5. TOUR OF CENTER FOR CAVE AND KARST STUDIES DYE TRACER LABORATORY, WESTERN KENTUCKY UNIVERSITY

This stop will include a tour of the new dye tracer laboratory recently donated to the Center for Cave and Karst Studies by Crawford and Associates, Inc. (Figure 24). It will include a discussion of dye tracer investigation techniques and protocols and a laboratory demonstration of dye analysis on a Shimadzu 5300 PC spectrofluorophotometer and computer analysis for multiple dyes using a non-linear, curve-fitting software program.

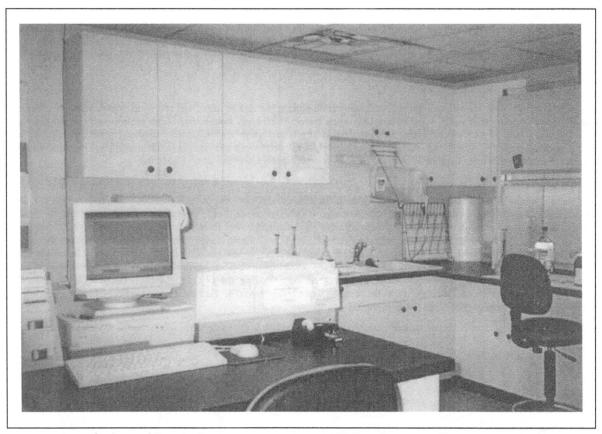

Figure 24: Center for Cave and Karst Studies dye tracer laboratory . Shimadzu 5300 spectrofluorophotometer at left, fume hood for activated charcoal dye receptor elution at right.

FIELD TRIP STOP 6. EGYPT PARKING LOT SINKHOLE FLOODING PROBLEM, WESTERN KENTUCKY UNIVERSITY: GEOPHYSICAL TECHNIQUES FOR LOCATING AND MAPPING CAVES FROM THE GROUND SURFACE

Introduction

Microgravity has been used extensively to investigate karst subsurface features in Kentucky and elsewhere by the Center for Cave and Karst Studies since 1985. Although other geophysical techniques have been tried, the best results have been obtained by using microgravity traverses to locate bedrock caves, voids in the regolith and to investigate sinkhole collapses.

Bouguer gravity can identify locations on the earth's surface that have relatively higher or lower gravity caused by lateral variations in subsurface density. Crawford has used microgravity extensively to locate bedrock caves from the ground surface. The lower density of the air, water, or mud within a cave compared to the surrounding carbonate rock results in a low-gravity anomaly. He has also used microgravity to locate voids in the regolith (all unconsolidated material above bedrock) that are potential sinkhole collapses. Since regolith is less dense than limestone bedrock, Bouguer gravity can also identify variations in depth to bedrock. In limestone areas, depth to bedrock is often very irregular, with limestone pinnacles that protrude upwards and cutters that extend downward. Cutters are V-shaped, regolith-filled crevices formed by solution of the limestone by soil water as it percolates down to the karst aquifer. Regolith arches form as regolith spalls into solutionally-enlarged voids in the bedrock. For these reasons, low-gravity anomalies indicate bedrock caves, voids in the regolith, or places where depth to bedrock is abruptly greater and therefore often indicative of places where regolith may be descending into bedrock crevices.

Detection of bedrock caves

Several researchers have demonstrated that gravity can be used to detect large bedrock caves (Omnes G. 1976; Kirk K.G. 1974; Butler D.K. 1983; and Smith, D.L. and Smith G.L. 1987). Figures 25 and 26 are examples where relatively large low-gravity anomalies along level traverses reveal the location of underlying cave passages. Although some studies had identified anomalies that were hypothesized to be caves, few, if any, wells had been drilled into these anomalies to confirm that they did in fact indicate cave passages previous to research performed by Crawford in 1985. Toxic and explosive vapors rising from contaminated cave passages into homes under Bowling Green, Kentucky in 1984 and 1985 resulted in an intensive effort to find the cave passages over a relatively large area (Crawford, 1989; Crawford, Webster, and Winter, 1989; Crawford, Lewis, Winter and Webster, 1999).

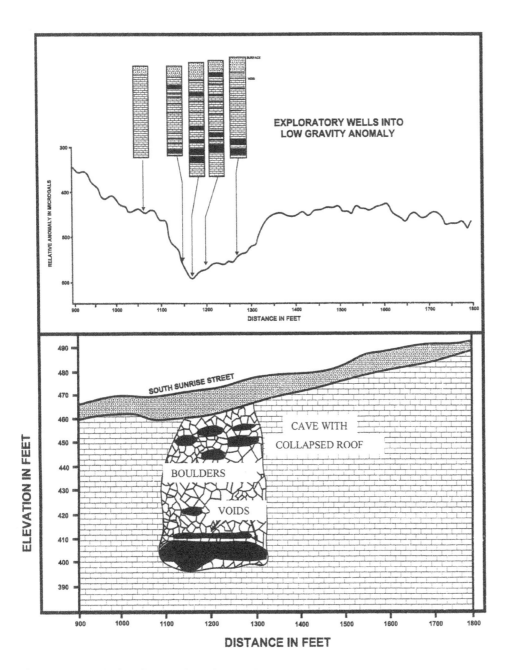

Figure 25: Four borings into this large low-gravity anomaly along South Sunrise Street in Bowling Green, Kentucky, intersected numerous voids and boulders indicative of a collapsed bedrock cave. There is no surface expression that might reveal the presence of the collapsed cave. Other borings along the traverse did not intersect voids.

Although several geophysical techniques for locating caves were considered and a few tried, the most successful was microgravity. The best results were obtained by taking microgravity measurements with a LaCoste and Romberg Model D Microgal Gravity meter or with a Scintrex CG-3M Autograv Microgravity meter at a ten-foot interval along traverses perpendicular to a hypothesized route of a cave stream. The hypothesized route was derived from topographic analysis, knowledge of local hydrogeology, dye traces and a detailed water table map. Voids existing beneath low-gravity anomalies were confirmed by exploratory drilling (Figures 25 and 26). By proceeding in a "leap frog" fashion with short parallel traverses, the route of the cave was established (Figure 27).

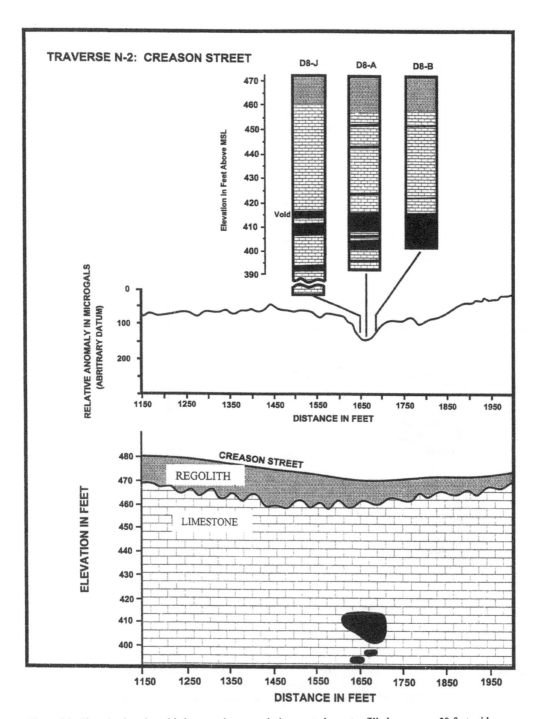

Figure 26: Three borings into this low-gravity anomaly intersected a water-filled cave over 50 feet wide.

Egypt Parking Lot sinkhole flooding problem

Storm water runoff from most of Western Kentucky University's campus has been directed into a large sinkhole basin that contains the Egypt Parking Lot, so named because it floods so frequently. Also within the sinkhole is a railroad underpass for Route 68/80 and Route 231. This sinkhole flooding problem has existed for more than forty years, and on one occasion it delayed fire trucks from reaching an apartment fire. On April 16, 1998, the sinkhole flooded, and over 80 student cars parked at the Egypt Lot were inundated. In 1999, the Center for Cave and Karst Studies at Western Kentucky University received a grant from the City of Bowling Green to use microgravity to investigate the flooding problem. The Center had used microgravity to locate caves from the ground

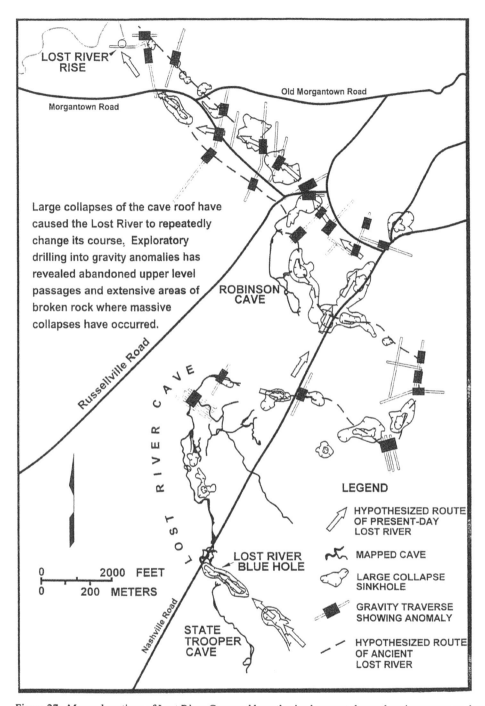

Figure 27: Mapped portions of Lost River Cave and hypothesized present-day and ancient routes as determined by microgravity (Crawford, 1986).

surface in various parts of Bowling Green since 1985 and had already located a large cave near the Egypt Parking Lot. Three wells drilled into a low-gravity anomaly along Creason Street revealed a large cave at least 15 feet high and over 50 feet wide (Figure 26). Additional traverses were made to trace the route to the cave stream under the Egypt Parking Lot. A large, low-gravity anomaly was detected under the parking lot, and a six-inch diameter well was drilled into the center of the low-gravity anomaly during the summer of 2000 (Figure 28). The exploratory boring intersected a large cave forty feet below the surface. The Center's down-hole camera revealed that the cave had three feet of air, two feet of water and twelve feet of silt. The diameter of the passage could not be estimated because the light on the camera could not expose the cave walls. Obviously, it is a wide cave passage.

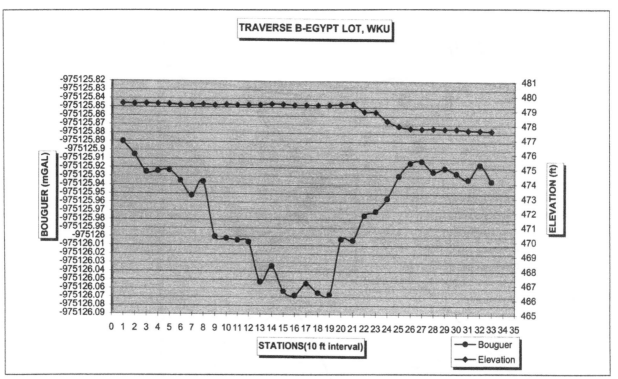

Figure 28: Egypt lot microgravity traverse B revealed a significant low-gravity anomaly. A well drilled into the anomaly penetrated 20 feet of regolith and 20 feet of bedrock before intersecting a large cave that contained 3 feet of air, 2 feet of water and 12 feet of sediments. A dye trace confirmed that the cave passage was part of the Lost River Cave system.

Figure 29 shows a traverse over the known cave using the Center's Sting resistivity meter with a 28 electrode dipole-dipole array. The resistivity data are modeled using two-dimensional resistivity tomography. Pressure transducers have been installed in the well into the cave stream and also on the surface at the bottom of the sinkhole in order to compare water levels during floods (Figure 30). The variation in water levels during floods will be used to evaluate the cave's capacity to accept additional storm water runoff. If the research indicates the cave has the capacity to receive additional runoff during floods, the Center will propose that a five to six-foot diameter well be drilled to permit storm water to drain rapidly into the large Lost River Cave and thus reduce or eliminate the flooding problem.

Research by the Center and other agencies has revealed that the "first flush" of storm water runoff contains most of the oil and grease, metals and other contaminants associated with urban storm water runoff. Presently, there are several drainage wells and even exposed bedrock crevices in the Egypt Lot and railroad underpass. Therefore, the "first flush" is presently going directly into the karst aquifer. As the capacities of these wells are exceeded, the cleaner storm water fills the sinkhole during floods. The Center intends to design a wetland treatment system so that the "first flush" would flow into a constructed wetland at the bottom of the sinkhole basin. As the water rises, the later runoff would then spill over into the large-diameter well. Therefore, the proposed mitigation would reduce or eliminate the flooding problem while also reducing the storm water runoff contaminant load that is presently flowing into the karst aquifer (Crawford, Fryer and Calkins, 2000). Figure 31 is a draft of the proposed storm water treatment facility designed by graduate students Joel Despain, Will Blackburn and Susan Marklin (2000).

Demonstration of karst research techniques

The Center's Scintrex CG-3M Autograv Microgravity meter and Sting Resistivity meter will be demonstrated at this stop. Also, a video log will be prepared of the well drilled into the cave stream using the Center's downhole video camera.

FIELD TRIP STOP 7. DISHMAN-McGINNIS ELEMENTARY SCHOOOL: USEPA HEALTH ADVISORY DUE TO TOXIC AND EXPLOSIVE VAPORS RISING FROM KARST AQUIFER INTO HOMES AND SCHOOLS. ALSO, DISCUSSION OF RADON GAS PROBLEM AND MITIGATION.

Toxic and explosive vapors

In January, 1984 fumes began to rise from contaminated cave streams into several homes and buildings in Bowling Green's Forest Park Subdivision. Initial investigations were made by the Bowling Green Fire Department and Kentucky Fire Marshal's Office. In May, 1984 Crawford was asked by the State Fire Marshal and by the Bowling Green City Commissioners to investigate and prepare a

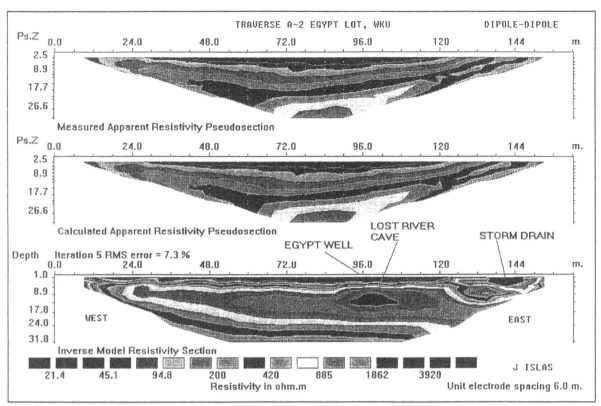

Figure 29: Traverse over the known cave using the Center's Sting resistivity meter and a 28 electrode dipole-dipole array. The resistivity data are modeled using two-dimensional resistivity tomography.

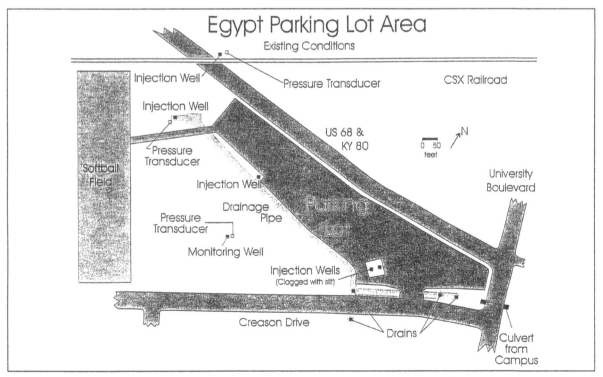

Figure 30: Map showing pressure tranducers – data loggers installed to study water levels during floods both on the surface and in the well drilled into Lost River Cave.

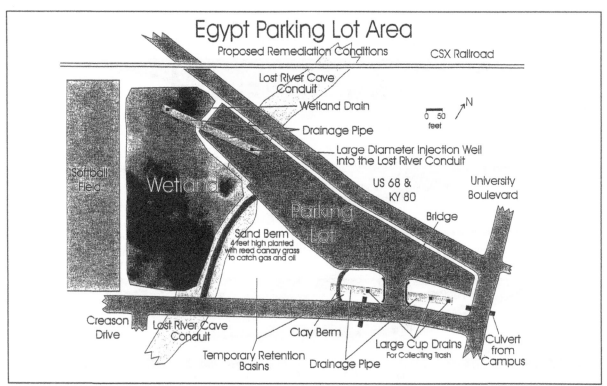

Figure 31: Draft of proposed storm water treatment and sinkhole flooding mitigation system designed by graduate students, Will Blackburn, Joel Despain and Susan Marklin (2000).

proposal for dealing with the problem. Crawford and his students (1984) identified the homes, buildings, drainage wells and sinkholes in the Forest Park area with fumes and recommended that the USEPA be asked to deal with the problem.

USEPA Region IV began a health investigation in cooperation with the U.S. Centers for Disease Control (CDC) in June, 1984. Air samples collected from bedrooms, crawl spaces, and basements of 125 homes in the Forest Park area were analyzed, as were blood samples and urine samples of residents from twenty homes. During the investigation the fume problem continued to worsen as fumes forced the partial evacuation of Parker-Bennett Elementary School in November, 1984 and Dishman-McGinnis Elementary School in January, 1985. Fumes were reported in over fifty homes, three commercial buildings, two schools, one church, eighteen drainage wells, four sinkholes, and nine caves. The most severe problems were in the Forest Park area above the Lost River Groundwater Basin. Another area of concern was in the Parker-Bennett School area above the Double Springs Groundwater Basin.

The USEPA and CDC health investigation lasted for over eight months. Meanwhile, the problem continued to worsen. After the initial investigation by Crawford and his students in May, Crawford recommended that a backhoe be used to excavate down to bedrock at the place where the fumes were the strongest at each affected home or building. He predicted that a bedrock crevice could be found leading down into the cave system at each location. A pipe could then be installed into the crevice, and a fan with an explosion-proof motor could then be used to pull the fumes from the crevice and vent them into the atmosphere above the home. The negative pressure created by the exhaust fan would prevent the fumes from entering the home or building. Unfortunately, no government agency was willing to provide the money or assume the responsibility until November, 1984, when fumes forced the evacuation of three classrooms in Parker-Bennett Elementary School and threatened closure of the entire school. The Bowling Green City School Board provided the money, and a crevice was found immediately as predicted. The ventilation system worked quite well, and the school did not miss a single day of classes (Figure 32). In January, 1985 fumes in Dishman-McGinnis Elementary School were quickly ventilated also (Figure 33)

After an extensive public health evaluation of the fumes by the CDC, a Health Advisory was issued in March, 1985 for the Bowling Green area. The advisory was issued on the grounds that explosive levels of fumes existed and that detectable levels of benzene, toluene, and xylene exceeded standards for "non-occupational settings". As a result of the Health Advisory, the USEPA, Region IV initiated a "Superfund" emergency response in April, 1985.

Crawford was hired by USEPA to assist with the problem, as were a team of his graduate students that included Chris Groves, Tony Able, John Hoffelt, and Jim Webster. The Center for Cave and Karst Studies Team explored and mapped caves, excavated entrances into caves, directed the drilling of two 30-inch diameter wells into caves that were located by using microgravity and exploratory borings, and with the help of the late Frank Reid and his magnetic induction transmission "cave radio", they directed the installation of monitoring wells into cave passages.

Figure 32: Pipe with an explosion-proof fan was installed over a bedrock crevice to vent fumes from the karst aquifer into the atmosphere at Parker-Bennett Elementary School.

Figure 33: The same procedure (Figure 32) was used to prevent fumes rising from the karst aquifer from entering Dishman - McGinnis Elementary School.

An extensive investigation to find the sources of the hazardous vapors was made by USEPA and the Kentucky Division of Water. A tank-testing program for underground storage tanks was initiated by USEPA, and four gasoline tanks in the Forest Park area were found to be leaking. The Kentucky Division of Water, assisted by USEPA, made a systematic investigation of factories in the South

Side Industrial Park and elsewhere. When violations were found, dye traces were performed by the Center for Cave and Karst Studies Team to determine the groundwater flow route of any contaminant being discharged into the ground. Seven factories were cited for discharging into the karst aquifer without a permit. Several factories had drains within their buildings that were connected to storm sewers, discharging their toxic waste directly into two wells it had drilled into the karst aquifer.

After the four leaking gasoline tanks were replaced and the factories stopped discharging toxic chemicals into the karst aquifer, the vapors rising into homes and buildings slowly decreased. However, the city continues to have sporadic problems. In August, 1999, gasoline vapors began to rise into the Greenwood Park Church of Christ and the Foundation Christian Academy. Again the Center for Cave and Karst Studies assisted with: 1) advice on the installation of fans to ventilate the building, 2) microgravity traverses to direct exploratory borings so that cave air could be drawn directly from the cave passage under the building, and 3) performing several dye traces to determine groundwater flow direction. The fans continue to prevent the gasoline vapors from entering the building as of January, 2001.

Radon Gas

The large fan installed at Dishman-McGinnis Elementary School in 1985 (Figure 33) has been removed and replaced with radon fans which are presently venting soil and cave air into the atmosphere to prevent radon gas from entering the building. In 1987 the Center for Cave and Karst Studies used a stratified, systematic, unaligned sample design to select 100 homes for radon testing. Activated charcoal canisters were used to perform initial screening test for radon. Residential radon levels averaged 10.4 pCi/l with several homes having levels over 100 pCi/l. The tests showed that 86% of the basement homes tested were above 4 pCi/l and 57% of the first floor homes (crawl space homes) tested were above 4 pCi/l. USEPA recommends that homes above 4 pCi/l be mitigated to reduce the levels below the 4 pCi/l action level (Webster and Crawford, 1989). In recent years, USEPA has come to realize that karst areas often have extremely high residential radon levels.

REFERENCES

Aley, T.J., and K.C. Thomson, (1981), Hydrogeologic Mapping of Unincorporated Green County, Missouri, to Identify Areas Where Sinkhole Flooding and Serious Groundwater contamination Could Result from Land Development, Green County Sewers District Report, Ozark Underground Laboratory, Protem, Missouri, p. 11.

Benson, R.C., Kaufmann, R.D., Yuhr, L.B. and D. Martin (1988). Assessment, prediction and remediation of karst conditions on I-70, Frederick, Maryland, 40th Highway Geology Symposium, pp. 1-15.

Blackburn, W.; Despain, J.; and S. Marklin (2000). Sinkhole flooding mitigation: Egypt Parking Lot, Western Kentucky University, Bowling Green, Kentucky. Field Methods paper directed by N. Crawford, Dept. of Geography and Geology. Western Kentucky University, Bowling Green, Kentucky, p. 11.

Booker, R.W. and Associates, Inc., (1978). Study of Sinkhole Flooding, Bowling Green and Warren County, Kentucky. Report prepared for Federal Insurance Administration.

Butler, K. K., (1983). "Micro-gravimetric and magnetic surveys, Medford Cave Site, Florida," Cavity Detection Research Report 1, Tech, Rep. GL-83-1. Vicksburg, MS: Army Engineer Waterways Station, p. 92.

Crawford, N.C., (1979). Grider Pond-Cave Mill Road interceptor project Phase I: dye tracing of septic tanks believed to be contributing to the impairment in water quality of the Lost River in Bowling Green, Kentucky, for G. Reynolds Watkins consulting Engineers, p. 19.

Crawford, N.C,. (1981). Karst Flooding in Urban Areas, Bowling Green, Kentucky; Proceedings of the Eighth International Congress of Speleology, Western Kentucky University, Bowling Green, Kentucky, p. 763-765.

Crawford, N.C., (1982). Hydrogeologic Problems Resulting from Development Upon Karst Terrain, Bowling Green, Kentucky, Guidebook for USEPA Karst Hydrogeology Workshop, Nashville, Tennessee, p. 34.

Crawford, N.C., (1984). Sinkhole Flooding associated with Urban Development upon Karst Terrain: Bowling Green, Kentucky. In Beck, B.F. (ed), Sinkholes: Their Geology, Engineering and Environmental Impact. Proceedings First Multidisciplinary Conference on Sinkholes. Rotterdam, Netherlands, A.A. Balkema, pp. 283-292.

Crawford, N.C., (1985). Map of Groundwater Flow Routes: Lost River Groundwater Basin, Warren County, Kentucky. Center for Cave and Karst Studies, Western Kentucky University, Bowling Green, Kentucky.

Crawford, N.C., (1988). Karst Hydrologic Problems of South Central Kentucky: Groundwater Contamination, Sinkhole Flooding, and Sinkhole Collapse. Field Trip Guidebook, Second Conference on Environmental Problems in Karst Terranes and Their Solutions. National Water Well Association, Nashville, Tennessee, p. 107.

Crawford, N.C. (1989a). Karst hydrogeology of Warren County, Kentucky, in Crawford (ed.) Karst Landscape Analysis, Warren County Comprehensive Plan, pp. 1-36.

Crawford, N.C., (1989b). Sinkhole flooding: Bowling Green-Warren County Storm Water management program, in Crawford (ed.) Karst Landscape Analysis, Warren County Comprehensive Plan, pp. 39-52.

Crawford, N.C., (1989c). Groundwater contamination in Warren County, Kentucky: toxic and explosive fumes rising from contaminated groundwater flowing through caves in Bowling Green, in Crawford (ed.) Karst Landscape Analysis, Warren County Comprehensive Plan, pp. 117-143.

Crawford, N.C., (1995). Microgravity techniques for detection of karst subsurface features, Site Investigations: Geotechnical and Environmental, Proceedings. Twenty-Sixth Ohio River Valley Soils Seminar. Clarksville, IN, pp. 1-24.

Crawford, N.C., (2000). Microgravity investigation of sinkhole collapses under highways, in Geophysics 2000 Proceedings, First International Conference on the Application of Geophysical Methodologies to Transportation Facilities and Infrastructures, St. Louis, Missouri, p. 13 (in press).

Crawford, N.C., Fryer, S.E., and C. R. Calkins, (2000). Geophysical techniques for locating and mapping caves from the ground surface: Microgravity subsurface investigation of the Egypt Parking Lot, Western Kentucky University. Proceedings of Mammoth Cave National Parks Eighth Science Conference, Mammoth Cave National Park, pp. 25-36.

Crawford, N.C., Groves, C.G., Tucker, R.B., Gorbis, S.K., Clauson, W.T., Erickson, K.E., and J.K. Patrick, (1995). Microgravity subsurface investigation at the site of a sinkhole collapse under the South Entrance Road to Mammoth Cave National Park, Kentucky, report for Mammoth Cave National Park, p. 112.

Crawford, N.C., Lewis, M.A., Winter, S.A., and J.A. Webster, (1999). Microgravity techniques for subsurface investigations of sinkhole collapses and for detection of groundwater flow paths through karst aquifers. In Beck, Pettit, and Herring (ed.) Hydrogeology and Engineering Geology of Sinkholes and Karst. Proceedings of the Seventh Multidisciplinary Conference on Sinkholes and the Engineering and Environmental Impacts of Karst, Harrisburg-Hershey, Pennsylvania, pp. 203-218.

Crawford, N.C., and J.L. Snead (1999). Groundwater contamination in Warren County, Kentucky: onsite sewage disposal systems, in Crawford (ed.) Karst Landscape Analysis, Warren County Comprehensive Plan, pp.145-172.

Crawford, N.C., Tucker, R.B., and M.A. Sumerlin, (1995). Microgravity subsurface investigation at the site of a sinkhole collapse under Interstate 65 at Elizabethtown, Kentucky. Prepared for Kentucky Department of Transportation, May 31, 1995, p. 105.

Crawford, N.C., and C.S. Ulmer, (1994). Hydrogeologic Investigations of contaminant movement in karst aquifers in the vicinity of a train derailment near Lewisburg, Tennessee. Environmental Geology V.213, pp. 41-52.

Crawford, N.C., and J.W. Webster, (1988). Microgravity Investigation of the proposed Dishman Lane Extension, Bowling Green, Kentucky, report for Law Engineering, Nashville, Tennessee, p. 20.

Crawford, N.C., Webster, J.W., and G. Veni, (1980). Sinkhole collapse problems in Warren County, in Crawford, N.C. (ed.) The Karst Landscape of Warren County, prepared for Bowling Green-Warren County Planning and Zoning Commission, Center for Local Government, Western Kentucky University, Bowling Green, Kentucky, pp. 71-115.

Crawford, N.C., Webster, J.W., and S.A., Winter, (1989). Detection of caves from the surface by microgravity followed by exploratory drilling: Lost River Groundwater Basin, Bowling Green, Kentucky, Prepared for the city of Bowling Green, Municipal Order No. 85-83.

Daugherty, D. L., (1976). Storm Water Management. City-County Planning Commission of Warren County, Kentucky.

Islas, J. L., (2000). Resistivity and microgravity techniques for locating and mapping the Lost River Cave from the ground surface, Field Methods paper directed by N. Crawford, Dept. of Geography and Geology, Western Kentucky University, Bowling Green, Kentucky, p. 32.

Kemmerly, P., (1981). The Need for Recognition and Implementation of a Sinkhole-Flood Plain Hazard Designation in Urban Karst Terrains. Environmental Geology, Vol. 3.

Kirk, K.G. (1974). Resistivity and gravity surveys applied to karst research, Proceedings of the 4th Conference on Karst Geology and Hydrology. West Virginia Geological Survey, pp. 61-71.

Lobeck, A. K. (1932). Atlas of American Geology, New York: The Geographical Press, Columbia University, Sheet No. 40.

Mace, C.E. (1921). Sewer system more then a million years old, Popular Mechanics Magazine, 35, 5, May 1921, p. 687.

McGrain, P., and D.G. Sutton, (1973). Economic Geology of Warren County, Kentucky, Kentucky Geoological Survey, Series 10, County Report 6, p. 28.

Mills, H.H., Starnes D.D., and K.D. Burden, (1982). Coping with Sinkhole Flooding in Cookeville, Tennessee, *Tennessee Technical Journal.*

Omnes, G. (1976). High accuracy gravity applied to the detection of karstic cavities, in Karst Hydrogeology (ed.). J.S. Tolson and F.L. Doyle, International Association of Hydrogeology Memoir 12. pp. 273-284.

Quinlan, J.F., and J. A. Ray (1981). Groundwater Basins in the Mammoth Cave Region, Kentucky, Occasional Publication #1, Friends of the Karst, p. 1.

Quinlan, J.F., and D.R. Rowe (1977). Hydrology and water quality in the Central Kentucky Karst, Phase I, University of Kentucky Water Resources Research Institute, Rep. 101, p.93.

Smith, D.L., and G. L. Smith, (1987).
Use of vertical gravity gradient analysis to detect near-surface dissolution voids in karst terrains, Second Multidisciplinary Conference on Sinkholes and the Environmental Impacts of Karst, (eds.) Beck, B.F. and Wilson, W.L. pp. 205-209.

Stroud, F,B.; Powell, G.W.; Crawford, N.C.; Rigatti, M.J., and P.C. Johnson (1986). Role of federal agencies in emergency response, to toxic fumes and contaminated groundwater in karst topography: a case study, Karst Environmental Problems in Karst Terranes and Their Solution Conference, National Water Well Association, Western Kentucky University, Bowling Green, Kentucky, pp. 197-225.

Webster, J.W. and N.C. Crawford (1989). Radon Levels in the Homes and Caves of Bowling Green, Warren County, Kentucky, Warren County Comprehensive Plan Technical Report, p. 15.

Geotechnical and Environmental Applications of Karst Geology and Hydrology, Beck & Herring (eds)
© 2001 Taylor & Francis, ISBN 90 5809 190 2

Field Trip Guide, Part 2: The Mammoth Cave karst aquifer

JOE MEIMAN   National Park Service, Mammoth Cave National Park, Mammoth Cave, KY 42259, USA,
joe_meiman@nps.gov

CHRIS GROVES   Hoffman Environmental Research Institute, Dept. of Geography & Geology,
Western Kentucky University, Bowling Green, KY 42101, USA, chris.groves@wku.edu

STOP ONE: PARK MAMMOTH OVERLOOK

Physiography of the Pennyroyal Plateau

We are standing on the platform (elevation about 250 meters) at the edge of the Dripping Springs Escarpment, looking down upon the Pennyroyal Plateau (Figure 1). Virtually everywhere within your view is within the park's karst groundwatershed – the exception being the area near the lone knob (Pilot Knob) the extreme southwest, which drains to the Barren River via the Graham Springs groundwater basin. The southern horizon comprises the headwaters of the Turnhole Spring groundwater basin. At nearly 250 km², it is the third largest karst watershed in Kentucky, and the largest draining into Mammoth Cave National Park.

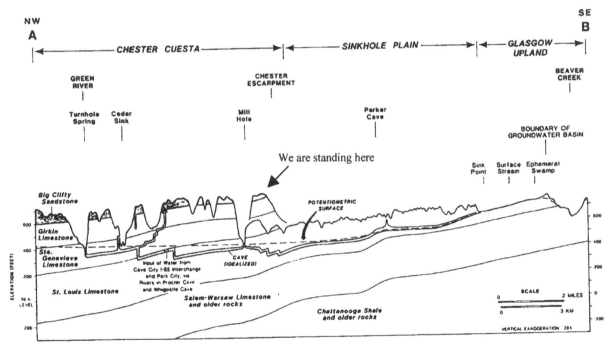

Figure 1. Cross-section of physiographic regions of the Mammoth Cave area (from Quinlan and Ewers, 1981).

The Pennyroyal Plateau is comprised of two distinct pysiographic regions: The Glasgow Uplands and the Sinkhole Plain. The former, ranging from three to six kilometers wide at an elevation ranging from 170 to 230 meters, is underlain by the argillaceous limestones of the lower portions of the St. Louis Limestone (Upper Mississippian) and is characterized by numerous sinking streams (Figure 1). These streams flow northward until they reach the more soluble beds of the upper portion of the St. Louis where they sink at discrete ponors. Note that although the regional dip is a gentle one to one and a half degrees to the northwest, the hydraulic gradient within the watershed (which is also generally towards the northwest) is even more subtle, thus as water flows downstream it is also flowing up-section.

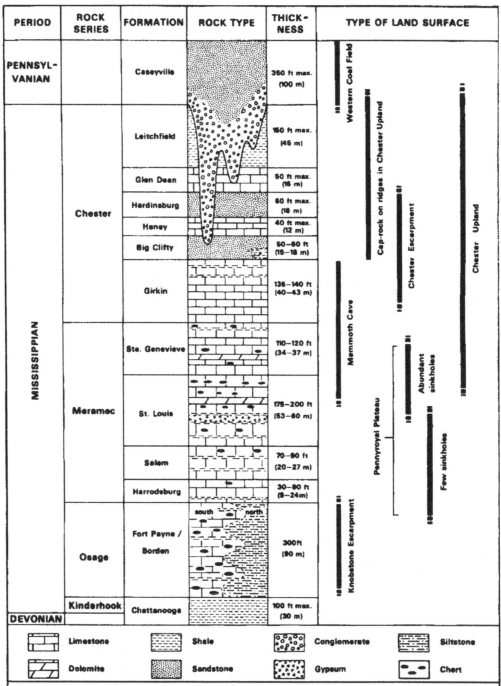

PERIOD	ROCK SERIES	FORMATION	ROCK TYPE	THICK-NESS	TYPE OF LAND SURFACE
PENNSYL-VANIAN		Caseyville		350 ft max. (100 m)	Western Coal Field
MISSISSIPPIAN	Chester	Leitchfield		150 ft max. (45 m)	Cap-rock on ridges in Chester Upland
		Glen Dean		50 ft max. (15 m)	
		Hardinsburg		60 ft max. (18 m)	Chester Escarpment
		Haney		40 ft max. (12 m)	
		Big Clifty		50–60 ft (15–18 m)	Chester Upland
		Girkin		136–140 ft (40–43 m)	Mammoth Cave
	Meramec	Ste. Genevieve		110–120 ft (34–37 m)	Abundant sinkholes
		St. Louis		175–200 ft (53–60 m)	Pennyroyal Plateau / Few sinkholes
		Salem		70–90 ft (20–27 m)	
		Harrodsburg		30–80 ft (9–24m)	
	Osage	Fort Payne / Borden	south north	300 ft (90 m)	Knobstone Escarpment
	Kinderhook	Chattanooga		100 ft max. (30 m)	
DEVONIAN					

	Limestone		Shale		Conglomerate		Siltstone
	Dolomite		Sandstone		Gypsum		Chert

Figure 2. Stratigraphic Section of the Southcentral Kentucky Karst (from Palmer, 1981).

Thus begins the Sinkhole Plain. Bounded by the Dripping Springs Escarpment to the north and the Glasgow Uplands to the south, the Sinkhole Plain (here, about five kilometers wide at an elevation of ranging from 170 to 210 meters) is entirely internally drained. The sinking streams of the Glasgow Uplands form the main trunk conduits carrying water through the karst aquifer and are fed by countless dolines of the Sinkhole Plain. In general, the "water table" (if such a thing exists in this aquifer) is approximately 50 meters beneath the surface of the Sinkhole Plain. The sinkhole ponds that you see are not the "water table, but are dolines that have been either naturally or anthropologically plugged and are perched above the aquifer. An extremely important groundwater recharge and storage mechanism is the epikarst that underlies the soils throughout this region. These solutionally-enhanced fractures and bedding planes, usually extending five to ten meters into the bedrock, provide a tremendous amount of readily-accessible stores that keep the cave streams flowing during times of extreme drought. They also can be extremely difficult to clean up following spills.

It is very easy, from this vista, to imagine the dissolutional geomorphic agents that shaped this land. For decades karst researchers have waxed on, sometimes almost poetically, about the formation of this classic karst landscape. How the Pennyroyal was lowered differentially with respect to the silisiclastic-capped uplands. How the Dripping Springs Escarpment has retreated to the north, leaving behind the knobs as outliers. It seems like normally rational individuals visit south-central Kentucky and don their karst-colored glasses, and attribute all they see to chemical dissolutional processes. We are now accumulating evidence of a totally different sort. We are hypothesizing that the south-central Kentucky Karst, including the Pennyroyal Plateau, is actually a fluvial landscape which has been relatively recently been modified by karst geomorphic processes. For the past twenty years Joe Ray has collected rounded chert stream gravel throughout the Pennyroyal. Joe suggests that the Pennyroyal was laterally planed by channelized flowing streams (Ray, 1996). This hypothesis may be correct, but then again, it could be wrong. The point being that regardless of what hypothesis proves accurate, one is currently being supported by actual data.

HAZMAP Project

The majority of the water flowing through the subterranean rivers of Mammoth Cave originates from private lands beyond park boundaries. These rivers, classified by the Commonwealth of Kentucky as Outstanding Resource Waters, support the most biologically diverse cave aquatic ecosystem in the world, including the Federally Endangered Kentucky Cave Shrimp. Groundwater flow properties of karst combine to create a hydrologic system in which the surface, and all surface activities, are highly integrated with the subsurface, and all aquatic ecosystems of the karst aquifer.

Across the recharge area, where surface water is immediately transferred to the cave below, travel tens of thousands of trucks and several dozen trains each day. It is typical to have as many as five spills of toxic material along these routes per year. If such a spill occurs at the wrong hydrologic location (a sinkhole or sinking creek) and/or during a rainfall event, the aquatic communities of the cave could be severely impacted.

Once a spill enters the karst aquifer there is little, if any chance for immediate mitigation. As the spill must be contained at the surface, both response time and prior knowledge of surface hydrology are critical. Before this project was completed, when a spill occurred, emergency responders arrived with little knowledge of the labyrinth of drainage ditches, culverts and sinkholes unique to the site and precious little time for their reconnaissance. The responders had to be equipped with a detailed hydrologic map that clearly indicates points of groundwater vulnerability relative to easily identifiable landmarks. The time saved by arriving on-site with the knowledge of where the spill is located with respect to a storm drain, sinkhole or sinking creek might prevent an ecological disaster.

Under the Mammoth Cave Area International Biosphere Reserve, Mammoth Cave National Park, in cooperation with the Kentucky Division of Water (KYDOW), and the Barren River Area Development District, created a set of groundwater hazard maps along the major transportation corridors (Interstate 65, 19 kilometers; the CSX Railroad, 22 kilometers; and the Cumberland Parkway, 8 kilometers) traversing the park's recharge basin (the Turnhole Spring Karst Groundwater Basin). These maps, which detail a 1/2 mile (0.8 km) section of roadway (or 1 mile [1.6 km] of railroad), display a host of landmarks (signs, mile-markers, guard rails, outcrops, and bridges), hydrologic features (paved ditches, streams, inlet boxes, pipe headwalls, and sinkholes), and potential hazards that may greatly alter the designed flow of surface waters (cracks, undercuttings, and collapses). Equipped with this map set, the emergency responder knows immediately where, hydrologically, the spill occurred and knows exactly where to deploy lines of defense for spill mitigation.

The map sets were completed and printed in summer of 1996. Training sessions were held and maps distributed for all area emergency responders. As long as hazardous materials are transported over karst lands there will be a need for rapid spill assessment and mitigation. It is the hope that through this effort of Mammoth Cave Area International Biosphere Reserve cooperators that spills will be dealt with quickly and effectively.

When the responder is called, locate the detailed map from the index map based upon highway mile markers (Figure 3). For example, if a spill occurs near mile marker 48 on I-65, the index map will tell the responder to look at map sheet I65-11 (Figure 4). The map sheet shows detailed landmarks, such as guardrails, signs, bridges, and outcrops, which will help determine the exact location of the spill. Once determined, the maps provide detailed surface hydrology, drainage basins, paved ditches, drop boxes, culverts, and sinkholes. The responder will know exactly where the spill is headed and thus know exactly where to position materials and personnel.

In order to display the relative speed at which a sinkhole will drain, a simple classification was developed. The following are in order from the least vulnerable (poorly drained) to the most vulnerable (well drained):

TYPE V   Soil covered sinkhole with no obvious sinkpoint. Poorly drained. May pond after rainfalls. May contain water tolerant plants.

TYPE IV   Soil covered sinkhole with no obvious sinkpoint. No signs of ponding. Fairly well drained.

TYPE III   Exposed bedrock sinkhole with no obvious sinkpoint. No signs of ponding. Fairly well drained.

TYPE II   Soil covered sinkhole. Obvious sinkpoint. Very well drained.

TYPE I   Exposed bedrock sinkhole. Obvious sinkpoint. Very well drained.

CLASS V INJECTION WELL   A sinkhole, crack, or fissure that has been modified or improved by man to allow efficient drainage of surface waters.

The map sheets also include other hydrologic features that may affect the movement of a spill.

SWALE   A shallow depression adjacent to a roadway with no obvious sinkpoint. Water drains diffusely into the ground.

STREAMS   Although not common, there are a few streams within the mapped area. All streams eventually sink into the ground. Ultimate sinkpoints are noted in map sheet text.

PONDS   Ponds are mapped as ephemeral and perennial. An ephemeral pond is a very poorly drained TYPE V sinkhole. It will usually contain water during the wetter parts of the year. A perennial pond contains water year-round, however, these ponds are subject to sudden failures, typically following heavy rains.

It is common that over the years many of the drainage features such as paved ditches, drop boxes, and culverts have experienced structural failure. These failures have been noted on the maps as potential hazards. Potential hazards will greatly affect the movement of water. For example, a crack in a paved ditch will cause surface water to enter the ground at that point, rather than its intended route. All potential hazards must be evaluated when a spill containment plan is developed. It will be of little use to position a sorbent boom at the foot of a paved ditch if the spill is diverted into a potential hazard further upstream. It must be noted that additional potential hazards may have developed after these maps were made. The responder must always be on the lookout for these new hazards.

STOP TWO: ENTRANCE ROAD COLLAPSE
On October 19, 1995 park maintenance workers mowing the roadside reported a soil collapse within a drainage ditch adjacent to the south entrance road (Figure 5). This seeming small, almost inconsequential collapse, as in many cases of sinkhole failure, was only the tip of the proverbial iceberg. To understand why subsidence occurred, a brief history of anthropomorphic alterations and a description of the geologic setting is necessary.

The collapse area is situated directly upon the contact of the Girkin Formation (the limestone that marks the upper limit of the thick carbonate sequence that is the Mammoth Cave Karst Aquifer) and the overlying Big Clifty Sandstone Member (the basal member of the Golconda Group) (Figure 2). It has long been known that along this contact are found the vast majority of vertical shafts in the cave system below (Brucker, *et al.*, 1972). Water flows across the ridge-capping sandstone in seasonal streams until it reaches the underlying carbonates, whereupon it sinks and enters the cave via shafts. If the sandstone capped ridges can be thought of as the "roof-top" of Mammoth Cave (preserving the conduit system from the erosional forces of water), then the seasonal streams are the "gutters" and the shafts the "downspouts".

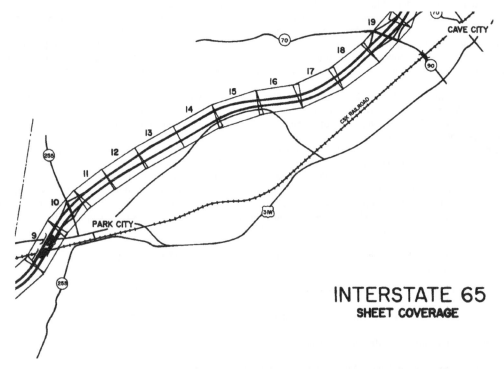

Figure 3. Portion of the index map sheet for the Groundwater Hazard Map of the Turnhole Spring Karst Groundwater Basin.

428

The history of anthropomorphic alterations to this area began in the late 1800's with the construction of the Mammoth Cave Railroad. A narrow-gauge railroad bed was constructed across the head of the small valley near the sandstone/limestone contact, at about the same point that the water was naturally sinking. This roadbed forced all runoff to sink at this point, where in the past a portion of the high-flow runoff (overflow) continued down-valley when the recharge capacity of the original sinkpoint was exceeded. This modification did not produce a profound effect as the surface catchment area of the sink was not greatly increased. Not to worry, this situation was remedied eighty years later. Better never than late in some cases.

In the early 1960's the current entrance road was constructed. The surface catchment area of the sinkhole, which originally was on the order of a few square meters, was suddenly increased to several thousand square meters. Where once a small area recharged a sink, which allowed excess water to overflow, now a large area is forced to drain all water through the subsurface. The result was the widespread collapse of October 19, 1995 (Figure 6).

Park management was more than a little concerned when it learned that the total vertical extent of limestone – thus the total possible extent of void space – between the surface subsidence and the base-level of the cave system was 120 meters. Upon closer inspection, not only had a small collapse in the drainage ditch occurred, there were noticeable cracks and subsidence in the roadway as well.

Crawford and Associates, Inc. was hired to perform an investigation of the site using microgravity traverses to map distributions of densities of earth materials, including potential voids, under the road and recommend remedial actions (Figure 7). It just so happened that there was a two-inch rotary drill rig in the park (used for exploratory drilling for sewer construction) at this time. We used the drill rig to make test bores along the microgravity traverses to aid in the interpretation of the geophysics (Figure 8). It is beyond the scope of the field guide to go into great detail of microgravity techniques. There are several fine overviews of this method,

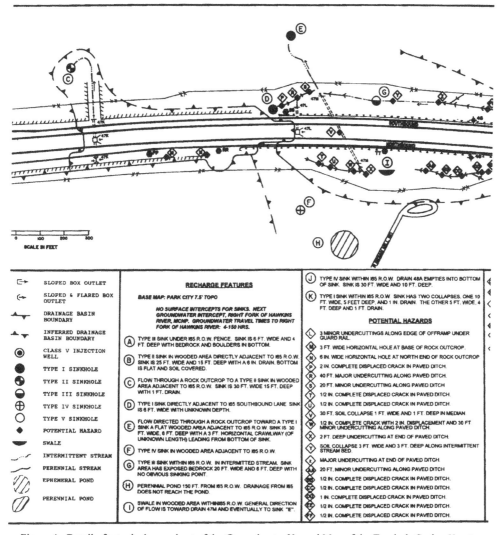

Figure 4. Detail of a typical map sheet of the Groundwater Hazard Map of the Turnhole Spring Karst Groundwater Basin.

Figure 5. Collapse of roadside drainage ditch as discovered in October 19, 1995.

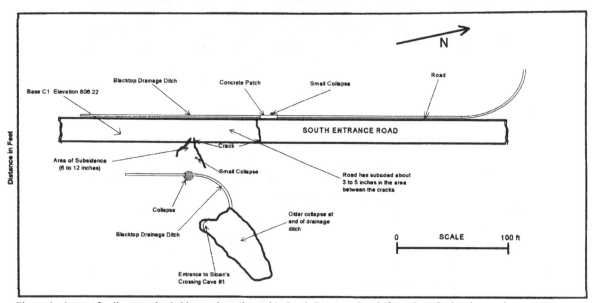

Figure 6. Areas of collapse and subsidence along the park's South Entrance Road (from Crawford and Associates, 1995).

including Nick Crawford's 1995 paper " Microgravity Techniques for Detection of Karst Subsurface Features" published in the proceedings of the Twenty-Sixth Ohio River Valley Soils Seminar, Clarksville Indiana.

An extremely large low-gravity anomaly was found beneath the south entrance road (Crawford and Associates, 1995). The void space beneath the road was estimated to be between 50 and 60 meters in diameter, and at least 30 meters deep – these estimates were supported by the test bores, where several bores fell through voids or loosely-compacted sand fill to a depth of greater than 30 meters (we only had 30 meters of drill steel on the rig!). With the abundance of vertical shafts located directly under the sandstone/limestone contact in the Mammoth Cave System, and with several of these shafts exceeding 50 meters in depth, it is easy to imagine a very large void beneath the road.

So, how was it fixed? At this point we had three options: 1.) Do nothing, and maybe one day a bus-load of tourists would discover one of the deepest shafts in the Mammoth System, 2.) Bridge the void with a series of geotextiles and layers of compacted material, or 3.) Move the road a few meters west onto competent bedrock. It was the opinion of Crawford and Associates and the park hydrogeologist to move the road. Park management and Park Service engineers opted for the second choice. Oddly, no one chose the first option.

The park chose to bridge the road over the void. After the asphalt surface was removed, the top two meters of loose sand fill was excavated. A large weight was dropped repeatedly (a process called "dynamic compaction") over the exposed sand fill. Next, a steel net was anchored into competent bedrock on either side of the void, spanning a distance of approximately 80 meters. Geotextile fabric was placed over the net followed by compacted fills of dense-grade gravel up to grade and paved with asphalt.

The longevity of this action remains to be seen, however the most important aspect to prevent further collapse also had to be addressed – drainage. Paramount to any efforts to repair the roadway is to, as close as possible, reestablish the natural surface hydrology of the area. Although it is not possible to route all surface water in its exact historic course, a few alterations were made to greatly increase the stability to this area. To simply fix the road and not address drainage alteration cures only the symptom, and not the disease. The disease in this case is the accelerated growth of the sinkhole caused by the past routing of the surface drainage. The following drainage modifications were made.

Figure 7. Crawford and Associates perform microgravity traverses using the LaCoste and Romberg Model D Microgal Gravity Meter.

A soil dike was constructed around the head of the main sinkhole east of the entrance road. The dike runs from up-slope of the collapsed paved ditch and ties into the existing railroad bed further down-slope. Any runoff between the entrance road and the dike is allowed to sheet off into the valley at the down-slope end of the dike. This modification restores the natural catchment area of the sinkhole.

A culvert was installed to take runoff from the unpaved ditch on the east-side of the entrance road to the west side of the road. Water is now routed from the intersection area to the west-side of the entrance road along its original course. Formerly the additional drainage from the intersection greatly increases the catchment of the sinkhole. As the culvert outfall area was well above the sandstone/limestone contact there was no need for additional paved ditches downstream from the headwall.

A catch basin or drop box was built near the upslope end of the soil dike and a culvert installed to take drainage from the existing east-side paved ditch, under the entrance road, to a paved ditch on the west-side. Drainage from this limited area now flows into the large sinkhole to the west of the entrance road. Although this action increased the runoff into the large sinkhole, the increase is relatively small and the sinkhole, which has received augmented flow via a paved ditch for decades and is over 100 feet from the road, shows no signs of instability. Although it would be desirable to route this water to another location this appears to be the only practical alternative.

The concrete patch in the paved ditch west of the road was excavated, compacted and repaired. The paved ditch was then extended an additional 40 meters up-slope to a point even with the road repair. This extension allows the paved ditch to begin over sandstone and keep water from sinking into the soil along side and beneath road.

STOP THREE: MAMMOTH CAVE, HISTORIC TOUR ROUTE

The Cave's Original Entrance

Much of the geological interpretation in the cave, especially the detailed stratigraphy we will be discussing, comes from the extensive work of Art and Peg Palmer, and Art's book *A Geological Guide to Mammoth Cave National Park* (1981) is very highly recommended to anyone interested in pursuing the subject further.

We will enter the cave through the original, Historic Entrance. Although the cave was reputed to have been "discovered" by a local hunter named Houchens in the late 1700's (either while chasing or being chased by a wounded bear, depending on who tells the story), it is clear that ancient residents of Kentucky used the cave as early as 4,000 years ago, according to dated artifacts from the cave system (Watson, 1969). These early visitors entered the cave for a variety of reasons, including the mining of sulfate minerals such as gypsum, epsomite, and mirabilite, using the cave for shelter, and perhaps even for sport spurred by the same curiosity that drives cave explorers today. According to archeologist Patty Jo Watson, who has made detailed studies of the cave, the Indians were terrific cave explorers. They visited several miles of passages, leaving an incredible museum of artifacts that have been preserved in the stable conditions in the upper passages. These artifacts have painted a fascinating picture of their culture, their activities, and even the food they ate, which may have even included their captured and killed enemies.

The cave was first shown commercially in 1816 and since then has been open for tourists on a continual basis. Sporadic exploration of the cave system occurred throughout the 1800's and early 1900's. All of the cave passages we will be visiting were known by about the 1870's, the ones beyond Bottomless Pit having been discovered about that time by the famous guide (and slave) Steven Bishop. Steven reportedly crossed over the gaping chasm on a cedar pole with a paying customer who had wanted to go where "no man had gone before." He went on to make a number of fabulous discoveries along this route, including Echo River with its bizarre eyeless fish, and the area he was to call his greatest find, Mammoth Dome.

Figure 8. Drill rig was used to verify microgravity data. The drill steel would typically drop, aided only by gravity, through the voids and unconsolidated material below. We only had 30 meters of steel, which did not reach the bottom in several holes.

The modern era of exploration in the park began in the late 1940's with a small group of cavers who started a systematic exploration of the caves on Flint Ridge, just to the east of Mammoth Cave. As time went on, these caves became integrated one by one and it was revealed that another great cave system, rivaling Mammoth Cave in extent, lay under that ridge. By and by the growing group of cavers formed the Cave Research Foundation, an organization devoted to the exploration and scientific study of the area's caves which was able to cooperate with the National Park Service, who until then had not supported cave exploration to any degree. By the late 1960's the Flint Ridge Cave system reached a surveyed length of over 150 kilometers, and a new challenge loomed before the explorers: if a way beneath the large karst valley separating the two great cave systems could be found, the two caves might possibly be connected making a cave system that would for all time be unrivalled as the world's longest. After a great deal of effort and difficult exploration by a number of dedicated cavers, a group of seven entered Flint Ridge on the morning of September 8, 1972. After a long, grueling trip through several miles of low, wet passages beneath Houchen's Valley, they emerged early the next morning from a low passage into Echo River and onto the tourist trail in Mammoth Cave. The "Everest" of speleology had been conquered, and since that day the Flint Ridge-Mammoth Cave system has been the world's longest. At the time of the connection the cave was 230 kilometers long--today the cave is over 560 surveyed kilometers in length, and no end is in sight.

Exploration continues today. In fact, a tremendous exploration breakthrough in nearby Fisher Ridge Cave occurred early in 1993, and that cave, now over 150 kilometers long by itself, comes to within less than 250 meters of the Mammoth Cave System. Although a connection is not imminent at this time, it is interesting to note that the combined length of the two caves exceeds 700 kilometers. Another major find during the summer of 1996 was Martin Ridge Cave, discovered by graduate students Alan Glennon and Jon Jasper from Western Kentucky University. Since then, they and their colleagues have explored over six kilometers of passages, and have found connections to Jackpot Cave (5 kilometers) and Whigpistle Cave (37 kilometers). This has resulted in a 48+ kilometer system, third longest in Kentucky and eleventh in the United States. Other large, nearby caves are being explored and surveyed, and will very possibly be integrated into the main system as time goes on. How long will the cave ultimately be? No one can say, but the once wild claim of an 800 kilometer long cave system is clearly not so wild after all.

As we descend the hill we will pass the contact between the Big Clifty Sandstone, and walk onto the highest of the three limestone units within which the cave is formed, the Girkin Limestone. It is within the lower Girkin that we enter the cave.

Rotunda

After passing through the entrance area known as Houchen's Narrows, the Rotunda is the first large room encountered in the cave, and is in fact one of the largest rooms in the system, although a few are considerably larger (Figure 9). This is also stratigraphically one of the higher passages in the cave. The walls here are carved primarily from the Paoli Member of the Girkin Limestone (Figure 10). The recessed niche of silty gray limestone near the floor towards Audubon Avenue (the large passage winding away to the right) is the P1 unit of the Paoli, which forms the base of the Girkin.

In the center of the room are the remains of a large saltpeter mining operation during the War of 1812. The fine cave sediments were leached for the compound calcium nitrate, which was then mixed with wood ashes to form potassium nitrate. This saltpeter was used in the manufacture of gunpowder. Although mining ended here just after the War of 1812, other caves in the southeast U.S. were utilized as a major source of saltpeter during the Civil War, when the Confederate Army was unable to get gunpowder from Europe. The artifacts here are completely original, as the cave has been preserved just the way it was at the end of the mining. What a horrifying experience it must have been for the miners in the cave during the New Madrid earthquakes of the winter of 1811-12! George and O'Dell (1992) have collected a series of stories handed down about the event, and although no deaths were reported in the cave, there was apparently considerable concern as the miners went running from the cave, screaming for their lives. The manager of the mining operation was unfortunately fired not long after the earthquake, because although he indicated his willingness to continue the job, he was never again willing to set foot in the cave. There is also evidence that the mining works were substantially damaged in the event.

In January of 1994, during the extreme cold snap that gripped the central U.S., a large slab of the Beaver Bend Member of the Girkin came loose from the roof, crashing down on the tourist trail and crushing part of the saltpeter works. The slab was about 20 meters long by 7 meters wide, and about 30 centimeters thick, which led to an estimate of close to 100 metric tons. Fortunately, the cave was closed at the time because of the winter storm outside, which had closed the entire park as well as Kentucky's highway system. The cause of the fall seems to be related to the cold weather, as it reached a low of -27°C during the period, with temperatures in the Rotunda falling well below zero with a strong wind blowing in through Houchen's Narrows. Speculation has suggested that either freeze-thaw wedging in the bedding plane above the slab, or contraction of the limestone slab itself, is responsible for the fall. This is the only very large rockfall to occur within the developed part of the cave during the 180 years of continuous show-cave operation here.

Booth's Amphitheater

As we wind down Main Cave to the left off of the Rotunda, we begin to slip down into the Ste. Genevieve Limestone, walking in the paleo-upstream direction. Note that we are only seeing the highest parts of these passages, which are filled with up to 25 meters of sediment in places (Palmer, 1981). Booth's Amphitheater has formed at the intersection of Main Cave with Gothic Avenue above. Gothic Avenue is the oldest known passage in Mammoth Cave Ridge. It began to form some time prior to two million years b.p. (Granger, et al., 2001), by draining water from the ancestral Houchen's Valley towards the Green River (Palmer, 1981).

The localized, dark black deposits near the ceiling have resulted from years of "torch throwing", where guides would fling tied bundles of kerosene soaked rags onto high ledges for unusual illumination. The practice was discontinued in 1991, for environmental reasons. The walls in this part of the cave, in fact, are rather dark in general, which may be the result of soot from thousands of years of cane-reed torches used by the aboriginal visitors to the cave (Watson, 1969) who mined sulfate mineral crusts in this area. Organic acids have been shown to be present in the coating as well (Quinlan and Traverse, 1967). The dark material seems to preferentially occur on gypsum crusts.

Giant's Coffin

At this point on the trip we will turn into a smaller passage on the right at Dante's Gateway, descending down through the Fredonia Members of the Ste. Genevieve. Giant's Coffin is the very large breakdown block behind which we will begin our descent. As we make our way down, we will pass through a more complex configuration of smaller passages, which formed during the early or middle Quaternary Period (Palmer, 1981). Erosion of the Mammoth Cave Plateau was by this time exposing limestone in new areas, so that water from the surface could enter the aquifer at many new discrete locations. These smaller yet more abundant active flowpaths have resulted in smaller yet more abundant cave passages, with an increase of passage complexity. The roughly horizontal elliptical tube passage that we travel along for a bit is Black Snake Avenue, and its tube shape suggests that it was formed largely under phreatic or pipe-full conditions.

Black Snake Avenue eventually winds close to the edge of Mammoth Cave Ridge, and in this area we will pass a number of dome-pit complexes. On the surface, at points along the edge of the Big Clifty Sandstone, water can make its way into the subsurface. Since this water is typically quite undersaturated with respect to limestone, it can bore these shafts, which only coincidentally intersect the horizontal passage along which we are moving. If the conditions are relatively wet, water can be seen at the bottom of Bottomless Pit. We are getting lower in the cave – the water at the bottom of this shaft is near the elevation of the Green River (128 meters), the local baselevel.

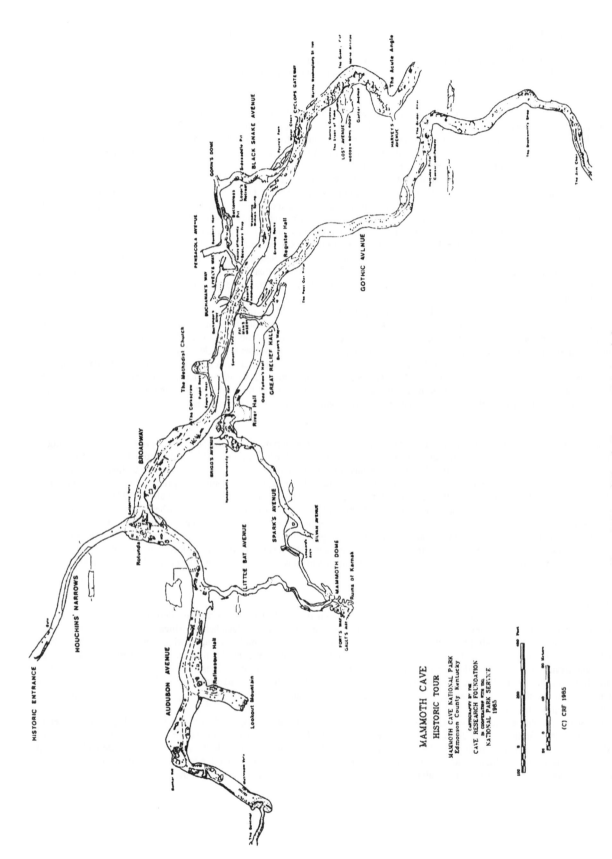

Figure 9. Map of Historic Tour route, Mammoth Cave

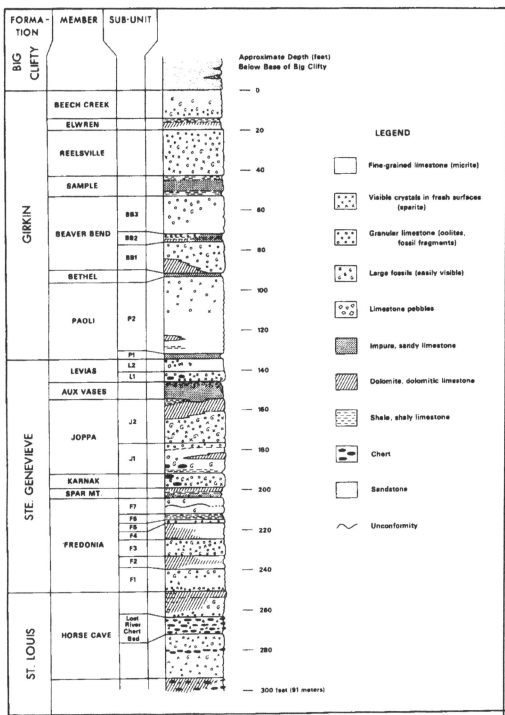

Figure 10. Detailed stratigraphic section of Mammoth Cave (from Palmer, 1981)

Great Relief Hall

After passing through Fat Man's Misery we reach Great Relief Hall. This will be a short rest and rest room break. Emerging from Fat Man's Misery we come into the passage that Steven Bishop called Relief Hall. Since the addition of rest rooms by the Park Service, it has been known as Great Relief Hall. This area was apparently once a popular spot for parties and dances, as can be seen by the various writings along the passage ceiling.

We are moving lower within the Ste. Genevieve and as we walk towards River Hall we eventually come into the top of the St. Louis. Great Relief Hall is formed within the Fredonia Member of the Ste. Genevieve. Lithologic detail, however, becomes obscured due to the frequent flooding and subsequent sedimentation in the cave's lower levels.

River Hall

At this point we reach the lowest major level of the Mammoth Cave System; it is at this level that the cave is still forming on a regular basis. At River Hall we will take a trip down the passage towards the left which leads to Echo River. In River Hall, the contact between the Ste. Genevieve and the underlying St. Louis is visible at the top of a prominent ledge near the ceiling (Palmer, 1981).

Dead Sea

The rather complex levels of passages we've traversed over the past hour and a half can also be seen at the current baselevel of the Mammoth System. The Dead Sea, representing the baselevel, continues River Styx Spring. Prior to rapid downcutting of the Green River during the Pleistocene, all flow was through River Styx. Following the entrenchment of the Green River, flow was diverted to Echo River Spring, and thereafter, reached River Styx only during flood events. Post-Pleistocene backfilling of the Green's channel (about 10 meters) caused an increased regional base-level, approximating that prior to the Pleistocene. Today, the pre-Pleistocene route of River Styx and the Pleistocene route of Echo River are both active as flow distributaries.

Mammoth Dome

As we work our way back to the higher levels of the cave, we will make up most of that elevation within a great vertical shaft complex known as Mammoth Dome. In the walls here almost the entire section of the Ste. Genevieve Limestone is exposed, from the basal contact with the St. Louis two meters below the floor of the lowest balcony (Palmer, 1981) to the uppermost part of the Joppa Member where we will emerge from the dome into Little Bat Avenue. Mammoth Dome is one of the largest of the hundreds known in the cave system. These shafts often provide routes connecting the different horizontal levels. Explorers and surveyors working in the cave system must become proficient at moving up and down ropes to negotiate these places. Sometimes the drains at the bottom of these shafts can be explored to lower levels, but quite often these drains are very wet and possibly filled with breakdown or silt.

Upon reaching the top of the dome via the fire-tower steps, we pass through Little Bat and eventually pop back into the Girkin Limestone at Audubon Avenue. A short hike to the right brings us back to the Rotunda, where we began our trip.

REFERENCES

Brucker, R.B., Hess, J.W., and White, W.B., 1972, Role of vertical shafts in the movement of groundwater in carbonate aquifers: Groundwater, V. 10, p. 5-13.

Crawford and Associates, 1995, Microgravity subsurface investigation at the site of a sinkhole collapse under the South Entrance Road at Mammoth Cave National Park, Kentucky: 68 p.

George, A.I. and G.A. O'Dell, 1992, The saltpeter works at Mammoth Cave and the New Madrid Earthquake: The Filson Club History Quarterly, V. 66, pp. 5-22.

Granger, D.E., Fabel, D., and Palmer, A.N., 2001, Plio-Pleistocene incision of the Green River, Kentucky, from radioactive decay of cosmogenic ^{26}Al and ^{10}Be in Mammoth Cave sediments: Geological Society of America Bulletin, in press.

Ray, J.A., 1996, Fluvial features of the karst-plain erosion surface in the Mammoth Cave Region: Proceedings of the Fifth Annual Mammoth Cave Science Conference, Mammoth Cave National Park, Kentucky, pp. 137-156.

Quinlan, J.F., and Ewers, R.O., 1981, Hydrogeology of the Mammoth Cave Region, Kentucky: Geological Society of America Field Trip Guide, American Geological Institute, V. 3, pp. 457-506.

Quinlan, J.F. and A. Traverse, 1967, Humic acid and humate deposits in Salt's Cave and Mammoth Cave, Kentucky: National Speleological Society Bulletin, V. 29, pp. 98-99.

Watson, P.J., 1969, The Prehistory of Salts Cave, Kentucky: Illinois Sate Museum, Reports of Investigations, No. 16, 86 p.

Author index

Printed and bound by CPI Group (UK) Ltd, Croydon, CR0 4YY

23/10/2024

01777679-0019